Stochastic Modeling and Geostatistics
Principles, Methods, and Case Studies

Edited by
Jeffrey M. Yarus
and
Richard L. Chambers

AAPG Computer Applications in Geology, No. 3

Published by
The American Association of Petroleum Geologists
Tulsa, Oklahoma, U.S.A.
Printed in the U.S.A.

ISBN: 0-89181-702-6

Association Editor: Kevin T. Biddle
Science Director: Richard Steinmetz
Publications Manager: Kenneth M. Wolgemuth
Special Projects Editor: Anne H. Thomas
Production: Custom Editorial Productions, Inc., Cincinnati, Ohio

Publisher's note: This volume contains frequent mention of trademarked hardware and software. The editors did not attempt to verify trademark registration; rather, they generally retained the authors' use of capitalization and the trademark registration symbol. By convention, capitalization is a more consistent indicator in text of trademark registration than is the presence or absence of a trademark symbol.

This and other AAPG publications are available from:
The AAPG Bookstore
P.O. Box 979
Tulsa, OK 74101- 0979
Telephone (918) 584-2555; (800) 364-AAPG (USA—book orders only)
FAX: (918) 584-0469; (800) 898-2274 (USA—book orders only)

AAPG
Wishes to thank the following
for their generous contributions
to

Stochastic Modeling and Geostatistics

Amoco Production Company

Amoco E & P Technology

Kansas Geological Survey

Marathon Oil Company

Mobil Oil Corporation

❖

Statoil

Contributions are applied against the production costs of the publication, thus directly reducing the book's purchase price and making the volume available to a greater audience.

Acknowledgments

The authors wish to acknowledge the contribution made by all of the participating authors. The time and effort they have put forth can only be appreciated by those of us involved in compiling this volume. We wish to thank the AAPG, specifically Cathleen Williams and Anne Thomas, without whose help and patience none of this would have been possible. In addition, we jointly wish to acknowledge Marathon Oil Company, GeoGraphix, and Amoco Production Company for supporting us in this endeavor.

Personal Acknowledgments from Jeffrey M. Yarus

I am indebted to the following individuals who have helped me throughout my career attain certain modest levels of proficiency in the field of mathematical geology. They are: Dr. Robert Ehrlich, my mentor who taught me to be a scientist and instilled me with confidence; Dr. John Davis, whom I admired from a distance for so many years and now inspires me as a colleague; Dr. Ricardo Olea, who has generously and patiently taught me the fundamentals of geostatistics; Dr. André Journel, who graciously brought me into his fold and remains a standard for my achievement as a geostatistician; R. Mohan Srivastava, whose guidance and wisdom has taught me the practical nature of geomathematics; and Dr. Gregory Kushnir, who has challenged me intellectually, and has allowed me to share in many of the mathematical developments in our mutual careers.

I would also like to thank my sons David and Jordan for allowing me to miss many of their soccer and hockey games, my daughter Rachel, who managed to be conceived and born during the course of this volume, and my wife Mary, for giving me all of the time, patience and encouragement necessary to complete this work. Finally, I thank my father Leonard W. Yarus, whom I miss, and mother Marilyn S. Yarus for instilling me with curiosity and tenacity.

Personal Acknowledgments from Richard L. Chambers

With encouragement from my parents, Bob and Aileen Chambers, I have been able to pursue my dreams, wherever they might lead, even in the direction they had not envisioned. When I entered college, I fully intended to prepare myself for a career in medicine and had given little thought to anything else. In my sophomore year, I enrolled in physical geology to learn about the rocks and landforms I saw while hiking and rock climbing. From that time on, I was hooked on geology, because it offered me the opportunity to combine my avocation with a fascinating profession. Drs. Sam Upchurch and Bob Ehrlich helped to develop my interest in statistics during my doctoral career at Michigan State University. Dr. Jean-Paul Marbeau provided me a deeper understanding of geostatistics and practical applications. I owe much to these scientists and thank them for their time and patience.

Throughout my career, my family has been very patient and understanding when so much of my time and energy was consumed by my day-to-day job. The additional time required to devote to the development of this book placed an additional burden on an already scarce commodity—my time with my family. My deepest gratitude goes to my wife Verda, and our three daughters, Kris, Kelli and Keri, for their love, support and patience.

IN MEMORY

This book is published in memory of Richard Leonard, son of Jay and China, brother of Ted.

About the Editors

Jeffrey M. Yarus was born in 1951 in Cleveland, Ohio. In 1969, he began his formal studies in geology at the College of Wooster, Ohio. During his undergraduate years, Yarus was awarded an opportunity to study geology at the University of Durham, England, on a scholarship for one year. Upon completion of the special program, he returned to Wooster and received his B.A. degree with honors in 1973. In September of that year, Yarus began his graduate studies at Michigan State University under the supervision of Robert Ehrlich. At Michigan State, Yarus first developed his interest in computer mapping and numerical and statistical analysis. In 1974, Yarus followed Ehrlich to the University of South Carolina where he finished his M.S. degree and continued through the Ph.D. program in geology. Yarus joined Amoco Production Company in New Orleans, Louisiana, in 1977 as a production geologist for the Gulf Coast region. In 1980, he left Amoco and moved to Denver, Colorado, where he worked in the independent oil business for eight years. As an independent, Yarus worked a variety of domestic basins in the Rockies, mid-continent, and Appalachian regions. In 1988, Yarus was hired by Marathon Oil Company's Petroleum Technology Center as a senior mathematical geologist. At Marathon, he played a major role in instituting desktop computer mapping and geostatistical technology. He was responsible for providing training and consulting in this area for the entire company. Yarus has written a variety of papers and taught courses on computer mapping and applied statistical methods. His professional contributions are many, and include an article in AAPG's first volume on computer applications (published in 1992), and an AAPG-sponsored course on computer mapping techniques. Yarus has been an adjunct professor at the University of Colorado, Denver, and is presently a visiting professor at the Colorado School of Mines. Today, he works for GeoGraphix, Inc., where he is the manager of technical services and support.

Richard L. Chambers was born in 1947 in Algona, Iowa. He earned B.A. and M. S. degrees in geology from the University of Montana, and then attended Michigan State University, earning a Ph.D. in geology in 1975, with an emphasis on quantitative sedimentology and statistical applications. In 1972, he joined the Great Lakes Environmental Research Laboratory (GLERL), National Oceanographic and Atmospheric Administration, U.S. Department of Commerce. The bulk of his work at GLERL involved sediment-water interface geochemistry, depositional processes, and mass balance studies in Lake Michigan. Chambers joined Phillips Petroleum Company in Bartlesville, Oklahoma, in 1980 as a senior geologist. From 1980 to 1988, his activities at Phillips included research in high-resolution P- and S-wave seismic imaging and regional basin studies to develop exploration plays and lease acquisition strategies. In 1988, he joined the Amoco Production Company Research Center in Tulsa, Oklahoma, as a senior research scientist to bring geostatistical technology to Amoco. His work at Amoco involves the application of geostatistical methods to exploration and production problems, with major emphasis on data integration, reservoir characterization, and uncertainty assessment. Chambers co-developed an in-house training course in geostatistics and has taught that course for the past five years. In 1991, he became the supervisor of the reservoir imaging and characterization group, where he now spends his time on less technical issues. He has written articles in sedimentology, geochemistry, geophysics, and geostatistical applications.

Table of Contents

Preface

Producing a viable book about some aspect of mathematical geology that geologists will use is not easy to do. Many geologists simply do not like to deal with math on anything but a rudimentary level, if that! This is particularly true in the United States where courses in mathematics have not been a strong component of a general earth science curriculum. This is not a criticism or an implication that a mathematical approach to geology is in some way superior, nor is it meant to imply that earth sciences do not require the use of important mathematical principles. It is, however, a statement of priority. Traditional geology is a highly qualitative science resting soundly on classification schemes and descriptions associated with physical observations.

Petroleum Geology is a good example of this qualitative science. Historically, when an exploration geologist correlates logs and interprets seismic sections, results are described using qualitative terms. Structural surfaces can be described as anticlines or synclines, depositional environments can be deltas or reefs, depositional features can be onlapping or offlapping, and lithologies can be sandstones or limestones. Although these terms are highly descriptive, they have great latitude in their meaning and are subject to individual interpretation. The quality of the interpretations rests largely on the experience of the interpreter, and we would expect different geologists to produce similar, yet slightly different results. The result of a geologic investigation is generally considered deterministic, and consists of maps and cross sections that depict a basin, or perhaps a portion of a basin, where potentially economic reserves of oil and gas reside.

This book is dedicated in part to understanding qualitative and deterministic geologic models in terms of statistics. The book is precisely concerned with the fact that given the same set of data, different geologists will generate different results, all of which may be valid interpretations. It is the premise of the authors that understanding the interpretive variations is far more important than identifying any one particular model as truth.

"Geostatistics" is the term applied to the set of tools used to solve various geologic problems like those identified in this volume. Those of us who attempt to use geostatistical techniques have become known as "geostatisticians." Geostatisticians are geologists, petrophysicists, geophysicists, or even engineers by training. Although I often wonder how some of us ended up in this peculiar area of the earth sciences, we are set apart from the rest of the geologic community in that we deal routinely with spatial statistics and computers. Although mathematical applications have become more popular as computers integrate with professional lives, many geologists resist taking the time to fully understand these quantitative techniques. Yet, in the business of petroleum, constructing mathematical models is a prerequisite for certain analyses, such as reservoir simulation. As consequence, this type of mathematical modeling is often left to engineers who are more comfortable with math and computers than geologists. Part of the geostatistician's job, regardless of professional origins, is to ensure that the geology does not get lost in this process.

To elaborate, many qualitative geologic models, by necessity, will undergo a transformation into mathematical models by someone other than a geologist. If the results were precise, the transformation would not be a problem. Unfortunately, all too often the final quantitative model is very different from the original. The changes have nothing to do with professional interpretation but are made for pragmatic reasons. Reservoir models are expensive to produce, and their cost is directly related to the size of the model, particularly the number of grid nodes. For example, a multiphase reservoir simulation with 50,000 grid nodes could cost roughly \$40,000 to \$50,000 on a Cray computer. Although geologists do not typically think in terms of grid nodes, their models can easily consist of millions of them. To execute a reservoir simulation with that resolution would be financially unacceptable. In order to reduce the cost of computer simulation, engineers will coarsen the geologic model by reducing the number of total nodes to a more affordable size.

There are ramifications to drastically reducing the size of a reservoir model in this way. Specifically, the heterogeneity, or complexity of the geology, is simplified. If the reservoir is not very complex to begin with, a coarser representation may suffice. However, when the reservoir is complex, simulation results can be misleading. To prevent this, history matching techniques are used to tune the coarser engineering model. Porosity, permeability, and other petrophysical parameters are "adjusted" until the fluid-flow simulations either match the observed performance of existing wells or the pressures and flow rates from production tests. If either or both these conditions are met, the assumption is made that the critical geology must be preserved. However, such may not be the case.

Although a reservoir simulation brings a certain closure to a study by providing the economics and plan for development, the ensuing production history is often disappointing. The original predictions for recoverable hydrocarbons are often over- or underestimated, individual wells do not perform as anticipated, and inappropriate drilling patterns or enhanced recovery methods are frequently implemented.

Geostatistics attempts to improve predictions by building a different kind of quantitative model. It attempts to construct a more realistic picture of reservoir heterogeneity using statistical methods which do not simply "average" important reservoir variables. Like the traditional deterministic approach, it provides mechanisms to retain indisputable "hard" data where it is known, and interpretive "soft" information where desired. Unlike deterministic modeling, however, geostatistics provides scientists with numerous plausible results. The degree to which the various outcomes differ is a reflection of what is not known, or a measurement of the "uncertainty." Some of the outcomes may challenge the prevailing geologic wisdom, and almost certainly provide a range of economic scenarios from optimistic to pessimistic. Although, having more than one result to analyze changes the paradigm of traditional analysis and may require multiple reservoir simulations, the benefits outweigh the additional cost. To paraphrase Professor André Journel of Stanford, ". . . it is preferable to have a model of uncertainty then an illusion of reality."

Those interested in applying geostatistics to solving practical problems will find this book useful. In particular, it is intended for frontline geoscientists and their managers as a source of information and guide. We hope that readers will be able to identify an application they wish to try or a case study they wish to use in their day-to-day work. The articles vary with respect to complexity. Some of the articles contain a fair amount of math and some contain no math at all. Those articles that seem mathematically challenging have been written in such a way that tenacious readers should be able to read around the math and still glean the essence of the authors' ideas.

The book is divided into four sections. Each section has its own short introduction. In general, **Section I: Getting Started** and **Section II: Principles** should be read first, in order. These sections give the reader both a philosophical approach to geostatistics as well as some of the fundamentals (including strengths and weaknesses). The chapters in **Section III: Methods and Case Studies** can be read in any order. **Section IV: Public Domain Software and Bibliography** contains a description of some of the software available through the public domain, and general bibliography for those newly interested in the field of geostatistics.

Jeffrey M. Yarus
Senior Editor

Introduction

In selecting papers for this volume, we have been strongly influenced by the need to provide a usable reference for front-line geoscientists, reservoir engineers, and those managers who wish to obtain an overview of geostatistics as it applies to the petroleum industry. Authors solicited to submit papers for **Section 2**, emphasizing general principles and techniques, were requested to minimize detailed mathematical and theoretical discussions. As you read chapters in that section, do not feel obligated to fully digest the occasional equation; it is possible to read through the mathematics and still capture the content of the author's message. All of the papers are original contributions, although some of them have been previously presented orally in informal settings.

In **An Overview of Stochastic Methods for Reservoir Characterization**, R. Mohan Srivastava presents a general discussion of the most commonly used stochastic methods for reservoir characterization: indicator simulation, Gaussian simulation, fractal methods, Boolean simulation, and annealing. He also provides the reader a sense of what each method requires in terms of time and resources, both human and computer.

♦

Chapter 1

An Overview of Stochastic Methods for Reservoir Characterization

R. Mohan Srivastava
FSS International
Vancouver, British Columbia, Canada

♦

INTRODUCTION

In the fall of 1988 the Society of Petroleum Engineers (SPE) held a forum on reservoir characterization in Grindenwald, Switzerland. Many of the authors who have contributed to this volume were at that forum discussing their ideas on stochastic methods for reservoir characterization. All of these ideas were then still quite new and largely untested; indeed, some of them had not been reduced to practice, but were merely the wild imaginings of creative and curious minds. At that time, there was still a fair bit of controversy over whether stochastic methods had any relevance to the practice of modeling petroleum reservoirs.*

It is a testament to the practical value of stochastic methods that three years later when the SPE held its next forum on reservoir characterization in Crested Butte, Colorado, all of the specific methodologies presented at the Grindenwald forum had found their way into practice through case studies on actual reservoirs. The controversy at the Crested Butte forum was no longer whether stochastic methods should be used, but which ones should be used.

Geologists, geophysicists, and reservoir engineers who observed this explosion in the use of stochastic methods now have a choice of several theoretically sound and practically tested approaches. This wide variety of methods is a double-edged sword—though choice entails the ability to be flexible, it also entails the need to make a decision. Choosing a specific approach can be a bewildering and daunting task for a newcomer to the field of stochastic methods. Technical articles tend to concentrate on the advantages of the author's method, leaving the disadvantages to be aired by other authors promoting their own favorite methods.

This overview attempts to provide a balanced accounting of the advantages and disadvantages of the more common and practically successful stochastic methods for reservoir characterization. For each method it also presents a brief outline that is not intended to give a complete description of the nuts-and-bolts details but rather to provide a skeleton on which to hang the discussion of the pros and cons.

WHAT DO WE WANT FROM A STOCHASTIC METHOD?

Before discussing various stochastic modeling methods, it will be useful to consider some of the possible goals of a reservoir modeling exercise because the appropriateness of any particular method depends, in large part, on the goal of the study. Stochastic methods vary considerably in their requirements for time, money, human resources, and computer hardware and software. Not all stochastic modeling studies need a Cadillac technique; some studies will do fine with the Volkswagen technique or with a solid pair of rollerskates. If we are considering stochastic modeling, we should give some thought to which of the following goals, either alone or in combination, are relevant.

Stochastic Models as Art

Stochastic models often are not used quantitatively; no hydrocarbon volumes or fluid flows are calculated from them. Even though such stochastic "artwork" is often dismissed as mere decoration for the walls of research centers, it may still have a useful role to play as a catalyst for better technical work. In an industry that has become too familiar with simple layer-cake stratigraphy, lithologic units that either continue to the next well or conveniently pinch out halfway, and contour maps that show

---

*One memorable exchange between an industry sedimentologist and an academic geostatistician:

Sedimentologist: Of what earthly value to me is any procedure that involves a random number generator when I know that the geologic processes I study are not, in fact, random but merely quite complicated?!

Academic: Any suitably complicated process can be simulated by probabilistic methods—even your sacred geology!

Sedimentologist: You, sir, are tremendously naive!!

graceful and gentle undulations, it is often difficult to get people to realize that there is much more interwell complexity than traditional reservoir models can portray. Stochastic models often have their greatest influence when they are used to challenge a complacent belief in the simplicity of a reservoir. When engineers and scientists familiar with the traditional approaches to reservoir modeling first see the results of a stochastic approach, their initial reactions often are surprise and skepticism. Their surprise is due to the fact that cross sections through stochastic models are much more visually complex than the cross sections they are familiar with; their skepticism is due to the fact that stochastic models provide many different renditions and they are accustomed to thinking of one "best" model. The initial surprise and skepticism often give way to curiosity as they realize that these many alternate renditions of the reservoir share a similar visual character and all somehow manage to honor the information at the well locations.

Once the results of a stochastic model are rendered on cross sections and hung on the walls of a meeting room, they often become a lightning rod for criticism when people point to a specific feature and argue that such a feature cannot exist. Their reasons range from geologic arguments based on depositional environment to geophysical arguments based on seismic stratigraphy to engineering arguments based on well test information. In all of these situations, the benefit of the stochastic model simply may be to focus the views and opinions of a wide variety of experts and, in so doing, to point to an improved reservoir model. Whether this new and clearer understanding is used to improve the stochastic model or whether it is used to make intelligent adaptations to more traditional models, the initial stochastic model still has played an important role in the evolution of an appropriate reservoir model.

Stochastic Models for Assessing the Impact of Uncertainty

Many engineers and scientists who forecast reservoir performance understand that there is always uncertainty in any reservoir model. They know that their performance forecasts or their volumetric predictions are going to be based on a specific model that is designated the "best" model, but they also want a few other models, more specifically a "pessimistic" model and an "optimistic" model, so that they can assess whether the engineering plan they have developed on the "best" model is flexible enough to handle the uncertainty.

When used for this kind of study, a stochastic approach offers us a variety of models, each of which is consistent with the available information. We can sift through these many possible renditions of the reservoir and select one that looks like a good low-side scenario and another that looks like a good high-side scenario.

Stochastic Models for Monte Carlo Risk Analysis

Stochastic modeling offers the ability to do Monte Carlo risk analysis because the various renditions it produces are not only plausible in the sense that they honor all of the information, but also they are equally probable in the sense that any one of them is as likely a representation of the true reservoir as the others. In such studies, hundreds, if not thousands, of alternate models are generated and processed to produce a distribution of possible values for some critical engineering parameter—breakthrough time, for example, or connected pore volume; these distributions are then used to optimize a decision by minimizing an objective function.

A critical aspect of this use of stochastic modeling is the belief in some "space of uncertainty" and that the stochastic modeling technique can produce outcomes that sample this space fairly. We know that we cannot possibly look at all possible outcomes, but we believe that we can get a fair representation of the whole spectrum of possibilities. When stochastic modeling techniques are used for this purpose, we hope that they do not have any systematic tendency to show optimistic or pessimistic scenarios because we are going to use all of the outcomes as equally probable representations of reality.

Though there is some overlap with these types of studies and the pessimistic/optimistic models described in the previous section, the important difference between the two is that Monte Carlo risk analysis involves the notion of a probability distribution, and the type of study described in the previous section does not. The previous section referred to studies where models are chosen by sifting through a large set and extracting two that seem to be plausible, but extreme, scenarios. No one particularly cares whether the optimistic case is the 95th percentile or the 99th percentile; it is simply an example of a case that produces a very optimistic result while still honoring the same basic data. In Monte Carlo risk analysis, however, we depend on the notion of a complete probability distribution of possible outcomes and on the hope that our stochastic approach is producing outcomes that fairly represent the entire distribution.

Stochastic Models for Honoring Heterogeneity

Though stochastic techniques are capable of producing many possible outcomes, many stochastic modeling studies use only a single outcome as the basis for performance prediction. When used in this way, stochastic techniques are attractive not so much for their ability to generate many plausible outcomes, but rather for their ability to produce outcomes that have a realistic level of heterogeneity.

Over the past decade it has become increasingly clear that reservoir performance predictions are more accurate when based on models that reflect the actual heterogeneity of the reservoir. Countless examples exist of reservoir performance predictions whose failure is due to the use of overly simplistic models. Most traditional methods for reservoir modeling end up with a model that is too smooth and continuous, gently undulating from one well to the next rather than showing the interwell variability known to exist. Such smoothness commonly leads to biased predictions and poor development plans—actual breakthrough times end up being much quicker than expected, for example, or sweepage is not as efficient as the smooth model predicted.

Though using a stochastic method to produce a single outcome is often viewed with disdain by those who would like to be pumping hundreds of possible outcomes through Monte Carlo risk analysis, we can argue that even a single outcome from a stochastic approach is a better basis for performance prediction than a single outcome from a traditional technique that does not honor reservoir heterogeneity.

As we will see when we discuss the various stochastic modeling alternatives, a key issue needs to be resolved if honoring heterogeneity is the principal goal: what heterogeneity to honor and how to describe it.

Stochastic Facies or Stochastic Rock Properties (or Both)?

As stochastic models have become more widely accepted for flow studies, workers have begun to recognize two fundamentally different aspects of a stochastic model of a reservoir. The first aspect is the architecture of the flow units, and the second aspect is the spatial distribution of rock and fluid properties within each flow unit.

The reservoir architecture commonly is the first priority in a stochastic reservoir model and is usually described in terms of the different geologic facies; in an eolian setting, for example, grain-flow sands might be modeled as one type of flow unit, wind-ripple deposits as a second flow unit, and interdune muds and shales as a third flow unit. Once the spatial arrangement of the different flow units has been modeled, we then must decide how to assign rock and fluid properties within each flow unit. In some reservoirs, the spatial arrangement of the flow units is the dominant control on reservoir performance and in such situations it is common to assign constant rock and fluid properties to each type of flow unit. To continue with the example of an eolian setting, we might decide to use three different permeabilities, one for the grain-flow sands, another for the wind-ripple deposits, and a third for the interdune muds and shales. The assignment of constant properties within flow units may not be appropriate, however, if there is considerable overlap in the distributions of rock and fluid properties within the different facies. The lower grain-flow permeabilities, for example, may be quite similar to the higher wind-ripple permeabilities. In such situations, we may need to carry the stochastic modeling exercise one step beyond the stochastic modeling of facies to also stochastically model the rock and fluid properties within each facies.

The important difference between modeling facies versus modeling rock and fluid properties is that the first is a categorical variable, whereas the second is a continuous variable. Categorical facies codes take on only certain discrete values and often do not have any significance to their ordering. In our eolian example, we might choose to assign 1 to the grain-flow sands, 2 to the wind-ripple deposits, and 3 to the interdune muds and shales. In this example, intermediate values, such as 2.5, have no meaning; furthermore, the ordering of the numerical codes does not necessarily reflect a spatial or depositional ordering—we don't necessarily have to go through facies 2 on our way from 1 to 3. Some of the stochastic modeling techniques discussed in following sections are well suited to categorical facies, and other techniques are better suited to continuous variables.

Though it has become conventional to assume that a lithofacies model is an appropriate model of reservoir architecture, it is worth considering whether this assumption is necessarily a good idea. Though the original depositional facies are easily recognized and described by geologists, these may not be the most important controls on flow behavior. If the permeability variations are due primarily to diagenetic and structural events that postdate the original deposition, then it is not at all clear why a model of the original depositional units is an appropriate model of the flow units. Other criteria, like the responses of a suite of petrophysical logs, might better indicate the arrangement of the flow units than would the more visibly obvious geologic core description.

Stochastic Models for Honoring Complex Information

The growing popularity of stochastic methods is due, in large part, to their ability to incorporate an increasingly broad range of information that more conventional techniques do not accommodate. Though many of the most recent developments in stochastic modeling could be applied to more conventional techniques, this has not (yet) been done. So, we may find ourselves using a stochastic method not because we are particularly interested in generating multiple outcomes or in honoring geologic heterogeneity, but simply because we want to integrate our seismic data with our petrophysical data, or to ensure that our model follows some basic geologic principles in terms of where certain facies can and cannot appear.

OUTLINE OF COMMON STOCHASTIC METHODS

The following sections present a thumbnail sketch of the stochastic methods in common use. Following these brief outlines is a discussion of the important practical advantages and disadvantages of each of the methods.

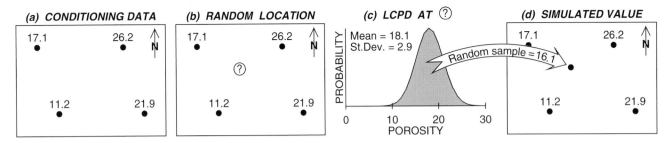

Figure 1. Sequential simulation of porosities in a layer.

Sequential Simulation

The family of "sequential" procedures all make use of the same basic algorithm shown in Figure 1:

1. Choose at random a grid node at which we have not yet simulated a value.
2. Estimate the local conditional probability distribution (lcpd) at that location.
3. Draw at random a single value from the lcpd.
4. Include the newly simulated value in the set of conditioning data.
5. Repeat steps 1 through 4 until all grid nodes have a simulated value.

The only significant difference, in terms of practice, between the various sequential procedures is the way in which the lcpd is estimated. Any technique that can produce an estimate of the lcpd can be used as the basis for sequential simulation. Multi-Gaussian kriging, for example, produces an estimate of the lcpd by assuming that it follows the classical bell-shaped normal distribution and estimating its mean and standard deviation; this was the approach used in the diagram in Figure 1. When multi-Gaussian kriging is used in a sequential simulation procedure, the algorithm is usually called sequential Gaussian simulation. Indicator kriging is an example of another technique that could be used to estimate the lcpd. With this procedure, no assumption is made about the shape of the distribution, which is estimated by directly estimating the probability of being below a series of thresholds or by directly estimating the probability of being within a set of discrete classes. When this method is used for sequential simulation, the algorithm is usually called sequential indicator simulation.

The fact that the sequential approach can accommodate any technique for estimating the lcpd has made it a very flexible and popular technique; many of the methods that use the sequential principle carry names that do not include the word "sequential." Markov-Bayes simulation and indicator principal components simulation, for example, are implemented within a sequential framework.

In addition to the conditioning data that we want to honor, our input to a sequential simulation depends on the procedure we use to estimate the lcpd. Most of these procedures require us to describe the spatial continuity, usually in terms of a variogram or correlogram. Those procedures that accommodate secondary data also may require a description of the spatial continuity of the secondary data and their relationship to the primary data; these relationships commonly are expressed in terms of secondary variograms and cross-variograms, but, for some algorithms, are simplified to fewer parameters that describe the strength of the correlation between primary and secondary data.

Boolean, Marked-Point Process, and Object-Based Simulation

Though their authors have not yet settled on a single, informative title, there is a family of methods that shares the idea that a reservoir model should be created from objects that have some genetic significance rather than being built up one elementary node or pixel at a time. To use such procedures we have to select for each lithofacies a basic shape that adequately describes its geometry; for example, we might decide that sand channels in a turbidite system look like half-ellipses in cross section, or that delta fans look like triangular wedges in map view. In addition to selecting a basic shape for each lithofacies, we need to choose what overall proportion of this shape we would like to see in the final model and we also need to choose distributions for the parameters that describe this shape. These parameters typically include such things as its size, anisotropy ratio, and orientation of its long axis. With some of the more sophisticated algorithms, we also get to choose rules that describe how the various shapes can be positioned relative to one another; these rules might dictate that certain shapes cannot occur within a certain minimum distance of other shapes, or that certain shapes must always overlap.

Once the distributions of parameters and positioning rules are chosen, the following simplified outline captures the essential steps in the procedure:

1. Fill the reservoir model with some background lithofacies.
2. Randomly select some point in the reservoir model.
3. Randomly select one of the lithofacies shapes and draw an appropriate size, anisotropy, and orientation.
4. Check to see if this shape conflicts with any conditioning information or with the other shapes that were previously simulated. If it does not

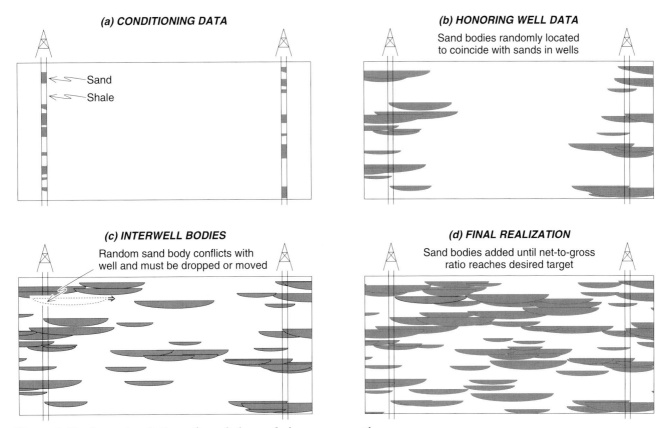

Figure 2. Boolean simulation of sand channels in a cross section.

conflict, keep the shape; if it does conflict, reject the shape and go back to the previous step.

5. Check to see if the global proportions of the various shapes have been reached; if not, go back to step 2.

If there are well data that must be honored, they are typically honored first, as shown in Figure 2, and, when the interwell region is simulated, care is taken to avoid conflicts with the known sequence of lithologies in the wells.

The many implementations of this approach differ largely in how they ensure that the relative positioning rules are honored. Some workers follow the outline given earlier and simply reject anything that does not meet the rules; other workers try to be a bit more efficient and check to see if a slight repositioning or resizing of the current shape or of previous shapes might allow the positioning rules to be met. The chapter by Hatløy in this volume gives a good example of how a Boolean approach can be used to model facies in a fluvial environment.

Though Boolean simulation usually is based on the shapes of sedimentary units (turbidite channels, mouth bars, overbank deposits, barrier islands, . . .), it also has been used for simulating the shapes of the surfaces between sedimentary units rather than simulating the shapes of the sedimentary units themselves. This approach embraces the idea that the flow behavior of certain reservoirs is controlled by the thin, low-permeability barriers between the

sand packages. Boolean simulation has also met with some success in simulating fracture patterns by viewing them as thin elliptical disks that are randomly positioned with specified length and orientation distributions.

Estimation Plus Simulated Error: Fractal and Turning-Bands Simulation

One of the earliest methods for stochastic simulation was a method known as turning bands, which tackles the simulation problem by taking a smooth model created by estimation and then adding an appropriate level of noise to create a model that still honors the original data, but also has an appropriate pattern of spatial variability. In the mid-1980s, when the petroleum industry first began to take an interest in fractals, the same basic idea was used to make fractal simulations honor control points at well locations. Though no single name has ever evolved for this family of methods, we refer to them as ESE procedures (for Estimation plus Simulated Error) for the purposes of this overview.

Figure 3 shows an example of an ESE procedure for the simulation of the porosity within a producing horizon. Figure 3b shows a smoothed map obtained by interpolating the known porosities in the four wells shown in Figure 3a. Figure 3c shows an unconditional simulation that has the same histogram and pattern of spatial continuity as the actual porosities, but is considered unconditional because it does not

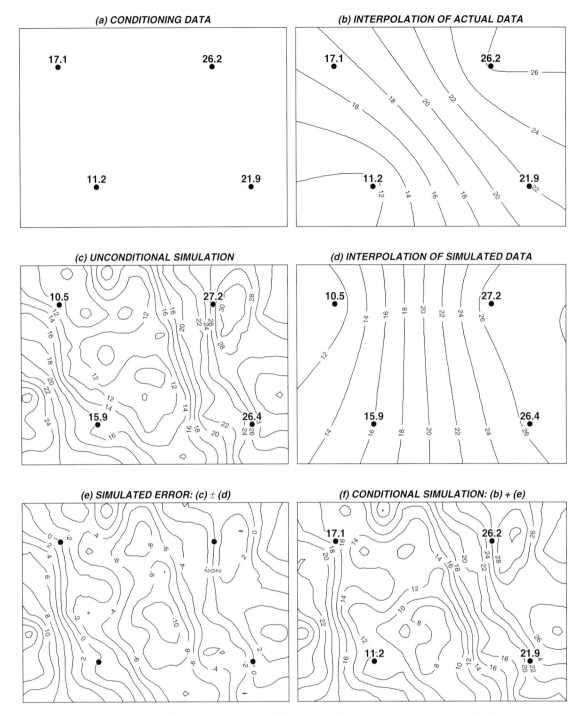

Figure 3. A stochastic model produced by adding simulated error to an interpolation.

honor the observed porosities at the well locations. This unconditional simulation can be generated in many different ways. In the turning-bands procedure, each value in the unconditional simulation is a weighted average of one-dimensional simulations produced on a set of radiating bands; in fractal procedures, each value is a weighted average of a large number of sine and cosine functions; in the example shown in Figure 3, the unconditional simulation was produced by a weighted moving average of white noise. Once this unconditional simulation is generated, it is sampled at

the locations where data actually exist. This sampling gives us a sample data configuration identical to the one used to do the original interpolation (Figure 3a); the sample data values, however, are different. We can interpolate these simulated samples and compare the resulting smoothed map back to the exhaustive unconditional simulation to see how much error we made at each location. Figure 3e, which shows the difference between the interpolation of the simulated samples (Figure 3d) and the exhaustive unconditional simulation (Figure 3c), is a map of the simulated error; it is zero at the locations where we have samples and tends to get larger as we move away from the actual sample locations. Figure 3f shows the result of adding the simulated error (Figure 3e) to the original interpolation of the actual data (Figure 3b); this is a conditional simulation that not only has the correct spatial variability, but also happens to honor the observed porosities at the well locations.

In addition to the conditioning information that we want to honor, an ESE simulation also requires us to describe the spatial variability of the phenomenon we are modeling; this spatial variability will be imposed on the unconditional simulation that we use to simulate the errors. In the turning-bands procedure, the spatial variability is typically described by a variogram or correlogram; in fractal simulations it is described by the fractal dimension.

All of the practical implementations of the ESE approach create the unconditional simulation through some kind of averaging process. The unconditionally simulated values thus tend to follow the bell-shaped normal distribution. Because this unconditional simulation is supposed to mimic not only the pattern of spatial continuity, but also the histogram of the original data, ESE procedures typically do not work with the original data values (unless they are normally distributed), but work instead with a transformed variable that is more symmetric. When simulating permeability, for example, it is unlikely that the original permeabilities will be normally distributed; their logarithms, however, might be closer to being normally distributed. There are more complicated mathematical functions called Hermite polynomials that accomplish the same purpose; there also are some simple graphical procedures that can transform the original data into an ideal normal distribution. Once the conditional simulation of the transformed values is produced, these transformed values need to be back-transformed to the original variable of interest.

The ESE procedures were historically the first to make extensive use of this idea of transforming to normality, simulating the normally distributed transform, and then back-transforming to the original variable of interest. Many of the later procedures, such as sequential Gaussian simulation and L-U* decomposition, also make use of the same approach.

Simulated Annealing

In simulated annealing the reservoir model is constructed by iterative trial and error. Figure 4 shows a simplified simulated annealing procedure for the problem of producing a sand/shale model that has a net-to-gross ratio of 70%, an average shale length of 60 m, and an average shale thickness of 10 m.

In Figure 4a the pixels in the model are randomly initialized with sands and shales in the correct global proportion. This initial model has the correct net-to-gross ratio, but because the sands and shales are

assigned randomly, its average shale length and thickness are too short. Figure 4b shows what happens after we perturb the initial model by swapping one of the shale pixels with one of the sand pixels. Such a swap will not affect the net-to-gross ratio because the global proportion of sand and shale remains unchanged. This swap, however, does make a tiny improvement in both the average shale length and the average shale thickness. Figure 4c shows the grid after a second swap. We are not quite as lucky with this second swap because it makes the average shale length and thickness a little worse, so we reject this second swap, go back to the previous grid shown in Figure 4b, and try again.

This swapping is repeated many more times and, after each swap, the average shale length and thickness are checked to see if we are any closer to the target values. If a particular swap does happen to get us closer to the length and thickness statistics we want, then we keep the swap; however, if the swap causes the image to deteriorate (in the sense that the length and thickness statistics are farther from our target) then we undo the swap, go back to the previous grid, and try again.

Figure 4d shows what the grid looks like after the 463rd swap, which happens to be a good one. At this point, 66 of the attempted 463 swaps have been good and were kept, but the remaining 397 swaps were bad and were not kept. Figure 4e shows the effect of the 464th swap, a bad one that was not kept because it causes the statistics to deteriorate. Figure 4f shows what the grid looks like after the 1987th swap. At this point there have been 108 good swaps and 1879 bad ones, and the shale length and thickness statistics are very close to our target values.

The key to simulated annealing is minimizing the deviation between the grid statistics and target values; the function that describes this deviation is usually called the energy or objective function. In the example shown in Figure 4, where we were trying to match the average length and thickness of the shales, the energy function was simply the sum of the absolute differences between the actual average dimensions and our target values:

$$E = |\text{Actual average length} - 60| \\ + |\text{Actual average thickness} - 10|$$

What simulated annealing tries to do is to drive the value of E to 0, thus decreases in energy are good and increases in energy are bad.

Though Figure 4 displays some of the key concepts in simulated annealing, it is a very simplified version and does not show some of the important aspects of practical implementations of simulated annealing. Most simulated annealing procedures do not recalculate the energy after each swap. What they do instead is perform several swaps before recalculating the energy. Initially, when the energy is high (the statistics of the grid are far away from the target values), many swaps are performed before the energy is recalculated; as the energy decreases and the grid statistics con-

*The L-U method takes its name from the Lower-Upper triangular decomposition that is done on the covariance matrix.

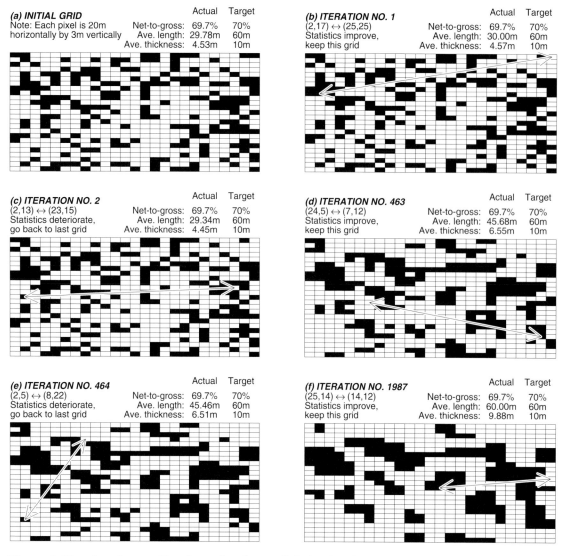

Figure 4. Simulated annealing for a simple sand/shale simulation.

verge to their desired target values, the number of swaps between recalculations is decreased.

The second important difference between most practical implementations of simulated annealing and the simplified example in Figure 4 is that they are somewhat tolerant of lack of progress. In the example shown in Figure 4, any swap that increased the energy (moved us farther away from our target statistics) was immediately undone. In the common practical implementations of simulated annealing, some increases in energy are tolerated. The chance of allowing the energy to increase is related to two factors: the magnitude of the increase and the number of iterations. Increases in energy are more tolerated early on in the procedure, and small increases are always more tolerated than large ones.

A third difference between the example given here and most practical implementations is that the example in Figure 4 does not have any conditioning information to honor. Conditioning information is quite easily honored in practice simply by setting the

appropriate pixels to the values observed in the wells and then ensuring that these conditioning pixels are never swapped later in the procedure.

The other important difference between most practical implementations of simulated annealing and the simplified example in Figure 4 is that the energy function is usually much more complex than the simple one used here. Simulated annealing is a very flexible approach that can produce stochastic models that honor many different kinds of information; the chapter by Kelkar and Shibli in this volume provides an example of the use of annealing for honoring the fractal dimension. For any desired features or properties that we can summarize numerically, we can use simulated annealing to honor these features by including appropriate terms in the energy function. In Figure 4, for example, we wanted to honor both the average shale thickness and the average shale length, so our energy function had two components. In more interesting practical examples, the energy function might have several more components. For example, we could have one component that describes how close the grid's variogram comes to some target variogram, a second component that describes how similar the synthetic seismic response of the grid comes to some actual seismic data, and a third component that describes how close a simulated tracer test on the grid comes to observed tracer test results.

Probability Field Simulation

In the sequential simulation procedures described earlier, the value drawn from the lcpd at a particular location is treated as if it were hard data and is included in the conditioning data set. This ensures that closely spaced values will have the correct short-scale correlation. If previously simulated values were not treated as hard data, then the random sampling of the lcpd based on only the original well data would result in too much short-scale variability. When based on only the original well data, and not including the previously simulated values, the lcpds at neighboring locations would be virtually identical, but the random sampling of the two very similar distributions might produce very different values, particularly if the distributions have a large spread.

The idea behind probability field, or P-field, simulation is to save computation time by basing the lcpd on only the original well data. To get around the problem previously described (too much short-scale variability when distributions are randomly sampled), the P-field approach controls the sampling of the distributions rather than controlling the distributions themselves as a sequential procedure would do.

Random sampling from a distribution is typically accomplished on a computer by first drawing a probability value using a uniform random number generator and then selecting the corresponding percentile from the distribution. For example, if the computer's uniform random number generator produced a probability value of 0.31, then we would take the 31st percentile of the distribution as our randomly sampled value. With sequential techniques, the random number or probability value used to sample the distribution at one location is completely independent from the one used to sample the distribution at a neighboring location. So we might, for example, use a very high probability value of 0.9 (the 90th percentile) from the lcpd at one location and a very low probability value of 0.1 (the 10th percentile) from the lcpd at a neighboring location. With the P-field approach, however, the probability values used to sample the local conditional probability distributions are spatially correlated. Thus, if 0.9 is the probability value used to sample the lcpd at a particular location, then the random numbers used to sample nearby lcpds have to be close to 0.9. The user controls the exact degree of correlation in the probability field, usually using the same pattern of spatial correlation as that of the variable being modeled.

The field of spatially correlated probability values can be produced by any technique for unconditional simulation. Any of the turning-bands, fractal, and moving average techniques that were discussed in the section on ESE methods could be used to generate spatially correlated fields of probabilities. The only slight modification that has to be made to the output from these procedures is that they have to be transformed from the bell-shaped normal distribution that these methods typically produce to the uniform distribution. This can be accomplished either by a graphical transformation, by ranking the values, or by using a polynomial approximation of the inverse of the cumulative distribution function for the standard normal distribution.

In this volume, the chapter by Srivastava shows how probability field simulation can be used for improving our ability to visualize uncertainty; the chapter by Bashore et al. shows a good example of how the P-field approach can be used to create an appropriate degree of correlation between porosity and permeability.

Matrix Decomposition Methods

A few stochastic modeling techniques involve matrix decomposition; the method of L-U decomposition is one example. With this approach, different possible outcomes are created by multiplying vectors of random numbers by a precalculated matrix. This precalculated matrix is created from spatial continuity information provided by the user, typically as a variogram or correlogram model.

Matrix methods can be viewed as sequential methods because the multiplication across the rows of the precalculated matrix and down the column vector of random numbers can be construed as a sequential process in which the value of each successive node depends on the values of the previously simulated nodes. They also can be thought of as ESE procedures because they can be written as an estimation plus a simulated error. For the purposes of this overview, however, they are treated separately because their use of a matrix decomposition gives them unique advantages and disadvantages.

Iterative Methods

Some iterative methods are similar to the simulated annealing procedure previously described in the sense that they iteratively update and improve a model that is initialized with no spatial correlation. These methods differ, however, in two important respects. First, they are not as flexible as annealing because the only statistics they try to honor are the histogram and variogram. Second, they do not update the grid by swapping, so they may not honor the histogram exactly.

Figure 5 shows an example of an iterative procedure used to simulate sand/shale lithologies. As with the annealing approach described earlier, the lithologies are originally assigned randomly and have no spatial correlation. The correlation is iteratively improved by visiting each grid node, discarding the value, estimating the local conditional probability distribution (lcpd) using the surrounding 24 values, and

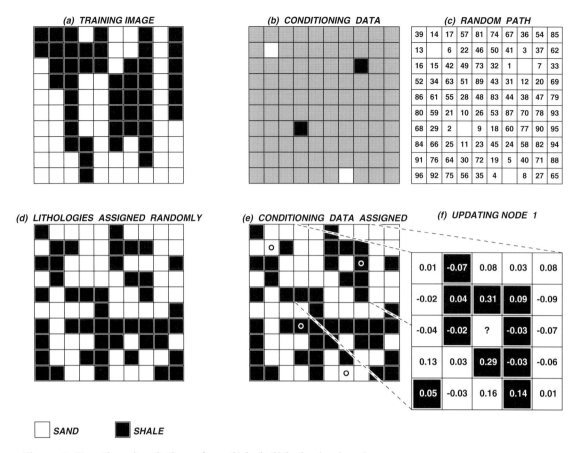

Figure 5. Iterative simulation of sand/shale lithologies in a layer.

then drawing a random sample from the lcpd to replace the previously discarded value.

In the example shown in Figure 5, the lithologies are known at four well locations shown in Figure 5b. Once the grid is randomly initialized with the correct net-to-gross ratio (Figure 5d), the lithologies at the conditioning locations are assigned their correct values. Once the lithology at any particular node is discarded, the local conditional probability distribution at that location has only two outcomes—sand or shale—and can be estimated using indicator kriging. The indicator kriging weights in Figure 5f are based on the exhaustive variogram from a training image (Figure 5a); the training image is not a necessary feature of an iterative technique, however, and the kriging weights could be based on a variogram model. Once the kriging weights have been calculated, the probability of encountering sand at the central location is the sum of the weights on the neighboring sand pixels and the probability of encountering shale is the sum of the weights on the neighboring shale pixels.

In this particular example, with strong north-south continuity in the training image, the indicator kriging weights are highest on the pixels immediately north and south. The local conditional probability distribution at the first node in the random path has a high probability of encountering shale because the pixels to the north and south are both shale; when this lcpd is randomly sampled, we will likely end up assigning

shale to the central pixel. As the updating procedure is repeated at the other locations in the random path (Figure 5c), the tendency is for the updated values to bear some similarity to their neighbors to the north and south, and the pattern of spatial correlation of the grid slowly acquires the north-south anisotropy of the training image.

Once the updating procedure has been run at each node in the random path, the pattern of spatial continuity will not be exactly what we want because we started from a very noisy image. The entire procedure has to be repeated several more times before the grid has an appropriate degree of spatial continuity. Even though the procedure has to be repeated several times, the kriging weights never change; thus, this is a remarkably fast procedure that involves repeated multiplication of a changing set of lithology codes by a fixed set of kriging weights.

Hybrid Methods

The practical success of the methods outlined above has been accompanied by the development of hybrid methods that take the best of two or more ideas. It is increasingly common to find reservoir models that are created using a combination of stochastic methods. For example, a Boolean technique may be used to create a model of the lithofacies within a reservoir, whereas a sequential Gaussian simulation is used to fill in the details of rock properties, such as porosity and permeability, within each lithofacies. Another type of hybrid starting to appear in the literature is the use of annealing as a postprocessor of models created by other techniques. The chapter by Murray in this volume provides an example in which the results of a sequential indicator simulation are used as the starting point for an annealing procedure; with this type of hybrid procedure, the simulated annealing step imparts statistical properties that are not directly honored by indicator simulation.

A different way of combining different techniques is to use one technique as the basis for developing parameters for another technique. In the study presented by Cox et al. in this volume, a deterministic approach is used to create plausible images of a sequence of eolian sediments. These computer-generated images are then used as the basis for calculating indicator variograms for a sequential indicator simulation procedure.

Though listing all of the possible hybrid methods is not particularly useful (and perhaps not even possible), it is worth emphasizing that many of the disadvantages discussed in the following section can be mitigated by combining the strengths of different methods into a hybrid procedure.

ADVANTAGES AND DISADVANTAGES

The following discussion of the advantages and disadvantage of the various methods is structured according to the goals discussed at the beginning of this chapter.

Stochastic Models as Art

If a stochastic model is being used as a visual catalyst for more critical thinking, then one of the primary criteria for choosing a method has to be the visual appeal of the final result.

For stochastic modeling of facies, the visual appeal of a good Boolean approach is hard to beat. Geologists like to be able to see familiar sedimentary features, and a Boolean approach gives them direct control of the shapes they would like to see. When used for facies modeling, pixel-based techniques, such as sequential indicator simulation, tend to produce images that, to a geologist's eye, are too broken up.

For stochastic modeling of rock or fluid properties, such as porosity, permeability, or saturation, a Boolean approach is unable to complete the job alone. Such an approach may be useful as a first step, but the artwork is often unconvincing if the rock properties within each facies are assigned randomly or assumed to be constant. The images look more plausible when the rock properties within a given facies have some variability and some pattern of spatial continuity. With most of the techniques for modeling continuous variables, the spatial continuity is described through a variogram model. Though experimental variograms of rock and fluid properties typically show a nugget effect, experience has shown that stochastic models of

continuous properties have more visual appeal if the variogram model used in the stochastic simulation procedure has no nugget effect.

For stochastic modeling of a continuous property, such as the thickness of an unfaulted reservoir, sequential techniques tend not to work well. The same variogram models that describe very continuous surfaces—the Gaussian variogram model, for example, or a power model with an exponent close to 2—also produce chaotic weighting schemes when some of the nearby samples fall directly behind closer samples. Because sequential procedures include each simulated value as hard data that must be taken into account when future nodes are simulated, and because they are typically performed on a regular grid, chaotic weighting schemes are virtually guaranteed whenever we are trying to simulate a very continuous surface. The result is often very unsatisfying visually. For such problems, one of the ESE techniques or P-field simulation would produce a more acceptable result.

Most of the stochastic modeling techniques for continuous variables give us the ability to honor a variogram or a set of indicator variograms. If we were trying to produce stochastic artwork that shows specific sedimentary features, such as a succession of fining-upward sequences or a fabric of cross-cutting features (neither of which is well described by a variogram), then annealing commonly is the most useful approach. As long as we can find a way of numerically summarizing the sedimentary features that we would like to see, then we can include appropriate terms in our energy function for annealing.

Stochastic Models for Assessing the Impact of Uncertainty

When the goal of a study is to produce a handful of models that capture a few extreme scenarios, as well as some intermediate scenarios, one of the practical issues that comes to the forefront is computational speed. This is especially true because it is usually not possible to steer a stochastic model to an optimistic realization or to a pessimistic one; the only way of finding such extreme scenarios is to produce many realizations and sift through them all.

Getting an objective ranking of the many different stochastic modeling algorithms is difficult because their respective authors believe that their own method is lightning fast. Though all of the methods discussed here are workable in practice, some require several days of run time on fast computers to produce a single realization despite their author's enthusiastic claims of speed. Such procedures cannot be a practical basis for producing many realizations.

As a general rule, sequential procedures are slow because they have to search through an ever-increasing set of hard data to find the nearby samples relevant for estimating the local conditional probability distribution. The notable exceptions to this are the sequential simulation procedures based on Markov random fields. These sequential procedures considerably accelerate the search for previously simulated data by using only the closest ring of data; some of these procedures accomplish greater speed by precalculating the kriging weights that eventually will be needed during the sequential simulation.

The fastest procedures are those that precalculate as much as possible and minimize the amount of recalculation needed to produce new realizations. Probability field simulation is generally faster than sequential approaches because the local conditional probability distributions are the same for all realizations; the only thing that needs to be updated is the field of spatially correlated probability values. The ESE methods have a similar efficiency because the kriging weights for any particular estimation do not change from one realization to the next and, therefore, can be precalculated and stored.

Matrix decomposition methods can be among the very fastest available because once the matrix decomposition is performed, the rest of the procedure is simply one of repeated multiplication. Unfortunately, these matrix decomposition methods are limited by the fact that it is difficult, even with very powerful computers, to do the matrix decomposition for more than a few thousand points. With many reservoir models containing millions of grid nodes, the matrix decomposition approach is often not feasible.

Like the matrix decomposition methods, iterative methods also accomplish tremendous speed by precalculating the necessary kriging weights and reducing the simulation exercise to one of repeated multiplications; however, unlike the matrix decomposition methods, they are not limited to small grids.

Annealing is a highly variable procedure in terms of its run time. For certain problems, annealing is very rapid; for other problems, it is abysmally slow. The key to rapid annealing is the ability to update rather than recalculate the energy function. If a global statistic has to be recalculated from scratch every time two pixels are swapped, then annealing is not likely to be speedy; if there is some convenient trick to updating the previous calculation rather than recalculating from scratch, then annealing can be very efficient.

Boolean techniques can be very quick if they do not have to deal with conditioning information in wells or with relative positioning rules. If these techniques do have to check for conflicts with wells or with previously simulated shapes, then they can be very slow.

Stochastic Models for Monte Carlo Risk Analysis

For studies that involve Monte Carlo risk analysis, there are two criteria that need to be considered when selecting a stochastic modeling technique. The first criteria is the computational speed issue discussed in the previous section; the second criteria is the issue of whether the stochastic method is fairly sampling the space of uncertainty.

This issue of the space of uncertainty is currently one of the thorniest theoretical problems for stochastic modeling. Some workers take the point of view that the space of uncertainty must be theoretically defined outside the context of the algorithm itself so that the user does have some hope of checking that a family of realizations represents a fair sampling of the space of uncertainty. Other workers, however, take the point of view that the space of uncertainty can be defined only purely through the algorithm and that it is, by definition, the family of realizations that is produced when all possible random number seeds are fed to the algorithm.

This issue of the equiprobability of the realizations and the fair sampling of the space of uncertainty comes to the forefront when annealing is used as a postprocessor of the results from some other technique. Though some theory backs up the claim that annealing fairly samples the space of uncertainty, this theory is entirely based on the assumption that the starting grid is random (or that the initial energy is very high). Though annealing offers tremendous flexibility in incorporating and honoring widely different types of information, there remains an awkward ambiguity about its implicit space of uncertainty and whether this space is being fairly sampled.

Stochastic Models for Honoring Heterogeneity

If the reason for producing a stochastic model is to respect the critical heterogeneities that control flow, then the choice of a stochastic modeling method should consider whether we actually get to control the critical heterogeneities or not.

One of the heterogeneous features that is usually difficult to model appropriately and yet is critical to flow behavior is the fracture network. As discussed, Boolean methods have often been used to model fractures as thin elliptical disks. Another approach to this problem is discussed in the chapter in this volume by Hewett, in which he describes the use of fractal methods for fracture characterization.

If the important control on the flow behavior is the spatial arrangement of the original facies, then a Boolean approach usually is best because it offers the user direct control over the size, shape, and orientation of the major flow units. Once we have a model of the architecture of the major flow units, this model can be used to guide the assignment of the extreme permeabilities. The chapter by Tyler et al. in this volume provides an example in which the sedimentological reservoir model is created using a Boolean approach and the petrophysical reservoir model is then created by simulating rock and fluid properties within each lithofacies.

If there is no facies architecture to guide the spatial positioning of the extreme permeabilities, then an indicator-based approach offers us considerable flexibility in describing the spatial arrangement of the critical extreme values. Most of the techniques that do not make use of indicators—all of the common ESE techniques, sequential Gaussian simulation, and matrix decomposition methods—do not offer the user direct control of the spatial continuity of the extreme values and implicitly assume that the most extreme values have poor connectivity. If this poor connectivity of extremes is contradicted by geologic knowledge (as is often the case), then such methods should not be used as the basis for modeling heterogeneities for flow simulation.

If some of the information about reservoir heterogeneity comes through well tests or from production information, then annealing is the technique most capable of directly accommodating this type of information. Annealing has been used, for example, to directly honor tracer test results; though the procedure is very slow (because the tracer test response cannot be readily updated when pixels are swapped), the resulting scenarios are tremendously realistic because they capture the connectivity of the extreme values that is implicit in the actual tracer test observations. The chapter by Deutsch and Journel presents the use of annealing for integrating permeabilities derived from well tests.

Stochastic Facies or Stochastic Rock Properties (or Both)?

As previously discussed, a model of the reservoir architecture often is best accomplished with a good Boolean approach. Once the positions of the major flow units have been modeled, some other approach can be used to fill in the details of the rock and fluid properties within each flow unit.

Though there are some very good implementations of Boolean methods for certain sedimentary environments, other environments do not seem to have been given as much attention. For example, though there are several good case studies of Boolean methods in turbidite and fluvial settings, there are very few (if any) case studies of Boolean methods in eolian or carbonate settings. In such situations, if a good Boolean approach is not available, then it may be necessary to tackle the facies modeling with some other type of approach.

Indicator methods often are well suited to facies modeling because they deal quite naturally with categorical facies codes. The chapter by McKenna and Poeter provides an example of the use of indicator methods for simulating facies for transport modeling in groundwater studies. Though most of the common implementations of indicator simulation define the indicator according to whether a continuous variable exceeds a given threshold, it is easy to modify these techniques so that they define the indicator according to whether some categorical code falls within a certain class.

The other stochastic modeling methods that deal with continuous variables—the ESE techniques, sequential Gaussian simulation, and matrix decomposition methods—can sometimes be used successfully to simulate categorical facies codes. As shown in the chapter by MacDonald and Aasen, facies models can be created by simulating a continuous variable that is then reclassified into a set of categorical codes according to whether the continuous variable is within a certain range of values. One of the practical problems with this approach is the issue of how to turn the original facies information into a continuous variable so that it can be used to condition the continuous simulation. Another practical problem is the issue of how to order the categorical codes to capitalize on the tendency of the continuous variable to pass through class 2 on its way from class 1 to class 3. If the depositional arrangement of the facies is such that the succession of facies is not a predictable sequence, then the use of a continuous variable as the basis for a facies simulation may induce some undesirable artifacts; however, if there is a predictable succession of facies, then the use of an indicator simulation may not capture the desired transitions between facies.

One of the advantages of the probability field simulation approach is that it offers some flexibility in the random sampling of the local conditional probability distributions of facies codes. Indicator kriging could be used to build the local probability distributions that provide the probability that the location belongs to any particular facies code. When the probability value from the spatially correlated P-field is used to sample these local distributions, it is possible to order the facies codes in such a way that certain transitions are more likely to occur than others. In this way, the P-field approach allows us to use indicator methods (and to avoid the problem of converting facies codes into continuous variables) while also honoring information about transitions between facies.

The chapter by Doveton provides a good overview of the statistical analysis of the transitions between facies and of the simulation of successions of facies in one-dimensional, two-dimensional, and three-dimensional models.

Stochastic Models for Honoring Complex Information

When the goal of a stochastic modeling exercise is to integrate measurements from wells with other types of information, such as seismic data, production information, and well test results, annealing becomes the most attractive of the various stochastic modeling alternatives. Even though the evaluation of the energy function is often slowed down considerably when it includes more complicated information, the annealing technique still eventually works.

Any of the stochastic modeling techniques that involve the explicit estimation of a local conditional probability distribution can easily accommodate secondary information available as a dense grid, as most seismic data would be. In sequential Gaussian simulation, for example, by switching from kriging to cokriging, we could honor the correlation between rock properties and seismic attributes; the same is true of the common ESE methods and P-field simulation. In this volume, the chapters by Almeida and Frykman and by Chu et al. provide examples of the use of the sequential Gaussian approach with cokriging to integrate seismic information. Another common adaptation of kriging that allows secondary information to be honored is the method known as kriging with an external drift; the chapter by Wolf et al. shows how this method can be used for integrating seismic data with sparse well data in a study of volumetric uncertainty.

If the additional information is geological rather than geophysical, then Boolean techniques are often the most attractive. Though most Boolean techniques have difficulty honoring dense seismic information, they excel at honoring information about the geometries and relative positioning of the various facies. Boolean techniques rarely make direct use of high-resolution three-dimensional seismic data, but use it to provide additional insight into geometric properties, such as channel widths and orientations. Boolean techniques also can incorporate insights that pressure transient analysis often provides on distances to permeability barriers.

Other Considerations

Though the goal of a study will usually be the most important consideration when choosing a stochastic modeling technique, some other factors will also constrain the choice.

The availability of data is an important issue when choosing between a parametric approach and a nonparametric, or indicator, approach. The basic philosophy of an indicator approach is to avoid unnecessary distributional assumptions and to let the data speak for themselves. While this usually sounds like a good idea, it should be quite clear that if we intend to let the data speak for themselves, then we need a fair bit of data to do the talking. If we have only a handful of data, then an indicator approach is likely to be frustrating. With parametric techniques, such as sequential Gaussian simulation, the assumed distributional model does the talking and we need only enough data to defend our choice of parameters for the distributional model.

Another factor that impinges on selecting an indicator method versus a method that deals with continuous variables is the need for resolution in the final model. In this volume, the stochastic structural modeling of an isopach variable in the chapter by Høye et al. would not have worked well with an indicator approach; the authors' choice of an approach that deals directly with continuous variables reflects the need for much higher resolution than an indicator method can provide with sparse data. When modeling the permeability in a reservoir, however, an indicator approach that breaks the permeability into several discrete classes is often quite acceptable because this achieves enough resolution for the purposes of flow

simulation; the chapter by Cox et al. describes a study in which permeabilities were simulated by using several indicators.

One of the important practical considerations with Boolean techniques is the spacing between well control relative to the size of the geometric shapes being simulated. If we have abundant well control and are trying to position large features, then most Boolean techniques will struggle with the repeated conflicts between the random shapes and the well control. Boolean techniques are generally much more successful with sparse well control.

A final consideration for any practical case study is the availability of good software, one of the issues raised by Coburn in his chapter. Certain techniques have been around long enough that there exist good implementations that are readily available in the technical literature. For example, there are several good public-domain implementations of the ESE technique of turning bands. Though somewhat younger than the veteran turning-bands procedure, sequential indicator simulation and sequential Gaussian simulation are also readily available. There are several good implementations of the basic Boolean procedure, and the details of honoring relative positioning rules can also be found in the technical literature. On the other end of the software maturity spectrum are the simulated annealing procedures. Though these simulated annealing procedures have tremendous potential for incorporating secondary information, there is still not enough readily available software and technical literature for the uninitiated user.

CONCLUSIONS

Though the discussion here has addressed some of the important shortcomings of all of the various methods, all of the methods discussed here have been successfully used in actual case studies. Unfortunately, much of the technical literature and many of the commercial software packages for stochastic modeling promote a single method as better than all others. There is no stochastic modeling method that is universally best for all possible petroleum reservoir problems. As stochastic modeling becomes more accepted in the petroleum industry, and as more stochastic modeling techniques are developed, the most successful case studies will be those that view the growing assortment of methods as a tool kit rather than as a set of competing methods.

SECTION I

Getting Started

The opening chapter, **Geostatistics and Reservoir Geology**, appropriately comes from André Journel. He is one of the world's leading experts in geostatistics and is largely responsible for its development and proliferation in petroleum geology. He presents geostatistics as a toolbox to analyze data and transfer the analysis and interpretation into the task of reservoir forecasting. He briefly addresses data integration, the importance of quantification and uncertainty assessment, and finally, the concept of stochastic imaging.

Several factors contribute to the recent acceptance of geostatistics in the petroleum industry. Declining oil prices, poor hydrocarbon recovery, and complex reservoirs have resulted in a need to find methods that can provide us with a better understanding of a reservoir. At the same time, we need to quantify the uncertainty of our reservoir description. T. C. Coburn presents a personal view on past barriers to the acceptance of geostatistical technology and its proliferation in the oil industry. Coburn, who was manager of quantitative geology for Marathon Oil Company, also points out key ingredients necessary to its successful implementation in future projects in **Reflections on the Proliferation of Geostatistics in Petroleum Exploration and Production**.

Chapter 2

Geostatistics and Reservoir Geology

André G. Journel
Department of Petroleum Engineering
Stanford University
Stanford, California, U.S.A.

INTRODUCTION

In geostatistics, the prefix "geo" clearly links geostatistics to the earth sciences. The geostatistical glossary (Olea, 1991) defines geostatistics as "The application of statistical methods . . . in the earth sciences, particularly in geology." Geostatistics provides a toolbox for the geologist to use in analyzing data and transferring such analysis and interpretation to the task of reservoir forecasting. A tool can never replace data, but it can help build an interpretation and the corresponding numerical model. Geostatistics is no substitute for the geologist's experience in formulating the model properties, but it may help in creating the model.

Geologists, at times, have become alarmed at the notion of geostatisticians treading over their turf and replacing some of their well-thought-out deterministic depositional models with "random numbers." Conversely, some engineers may have been wary of geostatisticians lending their numerical skill to geologists to impose upon them reservoir models with resolution far beyond that of traditional and comfortable layer-cake models, a resolution whose accuracy they question and which, in any case, their flow simulators cannot accommodate. Such concerns should be put to rest as reservoir modeling becomes truly interdisciplinary, and geostatistics is seen not as a self-centered discipline, but as a set of tools to be used jointly by geologists, geophysicists, and reservoir engineers.

DATA INTEGRATION

Because hard data at the correct scale are scarce (due to the expense), the single most important challenge in reservoir modeling is data integration. Relevant information comes from various sources at various scales with varying degrees of reliability. For example, geological data originates at many scales, from the microscopic image of a pore volume to the basin-scale interpretation of a depositional system; geophysical (seismic) data, often the most prevalent, span from the macroscale of well logs to the scale of the entire field. The former case provides resolution approaching the 10-ft (3-m) level; engineering data are more at the scale of the forecasting exercise, being from hundreds of feet for well tests to the entire field for production data. The challenge is to collect from that variety of data sets the information relevant to the final goal of the reservoir model. The information relevant to volumetrics (in situ) calculations need not be that relevant to recovery forecasting under a complex oil-recovery process.

The objective is not for each reservoir discipline (e.g., geology, geophysics, engineering) to maximize its contribution to the final reservoir model, rather the challenge is for each discipline to understand the specific forecasting goal and provide only the relevant information in a usable format and with some assessment of uncertainty. A facies distinction need not correspond to different flow units because of hydrodynamic continuity across facies due, for example, to later diagenesis or microfracturing. The major contribution of geostatistics to reservoir modeling lies in data integration; providing a formalism to encode vital, possibly nonnumerical information; combine different data accounting for uncertainty; and transfer such uncertainty into the final forecast.

IMPORTANCE OF QUANTIFICATION

Reservoir modeling, flow simulation, and recovery forecasting are eminently numerical exercises. Geological information will be accounted for only if it is made quantitative and is associated with some measure of uncertainty. As opposed to exploration geology, the qualitative understanding of the genesis of a formation may not be critical to performance forecasting as long as the hydrodynamic properties of that formation can be quantified from pore scale to reservoir scale. By participating directly in the numerical modeling with his or her own geostatistical skills, the geologist is better positioned to evaluate which geological information is relevant and control its proper integration. With the first computers, earth scientists had to rely on analyst programmers to harness computing power; later, digital computing became an additional tool controlled by the earth scientist. The same history applies to geostatistics, where tools and software are now in the public domain and widely available to all. As with any tool, the best geostatistics is that performed by whomever is comfortable with the tool, but most important, by those who understand the problem at hand and its controlling factors. In reservoir modeling, the problem is recovery forecasting, where the important controlling factors are aspects of the reservoir geology. Geostatistics and computers are tools to better analyze data, build numerical models, and then image the consequence of such models. Numbers, whether hard (single value) or probabilistic (distributions), are necessary vehicles for the transfer of information, in our case from geological interpretation to reservoir forecast.

IMPORTANCE OF ASSESSING UNCERTAINTY

Because data originate from such a variety of sources and at many scales, and because their relevance to the final goal of reservoir performance forecasting is not always clear before the fact, all data must be calibrated and the uncertainty of their information content assessed. Calibration means that some data are considered reference or "hard," whereas other, "soft" data are calibrated against them. Hard data are honored exactly by the numerical model; in geostatistical jargon they are "frozen." Soft data need not be honored as exactly; the degree to which they are honored depends on an assessment of their reliability, ideally provided by the calibration exercise. In many instances, however, there are not enough data to perform numerical calibration; in these cases calibration may be borrowed from another field, e.g., from an outcrop, data taken at another scale, lab data, or experience. In the latter case, all parties involved in the reservoir modeling must accept such uncalibrated information, which would then be frozen in the model without further question. Expert prior decisions are at the roots of any good reservoir model, and any such prior decisions should be stated clearly.

Two good reasons for calibration are that it

- Forces the proponent of a certain piece of information to document its origin and its relevance to the problem at hand, and
- Allows the assessment of the impact of that information on the final forecast through sensitivity analysis or, better, through the geostatistical tool of stochastic simulation (discussed in a following section).

Few geophysicists discuss the fact that seismic impedance values derived from a dense, three-dimensional seismic survey should be calibrated to well logs or better core data before being used to map porosity. Yet, many geologists would like their appreciation of reservoir continuity or compartmentalization to be taken at face value when such a decision is much more critical to predicting reservoir performance. Geophysicists may resent such lopsided treatment of their data. At the very least, some sensitivity analysis to the geological interpretation/decision should be performed. Again, geostatistics provides tools for such calibration or sensitivity analysis.

CONCEPT OF STOCHASTIC IMAGING

A stochastic simulation (imaging) algorithm is a mechanism that allows drawing alternative, equiprobable, spatial distributions of objects or pixel values. Each alternative distribution constitutes a stochastic image. There can be several stochastic images of the same phenomenon, for example, the distribution in space of reservoir units or facies. Each stochastic image/realization/outcome honors (1) specific statistics, such as a histogram, covariance, variogram, or correlation coefficient; for example, the simulated facies can be made to honor volume proportions, size distributions, aspect ratios, etc., and (2) hard and soft data at specific locations. Hard data, such as well data, are reproduced exactly by all realizations, whereas soft data are reproduced with some degree of tolerance.

From the constraining statistics and data honored, one cannot say that one realization is better than another. Yet, the various stochastic images differ one from another at locations away from the hard data locations. That difference provides a visual and quantitative measure of uncertainty about the properties being imaged.

Different equiprobable numerical reservoir models (different realizations) can be built, all honoring the same hard and soft data. These alternative models can be processed through the same flow simulator to yield a distribution of reservoir forecasts; thus, the transfer of uncertainty from the geological/petrophysical model to the final forecast has been achieved.

There is nothing uncertain about reality; it is our vision of that reality that is uncertain. Uncertainty, not being intrinsic to the phenomenon under study, cannot be estimated; rather, it is modeled. All uncertainty measures are models, all based on somewhat subjective decisions about what should be considered uncertain or soft and what should be considered unquestionable or hard or frozen. Better that we create a subjective model of uncertainty than an illusion of certainty.

REFERENCE

Olea, R., ed., 1991, Geostatistical glossary and multilingual dictionary: Oxford, Oxford University Press, 177 p.

◆

Chapter 3

Reflections on the Proliferation of Geostatistics in Petroleum Exploration and Production

T. C. Coburn
Marathon Oil Company
Littleton, Colorado, U.S.A.

◆

ABSTRACT

Use of geostatistics to address problems in the exploration and production segments of the petroleum industry is now steadily growing, due, in part, to increased awareness, incorporating better science in the development of contemporary techniques, greater access to advanced computer resources, and enhanced emphasis on interdisciplinary problem solving. This paper discusses some of the past barriers to the proliferation of the technology and enumerates key ingredients to the successful implementation of future geostatistical projects.

Geostatistics, as an art and science, is growing up within the petroleum industry, particularly within exploration and production. After approximately 10 years of determined effort by a relatively small number of individuals at a handful of industry and university research centers, geostatistics finally, although cautiously, seems to be catching on. In fact, because there has been such an infusion of interest in the early 1990s, signs of a small bandwagon effect are actually beginning to crop up!

What is driving this newfound acceptability is the realization that some very thorny problems, such as optimizing reservoir performance and recovery, seemingly cannot be fully addressed without probabilistic and numerical tools. Geoscientist and engineers are developing an appreciation for the use of geostatistics to characterize the variability associated with all sorts of studies of the unobservable subsurface, from the risking of competing exploration scenarios to the placement of development wells to the analysis of fracture, fault, and flow patterns.

It certainly has not hurt to have had an injection of better science in the day-to-day practice of geostatistics, or to have experienced an explosion in the growth and availability of the computing power necessary to crank through complex algorithms and process enormous volumes of data. It also has been fortuitous that geostatistics has helped bring representatives of diverse engineering and geoscience disciplines together for the common good of solving tough exploration and production problems—a highly desirable arrangement that previously has been paid a good deal of lip service, but not given much real encouragement.

As alluded to, in many circles of the petroleum industry, it is the mathematical constructs of geostatistics that are now being given so much credence. Since geologists began collecting measurements on rocks, geophysicists embarked on shooting seismic, and engineers started simulating and predicting reservoir performance, it has been practically impossible to combine the various sources of information because of differences in scale. Whereas geologists frequently collect measurements on formations at locations separated by small distances or conduct experiments on fairly small pieces of rock, engineers need information on the order of cubic miles and acre-feet because of the geographic expanse encompassed by most oil fields. The principles of geostatistics offer today's best hope of scaling up the geologists' microscale measurements to the macroscale of reservoir simulation and prediction in such a way as to effectively imprint the details of geological description on simulation outcomes.

Obviously, geostatistics has not always enjoyed as much nurturing as it is receiving today; frankly, there is still a long way to go before it is accepted as a routine technology. Although engineering professionals have become fairly avid supporters and are busily incorporating geostatistical methods and concepts into their studies, geoscientists still largely look at these ideas with jaundiced eyes.

The differences are age-old—engineers are comfortable with mathematical constructs, but geologists find them threatening. Engineers have a fair appreciation of the concepts of precision, error, and accuracy, but the work of geologists, by its very nature, is highly inexact. Even today, the most active geostatistics research programs (and there are some very well-funded ones) in academia are found imbedded in engineering programs. Although geology is recognized as an important and necessary component of geostatistical research and practice at these centers, it is often the weakest academic link. Sadly, the "geo" aspect of geostatistics continues to fight for survival in the petroleum industry today.

Given the ever-increasing acceptance of geostatistics by other disciplines and industries that deal in spatially- and geographically-oriented data, it is difficult to reconcile that the petroleum industry—the single industry that deals almost exclusively with these kinds of data—lags so far behind. The source of this quandary harkens back to the nonquantitative nature of geologists and their managers. Because these individuals are empiricists who live and breathe the physical earth that can be touched, seen, and felt, they have difficulty with the abstraction of equations.

Knowing this fact, it should be obvious to the proponents of geostatistics that they should package their products in such a way as to make them more palatable and nonthreatening to the geologists and geophysicists whose support they need. Unfortunately, just the opposite has been true. Consider the fact that the geostatistical literature is replete with algebraic abstraction and obtuse terminology that even professional statisticians find difficult to digest.

There are some additional sticking points. First, the community of geostatistical practitioners, both in this country and around the world, is somewhat cliquish, and many of the most prominent researchers have never personally worked on a real petroleum industry problem. Second, there has never been a clear, unapologetic distinction between statistics and geostatistics; those geoscientists who once learned about simple means and variances in college do not understand the relationship. Third, new geostatistical ideas and approaches are being promulgated more rapidly than they can be assimilated in everyday exploration and exploitation, leaving the impression (right or wrong) that validation against real data is not adequately undertaken. Fourth, exorbitantly priced computer packages are proliferating that promise the world, but deliver very little. Finally, many of the applications published to date leave the impression that data are being made up. Whereas engineers can tolerate the concepts of pseudo-wells, pseudo-functions, and even pseudo-data, geologists cannot! All of this has led to an unnecessary and unfortunate mystification of geostatistics that, in an early 1990s economic environment, is too risky for mainline petroleum explorationists to embrace.

With the lukewarm response of geologists and other explorationists to such misdirection, the upshot has been for some engineers in some oil companies, wanting to proceed; to simply do the geology, geophysics, and log analysis themselves. There could not be a more undesirable situation. This is not to say that such practice is universal; for there have been a number of excellent examples of cooperative, interdisciplinary, highly successful efforts that have virtually been glued together by the concepts and mechanisms of geostatistics. On the whole, though, there has been a definite trend for the two sides of the exploration and production business to go their separate ways on this technology. Perhaps what is now needed is simply to accept the foibles of the past as growing pains, focus more clearly on the mutual objectives to be achieved, and look more critically and positively at what it takes to implement and execute successful geostatistical projects in this business.

To these ends, one must recognize that successful geostatistical efforts require some key administrative ingredients. First and foremost, there needs to be a team of individuals committed to communication. In addition to being skilled in their individual disciplines, they must be highly flexible and willing to compromise for the overall good of the project. Unfortunately, geologists and engineers really do not speak the same language (and never will!), because their paradigms are different. For example, when a geologist engages in a description of reservoir variation or heterogeneity, she or he may choose to assess rapid changes in lithology, depositional setting, or porosity; but when an engineer talks about reservoir heterogeneity, she or he is usually interested in permeability changes. This intellectual clash is only exacerbated when numbers and equations (especially complex ones) are introduced, because suddenly, conflicting notions must somehow come together and be reduced to finite, concrete quantities.

Second, it goes without saying, that to have a successful geostatistical project requires unrestricted access to the knowledge and expertise of a trained geostatistician. Because it is easy for geostatistics to be misapplied and the results to be misinterpreted, attempts at geostatistical analysis by less-than-competent individuals usually are scoffed at by management because the analysis takes too much time, is too obvious or simplistic (the answer could have been obtained from other sources without all this work!), or eventually leads to results that are unhelpful, unbelievable, or blatantly wrong.

Third, there has to be a strong project leader. The leader must be someone who can be objective and can properly discern the project goals, the discipline-specific inputs, and the assumptions under which the geostatistical approaches are to be applied, all of which may conflict. In many cases, the most logical individual to serve in this role is actually the project geostatistician. The geostatistician, above all others, understands the data constraints of all parties and appreciates what can and cannot be accomplished with the tools available. However, if the geostatistician assumes this role, she or he must have exceptional consulting and negotiating skills and peer credibility, be a master of language and communication, and be broadly familiar with all disciplines involved in the project.

Having access to reliable, digital data in sufficient quantities is a must. As the old saying goes, "you can't get blood out of a turnip." As in all other types of geoscience modeling work, it is simply not possible to generate believable geostatistical results (reservoir characterizations, basin assessments, etc.) without adequate data. In today's environment, that data must be available in digital form. Someone, presumably the project leader, must be responsible for pulling together all the well logs, seismic sections, grid files, core measurements, and other data, and transforming them into a consistent format that can be easily imported by the software of choice. This, in turn, requires cooperation among all the sources or individuals (both internal and external) who supply information. In the end, geostatistical work is fairly tedious, and all of the data pieces need to fit together like a tightly constructed puzzle—one engineered by the geostatistician—to achieve good results.

Adequate computing resources also are a must. Geostatistical algorithms are complex, requiring lots of CPU cycles. In many exploration and production projects, the volumes of data to be processed are enormous. The computer resources needed to perform geostatistical computations are similar to those required in reservoir simulation. Adequate disk space, memory, processing speed, and network capacity, among other things, must be provided to complete the project on time. In addition, the right combination of software must be available (today's products are largely UNIX-based), and at least one project team member, usually the geostatistician, must be skilled in executing the code, as well as in operating the computer equipment.

Management must give its support if any geostatistical project is to succeed. In today's environment, managers and supervisors must have a certain amount of trust in the project staff simply to get a project started. Because geostatistical efforts are often interdisciplinary in nature and cross organizational bounds, the supervisors of all of the project's members must exhibit at least the same degree of cooperation as those individuals assigned to the team. They must commit to learning something about the basic principles and philosophy of geostatistics, and they must display a certain degree of patience and tolerance with problem-solving approaches that, to them, may seem out of the ordinary or unconventional. It goes without saying that adequate funding needs to be provided at all stages of a geostatistical project for it to be a success.

Finally, there must be a technical support staff. The geostatistician cannot do all of the work alone, even though she or he frequently contributes more than a fair share simply because the other project members are unlikely to have the same degree of numerical understanding. Having technicians who can collect information, input data, monitor the progress of computer-generated realizations, and do other similar tasks helps free up more of the geostatistician's time to design an effective modeling program.

Many of these same points can be applied to almost any kind of petroleum industry undertaking. However, they are especially crucial to geostatistical projects because of the mathematical complexity involved and the potential for misunderstanding. The seemingly natural wariness of geoscientists toward geostatistical techniques underscores the need for everything to work smoothly. Otherwise, the focus and criteria by which geostatistical projects are judged will inadvertently be shifted to peripheral operational issues.

The beauty of geostatistics is that it can bring order to chaos by logically aiding in incorporating ranges of mutually acceptable answers, explicitly honoring factual data, and weighting the professional intuition and experience of all parties involved. However, geostatistics should not be portrayed as a panacea, and it cannot be loosely and inappropriately applied. Rather, it should logically and intelligently be brought to bear on exploration and production problems where it can have a real impact. For the technology to proliferate beyond its current level of use, the geostatistical and geoscience communities must actively cooperate to specifically identify such situations, and they must jointly establish effective problem-solving strategies.

For their part, the geostatisticians must work to remove the mystique surrounding the technology that prohibits it from being more widely endorsed. This basically means their medium of communication must be altered. Instead of speaking the language of mathematics, geostatisticians must adopt the vernacular of geoscientists. They must effectively portray their ideas and concepts in nonthreatening ways using maps, cross sections, and other visual aids that geoscientists readily understand and accept.

With the adoption of all of these principles, it is easy to see how geostatistics may have a major impact on exploration and production problem solving in the next 10 years. If nothing else, considerable benefits will surely be derived from the data integration that geostatistical approaches foster. Proof of the technology's importance, however, will likely be realized only by validating its predictive capabilities against those of more conventional, deterministic methodologies. Hopefully, such an effort will not take more time and patience than the petroleum industry of the 1990s can afford.

SECTION II

Principles

The reader must grasp a number of basic principles to understand the strengths and weaknesses of geostatistics. In this section, principles regarding spatial modeling, sample support, and scale are discussed.

By their very nature, earth science data have an inherent spatial structure. That is, variables we measure, like elevation, lithology, or porosity, are tied to geographic positions on the earth. Observations taken in close proximity will be more strongly correlated than observations separated by more distance. Although this relationship is an accepted norm to geologists, it is not the case in most other sciences. Classical statistical analysis does not necessarily account for spatial relationships, and its use in the earth sciences can often produce misleading results. A prerequisite of any geostatistical study is the development of the spatial model. R. A. Olea provides an excellent overview of semivariograms, the tool used in geostatistics to estimate and model spatial variability, in the chapter entitled **Fundamentals of Semivariogram Estimation, Modeling, and Usage**.

The sample support is the area or volume over which a measurement is made. Understanding how one measurement relates to another in close proximity is critical because geostatistical methods are highly sensitive to the spatial controls on variability. This is particularly true with tensor variables like permeability, for which mathematical averaging over volumes of various dimensions tends to be a problem. Y. Anguy et al. discuss how image analytical data can be blended with physical measurements to provide information for modeling permeability at scales greater than the permeability plug in **The Sample Support Problem for Permeability Assessment in Sandstone Reservoirs**.

In the nineteenth century, Johannas Walther, a noted geologist, recognized the Law of Correlation of facies. The law, as paraphrased by Krumbine and Sloss (1963, Stratigraphy and Sedimentation, W. H. Freeman and Company), states that "in a given sedimentary cycle, the same succession of facies that occurs laterally is also present in vertical succession." Taken commutatively, geologists have often attempted to predict the lateral variability of facies from a vertical sequence of stratigraphy. In the chapter **Theory and Applications of Vertical Variability Measures from Markov Chain Analysis**, J. H. Doveton presents an interesting alternative to the geostatistical simulation approach of reservoir facies. He points out that Markovian statistics of vertical variability are applicable to selected problems of lateral facies prediction and simulation.

The final chapter in this section is **A Review of Basic Upscaling Procedures: Advantages and Disadvantages**. Upscaling is the term applied to the process of changing from a fine scale of resolution to a coarser scale of resolution. The process is often necessary when translating a digital geostatistical model consisting of perhaps millions of grid nodes to a model consisting of tens of thousands of grid nodes, which is a more economically viable size for reservoir simulation. All of the techniques presently available cause a severe reduction in vertical and horizontal resolution, and thus can present a problem for estimating reservoir performance and maintenance. Rather than trying to provide an exhaustive review of all practical details, J. Mansoori highlights permeability upscaling methods and provides an annotated bibliography of pertinent reference literature.

Chapter 4

Fundamentals of Semivariogram Estimation, Modeling, and Usage

Ricardo A. Olea
Kansas Geological Survey
Lawrence, Kansas, U.S.A.

ABSTRACT

The semivariogram is a measure of the rate of change with distance for attributes that vary in space. Semivariogram analysis is helpful in comparing such attributes and in designing their adequate sampling. In addition, the semivariogram is required in almost any geostatistical procedure for prediction away from control points.

Correct semivariogram estimation requires a thorough assessment of the sampling, variations in the mean, or geometry of the volume that may be associated to a measurement, and a sensitivity analysis of the semivariogram to variations in the azimuth. In the simplest case, the semivariogram is a scalar function of lag. Depending on the application, one may need to estimate the parameters for linear combinations of admissible semivariogram models, which is customarily done as a curve fitting to a tabulation by lag of estimated semivariogram values.

Numerical examples illustrate every stage in estimating and modeling a semivariogram.

INTRODUCTION

Moments are summary parameters that characterize statistical populations. Moments are employed in the indirect inference of population properties. Perhaps the most widely used moment is the mean, a parameter describing the central tendency of a population. For example, if we take a group of students from the same class of an elementary school and compute a mean age of six years and seven months, in such context this parameter indicates that most likely the students are first graders. Sole knowledge of the mean, however, is insufficient to fully characterize a population. The mean provides no information about the total number of individuals nor the variations among them. This is why there are moments other

than the mean. In the simplest cases, a few parameters can perfectly describe a large population. In most cases, however, the characteristics of a population cannot be fully described in terms of moments. Even so, the approximations that are possible depend heavily on the notion of some key moments.

The attributes of populations of geologic interest typically show distinct organized patterns that are the result of sedimentary, magmatic, metamorphic, or structural processes. A central difference between statistics and geostatistics is the recording and extracting of information related to the location of every observation associated with such processes. Attributes that have a location x associated with each observation, such as topographic elevation and porosity, are called regionalized variables.

Physically, in a similar way that the mean is a measure of central tendency, the semivariogram is a measure of the rate of variation of a regionalized variable with distance. This chapter demonstrates how to obtain a semivariogram and how to employ it for the comparison, sampling, and modeling of a regionalized variable.

SEMIVARIOGRAM AND COVARIANCE

As in statistics, geostatistics deals with two different realms: the complete collection of all individuals of interest, called the population, and population subsets, called samples. The entire population is usually unobservable or unmeasurable, whereas samples can be fully observed and measured. Given certain specifications, the population is unique, but the same population may be sampled in numerous ways. Statistics and geostatistics developed to a large extent due to the need to approximate populations through analyzing samples. One conceivably can interview all first graders in a particular school or core a well, but surveying all first graders in the nation or every cubic inch of an oil field is quite a different task.

Given two locations, x and $x + h$, inside the field of a regionalized variable $y(x)$, the semivariogram is a measure of one-half the mean square error produced by assigning the value $y(x + h)$ to the value $y(x)$. Given a sample of observations $y(xi)$, $i = 1,2,...,m$, provided that the mean is constant,

$$\hat{\gamma}(h) = \frac{1}{2n(h)} \sum_{i=1}^{n(h)} \left[y(x_i + h) - y(x_i) \right]^2$$

is an unbiased estimator of the semivariogram. The term $n(h)$ is the number of pairs of observations a distance h apart. One may notice that, unlike the mean, the semivariogram is not a single number, but a continuous function of a variable h called the lag. The lag is a vector, thus involving not only the magnitude of the separation, but the azimuth of the line through the pairs. An estimate of the semivariogram is also called an experimental semivariogram.

The formula for calculating the experimental semivariogram involves terms that depend on specific locations, namely xi and $xi + h$. The averaging, however, is supposed to cancel out the dependency on location and leave solely a dependency on the distance h. This is an assumption rather than a fact. Geostatistics does not have a test to verify the compliance of this assumption.

Users tend to overlook the constancy in the mean required for estimating a semivariogram. A gentle and systematic variation in the mean, such as the increase in temperature with depth, is called a drift. Proper semivariogram estimation practices require the removal of this drift. There is more than one way to eliminate the drift, but complications resulting in the analysis are beyond the scope of this introductory

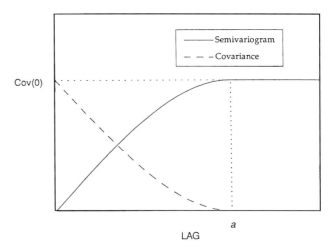

Figure 1. Idealized examples of covariance and semivariogram showing the location a of the range.

chapter. Let us only briefly mention that one method eliminates the drift by subtracting an analytical function—such as a polynomial—modeling the drift (Olea, 1975) and another filters the drift by differentiating in a manner completely similar to filtering of systematic variations in time series analysis (Christakos, 1992).

For those familiar with time series analysis, the definition of the semivariogram should resemble the definition of the covariance cov(h). When there is a covariance, the semivariogram and the covariance are related by

$$\gamma(h) = \text{cov}(0) - \text{cov}(h)$$

Even though both moments refer to the rate of spatial variation of an attribute, as the semivariogram increases the covariance decreases: the covariance measures similarity and the semivariogram dissimilarity. As shown in Figure 1, a typical semivariogram increases in dissimilarity with distance over short lags, then levels off. The lag a at which the semivariogram reaches a constant value is called the range. The value of the semivariogram beyond the range is referred to as the sill, which is equal to the variance of the exhaustive sampling when both such variance and the range are finite (Barnes, 1991).

Besides requiring a less-stringent assumption for its existence, estimating the semivariogram has an advantage over estimating the covariance in that the semivariogram does not require knowledge of the regionalized variable mean. Despite this and other important advantages in estimating the semivariogram over the covariance (Cressie and Grondona, 1992), use of the semivariogram is not widespread outside the field of geostatistics.

The semivariogram is called a second-order moment because its calculation involves squares of the observations. Other than practical rules of thumb, there are no statistical tests to assess how close an experimental semivariogram approximates the true semivariogram.

STRUCTURAL ANALYSIS

Structural analysis is the detection and numerical modeling of patterns of spatial dependence that characterize a regionalized variable, which to a large extent is synonymous with the study of the semivariogram.

Structural analysis is rarely the final purpose of a study. Instead, structural analysis is a preliminary step prior to the modeling of a regionalized variable using such techniques as kriging or conditional simulation (Olea, 1992), which involve solving systems of equations comprising matrices with semivariogram terms.

Generally, the semivariogram is a vectorial function of h. Because a discontinuous sampling precludes estimating the experimental semivariogram for every possible value of the lag, in practice the estimation is limited to a few points after sorting the pairs of variables not only by distance, but also by spatial orientation.

A favorable situation is when sampling is done at the nodes of a square grid, such as the one in Figure 2. Then, one can make estimates at fixed multiples, in directions other than that of the sampling, which is run in typically long one-directional transects, such as the columns. The symmetry of the configuration allows for investigation in other directions by taking successive readings from different rows and columns. For example, in a two-dimensional square grid oriented north-south and east-west, directions that can be analyzed in addition to, say, the north-south direction of sampling, include east-west, northwest-southeast and northeast-southwest. Notice that the minimum spacing possible in the northwest-southeast or northeast-southwest directions would be $\sqrt{2}$ times the spacing along the north-south or east-west directions. Somewhat less convenient than grids are linear traverses sampled at regular intervals. Calculations along a traverse work as well as calculations along a row or column in a grid, but the regularity of the sampling sites along directions other than the direction of sampling is lost.

Table 1 contains the results of an example calculation for what could be traverse data. For clarity in the mechanics of the calculation, I have selected a very short sequence. For reasons explained in following sections, a sequence this short cannot produce significant results.

Before starting any calculation one should inspect the sequence for systematic variations in the mean. In the previous example, there are six values above the mean and four below, with an initial increasing sequence of five values followed by a decreasing sequence of four values. One can safely assume then that the sequence does not have a systematic variation in the mean and calculation of the experimental semivariogram can proceed.

Note in this example in Table 1 the operation of a rule that applies to any semivariogram estimation: the number of pairs tends to decrease with increases in

Figure 2. A regular square grid.

the lag, up to a certain limit beyond which there are no more pairs of data. Because the accuracy of the estimates is proportional to the numbers of pairs, the greater the lag, the more unreliable the estimates. Fortunately, in the applications of the semivariogram to kriging and simulation, that portion near the origin is the one requiring the most accurate estimation for its higher influence in the results. An unsubstantiated practice is to limit the estimation to lags with a minimum of 30 pairs (Journel and Huijbregts, 1978; Hohn, 1988).

If one wants to estimate the semivariogram for a large lag close in magnitude to the diameter of the sampling area, then systematically the only pairs of observations that are that far apart are those comprising sampling sites located in opposite extremes of the sampling area, thus excluding the central points from the analysis. Such consideration is the justification for a second practical rule that advises one to limit the lag of the experimental semivariogram to one-half the extreme distance in the sampling domain for the direction analyzed (Journel and Huijbregts, 1978; Clark, 1979).

The worst configuration in terms of convenience in estimating the semivariogram is also the most common: irregularly located data. The practice in this case

Table 1. Estimation of an experimental semivariogram

x_i	$y(x_i)$	Δ^2_0	$\Delta^2_{0.5}$	$\Delta^2_{1.0}$	$\Delta^2_{1.5}$	$\Delta^2_{2.0}$	$\Delta^2_{2.5}$	$\Delta^2_{3.0}$	$\Delta^2_{3.5}$	$\Delta^2_{4.0}$	$\Delta^2_{4.5}$
7.0	3.2	0.0	1.21	3.24	10.89	22.09	24.01	18.49	16.81	12.25	3.76
7.5	4.3	0.0	0.49	4.84	12.96	14.44	10.24	9.00	5.76	2.25	
8.0	5.0	0.0	2.25	8.41	9.61	6.25	5.29	2.89	0.64		
8.5	6.5	0.0	1.96	2.56	1.00	0.64	0.04	0.49			
9.0	7.9	0.0	0.04	0.16	0.36	1.44	4.41				
9.5	8.1	0.0	0.36	0.64	1.96	5.29					
10.0	7.5	0.0	0.04	0.64	2.89						
10.5	7.3	0.0	0.36	2.25							
11.0	6.7	0.0	0.81								
11.5	5.8	0.0									
	$\Sigma\Delta^2_h$	0.0	7.52	22.74	38.97	50.15	43.99	30.87	23.21	14.5	6.76
	$n(h)$	10	9	8	7	6	5	4	3	2	1
	$\gamma(h)$	0.0	1.42	1.42	2.78	4.18	4.40	3.85	3.86	3.62	3.38

Mean: 6.23
Variance: 2.39

calls for the grouping of distances and directions into classes (David, 1977; Isaaks and Srivastava, 1989). A safe practice is to take a sampling interval equal to the average distance to the nearest neighbor and an angular tolerance of 22.5°.

For samples that badly miss these operational criteria, such as the example in Table 1, the experimental semivariogram may be an artifact. The user may better adopt the semivariogram from another area judged similar to the area of interest than to perform any calculations.

SEMIVARIOGRAM MODELS

The direct use of the experimental semivariogram can lead to singular matrices, multiple solutions, or negative mean-square errors in the kriging system of equations (Isaaks and Srivastava, 1989). The solution is to replace the tabulation obtained for the experimental semivariogram by admissible functions that have been proven to assure a unique solution to the normal equations and a nonnegative mean square error (Christakos, 1984). All in all, the models are a function of the vectorial lag h. In approximately decreasing frequency of use, for a given direction h of h, the most common models are as follows.

- The spherical semivariogram model:

$$\gamma(h) = \begin{cases} C\left[\dfrac{3h}{2a} - \dfrac{1}{2}\left(\dfrac{h}{a}\right)^3\right], & 0 \le h < a \\ \\ C, & a \le h \end{cases}$$

where C is the sill and a is the range. Near the origin, this semivariogram is close to a line in shape. The model is said to be transitive because it reaches a finite sill at a finite range. The tangent to this semivariogram at the origin intersects the sill at two-thirds of the range. The model is not admissible for a space with a dimension larger than three.

- The exponential semivariogram:

$$\gamma(h) = C\left(1 - e^{-\frac{3h}{a}}\right)$$

The model approaches the sill asymptotically. A practical definition of the range is the distance a, where the semivariogram is $0.95C$. Geometrically, a tangent at the origin intersects the asymptote C at lag $a/3$.

- The power semivariogram, which is the nontransitive model:

$$\gamma(h) = Ch^\alpha, \ 0 < \alpha < 2$$

The linear semivariogram is a special case of this model when $\alpha = 1$. The main advantage of the linear model is its simplicity. The linear semivariogram may be used to model an experimental semivariogram that is not transitive, or to model transitive semivariograms if the lag in the normal equations will be always smaller than the range, making the form of the semivariogram beyond the range immaterial. Notice, however, that the power semivariogram model does not have an equivalent covariance model.

- The Gaussian semivariogram:

$$\gamma(h) = C\left(1 - e^{-3\left(\frac{h}{a}\right)^2}\right)$$

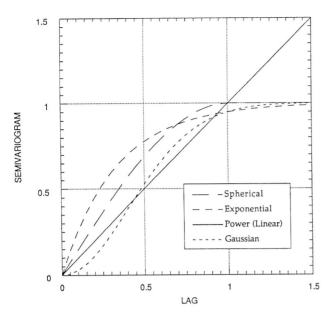

Figure 3. The most common semivariograms when the two model parameters are both equal to 1.

The sill is approached asymptotically. A practical rule is to consider the range to be the distance a, for which the semivariogram is $0.95C$. A graph of the model has a parabolic form near the origin.

Figure 3 is a graphical display of the most common models when both parameters are equal to 1. Near the origin, the exponential model is the steepest and the Gaussian is the flattest; for a fixed range, the transitive semivariograms can model three rates of deterioration of spatial continuity within short distances. The exponential and the Gaussian semivariogram cross at a lag equal to the range where, by definition, both are equal to 0.95 times the value of the sill; the linear and the spherical semivariogram cross for the same lag value of 1. The Gaussian semivariogram is the only one out of the four that has an inflection point.

A pure nugget effect model is a special degenerate case of a transitive semivariogram with an infinitesimal range—the semivariogram surges directly from 0 to a constant value. The experimental semivariogram of a sequence of random numbers follows a pure nugget effect model.

The sill is a critical parameter in all transitive models. Provided the observations are evenly distributed over the area of interest and the dimensions of such area are at least three times the range, the sample variance is a good approximation to the true sill (Barnes, 1991).

ADDITIVITY

Adding to the complexity of the modeling is the additivity property of admissible semivariogram models. Any linear combination of admissible models with positive coefficients is also admissible and is called a nested semivariogram. For instance, the sum of a pure nugget effect semivariogram model plus two spherical semivariograms with different parameters is an example of a nested semivariogram. By convention, a pure nugget effect model plus another model is not regarded as a nested model. Instead, it is referred to as a model with a nugget effect, such as an exponential semivariogram with a nugget effect (Christakos et al., 1991).

Single models are used for modeling experimental semivariograms that are close in shape to any of the basic admissible models, or for the approximate fitting of complex structural functions. Nested models are used to better fit complex structural functions.

PARAMETER ESTIMATION

Once the user selects a semivariogram model—simple or nested—the problem remains to estimate its parameters. Two different approaches can be used to estimate such parameters: maximum likelihood methods and curve fitting. Maximum likelihood estimates the model parameters directly from the data, skipping the calculation of an experimental semivariogram. The method is computational intensive, makes unrealistic statistical assumptions, and, according to McBratney and Webster (1986), is not practical for samples larger than 150.

The predominant practice is to calculate an experimental semivariogram first and then fit a model. The appeal comes from the possibility to visually compare the model to a few estimated points. Despite progress made in automation (Cressie, 1991), various difficulties account for the fact that fitting the model by eye still prevails. After obtaining the experimental semivariogram, the fitting is done by trial and error with the help of some computer software (Deutsch and Journel, 1992; Englund and Sparks, 1988). Increasing estimation variance with lag for the semivariogram estimator, the disproportionate influence of anomalous estimates, the nonlinearity of most models with respect to the parameters, and correlation among errors departing from normality preclude the use of ordinary least squares for automatic fitting. Generalized least squares is theoretically more satisfactory, but involves an iterative procedure that does not always converge (Gotway, 1991). Weighted least squares, intermediate in complexity between ordinary and generalized least squares, is claimed to provide a reasonable balance between simplicity, statistical rigor, and computational efficiency (Cressie, 1985; McBratney and Webster, 1986).

The choice of incremental and angular tolerance in the grouping of irregularly spaced data, and the insertion or deletion of a single observation in a small to medium sized data set, often make more difference in the experimental semivariogram than the analytical difference between alternative models. Parsimony should be observed to avoid overfitting with nested models.

Figure 4 shows a practical instance of semivariogram modeling using public domain data for the depth

32   Olea

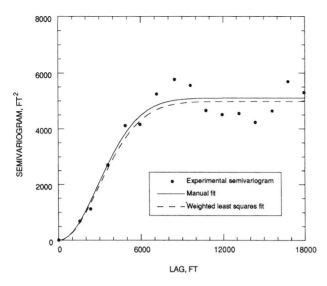

Figure 4. Omnidirectional semivariograms for an unconformity in the UNCF field. The dots denote the experimental semivariogram; the solid line the admissible Gaussian model with sill 5100 ft² (474 m²) and range 7150 ft (2179 m) fitted by hand; and the dashed line the admissible Gaussian model with sill 4975 ft² (462 m²) and range 7321 ft (2231 m) fitted by weighted least squares.

Table 2. Semivariogram fitting by weighted least squares, UNCF field

Model	Sill (ft²)	Range (ft)	AIC
Gaussian	4974.5	7321.3	-0.698
Spherical	4949.8	12167.4	-0.494
Exponential	7159.8	33085.7	-0.442

Model	Constant (ft²)	Exponent	AIC
Power	2.7	0.65	-0.125

to an unconformity in the UNCF field (Jones et al., 1986, Table B-3; Olea, 1992). The sample size is 70, with most of the observations recorded along traverses and a few taken from irregularly spaced wells. The average distance to the nearest neighbor is 1465 ft (445 m) and the lag increment is set to 1200 ft (365 m). The small sample size precludes an azimuth discrimination. The estimation of the experimental variogram for the first point at lag 1557 ft (475 m) has 24 pairs and is the only lag for which the number of pairs is below the recommended minimum of 30 pairs. The maximum distance between pairs is 30,000 ft (9140 m). The data set used to produce Figure 4 has a variance of 5480 ft² (509 m²) and a mean of 7912 ft (2412 m) devoid of drift.

I show simultaneously the manual and the automatic fit only for illustrative purposes. In practice, practioners attempt one or the other. Those in the majority who feel uncomfortable with automatic fitting or lack a program at hand go for manual fitting. The remainder minority go straight into automatic fitting. To make the exercise meaningful, I did the manual fitting first—once you know the answer there is no genuine trial and error. The slow semivariogram increase near the origin suggests that in this case the most convenient model to fit is the Gaussian. I employed the variance as a starting value for the sill and set the range equal to 7500 ft (2286 m), the smallest lag for which the experimental semivariogram reaches the value of the sill. It took me six trials to refine the approximation and decide that the model denoted by the solid line in Figure 4 was the best one. The fitting, especially for small lags, is good enough to discard the use of nested models.

For the automatic fitting I used a program prepared by Jian et al. (1995). The program provided in a fraction of a second the best fit for up to seven models using weights inversely proportional to the standard deviation of the $n(h)$ pairs of points used to estimate each $\hat{\gamma}(h)$ (see page 22). The results for the most common models are in Table 2. The Gaussian model is the best because its Aikake's Information Criterion (AIC) is the smallest, confirming the assumption I made in the manual fitting.

Besides speed, automatic fitting provides optimal parameters and an index to rank how well different models may fit an experimental semivariogram. Selecting and fitting a semivariogram manually remains closer to an art than a science.

ANISOTROPY AND DRIFT

An experimental semivariogram must be analyzed for as many different directions as feasible with the configuration of the data, with four directions being the bare minimum. The semivariogram is called isotropic when variations in the azimuth do not produce significant changes in the experimental semivariogram and is anisotropic otherwise. An anisotropic semivariogram is a true function of h rather than just of h. Directional analysis has the inconvenience of reducing the number of pairs available for the estimation, an inconvenience that may completely preclude a directional analysis for small data sets, as in the case of Figure 4.

In the presence of isotropy or in the absence of a large sample size, pairs can be pooled regardless of the azimuth. The resulting experimental semivariogram is said to be omnidirectional.

Figure 5 illustrates the most common and simple case of anisotropy—geometric anisotropy. In this situation, one model can be used to fit all experimental semivariograms by making the range a function of the azimuth. The common choice is an ellipse, in which case the user must find both the size and the orientation of the minor and the major axes.

A drift also makes the experimental semivariogram vary with the azimuth. In the presence of drift, as

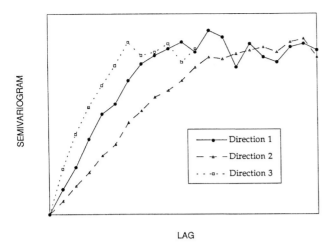

Figure 5. An example of semivariogram with geometric anisotropy.

testing of isotropy and the semivariogram modeling, particularly when the fitting is done by eye. Cross-validation is a procedure by which observations are dropped with replacement one at a time from a sample of size n, and for each discarded observation, an estimate is computed at the location of the discarded observation by using, at most, the remaining ($n - 1$) measurements. In principle, every parameter, such as those in the semivariogram model, should be reevaluated if the discarded point might have been used in the estimate. In practice, such parameters are presumed to be robust enough to be insensitive to the discarding of a single observation and one does not recalculate the parameters. A comparison among values estimated from the full sample and those estimated from the censored subsets is the basis for evaluating the overall quality of the estimation procedure; that is, the combined selection of the estimation method and its parameters. A search for better parameters is customarily done by trial and error.

Table 3 provides an ordinary kriging cross-validation for the same data employed to estimate the experimental semivariogram in Figure 4. I selected the mean square error of the discrepancies and the correlation coefficient between true and estimated values to measure the quality of the fitting.

The top of Table 3a is a sensitivity analysis to sill and range based on the results on Figure 4. When the changes are small, the mean square error and the correlation coefficient are not in agreement. For example, in terms of mean square error the second model is better than the third, but the correlation coefficient indicates the opposite. Regardless of the measure, neither of the models in Figure 4 gets the best mark. According to the mean square error, the best isotropic model is a Gaussian model with sill 4975 ft² (462 m²) and range 6900 ft (2103 m). The differences, however, are of secondary importance and for all practical purposes all three models are equiva-

shown in Figure 6, the experimental semivariogram in the direction of maximum dip has a slow increase near the origin, approximating a parabola, whereas in a perpendicular direction the experimental semivariogram is drift free. Such an experimental semivariogram in the drift-free direction is commonly used as a surrogate semivariogram approximating the one that would be obtained after removing the drift. The inconvenience of this approach is the impossibility to directly detect anisotropies in the semivariogram.

CROSS-VALIDATION

Geostatistics has no tests of significance to guide the user's decisions. Cross-validation is the only criterion available to measure the combined excellence for the selection of the kriging type, assumptions about the nature of the spatial variation of the mean, and

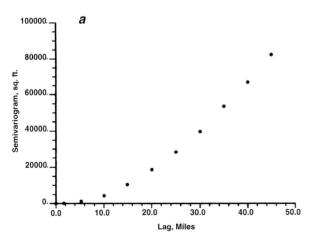

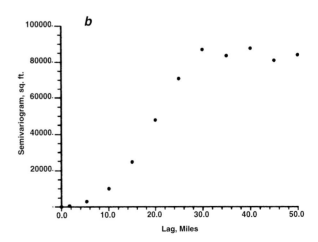

Figure 6. Semivariograms for the water table elevation in the High Plains Aquifer of Kansas during the winter 1980-81. Notice that the vertical scale in (a) is ten times that in (b). (a) Increasing semivariogram along the direction N80W of maximum dip for the linear trend . (b) Transitive semivariogram along N10E.

Table 3. Cross-validation for depth to unconformity, UNCF field

a) Isotropic semivariogram models

Model	Sill (ft^2)	Range (ft)	Mean square error (ft^2)	Correlation coefficient
Gaussian	4975	6900	411.8	0.962
Gaussian	4975	7150	414.4	0.958
Gaussian	4975	7321	418.4	0.961
Gaussian	4975	6500	417.7	0.960
Gaussian	5100	6900	411.8	0.961
Gaussian	5100	7150	414.4	0.958
Gaussian	5100	7321	418.4	0.961
Gaussian	5480	7500	424.6	0.962
Spherical	4950	12167	720.2	0.937
Exponential	7160	33086	757.5	0.934

Model	Constant (ft^2)	Exponent	Mean square error (ft^2)	Correlation coefficient
Power	2.7	0.65	899.2	0.927

b) Anisotropic Gaussian semivariogram models with a sill of 5100 ft^2

Range major axis (ft^2)	Major axis orientation	Axial ratio	Mean square error (ft^2)	Correlation coefficient
10000	N	0.65	207.1	0.982
9000	N	0.65	232.1	0.979
11000	N	0.65	216.7	0.981
10000	N10W	0.65	229.6	0.979
10000	N10E	0.65	231.9	0.978
10000	N	0.6	214.4	0.980
10000	N	0.7	222.2	0.980

lent. More relevant are the differences in the cross-validation for the models in Table 2. Now the ranking of the models is the same regardless whether the criterion used is the AIC, the mean square error, or the correlation coefficient.

Even more relevant are the results in Table 3b. Cross-validation is a convenient way to test for anisotropy in a situation such as the UNCF data in which the small sample size precludes testing the isotropy through estimation of experimental semivariograms along various directions. The results are particularly significant in terms of the mean square error, which shows a reduction of 49.7% relative to the best value for the isotropic models. Hence, according to cross-validation, the best model for the UNCF data is a Gaussian model with geometric anisotropy and a sill of 5100 ft^2 (474 m^2). The major axis of the range ellipse points north and is 10,000 ft (3048 m) long, while the minor axis in the perpendicular east-west direction is only 6500 ft (1981 m) long.

SUPPORT

Some of the geologic attributes of interest in geostatistics, such as formation thickness, are point measurements in the sense that there is a location associated to each measurement, but not a volume. Other attributes, such as permeability, are spatial averages and indeed have a volume associated with each measurement. In such a case, the location usually refers to the coordinates of the center of gravity of the analyzed volume. The technical name for such a volume is support, and its complete specification includes its geometrical shape, size, and spatial orientation.

A change in any of the characteristics of a support defines a new regionalized variable. At a given surface location and elevation, the permeability of a cylindrical core 3 in. (7.5 cm) long may be completely different from the permeability for the central 1 in. (2.5 cm) of the same core. Thus, an additional caution in structural analysis is to make sure that data for estimating the semivariogram relate to the same support. Generally, the effect of a larger support is to reduce variability, resulting in semivariograms with smaller sills and larger ranges. A different but related problem has to do with the heterogeneity of equipments, operating conditions, and experimental procedures used to obtain measurements. One commonly discovers that variations in the measurements are not related to actual spatial variation, but rather to changes in other confounding factors, such as the contractor who collected the data.

Analytical procedures account for the variations in the semivariogram due to changes in support (Clark, 1979; Journel and Huijbregts, 1978). One may have two reasons to worry about the support. When the sampling results in mixtures of supports, one needs to process different supports separately as different regionalized variables and use change of support transformations to consolidate the results.

A second and more common concern arises from the fact that kriging and conditional simulation require semivariogram models for supports other than points. Today, most computer programs rely only on a point semivariogram. Necessary transformations are computed internally by the software and are transparent to the user. This is the case, for example, in GSLIB (Deutsch and Journel, 1992) when preparing a numerical characterization for reservoir simulation. The user may provide a semivariogram based on core measurements, which, for the scale of the study, can be regarded as points. GSLIB performs the necessary upscaling to obtain the semivariogram for the reservoir simulator cells, which is the support of interest.

DIRECT APPLICATIONS

One can use reliable experimental semivariograms for comparative purposes and for sampling design. In a general way, similar semivariograms for the same type of regionalized variable indicate common origins, and significantly different semivariograms denote distinct geneses. Longer ranges and smaller sills relate to more continuous processes having smoother and less frequent fluctuations.

Good sampling of a regionalized variable calls for a first-round survey involving a few traverses that are sampled at a short sampling interval to allow for a good semivariogram estimation. Further densification of the sampling should be done following a regular pattern and at a density sufficient to ensure that, as a bare minimum, no location in the field is further away from the closest observation than a distance equal to the range. If this minimum is not followed, important structures can escape detection and the advantage of geostatistical estimations based on spatial correlation is lost. Geostatistics joins classical statistics to predict that the most likely value at those isolated locations is the sample mean.

In the presence of anisotropy, the sampling interval should be directional and vary proportionally to fluctuations in the range and inversely proportional to variations in the sill.

CONCLUSIONS

The semivariogram is a statistical moment assessing the average deterioration in similarity between two observations as their mutual distance increases. An estimated semivariogram is useful to compare regionalized variables and in sampling design. Semivariogram models play an important role in kriging estimation and conditional simulation. In essence, the prevailing practice for modeling a semivariogram requires the fitting of a valid curve to some estimated points.

In practice, both the estimation and the fitting are complicated by the attention one must pay to several important details: support must be homogeneous; fluctuations in the mean should not have a systematic variation; proper estimation requires at least 30 pairs for every lag smaller than one-half the maximum lag; provided the sample is large enough, the estimate should consider a sensitivity analysis to the azimuth; and the models are not linear with respect to their parameters. Modeling is necessary whenever the semivariogram is required for kriging or simulation, the most common situation. The selection is limited to linear combinations of admissible models, which can be tested through cross-validation.

The semivariogram's rich flexibility in the modeling is advantageous to experienced geostatisticians, but generally confusing to novices.

REFERENCES

Barnes, R. J., 1991, The variogram sill and the sample variance: Mathematical Geology, v. 23, no. 4, p. 673–678.

Christakos, G., 1984, On the problem of permissible covariance and variogram models: Water Resources Research, v. 20, no. 2, p. 251–265.

Christakos, G., 1992, Random fields models in the earth sciences: San Diego, Academic Press, 474 p.

Christakos, G., I. Clark, M. David, A. G. Journel, D. G. Krige, and R. A. Olea, 1991, Geostatistical glossary and multilingual dictionary: Oxford, Oxford University Press, 177 p.

Clark, I., 1979, Practical geostatistics: London, Applied Science Publishers, 129 p.

Cressie, N. A. C., 1985, Fitting variogram models by weighted least squares: Mathematical Geology, v. 17, no. 5, p. 563–568.

Cressie, N. A. C., 1991, Statistics for spatial data: New York, John Wiley, 900 p.

Cressie, N. A. C., and M. O. Grondona, 1992, A comparison of variogram estimation with covariogram estimation, in K. V. Mardia, ed., The art of statistical science: New York, John Wiley, p. 191–208.

David, M., 1977, Geostatistical ore reserve estimation: Developments in Geomathematics 2: Amsterdam, Elsevier Scientific Publishing, 364 p.

Deutsch, C. V., and A. G. Journel, 1992, GSLIB–geostatistical software library and user's guide: Oxford, Oxford University Press, 2 diskettes and 340 p.

Englund, E., and A. Sparks, 1988, Geo-EAS (geostatistical environmental assessment software)—user's guide: U. S. Environmental Protection Agency, 172 p.

Gotway, C. A., Fitting semivariogram models by weighted least squares: Computers and Geosciences, v. 17, no. 1, p. 171–172.

Hohn, M. E., 1988, Geostatistics and petroleum geology: New York, Van Nostrand-Reinhold, 264 p.

Isaaks, E. H., and R. M. Srivastava, 1989, An introduction to applied geostatistics: Oxford, Oxford University Press, 561 p.

Jian, X., R. A. Olea, and Y.-S. Yu, 1995, Semivariogram modeling by weighted least squares: Computers & Geosciences, in preparation.

Jones, T. A., D. E. Hamilton, and C. R. Johnson, 1986, Contouring geologic surfaces with the computers: New York, Van Nostrand Reinhold, 314 p.

Journel, A. G., and C. J. Huijbregts, 1978, Mining geostatistics: London, Academic Press, 600 p.

McBratney, A. B., and R. Webster, 1986, Choosing functions for semi-variograms of soil properties and fitting them to sampling estimates: Journal of Soil Science, v. 37, no. 4, p. 617–639.

Olea, R. A., 1975, Optimum mapping techniques using regionalized variable theory: series on spatial analysis no. 3: Lawrence, Kansas, Kansas Geological Survey, 137 p.

Olea, R. A., 1992, Kriging—understanding allays intimidation: Geobyte, v. 7, no. 5, p. 12–17.

◆

Chapter 5

The Sample Support Problem for Permeability Assessment in Sandstone Reservoirs

Y. Anguy
Laboratoire Energétique et Phénomènes de Transfert
L.E.P.T.-ENSAM
Esplanade des Arts et Métiers
Cedex, France

R. Ehrlich
C. M. Prince
V. L. Riggert
Department of Geological Sciences
University of South Carolina
Columbia, South Carolina, U.S.A.

D. Bernard
Laboratoire Energétique et Phénomènes de Transfert
L.E.P.T.-ENSAM
Esplanade des Arts et Métiers
Cedex, France

◆

ABSTRACT

All sandstone fabrics contain a characteristic complex of microstructures of a scale that is large with respect to the size of sample plugs conventionally used to measure permeability. A fundamental element of this microstructure consists of circuits composed of oversized pores and throats representing the sites of loose or flawed packing, and largely controlling permeability at the permeability-plug scale. Very small changes in the location of a permeability plug with respect to these circuits can result in significant variation in measured permeability. In the terminology of geostatistics, this is a classic problem of insufficient sample support, producing artificially heightened variance over small spatial scales as a result of a sampling volume that is too small. The varying pattern of the spatial variability of the microstructure dictates that permeability plugs must vary in size to contain enough of the microstructure to ensure adequate sample support. This is impossible in practice, but a combination of physical data and image analysis can yield permeability values representing rock volumes of the requisite size. A minimum sample size can be determined by measuring the scales of structural complexity using Fourier transforms of the image of the porous microstructure taken from thin sections. An adequate sample size is that which ensures local homogeneity (a measure of the scales of spatial variability within an image) and local stationarity (a measure of the variance between a set of mutually adjacent samples). A set of mutually adjacent samples that attains local stationarity defines a bed, wherein all samples at a scale of local homogeneity have either the same permeability or vary systematically. The physical significance of the structure can be determined by relating Fourier data to the relationship between pore type and throat size determined from petrographic image analysis. Methods of modeling permeability at the required scale can be based on the Darcy's law of empiricism or can be based on local volume-averaging theory and large-scale volume-averaging theory, wherein the permeability tensor arises naturally from first principles. The methodology discussed in this chapter represents a logical procedure for scaling up local observations and, in principle, can be operated validly over any scale range.

INTRODUCTION

Geostatistics is concerned with the spatial controls on variability. It is assumed that, in most earth systems, observations taken in close proximity to one another will be more strongly correlated than observations located more distantly from one another. The reduction in correlation as a function of distance can be measured or estimated using one or more related measures: the semivariogram, the autocorrelation function, or the autocovariance function. Empirical measurements are used either to explicitly define such functions or to support the general form of a relationship that changes as a function of spatial separation between measurements. In general, the higher the variance among closely spaced measurements, the shorter the distance within which observations can be correlated.

Permeability is a critical property in reservoir assessment and is the best scale-sensitive property. In principle, the permeability of a sandstone at various spatial scales should be measured on samples at each of those scales; however, operational problems dictate that highly precise estimates are made on very small core samples (permeability plugs) all of about the same size. It is common practice to combine such measurements in an attempt to estimate the permeability at larger scale. Such a derived estimate (e.g., the geometric mean) can be correct only fortuitously because it is uncommon to include data concerning the size and geometry of elements of contrasting permeability that may have been sampled by the permeability plugs. Use of an estimate, such as the geometric mean, carries with it an untested assumption that the units within the interval of interest have a random spatial arrangement. If the permeability plug is too small to be a valid sample, the problem of estimating permeability at a scale larger than the plug becomes intractable.

Permeability must be measured at a scale large enough to encompass a representative elementary volume (REV). At some scale, each adjacent sample that attains an REV will display similar values of permeability. The REV is larger than the textural equivalent of a unit cell of mineralogy because volumes much larger than unit cells are required for the volume to have representative physical properties. The volume of an REV most likely is not constant within or between sandstones. That is, considering the commonly observed progressions of bed forms, grain size contrasts, and discontinuities (such as shale drapes), no small sample can represent the entirety of a sandstone bed, say, 20 m thick; however, a local REV can exist whose properties can represent a volume of rock much larger than the volume of the REV. A sandstone therefore can be thought to consist of mesoscale geometric elements, the properties of each determinable by analyzing a sample larger than an REV, but smaller than an element. We can expect that this volume will vary from place to place within sandstone. In this chapter, we describe a strategy to accomplish this sort of reservoir description.

It is almost universally true that core-level measurements taken from permeability plugs do not scale up well at the level of reservoir simulation; production history matches are poor when simulation is based only on core measurements. Poor matches indicate either that core data are essentially irrelevant or that the small-scale permeability plugs are not providing an adequate estimate of the permeability in the portion of the reservoir represented by the core.

In this chapter, we demonstrate that the permeability-plug scale is smaller than an REV, and thus is inadequate for permeability measurement, even in sandstones that are visually uniform or massive. Inadequate sample size in such relatively homogeneous rocks means inadequacy for all sandstones. The reason for this deficiency is the presence of a universal substructure that occurs at scales near that of a permeability plug. A collection of permeability plugs of a certain size, taken in mutual proximity, will display more variance due to the interaction between the permeability plug and the underlying structure than variance displayed among samples of larger volume. That is, the smaller the permeability plug, the higher the variance between plugs taken in near proximity. This is a classic problem of sample support and must be dealt with before variograms or similar measures of spatial variability are calculated. A sample volume more relevant for permeability measurement should vary with the scale of spatial variability of the substructures.

We show how image-based data defining the porous microstructure can be combined with physical measurements to provide information for modeling permeability at scales greater than a permeability plug. First, we erect a formalism for the problem. We then document and discuss the image-based observations that demonstrate the existence of physically relevant structure. Finally, we outline two approaches for modeling permeability that eliminate most of the unwanted variance that arises from inadequate sample support. These approaches carry the seeds of a strategy for scaling up from small-scale observations.

NECESSITY OF A SAMPLE SUPPORT ALTERNATIVE

All sandstones, including massive sandstones, have a characteristic physically relevant substructure that is expressed at scales similar to or greater than the size of samples commonly used to measure permeability (Prince et al., in press.) Because of the presence of substructure, any small-scale estimate of the permeability does not represent a volume larger than itself and is related only to the local volume of the permeability plug V_β from which it is derived. In other words, physical measurements are commonly derived from sample volumes that are too small. Variability across such small-scale measurements therefore commonly is tainted with a high-frequency noise that may obscure and bias any statistical analysis of the data. The universal occurrence of such substructure in sandstones ensures that permeability values taken from samples smaller than the conventional 1-in. plugs (e.g., minipermeameter output) are even more variable and are less satisfactory for geostatistical analysis.

This problem requires developing a sample support procedure for sandstones. Practices developed in the mining industry, where the sample support concept originated, are inappropriate for attacking the sample support problems with respect to permeability. In the mining industry, increased variance due to insufficient sample support can be suppressed by physically homogenizing large volumes of rock (e.g., whole cores) and analyzing representative splits. In the case of permeable media, physical homogenization is out of the question. According to Isaaks and

Srivistava (1989) and David (1977), no "recipe" exists that permits estimation of the permeability variance at a large scale using only measurements taken at a scale where the samples are heterogeneous. What is required is a direct assessment of the spatial variability of the microporous structure to permeability at the permeability-plug scale.

The necessary link between fabric microstructure and flow physics measurements of permeability and formation resistivity factor at the micrometer to millimeter scale has already been established (Ehrlich et al., 1991a, b; McCreesh et al., 1991). Permeability measurements from permeability plugs, blended with structural information at relevant scales, can begin to solve the sample support problem for permeability.

In the following discussions, we demonstrate that fabric heterogeneities at or near scales of a permeability plug introduce undesired variance of permeability measurements. We first discuss criteria for establishing a scale defining local homogeneity of the microgeometry. Then, we discuss the concept of local stationarity. We assume that attainment of local stationarity of the microgeometry among a cluster of mutually adjacent samples implies achievement of local stationarity of the permeability estimates derived from this subset of contiguous samples, and vice versa.

DETERMINING LOCAL HOMOGENEITY AND LOCAL STATIONARITY

Local homogeneity is attained at that scale where the local fabric microstructure is either random or quasiperiodic. The appropriate measure of this condition is a Fourier transform of the thin section image where heterogeneities are defined in terms of spatial frequency. If the fabric is quasiperiodic, then spatial frequencies tied to relevant structural features should be included enough times to demonstrate that contrasting structures do not form clusters to produce another level of a structural hierarchy. Local homogeneity of the microstructure will reduce the variance among the physical properties of a cluster of adjacent local samples.

The condition of local stationarity occurs when two or more mutually adjacent, locally homogeneous samples yield similar values of the property of interest. Customarily, sampling is done systematically (e.g., every 10 cm), and modified as visually obvious changes in fabric occur. Systematic sampling always carries the possibility of aliasing (i.e., the sampling frequency goes in and out of phase with the spatial frequency of some contrasting fabric). For the purposes of this chapter, we will assume that aliasing is not a problem.

For all practical purposes, a large contrast in permeability between samples taken in near proximity indicates that the local homogeneity displayed by each sample was insufficient to produce local stationarity (i.e., minimal variance among adjacent samples). For instance, one sample may be taken from a sandstone and another from an adjacent conglomerate stringer in a core at a scale that ensures local homogeneity in each sample. The property of local stationarity is lacking, however, because of great contrast between samples. To achieve stationarity, samples expressing local homogeneity must contain many sandstone and conglomeratic stringer elements. The critical point is that without local stationarity, the physical properties of the sample have little bearing on the properties of the larger volume of rock that is of practical interest.

We define a bed as an interval in which local stationarity exists. Permeability measured among samples within a bed either will vary randomly or will express a simple gradient. In the latter case, the trend can be removed and the residual values represent the variance required for geostatistical analysis.

Beds represent the smallest flow units of practical significance with respect to permeability. The flow response of a sandstone is simply a function of the degree of permeability contrast between such flow units, the relative proportions of each kind of bed, and the size, shape, and arrangement of beds. Once bed-scale permeability is estimated correctly, then the existence of larger scaled (interbed level) contrasts can be readily determined by examining a spatial correlation function. As discussed in following sections, a comprehensive body of theory exists, based on fundamental principles, that permits assessment of a set of beds.

Small-scale permeability estimates appropriate for geostatistical analysis are those devoid of any high-frequency bias (i.e., estimates from samples that have attained both local homogeneity and local stationarity). Such conditions will be met when the local volumes V_β used to derive the permeability are geometrically consistent with some length-scale constraints of the type:

$$l_\beta, l_\sigma << r_0 << L_i \qquad (1)$$

where l_β, l_σ = characteristic length scales related to the mean pore size (l_β) and the mean grain size (l_σ), r_0 = the radius of the local volume V_β, and L_i = a characteristic length scale related to the volume-averaged variables appearing in Darcy's law governing flow within the local i^{th} bed.

The left side of inequality 1 states that many pores and grains must be accounted for to capture the essential characteristics of the pore fabric within each sample. Each sample, in turn, should enclose the fundamental length scale associated with features defined on a Fourier power spectrum or similar measure at least several times to comply with the condition of local homogeneity of the fabric microstructure.

An appropriate means to assess the characteristic length scales and character of spatial complexity is the Fourier transform. Local homogeneity implies that no significant power exists at spatial wavelengths approaching the size of the local sample V_β. Power, in the context of Fourier analysis, is the equivalent of the amount to variance associated with features that are expressed at any given wavelength. That is, if no relevant structural component occurs at wavelengths greater than, say, one-fourth of the domain, then local

homogeneity for the microstructure is declared. Achievement of local homogeneity defined in this way will result in a markedly lower intersample variance in either geometrical or physical properties than for samples of smaller size.

The right side of inequality 1 constrains the size of the local volumes V_β from a more geostatistical point of view. Significant variations of the porous structure should occur only at some length scales L_i bigger than the radii of the local samples V_β. Thus, local stationarity exists for scales less than that of L_i, which implies that the variance of the microstructure across a subset of adjacent local volumes should be less than the variance across the whole core. Local homogeneity will occur at variable length scales from place to place along a core. The intervals over which local stationarity exists we will define as local beds of length scale L_i.

Local stationarity can be determined by examination of the autocorrelation function, or its equivalent, the Fourier power spectrum, among permeability estimates derived from a suite of samples where each sample has achieved local homogeneity. Because each sample within a bed will be similar, variables that are a function of the microstructure can be analyzed as a spatial array, either through simple visual inspection or numerical analysis. Where the autocorrelation function decreases at a much lower rate (plateaus), intervals of local stationarity at scale L_i exist. Those specific beds can be identified by inversion of a filtered Fourier transform.

Declaration of local stationarity allows us to uniquely characterize the microstructure of a bed in terms of a set of N stationary, deterministic, or statistical parameters $\{P^i\}_{i=1,N}$ (e.g., the power spectrum and probability density function of the variable illustrative of pore or matrix on the basis of any local subsample) (Joshi, 1974; Quiblier, 1984). As an implicit function of the microstructure, permeability (K) can be expressed as

$$K = g(\{P^i\}_{i=1,N}) \qquad (2)$$

Variations of permeability all along the macroscopic system can no longer be due to sample support inadequacy, but to authentic contrasting beds of scale L_i, as determined from the autocorrelation function among the samples from a suite. Within a particular bed, the intersample variability is minimized. Let $V_\beta(x)$ and $V_\beta(x + \delta x)$, where $\delta x << L_i$, be any pair of local volumes within the ith bed. This yields

$$K(x + \delta x) - K(x) = dK = \sum_{i=1}^{N} \frac{\partial g}{\partial P^i} \, dP^i \approx 0 \qquad (3)$$

As a matter of fact, dK is caused only by $(dP^i)_{i=1,N}$, $(\partial g/\partial P^i)_{i=1,N}$ because it is an intrinsic function of the process used to derive K. The $\{P^i\}_{i=1,N}$ (i.e., the microstructure) being stationary over length scale L_i, they are statistically invariant over δx. That is

$$\{dP^i\}_{i=1,N} \approx 0$$

Equation 3 expresses that $\{dP^i\}_{i=1,N}$ does not contain high-frequency perturbations due to local interactions between sampling and microstructure. Such interactions would prevent use of $(\partial g/\partial P^i)_{i=1,N}$ as an intrinsic function of the problem, given the dependence of this term on boundary conditions. If such interactions exist, then the coupling between geometry and permeability, on which this whole undertaking relies, is no longer operational because of spurious interferences arising from the term $(\partial g/\partial P^i)_{i=1,N}$, obscuring the relation:

$$dK \propto \left\{dP^i\right\}_{i=1,N}$$

Thus, local stationarity of the microstructure supplies the local volumes V_β with operational significance. Local stationarity is a necessary condition to assess relevant physical properties of the macroscopic system at length scales of interest for modelers of petroleum reservoirs and groundwater contaminant transport.

EXISTENCE OF SANDSTONE MICROSTRUCTURE

Sandstones with homogeneous fabric over length scales of a few millimeters cannot exist. Graton and Fraser (1935) predicted, on the basis of experiments with monolayers of spheres, that mechanically assembled aggregates cannot achieve a global closely packed condition. Instead, in aggregates such as sandstones, which are assembled by the simultaneous sedimentation of large numbers of particles, many foci of close packing form simultaneously, but no process exists to produce mutual, parallel alignment of their internal fabrics. As the local well-packed areas grow toward one another, an intervening compromise zone of loose packing is formed. These internally loosely packed zones contain pores and throats that are oversized compared to those within the more closely packed zones. Loosely packed domains can arise spontaneously within a rapidly sedimenting mass or by sedimentation upon a cohesive substrate. The latter produces a zone of bedding-parallel loosely packed domains, as well as loosely packed domains radiating upward into the sedimenting mass (Prince et al., in press). Disruption along zones of dewatering produce domains oriented at high angles to bedding. Local mismatches in grain size also can produce loosely packed domains.

Generalization of this composite fabric into three dimensions is straightforward and has been done mainly by solid-state physicists who recognized that three-dimensional beadpacks are good models for the atomic structure of many crystals (Azaroff, 1960). In the physicist's lingo, the loosely packed domains are termed "packing flaws" or "defects." Once initiated, packing flaws propagate without limit through the solid, terminating only by intersecting with other flaws. Such a fabric in sandstones ensures that loosely packed domains

form preferred pathways for enhanced flow over long distances relative to grain size.

Zones of loose packing survive compaction and diagenesis in sandstones. Prince et al. (in press) demonstrated that in the case of both progressively sintered beadpacks and equigranular sandstones, loosely packed circuits are preferentially preserved relative to closely packed domains as porosity decreases. In the case of natural sandstones, Prince et al. (in press) observed grain dissolution occurring in loosely packed circuits whereas cementation occurred in closely packed domains.

EVALUATION OF SANDSTONE MICROSTRUCTURE

Correspondence Between Two-Dimensional and Three-Dimensional Data

The analysis of the fundamental length scales in sandstones is currently made from examining porosity exposed on the surface of planar sections cut perpendicular to perceptible bedding. Deductions of the three-dimensional porosity microgeometry may be made from such observations. The jump from two dimensions to three dimensions is aided by combining the sectioned porosity information with physical data taken from the remaining unsectioned sample (Ehrlich et al., 1991b; McCreesh et al., 1991). From a qualitative point of view, the procedure relies on the fact that the plane of section will pass through identical objects in different relative positions, yielding a distribution of sizes and shapes for each class of object, termed "pore type," that possesses a single size and shape in three dimensions (Ehrlich et al., 1991a).

For each thin section of a suite, a statistically representative number of patches of porosity (pores or connected clusters of pores) are geometrically characterized through an erosion-dilation procedure (Ehrlich et al., 1991a). Thin sections of a rock suite thus are characterized by a frequency distribution of sizes and shapes. These distributions are precise and unbiased because each section intersects large numbers of each class of three-dimensional objects (pore types) having orientations and positions that are random with respect to the plane of section. The lack of a common orientation of closely packed domains, illustrated by inspection of Fourier transforms of thin section images, provides evidence for the requisite randomness. This is not to say that sandstone fabric at all scales is random. The commonly observed fact that permeability measured parallel to bedding is usually greater than that measured perpendicular to bedding implies an anisotropy of the permeability tensor. Inequality in permeability measurements arises from the anisotropy in orientation and spatial density of the loosely packed circuits within the internal rock fabric.

Pores of a given type share a common size and shape. Derivation of pore types capitalizes on the fact that the relative proportions of pore types vary from thin section to thin section. These variations induce a characteristic web of correlations between the class intervals of the frequency distributions describing the sizes and shapes of porosity elements (porels) in the plane of section. The relative proportion of pore types in each thin section can be determined so that pore types can be placed into a mineralogical and diagenetic context (Ehrlich et al. 1991a). Most sandstones have fewer than seven pore types. Commonly, one pore type contributes a dominant portion of permeability in individual samples, with another pore type contributing a significant, albeit lesser, portion.

Multiple regression procedures link pore type abundance with capillary pressure data. Each pore type tends to be associated with a limited range of throat sizes (McCreesh et al., 1991). The relation between pore type and throat size is represented by a set of equations that predict the relative filling of each pore type as a function of pressure. In sandstones, the tendency of pore types to fill in mutually exclusive pressure intervals is a consequence of the proclivity of pores of the same type to be mutually adjacent, connected by similarly sized throats, which is a consequence of a heterogeneous fabric that results in a series of quasiparallel flow circuits. This pore type/throat size relationship has been used to model such basic reservoir properties as permeability and formation resistivity factor (Ehrlich et al., 1991b). Independent filling of pore types at different pressure intervals is largely due to the universal occurrence of circuits composed of loosely packed zones, juxtaposed against closely packed zones.

Consistency between calculated and experimental values of permeability has been shown for many formations (Ehrlich et al., 1991b). Variations in flow properties in a suite of samples have been established to result from variability in the relative abundance of pore types and changes in throat size tied to a particular pore type. Permeability values have a log-linear relation with the product of the number of pores per unit area and the fourth power of the throat diameter. This model is calculated as a function of unit cross sectional area, yielding the required dimensionality of permeability, L^2. The agreement between model and observation suggests that the independent flow paths tied to the different pore types show little tortuosity in the direction parallel to bedding. Results of permeability modeling provide convincing arguments that features observed in thin section are preserved statistically in the third dimension over length scales of at least the length of conventional permeability plugs (several centimeters). Such an inference is independent of the azimuth of the permeability plug with respect to the bedding plane, implying that the microgeometry is isotropic in planes parallel to bedding. Thus, thin sections are justified for understanding the three-dimensional structure of the porous medium because they are representative of the three-dimensional solid (Bowers, et al., in press.)

These results imply the validity of the inferences of Graton and Fraser (1935) that a universal characteristic structure exists in sandstones. Determination of the size and geometry of this structure is the key to scaling up physical measurements. Once the link between permeability and microstructure is established, then permeability at any scale is simply a function of the structure. In the next section we show how the fundamental structure of sandstones is determined and quantitatively assessed.

Fourier Analysis of Binary Images

Solving the sample support problem requires two steps: (1) local homogeneity of the fabric within local volumes must be achieved and (2) the existence of local stationarity must be demonstrated. The algorithm used in assessing local homogeneity carries the seed of a procedure to scale up permeability in a more robust fashion than has been done previously.

The first step is to search for local homogeneity in all samples of a suite. Recall that local homogeneity means that any structural element is repeated enough times in the sample so that permeability is not appreciably affected by small changes in the placement of the sample cylinder with respect to its position within a core. The homogeneity assessment is accomplished by analyzing the results of Fourier transforms of binary images taken from thin section.

Spatial organization of binary images can be analyzed using two-dimensional (2-D) Fourier transforms (Blackman and Tukey, 1958; Rosenfeld and Kak, 1982; Prince and Ehrlich, 1990; Prince, 1990). Binary images portray pixels of one class (in our case porosity) by a single intensity and the remainder (rock matrix) by another. Images of porous rock in thin section can thus be converted to images portraying pores by areas of black and matrix by areas of white through digitization (Crabtree, 1983; Crabtree et al., 1984). Areas covered by conventional high-resolution imagery (pixel size of less than 5 μm) are too small to address the question of local homogeneity of the structures in the context of laboratory measurements of permeability. To overcome the lack of relevance of such small areas, we have developed a postprocessing package of computer programs to merge multiple binarized views into one while preserving high resolution. Regardless of the magnification, we can generate mosaics of binarized images covering a thin section in its entirety (Figures 1A, 2A, 3A, 4A, 5A, 6A).

A double Fourier transform of such a binarized image is analogous to an x-ray diffraction pattern (Blackman and Tukey, 1958; Rosenfeld and Kak, 1982; Prince and Ehrlich, 1990; Prince, 1990). The degree of local homogeneity and the scales of local heterogeneity can be assessed easily from the transform (Prince, 1990; Prince and Ehrlich, 1990; Prince et al., in press). One measure of spatial order is the 2-D power spectrum that displays power [$F^2(u,v)$, the squared wave amplitudes] as a function of wavenumbers. $F^2(u,v)$ is a measure of the relative contribution of the elementary pattern, the x and y components of the spatial frequency of which are u and v, to the total sum (Rosenfeld and Kak, 1982). In other words, $F^2(u,v)$ is a measure of the contribution to the variance from spatial frequencies (u, v) (Blackman and Tukey, 1958).

Another way to display the information encompassed in the 2-D Fourier transform is the radial power spectrum, which displays variance integrated over rings of radii γ such that

$$\left[\left(u^2 + v^2\right)^{\frac{1}{2}} - 1\right] < \gamma \le \left(u^2 + v^2\right)^{\frac{1}{2}}$$

For ease of interpretation, the radial power spectrum is commonly expressed in terms of variance versus wavelength. We will use this convention in this chapter. Local homogeneity for both 2-D power and radial power spectra requires that little power arises from small wavenumbers (i.e., long wavelengths) relative to the total power.

The second step is to evaluate local stationarity. Attainment of local homogeneity for all individual plugs of a suite does not necessarily imply local stationarity and a solution to the sample support problem. Local stationarity implies a similarity of structure between a subset of mutually adjacent samples. Assessment of local stationarity can be done in several ways. One approach to analyzing the variance among large images is the use of Fourier power spectra. For practical purposes, local homogeneity is achieved in images which are four times larger than the smallest wavenumber associated with significant power.

Each image achieving local homogeneity can be subsampled (e.g., using every third pixel) and the variance in power among subsamples can be evaluated. Local stationarity can be assessed from a measure of the variance between spectra from large images compared to the variability contained within individual large images. That is, the variance among a set of adjacent samples should not be any greater than that arising from the subsampling of individual samples.

Another way to assess local stationarity is to model the permeability tensor K at the scale of local homogeneity and evaluate the variance across the simulated K, using an autocorrelation function. Still another way involves assessing the variance between porosity and pore-type proportions across samples. Such an approach can be taken as a first evaluation of local stationarity in that it is a necessary, but not sufficient, condition.

Failure to achieve local stationarity would indicate that local homogeneity, in fact, has not been attained and that the REV must be larger than the images used in the analysis. If this is the case, images must be collected at a still larger scale and the entire process repeated. If local stationarity is achieved, then measurement of permeability at that associated scale is free of a sample support problem.

We have concentrated most of our analyses on massive, relatively well-sorted, sandstones that are apparently structureless at the scale of micrometers to centimeters. Such sandstones, of course, are not visually homogeneous at scales greater than our sample volumes and areas. At the thin-section scale (permeability-plug scale), we have found that physically rel-

evant inhomogeneity is always present. The presence of structure in such "structureless" sandstones is strong support for the universal occurrence of structure in all sandstones.

Because the thin section area required to achieve local homogeneity varies between samples, the implication is that the volumes of permeability plugs also should vary. This is operationally impractical; however, knowing both measured permeability and microstructure of the permeability plug, with a set of reasonable assumptions one can assess the adequacy of the plug. If the plug radius is inadequate, the perturbations in variance are likely to have been caused by inadequate sample support.

We have verified, as predicted by Graton and Fraser (1935) and observed by Prince et al. (in press), that all first-order sandstone fabrics are composed of juxtaposed closely packed and loosely packed domains. Changes in the abundance or geometry of the domains or variations in grain size then produce second-order or higher level structures. Examples of these structures are shown in sandstones from the Jurassic of the North Sea and the Paleozoic of the United States mid-continent. We also have an example from the Mississippian Berea from Ohio, the sandstone used in laboratories worldwide because of its assumed homogeneity at the permeability-plug scale.

EVIDENCE OF SANDSTONE DOMAINAL FABRIC

For a well-sorted sandstone (Figure 1A), the existence of closely packed domains produces a well-defined peak on the radial power spectrum at a wavelength close to the mean grain size (Prince et al., in press) (Figure 1C). The better the sorting, the sharper the peak. In our samples from the Berea (Figure 2A) and mid-continent (Figures 3A, 4A, 5A), somewhat poor sorting is indicated by the more diffuse and lower amplitude grain size peak than that described by Prince et al. (in press). The grain size peak represents the center-to-center distances of pores within well-packed domains. A plateau adjacent to the grain size peak is the manifestation of loose packing, i.e., increased center-to-center distances between pores within loosely packed (flawed) domains (Figures 1C, 4C, 5C). The plateau can exhibit distinct modes, depending on the type of flaw or the variability in the roughness and smoothness of the flaws (Figures 2C, 3C).

The regions of close and loose packing can be visualized by filtering all wavelengths equal to or less than that of grain size from the 2-D Fourier transform and by inverting the filtered series into an image. Filtered images are displayed superimposed onto original images to show the relationships between them. In such synthetic images, porosity is not expressed in all portions of the image containing closely packed domains. In addition, nuances associated with pore or grain shape (the high frequency/short wavelength features) are removed from the zones of loose packing (stippled areas of Figures 1D, 2D, 3D).

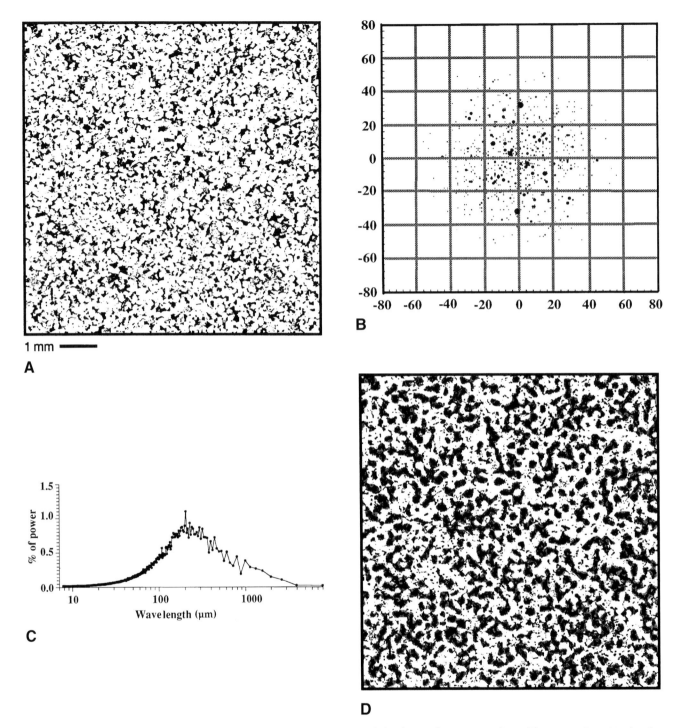

Figure 1. Fourier analysis of North Sea Jurassic sandstone. (A) Binary image produced by merging 20 single views (pixel size: 3.861 μm, image size 7.9 × 7.9 mm). Black indicates porosity. (B) Two-dimensional power spectrum of (A). Numbers on the axes represent wavenumbers (proportional to the inverse of wavelength). The size of the circles is proportional to the power. Data representing location of low power have not been plotted. Note the structural anisotropy in the northwest-southeast direction (concentration of large circles along that diagonal). Lack of power near (0,0) suggests attainment of local homogeneity. (C) Radial power spectrum of (A). Horizontal axis is calibrated in wavelength λ. Note the grain spike at $\lambda = 208$ μm and an adjacent plateau ($\lambda = 208$–349 μm) associated with loosely packed domains. Lack of power at long λ indicates local homogeneity. (D) Stippled pattern represents image after filtering all wavelengths equal to or less than grain size [porosity in original image (A) expressed by black]. Note that the stippled pattern selectively overlies the oversized pores associated with packing flaws—the lowest level in the structural hierarchy.

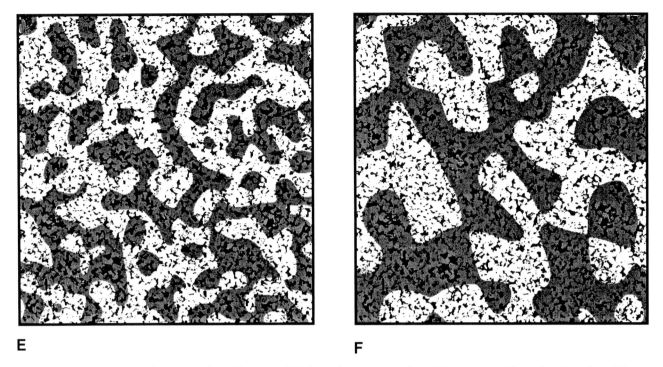

E   F

Figure 1 (continued). (E) Image after filtering all λ less than or equal to 659 μm, revealing circuits of packing flaws. (F) Similar to (E) except filtered at λ less than or equal to 1317 μm, illustrating a higher level in a structural hierarchy.

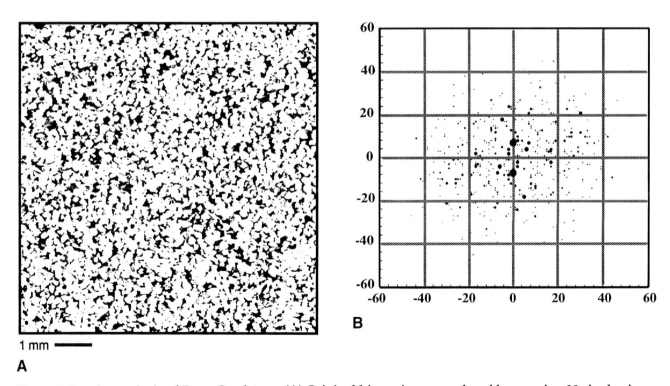

1 mm ———

A   B

Figure 2. Fourier analysis of Berea Sandstone. (A) Original binary image produced by merging 20 single views (pixel size: 3.861 μm, image size 7.9 × 7.9 mm). Black indicates porosity. (B) Two-dimensional power spectrum of (A). Notice the high concentration of power at wavenumber 7 (λ = 1129 μm). Local homogeneity is questionable because of the occurrence of maxima at wavenumber 7 and subpeaks at lower wavenumbers.

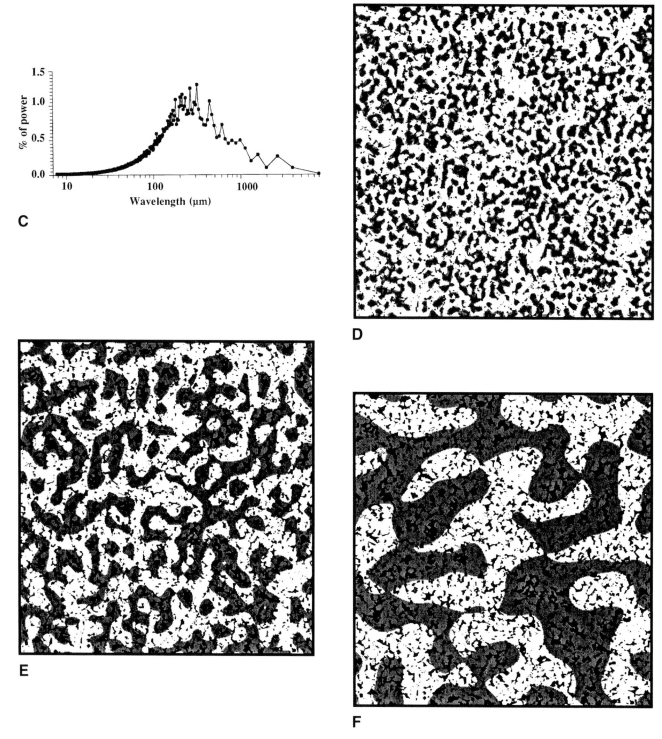

C

D

E

F

Figure 2 (continued). (C) Radial power spectrum of (A). Note the grain spike around $\lambda = 223$ μm and an adjacent plateau ($\lambda = 265$–316 μm) associated with loosely packed domains. (D) Image after filtering all wavelengths equal to or less than grain size; original image is overlain by stippled pattern (black). (E) Image after filtering all λ less than or equal to 439 μm revealing a circuit of packing flaws. (F) Similar to (E) except filtered at λ less than or equal to 1129 μm illustrating a higher level in a structural hierarchy.

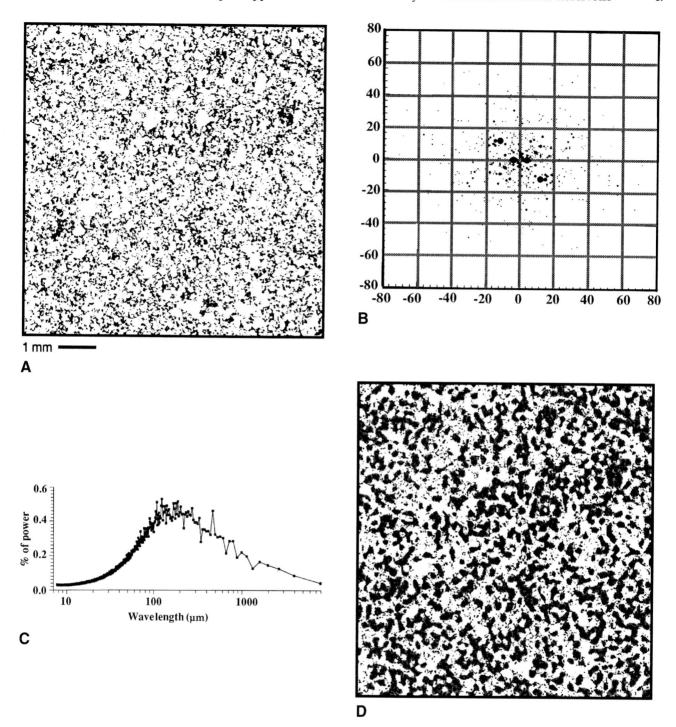

Figure 3. Fourier analysis of mid-continent sandstone, sample Mid 1. (A) Original binary image produced by merging 20 single views (pixel size: 3.861 μm, image size 7.9 × 7.9 mm). Black indicates porosity. (B) Two-dimensional power spectrum of (A). Local homogeneity is not achieved. (C) Radial power spectrum of (A). Note the grain spike at λ = 190 μm and an adjacent plateau (λ = 274–337 μm) associated with loosely packed domains. (D) Image after filtering all wavelengths equal to or less than grain size; original image is overlain by stippled pattern (black).

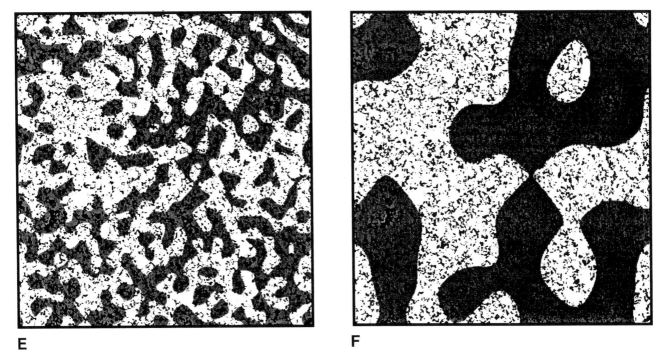

E

F

Figure 3 (continued). (E) Image after filtering all λ less than or equal to 465 μm, revealing a circuit of packing flaws. (F) Similar to (E) except filtered at λ less than or equal to 1976 μm, illustrating a higher level in a structural hierarchy.

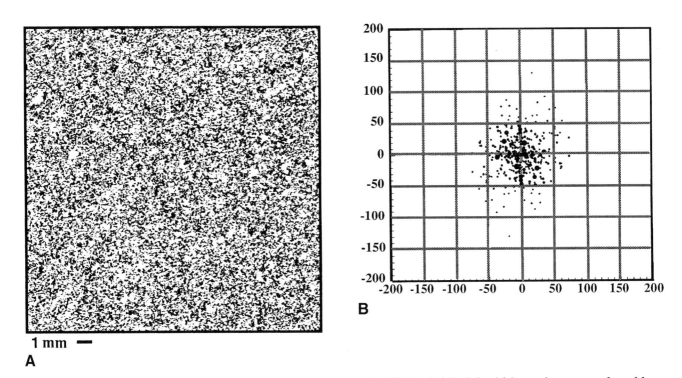

1 mm ▬

A

B

Figure 4. Fourier analysis of mid-continent sandstone, sample Mid 2. (A) Original binary image produced by merging 30 single views (pixel size: 9.524 μm, image size 19.5 × 19.5 mm). Black indicates porosity. (B) Two-dimensional power spectrum of (A). Local homogeneity is not achieved.

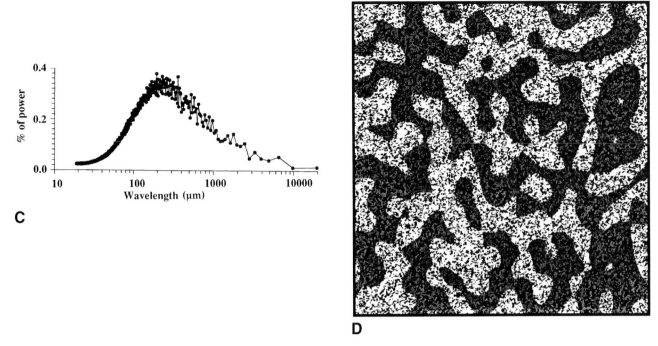

C

Figure 4 (continued). (C) Radial power spectrum of (A). Note the grain spike at $\lambda = 185$ μm and an adjacent plateau ($\lambda \approx 350$ μm) associated with loosely packed domains. (D) Image after filtering all λ less than or equal to 1773 μm, revealing a high level in a structural hierarchy; original image is overlain by stippled pattern (black).

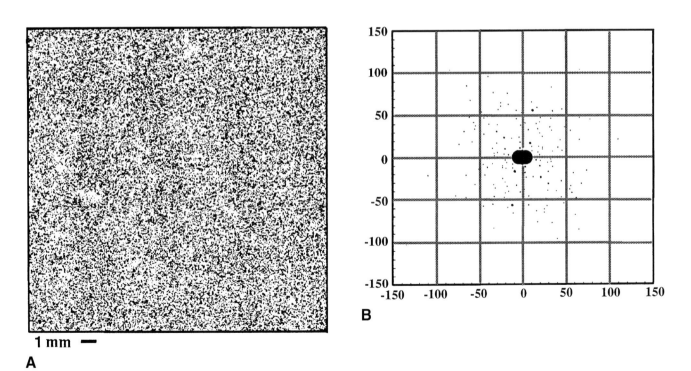

Figure 5. Fourier analysis of mid-continent sandstone, sample Mid 3. (A) Original binary image produced by merging 30 single views (pixel size: 9.524 μm, image size 19.5×19.5 mm). Black indicates porosity. (B) Two-dimensional power spectrum of (A). Local homogeneity is not achieved.

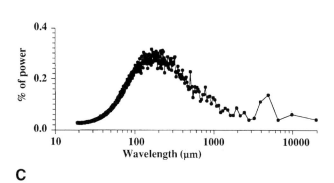

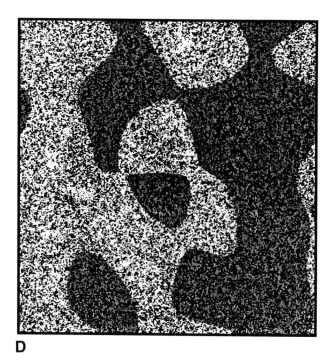

Figure 5 (continued). (C) Radial power spectrum of (A). Note the grain spike at $\lambda = 172$ μm and an adjacent plateau ($\lambda = 250$–319 μm) associated with loosely packed domains. (D) Image after filtering all λ less than or equal to 4876 μm, revealing a high level in the hierarchy of structures; original image is overlain by stippled pattern (black).

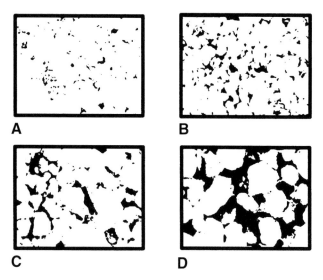

Figure 6. Pores types of the mid-continent suite of sandstones. Digitized views of porosity (black) are dominated by the pore types determined by petrographic image analysis. All pore types represent intergranular porosity. Each view is 1075 × 832 μm. (A) Pore type I has compact pores; mean diameter 13.6 μm, mean throat radius 0.3 μm. (B) Pore type II has compact to elongate pores; mean diameter 18.6 μm, mean throat radius 1.8 μm. (C) Pore type III has elongate pores; mean diameter 38.7 μm, mean throat radius 5.2 μm. (D) Pore type IV has coalesced elongate pores; mean diameter 79.3 μm, mean throat radius 5.2 μm.

Zones of loose packing (stippled areas of Figures 1D, 2D, 3D) represent the fundamental structure with respect to permeability in sandstones. They represent the continuous pathways in three dimensions that are characterized by the largest pore throats. Isolated blobs of packing flaws (loosely packed areas) observed in thin section represent slices through circuits penetrating the plane of section. Packing flaws propagate through the volume and define three-dimensional circuits of pores with relatively large throats. The spatial organization of packing flaws into larger structures (Figures 1E, 2E, 3E) constrains most of the permeability contribution to come from these areas.

DETERMINATION OF SPATIAL SCALE WITHIN DOMAINAL FABRIC

Closely and loosely packed domains are the fundamental textural elements in a hierarchy of structures. At a second level, these clusters coalesce into circuits of loosely packed domains penetrating a matrix of closely packed domains (Figures 1E, 2E, 3E). Higher order structures include various modes of clustering of these circuits, as well as contrasts due to differences in grain size (Figures 1F, 2F, 3F, 4D, 5D). All levels of structures above the fundamental level are expressed as additional peaks and plateaus at lower frequencies (longer wavelengths) on both the radial power spectrum and the 2-D power spectrum.

After removing spatial wavelengths less than the domain scale in the North Sea sandstone (Figure 1A), the second-order structures of closely packed domains (nonstippled) and loosely packed domains (stippled) are visible (Figure 1E). The packing flaw circuits (loosely packed domains) are not uniformly distributed, as shown by lower frequency (longer wavelength) peaks on both radial power and 2-D amplitude spectra. The peaks represent different clustering modes of circuits at one or more scales (Figure 1F). Because little power exists at wavenumbers less than 3 (wavelengths greater than 2635 µm) (Figure 1B), one might consider that local homogeneity has been achieved in this sample.

The types of structure shown in the North Sea sandstone are also displayed in the Berea (Figure 2A) and Mid 1 (mid-continent sample) sandstones (Figure 3A). Radial power spectra of these samples show a well-defined peak at wavelength 439 µm for the Berea (Figure 2C) and at 465 µm for the Mid 1 samples (Figure 3C). These peaks are related to the high concentration of power on the 2-D power spectra at wavenumber 18 for the former (Figure 2B) and 17 for the latter (Figure 3B). Structures associated with these peaks are packing flaw circuits (Figures 2E, 3E). Clusters of packing flaw circuits are associated with high power in the 2-D power spectrum at wavenumber 7 in the Berea (Figure 2B) and wavenumber 4 in Mid 1 (Figure 3B). Local homogeneity can be declared in the Berea (Figure 2A) because little power occurs at wavelengths greater than one-fourth of the image size. Occurrence of a major structure at radial harmonic 4, combined with a significant degree of power at wavenumber 2 (wavelength 3954 µm) (Figure 3B) suggests that local homogeneity has not been achieved in the Mid 1 sample.

Because the Mid 1 sample image (7.9 × 7.9 mm) did not achieve local homogeneity, we collected images of the sandstone at a larger scale in an attempt to do so. Images of samples Mid 2 (Figure 4A) and Mid 3 (Figure 5A) were collected at a scale of 19.5 × 19.5 mm. Clusters of circuits of flaws (stippled) are revealed in the Mid 2 (Figure 4D) and Mid 3 (Figure 5D) samples when the images are filtered for wavelengths of less than 1773 and 4875 µm, respectively. High power at very small wavenumbers in both the Mid 2 (Figure 4B) and Mid 3 (Figure 5B) samples, however, indicates that local homogeneity does not occur in the mid-continent sandstone, even at the 2-cm scale.

ASSESSMENT OF PHYSICAL RELEVANCE OF DOMAINAL STRUCTURE

We have demonstrated that structural heterogeneity at permeability-plug scale is a universal characteristic of sandstones. The physical significance of loosely packed circuits can be characterized as containing distinct pore type/throat size relationships using the procedures of Ehrlich et al. (1991a) and McCreesh et al. (1991). In other words, the spatial arrangement of pores of like type is

the same as that observed in filtered images, as discussed in the previous section. Most pore types contributing to permeability thus are intergranular pores associated with loosely packed domains. If a sample has variable grain size, a pair of pore types commonly is required for each significant change in size. In most sandstone reservoirs, three to six pore types are required to characterize the porosity. One or two of those pore types are frequently moldic or intragranular porosity that is usually unrelated to permeability.

We have observed that in most sandstones one or two pore types account for over 80% of the permeability. As determined by the procedure outlined in Ehrlich et al. (1991a), four pore types are present (Figure 6) in the sample suite from the mid-continent sandstone. In the Mid 1 sample (Figure 3A), pore type III (Figure 6C) and pore type IV (Figure 6D) are identical to the pores comprising the circuits of packing flaws (stippled area, Figure 3E). Pore type III (36.0% of the optical porosity) represents most of the porosity in the circuits of packing flaws and is in series with a small amount of pore type IV (4.8% of the optical porosity). Both pore type III and pore type IV have effective throat radii of about 5.2 µm, based on the analytical procedure described by McCreesh et al. (1991). Pore type II (58.0% of the optical porosity) and pore type I (1.2 % of the optical porosity) occupy the closely packed portions of the fabric (nonstippled area, Figure 3E) and have effective throat radii of 1.8 and 0.3 µm, respectively. The permeability model described by Ehrlich et al. (1991b) predicts a permeability of 26.3 md in this sample, which favorably compares to the measured value of 26.4 md. Pore types III and IV account for all but 2.8 md of the permeability—essentially all of the permeability.

In the mid-continent samples, the relative abundance of loose packing controls permeability so that permeability measured at too small a scale will yield high variance. The spatial density of packing flaw circuits in Mid 1 (Figure 3A) is not uniform at the imaged scale (7.9 × 7.9 mm; Figure 3F). Local homogeneity therefore is not achieved. Because there is a paucity of permeable circuits associated with pore types III and IV (stippled areas) in the upper left quadrant of the image (Figure 3F), Mid 1 is heterogeneous with respect to discharge at a scale approaching that of the image.

CALCULATION OF THE PERMEABILITY AT THE RELEVANT SCALE

We have documented the relevance of the conclusions of Graton and Fraser (1935) that permeability is a function of the three-dimensional microstructure. A measure of similarity of the porous microstructure within sample volumes (local homogeneity) and among sample volumes (local stationarity) of a suite is required to reduce the sample support problem to make the data suitable for geostatistical analysis.

Our observations suggest that packing is the source of most structural contrasts at permeability-plug scale. Fourier analysis suggests that the next several levels in the sandstone structural hierarchy express a nonuniform distribution of loosely packed domains. Samples at a scale where local homogeneity and local stationarity occur are the correct size for relevant permeability measurements. As demonstrated, this scale fluctuates with the fluctuating character of the underlying microstructure pattern. Thus, the sample support problem can be largely overcome by requiring a permeability plug of variable volume. This, of course, is operationally impractical; however, given image data and capillary pressure information, one can calculate the permeability at the scale of the image using the relationship between pore type and throat size (Ehrlich et al., 1991a). The procedure is a variant of the Hagen–Poiseuille model and has been shown to be valid over length scales of a few centimeters in a bedding parallel direction, as shown in Mid 1, Bowers et al. (in press), and Coskun and Wardlaw (1993). The success of the permeability model must result from the continuity of loosely packed circuits in the bedding-parallel direction. An advantage of the Hagen–Poiseuille model is that permeability is apportioned among pore types, a method effective for most reservoir sandstones. Such a model may not be adequate, however, for certain classes of sandstones, such as complexly laminated eolian sandstones, where locally permeable elements may be isolated behind cemented laminae.

Although the permeability model may be sufficient if local homogeneity occurs over a relatively small scale, it is not known what the limiting scale is. Use of this approach for the assessment of flow efficacy with tubelike models might be inappropriate for the structural hierarchy at a larger scale. On these grounds, one might need a procedure relying on the knowledge gained from the synthesis of thin section data and physical measurements (pore typing), but devoid of the assumptions tied to tubelike models. This criticism can be met by using volume-averaging theory (Anderson et al., 1967; Marle, 1967; Slattery, 1967; Whitaker, 1967; Raats et al., 1968; Whitaker, 1986), which derives the macroscopic and smoothed transport equation at the scale of interest (Darcy's law):

$$\langle v_\beta \rangle = -\frac{1}{\mu_\beta} K \left(\nabla \langle p_\beta \rangle^\beta \right)$$

The theory develops a local closure form allowing the explicit prediction of the permeability tensor K as an implicit function of the microstructure. It also provides the dependent variables $\langle v_\beta \rangle$ and $\nabla \langle p_\beta \rangle^\beta$ appearing in Darcy's law with operational significance to permit comparison between theoretical calculations and laboratory measurements.

Derivation of K through the volume-averaging theory implies an upward scaling of a microscopic field. This is physically satisfying only at an appropriate scale that contains an REV of the macroscopic system, as determined by Fourier analysis. Application of the volume-averaging formalism to the case of sandstone reservoirs raises the possibility of three-dimensional, fully determined porous media (i.e., the coordinates of a point determine whether the point belongs to pore or matrix). This problem can be removed in

building a mathematical simulation characterized by statistical moments, such as the probability density and the inverse Fourier transform of the power spectrum of the microstructure.

As a matter of fact, the not fully deterministic nature of sandstone reservoirs allows description of a porous medium through a set of random variables $Z_{i,j,k} = F(S_{i,j,k})$ taking values at distances $S_{i,j,k}$ (from arbitrary origin and axes), according to a probability density function indicating the state (pore or solid) of every point within the medium (Joshi, 1974). Local stationarity of the microgeometry within i^{th} bed (scale L_i) means that the probability density function tied to any set of distances $S_{i,j,k}$ depends only upon distance lags (Joshi, 1974):

$$Lag(r,s,t) = S_{i+r,j+s,k+t} - S_{ijk}$$

The stochastic process, of which the time series $Z_{i,j,k} = F(S_{i,j,k})$ for the medium is a particular realization, can be described by moments such as the inverse Fourier of the power spectrum of the phase function and density probability function. Such moments, as shown by the plateaus on the autocorrelation function among samples of a suite, might be obtained from any local volume within a specific i^{th} bed (L_i). Relevance of 2-D structural features to the volume allows description of the stochastic process in terms of inverse Fourier transform of the 2-D power spectrum of the phase function and density probability function, measured into the plane of section enclosing most of the variance of the microgeometry (i.e., cut perpendicular to apparent bedding).

An algorithm can be built transforming a stochastic process with Gaussian density probability (mean 0 and variance 1) into a stochastic process at the origin of the microstructure within the ith bed (L_i) (Joshi, 1974). Desired autocorrelation and probability density functions are achieved at the end of the process by passing pure white noise through a linear filter and by applying an inverse function for probability control to the resulting linear process. This approach is justified because moments such as 2-D autocorrelations preserve relative phase information, but lose absolute phase information. In other words, if one repeatedly samples a large-scale fabric, location of the fabric can vary with respect to the boundaries of the sampling area. All such images can have identical power spectra and, therefore, one can generate a family of synthetic porous media from initially random fields through use of the stochastic process discussed.

Synthetic, petrographically relevant 2-D and three-dimensional (3-D) media generated by such stochastic modeling can serve as an input for 2-D and 3-D finite elements and/or finite volume numerical schemes of the local closure form tied to the averaging theory (Anguy et al., accepted). Having done so, one is supplied with the tools needed for simulating the permeability tensor at progressively greater scales. The models developed at present are capable of encompassing the microstructure, as well as higher levels of the structural hierarchy. Such models yield permeability tensor as a natural consequence and describe nuances of flow through the medium. By including or removing specific levels of structure from the synthetic media, one can assess the relevance to flow of various levels of the structural hierarchy from observed changes in the permeability tensor with respect to volume.

At scales greater than the scale investigated (micrometer to centimeter), the probable occurrence of heterogeneities delineated by these bed boundaries can be handled in terms of the large-scale averaging process (Quintard and Whitaker, 1987, 1988). The large-scale process must be differentiated from the local volume-averaging which aims at incorporating into the averaged equations the local boundary conditions tied to the description of the flow within a single bed. The "large-scale averaging process" is less complex in that it just incorporates into the averaged equations the heterogeneities.

CONCLUSIONS

We have documented here and elsewhere (Prince and Ehrlich, 1990; Prince et al., in press) that the sample support problem within sandstones beds arises from interactions between sample size and a fabric containing juxtaposed closely packed domains and loosely packed domains. This problem exists in all sandstones because packing is fundamental to all sedimented aggregates. Circuits defined by loosely packed domains are commonly responsible for 80% or more of the flow within a sample. Besides its importance in the sample support problem, this ubiquitous fabric should have an important control on the behavior of multiphase flow within the medium.

We can calculate the permeability at scales greater than the permeability plugs using the procedure described by Ehrlich et al. (1991b), but the practical implementation of the local volume-averaging technique, including proof of concepts (Anguy et al., accepted) and the large-scale volume-averaging technique (Quintard and Whitaker, 1987, 1988), provide the tools for modeling the medium at any appropriate scale with a full characterization of the permeability tensor.

All of our discussions have centered on proper assessment of core data. Our undertaking, of course, is only one component of the overall effort pursued by all the contributors to this volume to understand flow properties of sandstones at scales far greater than those discussed in this chapter.

ACKNOWLEDGMENTS

The authors wish to thank Elf Aquitaine Production (E.A.P.), specifically J. P. Tricart and O. Brevart, for advice and financial support.

REFERENCES

Anderson, T. B., and R. Jackson, 1967, A fluid mechanical description of fluidized beds: Ind. Eng. Chem. Fund., v. 6, p. 527–538.

Anguy, Y., 1993, Application de la méthode de prise de moyenne volumique à l'étude de la relation entre le tenseur de perméabilité et la microgéométrie des milieux poreux naturels: Thése Universite Bordeaux I, France, Spécialité Mécanique, 170 p.

Anguy, Y., D. Bernard, and R. Ehrlich, accepted, The local change of scale method for modeling flow in natural porous media I: numerical tools: Advances in Water Resources.

Azaroff, L. V., 1960, Introduction to solids: New York, McGraw-Hill, 460 p.

Blackman, R. B., and J. W. Tukey, 1958, The measurement of power spectra from the point of view of communications engineering: New York, Dover, 190 p.

Bowers, M. C., R. Ehrlich, J. J. Howard, and W. E. Kenyon, in press, Determination of porosity types from NMR data and their relationships to porosity types derived from thin section, Journal of Petroleum Science and Engineering.

Coskun, S. B., and N. C. Wardlaw, 1993, Estimation of permeability from image analysis of reservoir sandstones: Journal of Petroleum Science and Engineering, v. 10, p. 1–16.

Crabtree, S. J., 1983, Algorithmic development of a petrographic image analysis system: Ph.D. dissertation, University of South Carolina, Columbia, South Carolina, 116 p.

Crabtree, S. J., R. Ehrlich, and C. M. Prince, 1984, Evaluation of strategies for segmentation of blue-dyed pores in thin sections of reservoir rocks: Computer Vision, Graphics, and Image Processing, v. 28, p. 1–18.

David, M., 1977, Geostatistical ore reserve estimation: Amsterdam, Elsevier, 364 p.

Ehrlich, R., S. J. Crabtree, K. O. Horkowitz, and J. P. Horkowitz, 1991a, Petrography and reservoir physics I: objective classification of reservoir porosity: AAPG Bulletin, v. 75, no. 10, p. 1547–1562.

Ehrlich, R., E. L. Etris, D. Brumfield, L. P. Yuan, and S. J. Crabtree, 1991b, Petrography and reservoir physics III: physical models for permeability and formation factor: AAPG Bulletin, v. 75, no. 10, p. 1579–1592.

Graton, L. C., and H. J. Fraser, 1935, Systematic packing of spheres—with particular relation to porosity and permeability: Journal of Geology, v. 43, no. 8, pt. 1, p. 785–909.

Isaaks, E. H., and R. M. Srivistava, 1989, An introduction to applied geostatistics: New York, Oxford University Press, 561 p.

Joshi, M. Y., 1974, A class of stochastic models for porous media: Ph.D. dissertation, University of Kansas, Lawrence, Kansas, 151 p.

Marle, C. M., 1967, Ecoulements monophasiques en milieux poreux: Revue Institut Français du Pétrole, v. 22, p. 1471–1509.

McCreesh, C. A., R. Ehrlich, and S. J. Crabtree, 1991, Petrography and reservoir physics II: relating thin section porosity to capillary pressure, the association between pore types and throat sizes: AAPG Bulletin, v. 75, no. 10, p. 1563–1578.

Prince, C. M., 1990, An analysis of spatial order in sandstones: Ph.D. dissertation, University of South Carolina, Columbia, South Carolina, 139 p.

Prince, C. M., and R. Ehrlich, 1990, Analysis of spatial order in sandstones I: basic principles: Mathematical Geology, v. 22, no. 3, p. 333–359.

Prince, C. M., R. Ehrlich, and Y. Anguy, in press, Analysis of spatial order in sandstones II: grain clusters, packing flaws, and the mesoscale structure of sandstones: Journal of Sedimentary Research, section A, v. 64(4).

Quiblier, J. A., 1984, A new three-dimensional modeling technique for studying porous media: Journal of Colloid Interface Science, v.98, p. 84–102.

Quintard, M., and S. Whitaker, 1987, Ecoulement monophasique en milieu poreux: effets des hétérogénéotés locales: Journal de Mécanique Théorique et Appliquée, v. 6, no. 5, p. 691–726.

Quintard, M., and S. Whitaker, 1988, Two-phase flow in heterogeneous porous media: the method of large-scale volume averaging: Transport in Porous Media, v. 3, p. 357–413.

Raats, P. A. C., and A. Klute, 1968, Transport in soils: the balance of momentum: Soil Science Society of America Proceedings, v. 32, p. 161–166.

Rosenfeld, A., and A. C. Kak, 1982, Digital picture processing (2d ed.), v. 1: New York, Academic Press, 435 p.

Slattery, J. C., 1967, Flow of viscoelastic fluids through porous media: AIChE, v. 13, p. 1066–1071.

Whitaker, S., 1967, Diffusion and dispersion in porous media: AIChE, v. 13, p. 420–427.

Whitaker, S., 1986, Flow in porous media I: a theoretical derivation of Darcy's law: Transport in Porous Media, v. 1, p. 3–25.

Chapter 6

Theory and Applications of Vertical Variability Measures from Markov Chain Analysis

John H. Doveton
Kansas Geological Survey
Lawrence, Kansas, U.S.A.

ABSTRACT

Finite Markov chain analysis has been used widely by sedimentologists in the search for fundamental patterns of lithological repetition that are statistically significant. The probability structure of a Markov model describes the relationship between adjacent events in a first-order process, but can be expanded to incorporate higher order memories. Simulations of stratigraphic successions from transition probabilities often are effective provided that any ancillary long-term trends also are accommodated. Markov stratigraphy can be used to produce multiple realizations of the internal structure of hydrocarbon reservoirs for use in fluid flow models. In addition, Markovian sequences have been modeled by synthetic seismograms. The discrimination of reflection frequency characteristics between synthetic seismograms from known facies types allows a lithostratigraphic classification of field seismic records. The Markovian statistics of vertical variability are applicable to selected problems of lateral prediction and simulation. The switch from the vertical to lateral direction is made possible by Walther's law, which states that lithologies that overlie one another must also have been deposited in adjacent tracts. Exceptions to Walther's law are caused by erosional breaks, but these are absorbed as a noise term within the probability model. Simulation of two- and three-dimensional models from Markovian vertical transitions must take into account the marked differences in scale and orientation that exist between the vertical and horizontal dimensions; however, some initial experiments indicate that results may be useful in applications ranging from pore network and rock fabric simulation to the modeling of local and regional geology. A finite Markov chain necessarily limits a simulation to a discretely stepped presentation of stratigraphic architecture; however, the discrete structure allows effective representations of bed boundaries and other sharp discontinuities. Geostatistical random functions can then be used to model internal variability of the Markovian events to refine the simulation.

PRELIMINARIES

In many situations, a sequence of events in either time or space is observed as a succession of states taken from a limited set of alternatives. Geological examples include traverses across a thin section in which each event corresponds to the mineral recorded at each point and stratigraphic successions where observations spaced at constant intervals record the occurrence of rock types. The relationship between adjacent events may be summarized by a transition tally matrix in which each cell sums the number of times that one state (identified by the row) is succeeded by another (identified by the column). For example, a succession consisting of lithologies A, B, C, and D may be summarized as a transition tally matrix by taking observations at successive 1-ft intervals and accumulating transition totals in the appropriate cells. This might appear as

$$
\begin{array}{c c c c c}
 & A & B & C & D \\
A & 8 & 4 & 3 & 1 \\
B & 4 & 5 & 2 & 1 \\
C & 1 & 2 & 4 & 5 \\
D & 3 & 1 & 3 & 3
\end{array}
$$

where for example, the number of times C succeeds A is 3. It can be seen that the ith row total is equal to the ith column total, which is a property of all transition matrices of this type because every lithology that is entered is also left (with the exception of the initial and terminal events). The row (or column totals) may be written as the vector

$$[16 \quad 12 \quad 12 \quad 10]$$

which represents the number of times that the succession observations are in each of the four states.

Division of the tally matrix by each of the row totals leads to a transition probability matrix, P

$$
\underline{P} = \begin{array}{c}
\begin{array}{c c c c}
 & A & B & C & D
\end{array} \\
\begin{array}{c}
A \\
B \\
C \\
D
\end{array}
\begin{bmatrix}
0.50 & 0.25 & 0.19 & 0.06 \\
0.33 & 0.42 & 0.17 & 0.08 \\
0.08 & 0.17 & 0.33 & 0.42 \\
0.30 & 0.10 & 0.30 & 0.30
\end{bmatrix}
\end{array}
$$

Similarly, division of the totals vector by the grand total results in an estimate of the fixed probability vector

$$[0.32 \quad 0.24 \quad .024 \quad .020]$$

which expresses the proportions of each lithology in the entire sequence. Readers unfamiliar with these basic concepts may find it helpful to work through the example shown in Figure 1. The counting of transitions in the short sequence of events is tabulated as the transition tally matrix. The tally matrix is then

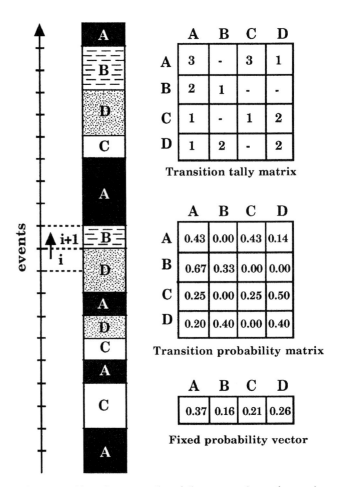

Figure 1. Simple example of the extraction of transition properties of adjacent events from a short sequence made up of four states.

converted to the transition probability matrix and fixed probability vector by the same operations described in the preceding paragraphs.

Because A, B, C, and D are mutually exclusive events, the probability that one state is followed by another is either a conditional or an unconditional probability. $P(B/A)$ is the notation that A will be followed by B, given that A has occurred as the previous event. In the unconditional case

$$P(B_{i+1} / A_i) = \frac{P(A \text{ and } B)}{P(A)} = \frac{P(A)\,P(B)}{P(A)} = P(B)$$

as opposed to the conditional alternative where

$$P(B/A) \neq P(B)$$

If all transitions are unconditional, then

$$P(B/A) = P(B/B) = P(B/C) = P(B/D)$$

and the model is one of independent events. The expected transition probability matrix, A, for independent events consists of rows of the fixed probability vector. For the example:

$$\underline{A} = \begin{bmatrix} 0.32 & 0.24 & 0.24 & 0.20 \\ 0.32 & 0.24 & 0.24 & 0.20 \\ 0.32 & 0.24 & 0.24 & 0.20 \\ 0.32 & 0.24 & 0.24 & 0.20 \end{bmatrix}$$

A matrix of expected tallies can be computed by multiplying the probabilities by the row totals of the observed tally matrix. The null hypothesis of independent events may be tested as a chi-square contingency with $(m - 1)^2$ degrees of freedom. If the null hypothesis is rejected, then the alternative model is accepted of a partial dependency between adjacent events and is known as a finite Markov chain. When the sequential pattern is controlled entirely by relationships between adjacent events, then the chain is of first order. If the sequence is shown to have a Markov property, then the structural form of the Markov chain may be deduced from a comparison of the observed transition frequencies with those expected from the independent events predictions. Transition types that occur more frequently than expected can be linked together as a preferred transition path that is a graphic description of the chain. In geological applications, the pattern of preferred transitions reflects depositional changes that are elements of sedimentary facies models. The pattern also can be interpreted readily in terms of repetitive motifs, such as cycles or rhythms, because of the limited number of states in the Markov chain.

Named after its discoverer A. A. Markov, whose inspiration was the alternation of vowels and consonants in Pushkin's poem *Onegin*, Markov chain models are an example of a stochastic process model. Markov chain models occur in the range between the extremes of determinism, where every event is exactly specified by its predecessor, and independent events, where there is no relationship between successive events.

If the matrix P is squared

$$P^2 = P.P$$

the resulting matrix is the expectation of the probability of the $(i + 2)$ event given that of the ith event, as predicted by the first-order Markov chain. If the matrix differs significantly from that observed in the sequence (as judged by a chi-square test), then the sequence has second-order Markov properties. In general, if the matrix P is successively powered to the limit, the matrix converges to a matrix of equilibrium proportions of the states, whose rows match the fixed probability vector. When a sequence is shown to have a first-order Markov property, then the transition probability matrix is a significant descriptor of the transition properties; however, the description is constrained to the relationship between immediately successive events and independent of all prior events. The type of dependency is often known as the memory of the process, which in this instance, has a length

of one step (the distance between adjacent events). In a second-order Markov model, the state event immediately preceding the pair of adjacent events is incorporated as part of the conditional probability. The memory contained within the transition probabilities is now extended to two steps. The concept can be generalized to n^{th}-order Markov models with memories of n steps. At one extreme, a memory of no steps is an independent-events situation where past events have no influence on the present event and the succession is random. Models with long memories suggest complex patterns of ordering with systematic, long-term components that are relatively free from short-term disruptions. In practice, many rock sequences can be represented adequately by low-order Markov models (Doveton, 1971). A first-order Markov model is often a sufficiently useful representation for descriptive purposes; however, the simulation of entire successions from low-order transition probability matrices involves scaling-up considerations, because the model is controlled by a short-term memory. If there are systematic longer term mechanisms, they will not be captured by the transition probabilities. This situation commonly occurs when the succession is markedly nonstationary because of long-term trends. The transition probabilities will then change in value over the length of the succession. The problem, however, can be accommodated by using several sets of transition probabilities keyed to subsequences or by linking the probabilities with position within the succession.

In practice, the Markov chain model has been found to be a simple and powerful method for both describing and simulating sedimentary sequences. Following the pioneer paper of Vistelius (1949), numerous stratigraphic applications have been published including Allegre (1964), Carr et al. (1966), Krumbein (1967), Gingerich (1969), Read (1969), Doveton (1971), and Ethier (1975). For the most part, these studies were concerned with attempts to isolate systematic transition patterns that could then be interpreted in terms of sedimentological processes or cyclic phenomena. Some authors also experimented with modeling sedimentary successions from transition probability matrices. Simulation examples ranged in scale from the internal structure of fluvial sandstones (Potter and Blakely, 1967), through simulation of the Pennsylvanian rock sequence (Schwarzacher, 1967), the lateral migration of transgressive-regressive strandline deposits (Krumbein, 1968), to marine sedimentation in space and time (Harbaugh, 1966). At the time, these simulation studies had intrinsic academic interest, but they also carried some expectations that they would find practical applications in a future of more powerful computers and greatly expanded databases. The comparatively recent emergence of stochastic reservoir models (Haldorsen and Lake, 1984) has stimulated renewed interest in the potential of Markov chains to model the internal structure of reservoirs. In addition, extensive improvements in the recording and processing of seismic records has been paralleled by new thinking in seismic modeling. Synthetic seismograms can be computed from the deterministic record of a sonic log from a selected well;

however, they also can be modeled from stochastic expectations of a succession generated by a Markov chain. These more recent developments are the focus of the remainder of this chapter. Applications of simple Markovian models are described in successively one, two, and three dimensions.

MARKOV CHAIN ANALYSIS AND SIMULATION OF SEDIMENTARY SEQUENCES IN ONE DIMENSION

The fundamental Markovian descriptor of a sequence is the transition probability matrix, p, which takes the form

$$\begin{bmatrix} p_{11} & p_{12} & p_{13} \\ p_{21} & p_{22} & p_{23} \\ p_{31} & p_{32} & p_{33} \end{bmatrix}$$

for a first-order Markov chain with (in this example) three states. In general applications, observations of the state of a process are taken at equal increments of time. The time unit for each step is chosen so as to be small enough so that all events of significance are recorded in the sequence of transitions. In the case of stratigraphic data, we do not generally have the luxury of a record measured in time units. Instead, the state of events (lithologies, facies, or other characteristics) is recorded at equal increments of depth. The substitution of depth for time simply changes the reference framework from time to space; however, the selection of the length of the interval between successive events must be considered carefully. If too small an interval is chosen, then the number of transitions of states to themselves becomes extremely large. Statistical tests for a significant Markov property will then reveal the trivial fact that successive observations tend to be repetitions of the same state. If too large an interval is used, many thin-bed events are missed altogether and the chain is biased toward states that tend to have thicker beds.

The off-diagonal elements of the transition probability matrix capture the transition characteristics between states and will be the same for any sampling interval finer than the thinnest bed. The transition probabilities of a state to itself occur on the main diagonal of the transition probability matrix. They dictate the statistics of the distribution of thicknesses of each state and will vary with the length of the interval used. The mean thickness, m, and its variance, v, of the state i are easily computed from the equations

$$m = \frac{1}{\left(1 - p_{ii}\right)}$$

and

$$v = \frac{p_{ii}}{\left(1 - p_{ii}\right)^2}$$

where the units are in number of interval spacings (Kemeny and Snell, 1960). The equations of these parameters show that thicknesses implied by Markov transition probabilities follow a geometric distribution (Krumbein and Dacey, 1969). This becomes an important consideration if the Markov transition probability matrix is used to model a synthetic stratigraphic succession. Studies of sedimentary bed thicknesses have generally concluded that they are lognormally distributed (see, for example, Pettijohn, 1957; Potter and Siever, 1955); however, this identification comes from the pragmatic observation that bed thicknesses are fitted approximately by straight lines when plotted on a log probability grid. A theoretical model of bed sedimentation based on simple probability will generate a geometric thickness distribution as pointed out by Krumbein and Dacey (1969). Of course, such a model is intrinsically Markovian.

If a transition probability matrix is used to model a synthetic succession, then systematic comparisons should be made with the real succession to verify that desired features are replicated by the simulation. The transition matrix of a first-order chain represents a very myopic view of the succession because it is restricted to the conditional relationships of adjacent events. Checks on long-term statistics will confirm whether the Markov simulation is a satisfactory match. If not, then a higher order Markov chain may be appropriate or additional longer term elements should be incorporated to supplement the Markov model.

The real succession should first be analyzed for stationarity. Do the transition probabilities stay effectively constant at all positions in the succession? If they do not, then the succession may have to be subdivided into segments that are effectively stationary. Alternatively, the transition probabilities can be modified to incorporate a drift component that reflects relative depth position. Is a first-order Markov chain an adequate description of the transition properties of the sequence? A one-step memory may be insufficient to capture systematic transitional behavior, particularly if there are distinctive sequential patterns that pick up phenomena such as fining-upward or coarsening-upward trends. These alternatives can be checked by comparing first-order Markov predictions of two-step transition types with their actual frequencies (Doveton, 1971). If the first-order Markov prediction fails to be adequate, then the system can be expanded to a second-order (or higher) Markov model. So, for example, Schwarzacher (1967) elected to use a double-dependence Markov chain in his simulation of the Pennsylvanian sequence of Kansas.

Some of the longer term properties of a succession will be honored automatically by the Markov simulation if the transitions are counted from the succession. At the most fundamental level, the proportions of the lithologies in the total succession are registered by the row and column totals of the transition tally matrix and these proportions are transferred into the transition probability matrix. In addition, Matalas (1967) pointed out that the first three moments of a succession will be preserved in a Markovian simulation.

The means and variances of the thicknesses of the lithologies in the succession should be compared with those calculated from the transition probability matrix. If they are radically different, then the actual dispersion of thicknesses appears to be represented poorly by a geometric distribution. The most common solution to this problem is to adopt a modified model, the embedded Markov chain (Gingerich, 1969). Rather than make observations at fixed intervals, the observation matrix is used to record only transitions between states. Transitions of a state to itself are now precluded and the main diagonal of the transition probability matrix has zero values. Each step of a simulation of a sequence based on this matrix now becomes a separate state. The thickness of each event can be generated by a random selection from the actual distribution of thicknesses of the corresponding state in the real succession.

Does the occurrence of the states throughout the synthetic succession show comparable characteristics to those observed in the real succession? Useful measures for this comparison are those of the first-passage time statistics (Doveton and Duff, 1984). The mean first-passage time expresses the average number of events that occurs after leaving a state before the same state is reentered. The variance measures the relative dispersion of passage time about the mean value. The expected passage time statistics can be calculated from the transition probability matrix as a matrix, M, of mean values, and W, of their variances. A fundamental matrix, Z, may be defined as

$$Z = (I - P + A)^{-1}$$

Then

$$M = (I - Z + EZ_d)D$$

and

$$W = M(2Z_dD - I) + 2[ZM - E(ZM)_d]$$

where E is an $m \times m$ matrix of unit values; Z_d is the diagonal matrix of Z; D is a diagonal matrix whose elements are the reciprocals of the fixed probability vector; and m is the number of states. Interested readers are referred to Kemeny and Snell (1960) for additional details on the derivation of these matrix algebra relationships.

The passage time statistics are the Markovian prediction of the mean and variance of the number of lithologies that intervene between successive occurrences of any given lithological state. These measures are particularly useful when applied in successions believed to have been deposited as cyclic sequences. If one of the states can be identified with the initiation or conclusion of a cycle, then the state forms a reference marker and the passage time statistics of this state are Markovian descriptors of the cycle. Comparisons with actual cycles extracted from the real succession will verify whether the Markov simulation is adequate in their representation.

The success of a Markovian simulation has often been judged in a seemingly cursory manner. So, for example, Krumbein and Dacey (1969) considered whether "simulations from empirical transition probability matrices 'look like' real-world stratigraphic sections."

Realistically, the answer to this question is dictated by the purpose of the simulation. If the intent is to understand the mechanisms of sedimentation at all possible scales, then a closest possible match is desirable between the simulation and reality. Hopefully, insights into sedimentary processes will be gained at each successive adaptation of the simulation to a better representation. At times, a simulation will be considered to be successful when it closely emulates certain characteristics while approximating other, less important features. The simplicity and speed of one algorithm may prove more cost effective than more expensive algorithms that result in only marginal improvements in representation. This situation can occur when there is a specific, limited, and pragmatic purpose for the simulation. Is the simulation to be used for the modeling of a single reservoir or an entire basin? Will the simulation be used to model fluid flow or the passage of seismic waves? The scale of the key elements and the physical properties of the system are important issues to be considered. If the results of a Markov simulation are judged to be acceptable, then multiple synthetic realizations often can be generated relatively rapidly. This property can be extremely useful in situations such as reservoir studies where the measurement of the equivalent number of real sections would be prohibitively expensive. Potter and Blakely (1967) experimented with the Markov simulation of the vertical profile of a fluvial sandstone body. They allocated bedding types between five states: cross-beds, massive beds, parting lineation, ripple beds, and mudstones. The channel sandstone was divided into a lower, middle, and upper section, corresponding to deposits of the thalweg and point bars, inner flood plain, and backwater environments, respectively. Each section was modeled by a different transition probability matrix to honor the marked proportional changes in the states and their transitions. Every step in the Markov simulation represented a single bed rather than a lithological state. Consequently, transitions of a state to itself resulted in stacked sets of the same bedding type. At each step, the thickness of the bed was chosen randomly from a lognormal thickness function modeled for each state. The tendency for beds to thin at higher levels in the channel was incorporated in the simulation by an equation that reduced the selected thickness as a function of height. Potter and Blakely (1967) noted the general relationship between sandstone bedding thickness and both grain size and permeability. Using linear regression equations developed from field data, they computed estimates of grain size and permeability for each bed of the simulation. At each step random samples from the standard errors of the regression estimates of grain size and permeability were added to incorporate simulated variability about the estimate. An example of one of these channel simulations created by Potter and Blakely (1967) is shown in Figure 2. The overall pattern shows a good generalized similarity with profiles of actual channel reservoir units. Other variables, such as porosity, also could be simulated by the Markovian model. The specific characteristics of the simulation shown is only one of a wide range of possibilities. Repetition of the simulation would result in an extended set of synthetic sections, which would differ in detail, but would maintain the same proportions of bed occurrence and transition properties. These separate realizations could be made the subjects for fluid flow simulations in dynamic reservoir models for comparison with data from real field tests. Alternatively, the transition probabilities of bedding structure and regression functions used to predict reservoir properties could be changed progressively to reflect lateral changes in the framework and hydraulic properties of the channel. At each stage, measures of the internal geometry and reservoir properties of simulated profiles would be checked carefully to verify that they provided reasonable matches with real sections. Where necessary, adjustments to the input parameters would be made to improve the quality of the synthetic profile.

In a second example, a Markov model can be used to produce synthetic successions for the generation of synthetic seismograms for comparison with field seismic records. A first-order Markov simulation of a sandstone-shale sequence is shown in Figure 3, together with its corresponding synthetic seismogram (Doveton, 1994). The simulation could be thought of as representing a turbidite succession in which the relative thickness of sandstones and shales can be controlled by changing the probabilities to emulate more proximal or distal facies. Any individual simulation holds no particular interest in and of itself, but an extended family of repeated simulations yields useful information on generalized seismic characteristics to be expected for successions of this type.

Sinvhal and Khattri (1983) applied this concept in a practical case study of facies discrimination from seismic reflection records. They compiled transition probability matrices for two types of successions logged in wells from a sedimentary basin in India. One facies was considered to have been deposited in shallow-marine to littoral environments resulting in thick sandstones and coals interbedded with shales. The other facies represented deeper marine deposition dominated by shales with little coal and a reduced sandstone content. The authors used the transition characteristics of each facies to generate multiple simulations of synthetic successions. Reflection coefficient series calculated for each sequence were convolved with a source wavelet to produce sets of synthetic seismograms for each facies type.

Seventeen potentially useful descriptors of the seismic signatures of the two facies were drawn from their autocorrelation functions and power spectra. The descriptors were used in a linear discriminant function analysis to find the most effective linear combination to discriminate between the two facies. Calibration checks of the resulting discriminant function resulted in a 75% success rate of correct classification. The function was then applied to the classification of field seismograms in a carefully controlled study that suggested that the method was an effective way to classify lithostratigraphy from seismic data.

MARKOV CHAIN SIMULATION IN TWO AND THREE DIMENSIONS

One-dimensional simulations from transition probability matrices are comparatively easy to generate.

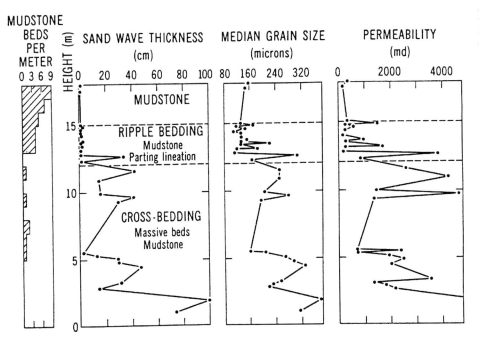

Figure 2. Vertical profile of a fluvial sandstone body simulated from transitions of embedded first-order Markov chains linked with sampling from bed thickness distributions. From Potter and Blakely (1967).

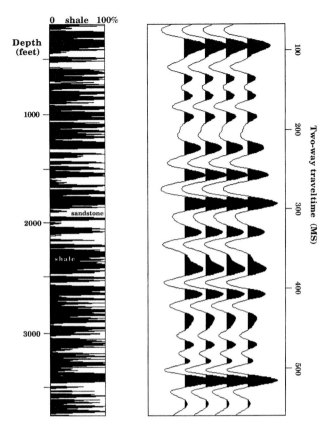

Figure 3. Markov chain simulation of a sandstone-shale sequence (left) convolved with a 30-Hz Ricker wavelet to produce a synthetic seismogram (right). The seismogram has been depth shifted to match maximum energy of the wavelet with the reflecting surface.

models using a theorem proposed by Switzer (1965). Switzer's theorem postulates that a finite-state random process can be represented on a plane where the transitions between states along any straight line are Markovian. The theorem can be used as the basis for a simulation method. Random lines are generated on a planar figure (such as a circle). The number of lines is controlled by a parameter λ, which is determined by transition probabilities measured on the process to be modeled. The intersection of the lines subdivides the plane into polygonal areas. The polygons are labeled with identification numbers and a state randomly assigned to each polygon. Adjacent polygons with the same state are fused together as single entities.

Lin and Harbaugh (1984) provided several geological examples using this method, of which the most convincing was a simulation of pore structure within a sandstone (Figure 4). The most appropriate applications appears to be for subjects where actual spatial arrangements take the form of mosaic structures. A representation by polygonal forms might be useful in cases such as the modeling of rock fabrics; however, it is a poor medium to model reservoir architectures or basin geometries. In these instances, a layered structure is a preferable template to produce a reasonable facsimile of actual spatial relationships. Rather than assign states to a preexisting polygonal subdivision, a planar area may be progressively filled by contiguous rectangles whose states are assigned by transitions from the states of preceding triangles. In stratigraphic applications, the plane would represent either a transect section of beds or a plan view of lateral relationships. In either case, transition probabilities computed from vertical sequences should have a strong predictive power for the nature of lateral transitions. This assertion is a natural consequence of both Walther's law and the repetitive nature of sedimentary units. Walther (1894) was one of the first to recognize that facies that succeed one another conformably must have been deposited in adjacent environments. This concept is known as Walther's law and is summarized by Selley (1976) as "a conformable vertical sequence of facies was generated by a lateral sequence of environments."

The concept is intuitively attractive, but is difficult to put into practice. The vertical dimension is handled easily with thicknesses usually assigned from samples

Computer programs for this purpose are listed by several authors, including Harbaugh and Bonham-Carter (1970). The extension of the same ideas to two and three dimensions is more difficult. In the case of two dimensions, the simulation must be designed such that the areas of the states must correspond to their relative proportions of occurrence. At the same time, the length of their common boundaries should be proportional to their transitional probabilities.

Lin and Harbaugh (1984) experimented with the production of two- and three-dimensional Markov

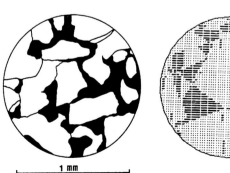
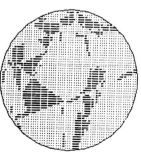
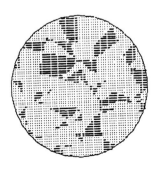

Figure 4. Sketch of a thin-section of an upper Logan Canyon Sandstone sample (left) compared with computer printouts from two Markov two-dimensional simulations of a porous sandstone (middle and right). From Lin and Harbaugh (1984).

of measured distributions. The lateral dimensions must be satisfactory representations of areal extents seen from outcrop and detailed mapping. Furthermore, the overall shapes of these simulated states should be consistent with their geology: channels should look like channels; sheet bodies should look like sheets. Finally, the overall geographic grain of the system should be reproduced. Some sedimentary systems may be isotropic, but most have distinctive axes of depositional dip and strike.

Moss (1990) described an interesting Markov simulation in which he attempted to model the internal structure of the Statfjord Formation in the Statfjord field of the Norwegian North Sea. He used stochastic lithotypes as states in a Markov chain analysis to control lateral relationships based on vertical transitions seen on logs. Clastic lithotypes were identified from the analysis of wireline logs from wells in the Statfjord Formation. Roe and Steel (1985) considered the formation to have formed in a coastal fan and fan delta complex on a low-energy coastline during the Late Triassic and Early Jurassic. Lithotypes were linked with electrofacies defined from compensated neutron, calculated effective porosity, corrected gamma-ray, and true resistivity logs. Four fundamental types were isolated and interpreted. These were lagoon-bay shales, distal fan and flood-basin bay siltstones, fan-plain medium-grained sandstones, and coarse-grained channel sandstones. An initial experimental model was produced using these four types. Moss (1990) then expanded the lithotype set to two types each of siltstone, medium sandstone, and coarse sandstone to obtain a more realistic simulation.

The integration of log measurements with core observations has several important advantages. If distinctive electrofacies can be linked with lithotypes in cored wells, then the observational database can be expanded to all logged wells, rather than to the restrictive number that have been cored. Secondly, the sensitivity of the common logs to porosity and shale content means that electrofacies are likely to be particularly useful as the building blocks of a simulation used to model fluid flow. The synthetic reservoir was produced by successive sampling from a Markov chain using transition probabilities between the lithotypes observed on the logs. The geographic axes of the model were aligned with the depositional dip and strike of the Statfjord Formation. The model orientation resulted in lithotype widths being drawn either from dip or strike distributions of lithotype elongations. So, for example, siltstones were modeled with equal dimensions in both directions, but medium-grained fan plain sandstones were elongated parallel to strike, and coarse channel sandstones were lengthened parallel to dip. In assigning wedges of lithotypes to coordinates in the model, overlaps were handled by a protocol of erosional rules. Coarse-grained sandstones eroded all other lithotypes; medium-grained sandstones eroded siltstones and shales. A completed model run was a realization of a stochastic reservoir architecture of the Statfjord Formation such as the strike and dip sections shown in Figure 5. In his paper, Moss (1990) empha-

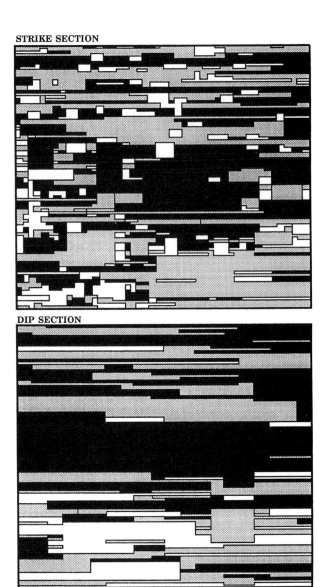

STRIKE SECTION

DIP SECTION

□ coarse ss   ▨ medium ss   ■ siltstone

Figure 5. Strike and dip sections of a Markovian reservoir model of the North Sea Statfjord Formation. Modified from Moss (1990).

sized that the modeling procedure was in a provisional state of development; however, the basic strategy shows a way forward in stochastic modeling in its use of both multiple stochastic facies and vertical transition probabilities to direct the building of the model. A three-dimensional model would involve an extension of the basic principles of this two-dimensional example. The step up in dimensionality, however, would be matched by a significant increase in complexity. The design algorithm would be required to assign lithotypes in spatial arrangements that reproduced the essential shapes, orientations, and contiguities of real reservoirs. The goal of an effective simulation is not unreasonable because actual profiles of reservoir properties are generally restricted to cored wells that are typically few in number. At the same time, modern fluid-flow simulators are designed to handle moderately complex representations of reservoir architecture. Consequently, methods to produce synthetic renditions of reservoir units have practical applications provided that they can be demonstrated to emulate the critical controlling variables in a realistic manner. Finite Markov models are one possible approach to the problem that can provide reasonable results, providing one pays careful attention to the longer term properties of the succession.

DISCUSSION

A major competing paradigm is that of models based on fractals. For example, Stolum (1991) claimed that a Markov process is incapable of producing a hierarchical pattern unless an unwieldy model is used with an extended memory of many steps. The fractal technique has an advantage in that its central axiom of self-similarity over a wide range of scales means that, if true, fractal simulations do not have the potential scaling-up problems of Markov models. At the same time, self-similarity at all scales implies a seamless continuum of variability that is difficult to reconcile with a notion of hierarchical organization. Scales of heterogeneity are commonly recognized in reservoir structure and stratigraphic sequences in general. Markov simulations, therefore, have a useful role to play when supplemented by procedures that model higher scaled features.

A finite Markov chain must limit the form of any resulting model to that of a discretely stepped representation of a reservoir. While discontinuities between lithotypes will be maintained explicitly, the reservoir properties within each event will remain homogeneous unless modified by an additional algorithm. Deutsch and Journel (1992) noted that stochastic simulation generated from geostatistics has the reverse problem. By their very nature, random function models cannot generate the sharp breaks that occur when crossing boundaries between distinctive lithotypes, but will often model internal variability quite successfully. Consequently, Deutsch and Journel (1992) recommended that "the spatial architecture of major lithofacies . . . should be represented first" and "then the distribution of petrophysical properties within each facies . . . can be simulated." They further noted that, "a two-step approach to simulation is not only more consistent with the underlying physical phenomenon (e.g., geology), it also avoids stretching the stationarity/homogeneity hypothesis underlying most continuous RF models." Finite Markov models provide the necessary tools for the first step, and the second step is fulfilled by geostatistical methods described at length in other chapters of this volume.

REFERENCES

Allègre, C., 1964, Vers une logique mathématique des séries sédimentaires: Bulletin de Société Géologique de France, Series 7, v. 6, p. 214–218.

Carr, D. D., A. Horowitz, S. V. Hrarbar, K. F. Ridge, R. Rooney, W. T. Straw, W. Webb, and P. E. Potter, 1966, Stratigraphic sections, bedding sequences, and random processes: Science, v. 28, no. 3753, p. 89–110.

Deutsch, C. V., and A. G. Journel, 1992, GSLIB: Geostatistical Software Library and User's Guide: New York, Oxford University Press, 340 p.

Doveton, J. H., 1971, An application of Markov chain analysis to the Ayrshire Coal Measures succession: Scottish Journal of Geology, v. 7, no. 1, p. 11–27.

Doveton, J. H., 1994, Geologic log analysis using computer methods: AAPG Computer Applications in Geology No. 2, Tulsa, AAPG, 169 p.

Doveton, J. H., and P. McL. D. Duff, 1984, Passage-time characteristics of Pennsylvanian sequences in Illinois: Ninth ICC Congress Compte Rendu, v. 3, p. 599–604.

Ethier, V. G., 1975, Application of Markov analysis to the Banff Formation (Mississippian), Alberta: Mathematical Geology, v. 7, p. 47–61.

Gingerich, P. D., 1969, Markov analysis of cyclic alluvial sediments: Journal of Sedimentary Petrology, v. 39, no. 1, p. 330–332.

Haldorsen, H. H., and L. W. Lake, 1984, A new approach to shale management in field-scale models: SPE Journal, v. 29, no. 4, p. 447–457.

Harbaugh, J. W., 1966, Mathematical simulation of marine sedimentation with IBM 7090/7094 computers: Kansas Geological Survey Computer Contribution 1, 52 p.

Harbaugh, J. W., and G. F. Bonham-Carter, 1970, Computer simulation in geology: New York, Wiley-Interscience, 575 p.

Kemeny, J. G., and J. L. Snell, 1960, Finite Markov chains: Princeton, Van Nostrand, 210 p.

Krumbein, W. C., 1967, FORTRAN IV computer programs for Markov chain experiments in geology: Kansas Geological Survey Computer Contr. 13, 38 p.

Krumbein, W. C., 1968, FORTRAN IV computer program for simulation of transgression and regression with continuous time Markov models: Kansas Geological Survey Computer Contr. 26, 38 p.

Krumbein, W. C., and M. F. Dacey, 1969, Markov chains and embedded Markov chains in geology: Mathematical Geology, v. 1, no. 1, p. 79–96.

Lin, C., and J. W. Harbaugh, 1984, Graphic display of two- and three-dimensional Markov computer models in geology: New York, Van Nostrand Reinhold, 180 p.

Matalas, N. C., 1967, Some distribution problems in time series simulation: Kansas Geological Survey Computer Contr. 18, p. 37–40.

Moss, B. P., 1990, Stochastic reservoir description: a methodology, in A. Hurst, M. A. Lovell, and A. C. Morton, eds., Geological applications of wireline logs: Geological Society of London, Special Publication 48, p. 57–75.

Pettijohn, F. P., 1957, Sedimentary rocks: New York, Harper, 718 p.

Potter, P. E., and R. F. Blakely, 1967, Generation of a synthetic vertical profile of a fluvial sandstone body: Journal of the Society of Petroleum Engineers of AIME, p. 243–251.

Potter, P. E., and R. Siever, 1955, A comparative study of upper Chester and Lower Pennsylvanian stratigraphic variability: Journal of Geology, v. 63, no. 5, p. 429–451.

Read, W. A., 1969, Analysis and simulation of Namurian sediments in central Scotland using a Markov process model: Mathematical Geology, v. 1, p. 199–219.

Roe, S. L., and R. Steel, 1985, Sedimentation, sea-level rise and tectonics at the Triassic–Jurassic boundary (Statfjord Formation), Tampen Spur, Northern North Sea: Journal of Petroleum Geology, v. 8, no. 2, p. 163–186.

Schwarzacher, W., 1967, Some experiments to simulate the Pennsylvanian rock sequence of Kansas: Kansas Geological Survey Computer Contr. 18, p. 5–14.

Selley, R. C., 1976, An introduction to sedimentology: London, Academic Press, 408 p.

Sinvhal, A., and K. Khattri, 1983, Application of seismic reflection data to discriminate subsurface lithostratigraphy: Geophysics, v. 48, no. 11, p. 1498–1513.

Stolum, H.-H., 1991, Fractal heterogeneity of clastic reservoirs, in L. W. Lake, H. B. Carroll, Jr., and T. C. Wesson, eds., Reservoir characterization II: San Diego, Academic Press, p. 579–612.

Switzer, P., 1965, A random set process in the plane with a Markovian property (note): Annals of Mathematical Statistics, v. 36, p. 1859–1863.

Vistelius, A. B., 1994, On the question of the mechanism of formation of strata: Doklady Akademii Nauk SSSR, v. 65, p. 191–194.

Walther, J., 1894, Einleitung in die Geologie als Historische Wissenschaft, Ed. 3, Lithogenesis der Gegenwart: Fischer, Jena, p. 535–1055.

Chapter 7

A Review of Basic Upscaling Procedures: Advantages and Disadvantages

John Mansoori
Amoco Production Company
Tulsa, Oklahoma, U.S.A.

BACKGROUND

Predicting hydrocarbon recovery from oil and gas reserves generally involves the following three steps: (1) understanding the role and importance of reservoir geology, (2) quantification of geologic and other relevant data into a system of numerical grids, and (3) performing computer simulation of the displacement process. An accurate implementation of the current technology in each of these steps can contribute to an improved prediction of reservoir performance.

In the course of geologic events leading to the present state of a reservoir, permeability distribution plays the most important role. Subsurface flow behavior is predominantly governed by such a permeability distribution. Fluctuations in magnitude and direction of permeability over the extent of a reservoir or aquifer define the degree of anisotropy, and additional changes in lithology give rise to reservoir heterogeneity. Hence, an adequate knowledge of the spatial distribution of permeability is critical in predicting reservoir depletion performance by any recovery process.

A quantitative measure of both anisotropy and heterogeneity depends strongly on the scale at which such measurements are made. On the large or macroscale, heterogeneity can manifest itself via faults, fractures, cross-bedding, contrasting lithology, and other sedimentological complexities. On the small or microscale, the use of the scanning electron microscope has revealed much information on pore size distribution and diagenesis. In addition, field observations have shown that heterogeneity is present to varying degrees at any arbitrary scale in porous medium. Hence, a successful prediction of reservoir behavior under any depletion scheme hinges upon our ability to (1) provide a realistic description of reservoir geology at the scale at which important variations of rock properties can be captured, and (2) use a numerical simulation tool for modeling subsurface fluid flow with sufficient accuracy and computational efficiency.

In addressing the first issue, we have seen an overwhelming effort in developing geostatistical methods to improve our ability in data integration of various sources and scales of measurement for developing realistic reservoir descriptions. Such a stochastic approach, contrary to the conventional (deterministic) approach, allows for quantification of uncertainty and provides for a more accurate economic model for better reservoir management.

Although geostatistical techniques are capable of generating reservoir descriptions at any arbitrarily small scale, flow simulations at such small scales require computing power not yet readily available to the general reservoir engineering community. Hence, we need a methodology where small-scale geologic variability represented by a very large number of numerical grid blocks can be upscaled to levels at which fluid flow simulations become practical. In adopting an upscaling procedure, however, you must consider the balance between the loss of accuracy due to the averaging process and the gain in computer speed due to a fewer number of grids.

In formulating conservation laws to describe macroscopic subsurface flow, most models are based on a continuum approach. In this methodology, the physical properties of the medium, such as porosity and permeability, are assumed to belong to a representative elemental volume (REV) that varies continuously in space. The partial-differential equations governing fluid flow are then derived by expressing the conservation laws at the scale of the volume averaged variable (Bear, 1972; Baveye et al., 1984). The strongly nonlinear nature of the resulting equations, however, makes the numerical solution approach favored over the approximate analytical methods. In such numerical methods, the flow domain is divided into grid blocks and rock properties are assigned to each grid. Ideally, the resolution of such a gridded network should be at the scale at which rock property measurements and lithofacies identification have been made. As mentioned, recent geostatistical development and stochastic simulation allow for generating high-resolution reservoir descriptions by integrating data from core measurements, well logs, seismic, and geologic features covering a broad range in the scale of measurement. The process of upscaling involves assigning average rock and/or flow properties to coarser grids that may contain up to several thousand smaller grids defined via the conditional simulation.

The intent of this chapter is not to present a detailed technical review of upscaling methods, but to provide a brief review of such techniques and to serve as a source of references for additional reading in both petroleum engineering and hydrology literature. I hope that future advances in computing technology precludes the need for upscaling.

IMPORTANT CONCEPTS IN UPSCALING

The main objective of an upscaling process is to use information at one length scale in order to compute equivalent or effective properties at a larger scale. The accuracy of such effective properties as applied to flow through porous media is usually judged by how well fluid-flow predictions made at the coarser (macroscale) level mimic predictions made at the finer (microscale) level.

In the most general context, due to a nonlinear coupling of rock and fluid properties in a multiphase flow simulation, upscaling should be applied not only to porosity and absolute permeability, but also to relative permeability, capillary pressure and dispersion in miscible displacements. Furthermore, if at all possible, the upscaling methodology must not be directly tied to the solution at the fine scale because the main purpose of an upscaling process is to avoid conducting such time-consuming flow simulations. The work reported in the literature in the area of multiphase upscaling is relatively sparse due to difficulty of analyzing the complex, nonlinear, coupling between rock properties and fluid-flow effects. Such derivations usually result in determining a set of pseudofunctions that are problem specific and cannot be easily generalized to other flow geometries. The review presented here does not cover the subject of upscaling in multiphase flow problems.

Avoiding the direct use of simulation results achieved at one scale to estimate flow behavior at a coarser scale requires certain approximations whose validity depends greatly on the original objectives of the problem. In increasing order, the first level of approximation involves replacing the original multiphase system with one in which the averaged properties of a coarse grid are obtained by solving the flow problem within the coarse grid using local boundary conditions. The sensitivity of results to the choice of boundary condition can be mitigated by considering several different boundary conditions and taking the average of upscaled properties calculated from flow simulation results.

The next level of approximation replaces the multiphase system with a single phase, steady-state flow field in which the local boundary condition assumption is applied to upscale the absolute permeability or block interface transmissibilities. Either of these properties can be used to obtain an approximate solution for multiphase flow problems using coarse grid simulations. Presently, the majority of work reported in the literature discusses the problem of upscaling under the assumption of single-phase, steady-state flow. These methods attempt to derive an effective absolute permeability (or transmissibility) in petroleum engineering applications or, alternatively, effective hydraulic conductivity in groundwater hydrology. The application of such single-phase derivations to multiphase flow problems is purely heuristic and its validity should be thoroughly checked under extreme upscaling conditions. The selection of a flexible gridding algorithm to better represent the variations in reservoir heterogeneity and the treatment of permeability as a full tensor are among the many methods developed to reduce inherent errors associated with these approximations.

CLASSIFICATION OF UPSCALING SOLUTION METHODS

The majority of work in the literature addresses the upscaling issues in the context of either permeability or hydraulic conductivity under single-phase flow conditions. Two distinct methods based on analytical or numerical techniques have been proposed. The numerical methods are much more general and do not suffer from limitations associated with the analytical techniques; however, under certain conditions, analytical methods can provide quick estimates of the effective properties. Analytical derivations are typically based on the application of stochastic differential equations to fluid flow in porous media whereas numerical methods are applied in the framework of Monte Carlo simulation.

The literature discussed in this chapter covers a rather broad range of property distribution. Properties can be spatially distributed either deterministically or by stochastic methods. Properties can include simple systems, such as permeability beds, becoming gradually more complex in the case of stochastic shales and uncorrelated and correlated spatial distributions. Depending on the spatial arrangement of permeability, several analytical methods can be used to solve for the effective properties. Many of these methods are based on the concept of stochastic differential equation. Finally, I discuss the relation between the geologic feature of sedimentary environment and the average permeability.

Numerical Methods

The effective permeability, determined using a coarse grid composed of several smaller grids, can be estimated using numerical simulation methods. Assuming a steady-state single-phase flow, rate or pressure boundary conditions are applied to various surfaces of the coarse grid to establish a flow path through the fine-grid heterogeneous system. The effective permeability of the network is then obtained by dividing the total volumetric flux across a section perpendicular to the flow direction by the imposed pressure gradient.

In a pioneering work presented by Warren and Price (1961), permeability fields at the fine scale were generated using several statistical distribution functions such as lognormal, exponential, skewed lognormal, and linear and discontinuous distributions. The spatial arrangement of permeabilities, however, was purely random with no correlation. Finite difference approximations were applied to the flow equations to solve for the steady-state pressure field. The effective permeability was then calculated. Using a Monte Carlo technique, Warren and Price (1961) showed that regardless of the type of permeability distribution, the effective permeability can best be approximated by the geometric mean of the distribution.

In a work similar to Warren and Price (1961), Bower (1969) used analog simulations using a two-dimensional flow field. He also concluded that the heterogeneous medium could be replaced by a homogeneous system whose effective permeability is the geometric mean of the distribution.

Freeze (1975) used a Monte Carlo technique to analyze both steady-state and transient one-dimensional groundwater flow in nonuniform, homogeneous, bounded media. Similar to the other investigators, he considered only a random distribution of hydraulic conductivities at the fine scale. The study was limited to one dimension. Freez observed that under steady-state conditions, the effective hydraulic conductivity is better approximated using a harmonic mean of the distribution.

Smith and Freeze (1979a, b) extended this study to two and three dimensions where hydraulic conductivities were spatially correlated using a set of joint probability density functions. All simulations were done for statistically homogeneous media using isotropic covariance functions. For the case of one-dimensional flow, the geometric mean accurately predicted the average behavior, with results becoming less accurate for the two dimensional problems. These authors concluded that the effective conductivity is a function of both spatial distribution and system dimensionality.

El-Kadi and Brutsaert (1985) conducted simulations of unsteady gravity drainage from a large unconfined aquifer. Hydraulic conductivities were assumed to be lognormally distributed with an exponential covariance of the log values. These authors showed that the head variance was an increasing function of both the correlation length and the hydraulic conductivity variance.

Journel et al. (1986) proposed an indicator approach to generate realizations of sand/shale environments. This technique can be used to obtain permeability fields that are highly variable, highly anisotropic, and whose spatial correlation covers multiple scales of variability. A Monte Carlo approach was again used to derive an effective permeability for various sand/shale proportions. The correlation between the effective permeability and shale proportion contained in a coarse grid was fitted with a power average model. In such a model, the power exponent takes the value of –1 for the harmonic average, +1 for the arithmetic average, and 0 for the geometric average. These authors showed that a power exponent value of 0.57, a value between the arithmetic

and geometric means, best characterized horizontal flow in a shaly sand environment. For vertical flow, the best fit was obtained with an exponent value of 0.12. These results, however, are quite sensitive to such factors as shale geometry, dimensions of the block relative to the correlation range, and the nature of multimodal distribution.

In a similar study, Desbarats (1987) used a numerical technique to estimate the effective permeability in sand/shale formations under saturated, steady-state flow conditions. Horizontal and vertical variograms from actual Assakao fluvial sandstone outcrops located in the Tassili region of the central Sahara were modeled and used within a turning bands algorithm (Mantoglou and Wilson, 1981) to generate the stochastic distribution of sand and shale sequences. Using a Monte Carlo approach, the effective permeability in a finite flow field was correlated with the shale volume fraction, the spatial correlation structure, and the flow field dimensions.

Deutsch (1989) compared the power average model and a percolation model to correlate the relationship between the effective permeability, the volume fraction of shale, and the shale anisotropy. A simple graphical procedure was developed for determining the power exponent from shale aspect ratio, shale volume fraction, and the shale and sand permeabilities. Both models were equally suitable for fitting the observed correlation; however, the power averaging method was deemed superior due to its simplicity and in its requiring only a single parameter to be determined as opposed to three parameters needed for the percolation model.

Desbarats (1988) presented the results of a numerical study using actual permeability data from the lower Stevens Sandstone formation of the Paloma field in Kern County, California. Measured permeabilities exhibited a lognormal distribution. Permeability fields were generated using univariate (histogram), spatial statistics (variogram), and turning bands simulation method. Results based on Monte Carlo technique showed that the effective permeability was a function of the spatial correlation structure, and that this dependency was reduced with increasing nugget value of the variogram model.

Gomez-Hernandez and Gorelick (1989) applied a stochastic simulation approach to investigate the influence of the spatial variability of aquifer hydraulic conductivity on the hydraulic head under a steady-state flow condition. Conductivity realizations were generated assuming isotropic porous media, an exponential covariance model, and a second-order stationarity. Monte Carlo simulations were performed to estimate mean and standard deviation of the hydraulic head. Results for both conditional or unconditional simulations were observed to fit best with a power averaging model using an exponent between the harmonic and the geometric means of the distribution. The effective hydraulic conductivity was greatly influenced by the distribution type, anisotropy, correlation length, and boundary conditions.

Analytical Methods

Compared to the numerical methods, which are quite general and have been successfully applied to systems with complex heterogeneity and anisotropy, analytical methods of calculating effective properties have a much more limited applicability. These methods are developed within either a deterministic or probabilistic framework. In a deterministic approach, the spatial distribution of permeability field is assumed to be known at a given scale of heterogeneity. Under the assumption of single-phase, steady-state flow, the continuity equation is combined with Darcy's law to arrive at the pressure equation, which is then solved analytically for an arbitrary boundary condition. Results are then used to calculate the effective permeability. Exact solutions, however, can be derived only for cases of simple heterogeneity and anisotropy.

In a probabilistic approach, the permeability field is modeled as a random variable with a known joint probability distribution that defines its spatial structure. The governing flow equation is solved directly in the framework of a stochastic differential equation. The solution to the stochastic system of equations is itself a random function characterized by a probability density function (pdf). In the Monte Carlo approach, the entire pdf of the output random variable can be obtained as the number of simulations performed approaches infinity. For analytical methods using stochastic approach, one can derive only a few parameters of the output random variable's pdf, such as its mean and covariance.

In the following sections, I first present a review of the upscaling methods for calculating the effective properties for simple, deterministic heterogeneities. I then review the methods based on stochastic approach.

The analytical techniques used to derive effective properties for porous media containing sand/shale sequences are generally referred to as streamline methods. Using a simple geometry representative of the geometric distribution of the sand/shale strings, these methods provide an estimate of the increase in flow path tortuosity due to the presence of shales.

Prats (1972) calculated an effective vertical to horizontal permeability ratio (K_v/K_h) for the case of a large number of very thin, impermeable horizontal shale strings with uniform distribution. By assuming that barriers were strips of infinite length, the problem was conveniently treated in a two-dimensional geometry. The analytical expression derived by Prats related the effective permeability to the width, vertical spacing, and degree of overlapping of the shale strings.

Weber (1982) examined various depositional environments where barriers to flow resulted from such geologic features as discontinuous clay, silt intercalations, and cross-bedded sand. Weber's analytical treatment was very similar to Prats's (1972), but with less stringent assumptions in that such barriers were allowed to process finite thicknesses and nonzero permeabilities. Furthermore, the sand body itself was treated with general permeability anisotropy. The analytical expression derived for the effective perme-

ability showed that, for the same potential difference in the sand body, the ratio of total flux in the presence of shale barriers to that of the sand body only (in the absence of shale) is a strong function of the sand's vertical/horizontal permeability ratio (K_v/K_h).

Richardson et al. (1978) studied in detail the geology of various depositional environments and were able to quantify the size and spatial distribution of several types of potential barriers to fluid flow. Using a simple analytical method, they then calculated the rate of drainage of a layer of oil accumulation over a barrier. This allowed them to derive equations for estimating the oil layer thickness as a function of time. The time required for oil to be drained over a barrier was a function of barrier geometry, oil viscosity, and horizontal permeability. These simple analytical results agreed very favorably with detailed numerical simulation of the oil drainage process.

Haldorsen and Lake (1984) combined the analytical method with statistical information on shale lateral continuity and spatial disposition to estimate the effective permeability of sand/shale depositional environment. Shale statistics were obtained from well logs and outcrop studies. Effective permeabilities in both horizontal and vertical directions were derived by first calculating an effective cross sectional area open to flow calculated from the stochastic distribution of shale strings within a coarse grid block. This required developing a very efficient computational algorithm. The Haldorsen and Lake technique, however, was restricted to two-dimensional geometries, assuming a grid-block length/height ratio greater than ten. Further assumptions included impermeable shales and a homogeneous sand body.

The method presented by Haldorsen and Lake (1984) relied heavily on a deterministic knowledge of the spatial distribution of shale within a homogeneous sand body. Begg and King (1985) presented a statistical technique by directly calculating the effective permeability of the medium using a histogram of shale lengths and volume fractions. Furthermore, restrictions on dimensionality and the grid-block aspect ratio associated with Halderson and Lake (1984) were removed; however, the assumption of zero permeability for the shale strings could not be relaxed. Begg and King's (1985) derivation showed a strong dependency of the effective permeability to system dimensions, and the density and thickness of shale barriers.

The statistical method of Begg and King (1985) was further generalized by Begg et al. (1985) for layered medium in which sand permeability anisotropy and shale frequencies and dimensions vary from layer to layer. Although the method was strictly applicable to impermeable shales, it could be extrapolated with reasonable accuracy to cases where the effective permeability was at least two times less than that derived for the impermeable shales. These authors showed that under similar shale geometry and spatial distribution, the effective permeability in three dimensions is always greater than that calculated in two dimensions. This difference, however, became increasingly negligible as the vertical effective permeability was reduced.

The streamline technique was applied by Martin and Cooper (1984) to model a steam-drive pilot project in a complex clastic reservoir with low-permeability barriers. Hazeu et al. (1986) used the method in a three-dimensional simulation study of the Frigg field in the North Sea.

For a simple case of a heterogeneous formation consisting of parallel beds of uniform permeability, the analytical solution of Darcy's flow equation yields effective permeability equal to the arithmetic mean when flow direction is parallel to the bedding plane, and a harmonic mean when flow is perpendicular to bedding plane. Dagan (1979) showed that in the general case of a heterogeneous formation with an arbitrary spatial arrangement of permeabilities, the effective permeability of the medium must lie between the harmonic and arithmetic mean. As previously stated, Warren and Price (1961) showed that for the case of purely random, uncorrelated distribution, the geometric mean is statistically the best average for the effective permeability.

Additional analytical techniques also have been developed based on perturbation expansion methods, the effective medium theory, and homogenization or multiple scale theory. Matheron (1967) and, more recently, Bakr et al. (1978), Gutjahr et al. (1978), and Tang and Pinder (1977) investigated the flow in a heterogeneous medium using a small perturbation approximation of flow equation. Effective conductivities for various formations were derived based on statistical models of local-scale conductivity variability. These studies, however, were limited to infinite domains and single-phase, steady-state flow. An important assumption behind the small perturbation technique was that the standard deviation of permeability should be small compared to its mean value.

For unsteady flow, Dagan (1982) found that the effective hydraulic conductivity was time dependent and showed a deviation from the arithmetic mean at early times. Dagan (1979, 1981) also considered more realistic permeability distributions and presented a self-consistent method for deriving the effective statistical properties, such as mean and head variance, for one-, two-, and three-dimensional groundwater flows. In this approach, the medium is represented by a set of blocks with random conductivity, size, and position. Fluctuation of the head field induced by each heterogeneous block is computed by assuming that it is surrounded by a homogeneous matrix whose average properties are obtained by the self-consistency requirement. These average properties are equal to the effective properties of the medium as a whole. The treatment of reservoir heterogeneity by this approach does not require an assumption on the smallness of the variance of medium properties, as was the case for perturbations methods.

Delhomme (1979) used a geostatistical technique and showed how the estimated variance of the hydraulic head can be reduced by conditioning the conductivity field to known point values.

King (1987) treated the pressure equation for single-phase, steady-state flow as a stochastic differential equation and used the field theoretical methods to include higher order terms in the perturbation expansion for estimating the effective permeability, mean pressure, and pressure variance. Diagrammatic methods were used to sum parts of an infinite perturbation series.

A new method for deriving the effective properties of random fields based on the effective medium theory (EMT) was suggested by Kilpatrick (1973). This method was originally applied to estimate the transport coefficients, such as electrical conductivity and diffusion, constant in composite media. The theory is based on the idea of replacing the inhomogeneous medium by an equivalent homogeneous medium such that the fluctuations induced by restoring the heterogeneity average to zero (Kilpatrick, 1973; McGill et al., 1991; King, 1989).

The original EMT developed by Kilpatrick (1973) assumed isotropic conductance (or permeability). This restriction was removed by Bernasconi (1974) who generalized Kilpatrick's single-bond EMT to an anisotropic heterogeneous network.

Harris (1990) presented a generalized effective-medium theory applicable to networks of arbitrary topology. The generalization incorporated anisotropic networks developed by Bernasconi (1974). Harris's (1990) method was applied to a fractured network to obtain estimates of conductivity tensor.

An important assumption behind the effective-medium theory is that the mean fluctuations in the output variable (i.e., hydraulic head or pressure) should be small and average to zero. This assumption may not be valid as the fluctuations of the input variable (i.e., hydraulic conductivity or permeability) increase in porous media. King (1989) proposed a new real-space renormalization approach for calculating the effective permeability of a heterogeneous medium to relax the stringent assumption on the field properties. In this approach, a permeability field whose scale length was representative of the original data sample size was first grouped into blocks of four in two dimensions or eight in three dimensions. The effective permeability of these blocks was calculated using an equivalent resistance network approach. This process was continued, using successively larger regions, until a single, 2×2 (in two dimensions) or 8×8 grid (in three dimensions) was achieved. The final averaging process resulted in one value for the effective permeability of the entire field. King (1989) showed that the method can be applied even when the permeability fluctuations are large, and that the method compared favorably with the numerical simulation technique at a fraction of computation time.

Tensor Methods

Generally, the analytical methods discussed can be used for calculating the effective permeability for mild cases of heterogeneity and anisotropy. The most general treatment of effective properties, however, demands that these properties be treated as a full tensor whose various terms reflect spatial variations of the property in both magnitude and direction within the porous media. For example, in a porous medium, such as sandstones, the permeability parallel to sedimentary layers is generally greater than permeability in the perpendicular direction (Greenkorn et al., 1964). For a medium with an arbitrary orientation with respect to the coordinate system, the off-diagonal terms are needed to correctly account for the fluid flow in directions perpendicular to the direction of main pressure gradient. The permeability anisotropy and development of heterogeneity in porous media are now believed to be a direct result of variations in shape, orientation, and textural differences of grains and other building blocks of reservoir rocks. Thus, before selecting an upscaling technique, one should carefully examine the characteristics and the degree of complexity of the medium to ensure reliability of such calculations. The geologist can play a significant role in this process.

Leung (1986) presented a systematic combination of permeability anisotropy and heterogeneity in an increasing order of complexity. A porous medium can be classified as (1) anisotropic and homogeneous, (2) anisotropic and heterogeneous (limited anisotropy), or (3) fully anisotropic and heterogeneous. In such complex situations, Dullian (1979) elaborated on the necessity of treating permeability as a full tensor based on a mathematical analysis of fluid flow in an anisotropic porous media. For a porous medium with three mutually orthogonal, principal axes (orthotropic), the permeability tensor is symmetric; furthermore, the alignment of permeability principal axes with the coordinate system results in a simple diagonal matrix. In both anisotropic and homogeneous, and anisotropic and heterogeneous mediums, the coordinate axis can be aligned with the principal permeability direction for fluid-flow modeling; however, in the fully anisotropic and heterogeneous medium, both magnitude and direction of the principal permeability axis can vary spatially. This requires that the full permeability tensor be considered for calculation of effective properties.

Kasap and Lake (1990) derived analytical expressions for both diagonal and off-diagonal terms of the permeability tensor in parallel beds and beds in series. For isotropic permeability, these equations yielded the familiar harmonic average for serial beds and the arithmetic average for parallel beds. These authors also provided approximate calculations for the case of incomplete layering, which typically occurs in reservoirs with discontinuous shales and cross-beds. Good comparison was observed between the analytical results and those obtained via numerical simulations.

White and Horne (1987) presented a numerical simulation technique to calculate the effective transmissibility tensor using information readily available at the small scale. In this method, multiple flow simulations using several kinds of boundary conditions were performed on a coarse grid under the assumption of steady-state, incompressible flow. The resulting

microscale pressures and fluxes were then used to calculate macroscale properties. The components of macroscale transmissibility tensor were obtained by minimizing the error in microscale fluxes using a least square procedure.

Aasum (1992) presented a more rigorous analytical technique for calculating full permeability tensor for a heterogeneous, anisotropic grid block. Full permeability tensor is requited when permeability is anisotropic and oriented at an angle not aligned with a major coordinate axis, such as cross-bedding and laminations, and when heterogeneities are not centered in the grid block, such as the case of nonuniform shale distribution within a sand body. The procedure consisted of dividing the grid block into four quadrants, then determining the effective permeability tensor in each of the four quadrants. The effective permeability tensor for the entire grid block was obtained by combining the effective permeability tensors from each of the four quadrants. Dividing the grid block into four quadrants allowed for a more rigorous treatment of the location of major heterogeneities within the grid block. Excellent agreement was observed between the analytical and numerical methods.

Samier (1990) described a finite element technique for calculating components of the transmissibility tensor for a heterogeneous grid block based on single-phase flow. The procedure was applied for calculating the off-diagonal transmissibility terms for a simulator grid block.

Holden et al. (1990) observed that the iterative method used to solve the steady-state, incompressible single-phase flow equations for a heterogeneous block converged much faster for the effective permeability than it did for pressure calculation. Hence, a significant saving in computer time was realized by terminating the iterative solver before reaching convergence on pressure. Off-diagonal elements were calculated by rotating the grid block and repeating the procedure.

Gomez-Hernandez and Journel (1990) used the standard numerical simulation technique similar to that of White and Horne (1987) to calculate components of the effective permeability tensor. They also elaborated on some techniques for mitigating the effect of boundary condition on the calculated effective permeability.

Begg et al. (1989) pointed out that in the case of large heterogeneities (in the order of block dimensions), the effective permeabilities obtained using numerical simulation on the fine scale showed some sensitivity to the boundary conditions imposed on the surfaces of the coarse grid block. Gomez-Hernandez and Journel (1990) argued that if the range of permeability covariance was finite, a minimum block size existed beyond which the block effective permeability became independent of the boundary condition. These investigators applied several different boundary conditions half a grid block away from the boundary of the grid block for which interface permeability tensors were to be determined. This technique was successfully used even for blocks having a size approaching the scale of large heterogeneities. In a subsequent development, these authors used the results of one set of effective permeability calculations on a synthetic training image to infer a statistical model of the joint spatial variability of the upscaled permeability tensors. This model was then used to generate multiple realizations on the coarse scale, eliminating the need for the often costly upscaling process.

Several investigators have applied homogenization theory to determining the effective permeability (Bourgeat, 1984; Mei and Auriault, 1989; Saez et al., 1989; Durlofsky and Chung, 1990). In this approach, the partial differential equations governing fluid flow in porous media are applied at two different scales of heterogeneity related via an arbitrary scale factor defined as the ratio of characteristic lengths at the two scales. Saez et al. (1989) presented the detailed derivation of the macroscopic flow equation for a single-phase, slightly compressible fluid. The dependent variable (i.e., pressure in a single-phase flow problem) was considered to be a function of time and position in both scales. A perturbation expansion was applied to the pressure using the scale ratio as the perturbation parameter. The results were substituted into the governing partial differential equations and the coefficients of powers of scale factor were set equal to zero. By assuming spatially periodic heterogeneity for the porous medium, a macroscopic partial differential equation resulted in which the macroscale effective permeability was related to the microscale permeability tensor and a microscale tortuosity tensor.

Otero et al. (1990) applied the homogenization technique to a porous medium with fractures and a porous medium containing stochastic shales. Results of this study showed that for a porous medium containing shale barriers, two-dimensional models significantly underpredicted the effective vertical permeability as compared to more realistic three-dimensional representations; however, two-dimensional models were adequate for estimating effective permeabilities for fractured porous media.

Durlofsky and Chung (1990) showed that the assumption of periodic boundary conditions always yielded a symmetric effective permeability tensor.

GEOLOGIC CONSIDERATIONS IN PERMEABILITY UPSCALING

A good discussion on identifying, classifying, and predicting reservoir nonuniformities has been provided by Hutchinson (1959) and Hutchinson et al. (1961). For instance, they showed that geologic features, such as cross-bedding, textural qualities, and cement content in sandstone formations inferred from well cores, could be used to assess the size and permeability contrast of nonuniformities. A general review of various types of reservoir heterogeneity and their relations to vertical and horizontal permeabilities is also given by Craig (1971).

A quantitative estimate of well productivity or injectivity requires the knowledge of the average permeability within the well drainage area. This can be achieved either by averaging core permeabilities measured at several depths representing a fair statistical variation of rock properties along the pay thickness, or by well testing and analyzing pressure transient results (Matthews, 1967). Using core permeabilities, the averaging technique used should be consistent with reservoir geology determined by the original depositional environment, as well as by secondary alterations of reservoir rock. The proper method of averaging should depend on the nature of heterogeneity in the primary direction of flow. Lishman (1970) argued that a false permeability anisotropy may be estimated due to geometric effects in core plugs that may distort true permeability anisotropy under actual reservoir dimensions. Average permeabilities calculated for series and parallel flows showed a distinct dependency on the length of the flow path. This effect was more pronounced for vuggy carbonate reservoirs than sandstone formations. If the permeability distribution shows lateral continuity, thickness-weighted harmonic averaging should be used for estimating the average permeability from core plugs. This has been shown to be applicable for marine-delta complexes and braided stream deposits (Mattax and Dalton, 1990). Warren and Price (1961), however, showed experimentally that the most probable behavior of a heterogeneous system approaches that of a uniform system having a permeability equal to geometric mean. This suggests that for reservoirs deposited within small meandering streams, geometric averaging of core plug permeabilities may be more suitable.

Well testing is probably the most common method of measuring the average permeability thickness for a section of a reservoir (Matthews and Russell, 1967). It is often a good engineering practice to compare the average permeability obtained from rock samples and that derived from well testing. A noticeable lack of agreement between the two averaging processes may reveal the presence of fractures, significant cross-bedding, or permeability barriers that were probably missed during core sampling. For instance, Willcox and Riley (1975) noticed that core permeabilities in the North Sea Leman field are typically much higher than the well-test permeabilities. This was attributed to extensive cross-bedding and a widespread presence of thin, highly permeable streaks. A similar disagreement was also observed in the Judy Creek carbonate field, but core data from the Jay–Little Eskambia Creek field in Florida exhibited good agreement between pressure buildup data and arithmetic averages of core permeabilities (Mattax and Dalton, 1990).

FINAL REMARKS

With diminishing hope of finding very large reserves, and as most domestic and international hydrocarbon reservoirs approach maturity, sound reservoir engineering practices for better reservoir management and future forecasting have become increasingly important. A new trend of a multidisciplinary team consisting of geoscientists and reservoir engineers for conducting reservoir studies has gained much popularity in the oil industry. New advances in geostatistics have now made it possible to incorporate many subtle details of reservoir heterogeneity into a reservoir description based on information obtained over a broad range of measurement scales.

Unfortunately, despite significant advances in computer technology, both in terms of increased storage capacity and faster cycles, the number of grid blocks needed to represent reservoir heterogeneity at a fine scale render flow simulations impractical. Until significantly faster computers become available, the only practical solution for making future performance predictions of a hydrocarbon recovery process is in applying upscaling methods to reduce the number of grid blocks of a reservoir model.

In this process, two considerations must be given the highest priority. First, the upscaling method used must be consistent with the degree of geologic complexity of the formation of interest. For simple heterogeneities, calculating effective permeability may not require full tensorial treatment. This, however, may become extremely important for complex cases of high anisotropy and heterogeneity over multiple scales. Second, the degree of upscaling should be carefully chosen such that an optimum balance is achieved between flow simulation time on the coarse grid and preservation of important geologic features in the fine grid reservoir model.

Most of the techniques discussed in the literature consider only absolute permeability (or hydraulic conductivity) as the parameter to be upscaled for flow calculations. Note that for cases where changes in rock properties in a heterogeneous medium have a significant effect on relative permeability, capillary pressure, and rock dispersivity, the upscaling of absolute permeability under a single-phase flow assumption may not be adequate to ensure satisfactory agreement between the fine- and the coarse-grid multiphase flow simulation results. The results of flow simulations in an upscaled system should be checked for excessive numerical dispersion, proper rate allocation for injection and production wells, and performance predictions inconsistent with major geologic features. The use of an adaptive gridding system to better reflect various scales and types of heterogeneity may provide more reliable simulation results. Finally, future research should aggressively address the issue of upscaling methods for multiphase flow in porous media.

REFERENCES

Aasum, Y., 1992, Effective properties in reservoir simulator grid blocks: Ph.D. Dissertation, The University of Tulsa, Tulsa, Oklahoma, 228 p.

Bakr, A. A., L. W. Gelhar, A. L. Gutjahr, and J. R. MacMillan, 1978, Stochastic analysis of spatial variability in subsurface flows 1. Comparison of one- and three-dimensional flows: Water Resources Research, v. 14, no. 2, p. 263–271.

Baveye, P., and G. Sposito, 1984, The operational significance of the continuum hypothesis in the theory of water movement through soils and aquifers: Water Resources Research, v. 20, no. 5, p. 521–530.

Bear, J., 1987, Dynamics of fluids in porous media: New York, Elsevier, 784 p.

Begg, S. H., and P. R. King, 1985, Modeling the effects of shales on reservoir performance: calculation of effective vertical permeability: SPE paper 13529, presented at the 1985 SPE Reservoir Simulation Symposium held in Dallas, TX, February 10–13, 1985.

Begg, S. H., D. M. Chang, and H. H. Haldorsen, 1985, A simple statistical method for calculating the effective vertical permeability of a reservoir containing discontinuous shale: SPE paper 14271, presented at the 60th Annual SPE Technical Conference held in Las Vegas, NV, Sept. 22–25, 1985.

Begg, S. H., R. R. Carter, and P. Dranfield, 1989, Assigning effective values to simulator gridblock parameters for heterogeneous reservoirs: SPE Reservoir Engineering, v. 4, no. 4, p. 455–463.

Bernasconi, J., 1974, Conduction in anisotropic disordered systems: effective-medium theory: The Physical Review, v. B9, p. 4575–4579.

Bourgeat, A., 1984, Homogenized behavior of two-phase flows in naturally fractured reservoirs with uniform fractures distribution: Computer Methods in Applied Mechanics and Engineering, v. 47, p. 205–216.

Bower, H., 1969, Planning and interpreting soil permeability measurements: Journal of Irrigation and Drainage Engineering, IR3, p.391-402.

Craig, F. F., Jr., 1971, The reservoir engineering aspects of waterflooding: SPE Monograph, v. 3, Henry L. Doherty Series.

Dagan, G., 1979, Models of groundwater flow in statistically homogeneous porous formations: Water Resources Research, v. 15, no. 1, p. 47–63.

Dagan, G., 1981, Analysis of flow through heterogeneous random aquifer by the method of embedding matrix, 1, steady flow: Water Resources Research, v. 17, no. 1, p. 107–121.

Dagan, G., 1982, Analysis of flow through heterogeneous random aquifers, 2, unsteady flow in confined formations: Water Resources Research, v. 18, no. 5, p. 1571–1585.

Delhomme, J. P., 1979, Spatial variability and uncertainty in groundwater flow parameters: a geostatistical approach: Water Resources Research, v. 15, no. 2, p. 269–280.

Desbarats, A. J., 1987, Numerical simulation of effective permeability in sand-shale formation: Water Resources Research, v. 23, no. 2, p. 273–286.

Desbarats, A. J., 1988, Estimation of effective permeabilities in the lower Stevens Formation of the Paloma field, San Joaquin Valley, California: SPE Reservoir Engineering, v. 3, no. 4, p. 1301–1307.

Deutsch, C., 1989, Calculating effective absolute permeability in sandstone/shale sequences: SPE Formation Evaluation, v. 4, no. 3, p. 343–348.

Dullien, F. A. L., 1991, Porous media fluid transport and pore structure: Academic Press, 574 p.

Durlofsky, L. J., and E. Y. Chung, 1990, Effective permeability of heterogeneous reservoir region: 2nd European Conference on the Mathematics of Oil Recovery, Proceedings, p. 57–64.

El-Kadi, A. I., and W. Brutsaert, 1985, Applicability of effective parameters for unsteady flow in nonuniform aquifers: Water Resources Research, v. 21, no. 2, p. 183–198.

Freeze, R. A., 1975, A stochastic conceptual analysis of one-dimensional groundwater flow in nonuniform homogeneous media: Water Resources Research, v. 11, no. 5, p. 725–741.

Gomez-Hernandez, J. J., and S. M. Gorelick, 1989, Effective groundwater model parameter values: influence of spatial variability of hydraulic conductivity, leakance, and recharge: Water Resources Research, v. 25, no. 3, p. 405–419.

Gomez-Hernandez, J. J., and A. G. Journel, 1990, Stochastic characterization of grid-block permeabilities: from point values to block tensors: 2nd European Conference on the Mathematics of Oil Recovery, Proceedings, p. 83–90.

Greenkorn, R. A., C. R. Johnson, and L. K. Shallenberger, 1964, Directional permeability of heterogeneous anisotropic porous media: SPE Journal, v. 4, p. 124–132.

Gutjahr, A. L., L. W. Gelhar, A. Bakr, and J. R. MacMillan, 1978, Stochastic analysis of spatial variability in subsurface flows 2. Evaluation and application: Water Resources Research, v. 14, no. 5, p. 953–959.

Haldorsen, H. H., and L. W. Lake, 1984, A new approach to shale management in field-scale models: SPE Journal, v. 24, p. 447–457.

Harris, C. K., 1990, Application of generalized effective-medium theory to transport in porous media: Transport in Porous Media, v. 5, p. 517–542.

Hazeu, G. J. A., O. S. Krakstad, D. T. Rian, and M. Skaug, 1986, The application of new approaches for shale management in a 3-D simulation study of the Frigg field: SPE paper 15608, presented at the 61st Annual SPE Technical Conference held in New Orleans, LA, October 5–8, 1986.

Holden, L., J. Hoiberg, and O. Lia, 1990, An estimator for the effective permeability: 2nd European Conference on the Mathematics of Oil Recovery, Proceedings, p. 287–290.

Hutchinson, C. A., Jr., 1959, Reservoir inhomogeneity assessment and control: Petroleum Engineering, v. 31, p. B19–B26.

Hutchinson, C. A., Jr., C. F. Dodge, and T. L. Polasek, 1961, Identification, classification and prediction of reservoir non-uniformities affecting production operations: Journal of Petroleum Technology, v. 13, p. 223–230.

Journel, A. G., C. Deutch, and A. J. Desbarats, 1986, Power averaging for block effective permeability: SPE paper 15128, presented at the 56th California Regional Meeting of the SPE held in Oakland, CA, April 2–4, 1986.

Kasap, E., and L. W. Lake, 1990, Calculating the effective permeability tensor of a gridblock: SPE Formation Evaluation, v. 5, no. 2, p. 192–200.

Kilpatrick, S., 1973, Percolation and conduction: Reviews of Modern Physics, v. 45, p. 574–614.

King, P. R., 1987, The use of field theoretic methods for the study of flow in a heterogeneous porous medium: Journal of Physics/Section A Mathematical & General, v. 20, p. 3935–3947.

King, P. R., 1989, The use of renormalization for calculating effective permeability: Transport in Porous Media, v. 4, p. 37–58.

Leung, W. F., 1986, A tensor model for anisotropic and heterogeneous reservoirs with variable directional permeabilities: SPE paper 15134, presented at the 56th California Regional Meeting of the SPE held in Oakland, CA, April 2–4, 1986.

Lishman, J. R., 1970, Core permeability anisotropy: Journal of Canadian Petroleum Technology, v. 9, p. 79–84.

Mantoglou, A., and J. L. Wilson, 1981, Simulation of random fields with the turning band method: Reprint 264, Ralph M. Parsons Laboratory, Department of Civil Engineering, Massachusetts Institute of Technology, Cambridge, Massachusetts, 199 p.

Martin, J. H., and J. A. Cooper, 1984, An integrated approach to the modeling of permeability barrier distribution in a sedimentologically complex reservoir: SPE paper 13051, presented at the 59th Annual Technical Conference held in Houston, TX, September 16–19, 1984.

Matheron, G., 1967, Elements pour une theorie des milieux poreux: Paris, Masson et Cie, 168 p.

Mattax, C. C., and R. L. Dalton, 1990, Reservoir simulation: SPE Monograph 13, 173 p.

Matthews, C. S., and D. G. Russell, 1967, Pressure buildup and flow tests in wells: SPE Monograph 1, 167 p.

McGill, C., P. Kilg, and J. Williams, 1991, Estimating effective permeability: a comparison of techniques: poster presented at the 3rd International Reservoir Characterization Technical Conference held in Tulsa, OK, November 3–5, 1991.

Mei, C. C., and J.-L. Auriault, 1989, Mechanics of heterogeneous porous media with several spatial scales: Proceedings of the Royal Society of London, v. A 426, p. 391–423.

Otero, C., A. E. Saez, and I. Rusinek, 1990, Effective permeabilities for model heterogeneous porous media: In Situ, v. 14, p. 229–244.

Prats, M., 1972, The influence of oriented arrays of thin impermeable shale lenses or of high conductive natural fractures on apparent permeability anisotropy: Journal of Petroleum Technology, v. 24, p. 1219–1221.

Richardson, J. G., D. G. Harris, R. H. Rossen, and G. van Hee, 1978, The effect of small discontinuous shales on oil recovery: Journal of Petroleum Technology, v. 30, p. 1531–1537.

Saez, A. E., C. J. Otero, and I. Rusinek, 1989, The effective homogeneous behavior of heterogeneous porous media: Transport in Porous Media, v. 4, p. 213–238.

Samier, P., 1990, A finite element method for calculating transmissibilities in n-point difference equations using a non-diagonal permeability tensor: 2nd European Conference on the Mathematics of Oil Recovery, Proceedings, p. 121–130.

Smith, L., and R. A. Freeze, 1979a, Stochastic analysis of steady-state groundwater flow in a bounded domain, 1. One-dimensional simulations: Water Resource Research, v. 15, no. 3, p. 521–528.

Smith, L., and R. A. Freeze, 1979b, Stochastic analysis of steady-state groundwater flow in a bounded domain, 2. Two-dimensional simulations: Water Resource Research, v. 15, no. 6, p. 1543–1559.

Tang, D. H., and G. F. Pinder, 1977, Simulation of groundwater flow and mass transport under uncertainty: Advances in Water Resources, p. 25–30.

Warren, J. E., and H. S. Price, 1961, Flow in heterogeneous porous media: SPE Journal, p. 153–169, Trans. AIME, 222.

Weber, K. J., 1982, Influence of common sedimentary structures on fluid flow in reservoir models: Journal of Petroleum Technology, v. 34, p. 665–672.

White, C. D., and R. N. Horne, 1987, Computing absolute transmissibility in the presence of fine-scale heterogeneity: SPE paper 16011, presented at the 9th SPE Symposium on Reservoir Simulation held in San Antonio, TX, February 1–4, 1987.

Willcox, P. J., and H. G. Riley, 1975, Performance matching for a North Sea gas field: SPE paper 5535, presented at the 50th Annual Fall Meeting held in Dallas, TX, September 28–October 1, 1975.

SECTION III

◆

Methods and Case Studies

The chapters assembled for this section represent applications of a wide spectrum of methods, and case histories covering a great diversity of geological settings, ages, and depositional environments. Methods include stochastic simulation, kriging, cokriging, external drift, Boolean, and fractals. Variables are treated either as indicator, Gaussian, or truncated Gaussian data. Some chapters illustrate the integration of soft data (seismic) as ancillary data to aid in spatial estimation and uncertainty reduction.

K. Tyler et al. apply an object- (Boolean-) based method to model fluvial system heterogeneities and show the impact of the model on production profiles in **Modeling Heterogeneities in Fluvial Domains: A Review of the Influence on Production Profiles**.

Modeling of shoreface reservoirs is used as an example to illustrate the link between geologic and stochastic models. A. C. MacDonald and J. O. Aasen illustrate the use of truncated Gaussian fields to describe the distribution of discrete or categorical variables, such as facies or facies tracts. Shoreface facies are simulated with a linear trend to account for the direction of progradation and the degree of contemporaneous aggradation. Their chapter is entitled **A Prototype Procedure for Stochastic Modeling of Facies Tract Distribution in Shoreface Reservoirs**.

A. S. Hatløy discusses the use of an object- (Boolean-) based method, combining empirical relations, probability distributions, and (geo)logical rules to model fluvial facies in **Numerical Facies Modeling Combining Deterministic and Stochastic Methods**.

Indicator kriging was used by M. E. Hohn and R. R. McDowell to assess heterogeneity and trends in initial potential and cumulative production data. They believe that the use of multiple cutoffs was instrumental in revealing underlying geologic features that control production. Follow their discussion in **Geostatistical Analysis of Oil Production and Potential Using Indicator Kriging**.

Recently there has been an increased effort to integrate production data in geostatistically based reservoir models. C. V. Deutsch and A. G. Journel present their approach, based on simulated annealing, to integrate well test-derived effective permeability into reservoir models in **Integrating Well Test-Derived Effective Absolute Permeabilities in Geostatistical Reservoir Modeling**.

Five chapters illustrate the use of soft or secondary data (seismic data) to aid in interwell estimation and reduction in model uncertainty. R. L. Chambers et al., in their chapter, **Constraining Geostatistical Reservoir Descriptions with 3-D Seismic Data to Reduce Uncertainty**, use seismic data inverted to acoustic impedance to aid in estimating interwell porosity. They use the external drift data integration method and compare and contrast kriging and simulation methods. In their chapter, **Importance of a Geological Framework and Seismic Data Integration for Reservoir Modeling and Subsequent Fluid-Flow Predictions**, W. M. Bashore et al. show how the choice of the geological model (lithostratigraphic or chronostratigraphic) can impact the character of the reservoir model and ultimate fluid-flow prediction. They also illustrate the use of seismic data in constraining the reservoir description.

D. J. Wolf et al. illustrate the use of kriging with external drift to integrate seismic reflection amplitude to aid in estimating interval porosity thickness. They also introduce the use of risk maps to assess uncertainty in **Integration of Well and Seismic Data Using Geostatistics**. J. Chu et al. present in some detail a three-dimensional case study of the reservoir described by Chambers et al. In their chapter, **3-D Implementation of Geostatistical Analyses—The Amoco**

Case Study, they illustrate the difficulties in building a 3-D description, especially when one is attempting to integrate seismic information into the model and present some solutions.

The chapter by T. Høye et al. details the use of seismic reflectors (top, base, and intrareservoir reflectors) into stochastic models to aid in constructing a structural model of the reservoir. Their objective was to highlight a target well location and assess uncertainty in the modeled results. Depth conversions were aided by modeling surface velocity as a Gaussian random field conditioned on the well locations. Follow their procedures in **Stochastic Modeling of Troll West with Special Emphasis on the Thin Oil Zone**.

S. A. McKenna and E. P. Poeter illustrate the use of an indicator simulation technique to incorporate soft (imprecise) data into a simulation of facies to help reduce model uncertainty. The soft data used in their study are synthetic data (real data plus noise). Their chapter, **Simulating Geological Uncertainty with Imprecise Data for Goundwater Flow and Advective Transport Modeling**, is applicable to the petroleum industry.

In **Fractal Methods for Fracture Characterization**, T. A. Hewett illustrates the concept of fractals and self-similarity and scaling laws using fracture networks. He points out that the most useful fractal sets for describing natural property distributions are those displaying self-similarity.

Description of Reservoir Properties Using Fractals, by M. Kelkar and S. Shibli, rounds out the discussion of fractals with a case study. This work investigates the utility of fractal geometry for describing the spatial correlation structure of rock properties in carbonate and clastic rock types.

A. S. Almeida and P. Frykman obtained stochastic images of porosity and permeability in a Maastrichtian North Sea chalk reservoir. In **Geostatistical Modeling of Chalk Reservoir Properties in the Dan Field, Danish North Sea**, they use a Gaussian collocated cosimulation algorithm, built on a Markov-type hypothesis, to perform direct cosimulation of spatially interrelated variables. Log-derived porosity was used as soft data during the cosimulation.

Innovative modeling techniques were combined to generate a realistic characterization of a complex eolian depositional environment for a multicompany reservoir management problem. D. L. Cox et al. first model the eolian bedding geometries and dimensions of four different stratification types. Permeability fields were generated within the conditional simulation of reservoir model, scaled-up for flow simulation and compared to historical field production data in **Integrated Modeling for Optimum Management of a Giant Gas Condensate Reservoir, Jurassic Eolian Nugget Sandstone, Anschutz Ranch East Field, Utah Overthrust (USA)**.

C. J. Murray uses a variety of methods, including cluster analysis, discriminant function analysis, and sequential indicator simulation (SIS) to identify and model petrophysical rock types. Simulated annealing was used to postprocess the SIS images such that the images honor rock type transitional frequencies in well data. Follow this process in **Identification and 3-D Modeling of Petrophysical Rock Types**.

Uncertainty is a statistical term used to describe what we do not know. The results of a geostatistical model include a number of possible outcomes, each equally likely, and each physically portraying, in a variety of ways, the portion of the model we do not know. Thus, for example, a channel may vary in size, shape, and location from outcome to outcome in an area where there is no hard data. Tracking these variations and developing a sense of how variable the model is can be very important. This section's final chapter, by R. M. Srivastava, nicely illustrates the topic of visualizing uncertainty. In **The Visualization of Spatial Uncertainty**, Srivastava demonstrates a unique way of looking at the various possible outcomes of a given geostatistical model.

Chapter 8

◆

Modeling Heterogeneities in Fluvial Domains: A Review of the Influence on Production Profiles

Kelly Tyler
Adolfo Henriquez
Tarald Svanes
Statoil
Stavanger, Norway

INTRODUCTION

The goal in reservoir characterization for heterogeneous formations is to establish a reservoir model based on explicitly modeling the known heterogeneities (conditioning to well observations) and using mathematical and statistical algorithms to systematically simulate the spatial distribution at interwell locations. Modeling of the unknown requires geological expertise acquired from experience with outcrop studies, well data (cores and logs), seismic, well test and production data. For many years, geologists have been developing their understanding of the sedimentological environments and production behavior of heterogeneous reservoirs, but without the use of current statistical techniques and computer hardware capable of handling detailed three-dimensional (3-D) modeling algorithms. Until recently, reservoir models were limited to layer-cake models, which are known to be overly simplistic for most sedimentological environments.

The fluvial environment is an example of a class of reservoirs requiring modeling of the interwell region with tools other than the layer-cake model. Fluvial environments, as observed at outcrop analog scales, comprise a labyrinth of sandstones, shales, crevasse splays, coal, and calcite layers. Many of the North Sea reservoirs are dominated by fluvial reservoir sequences and, therefore, it is no coincidence that companies operating in the North Sea have pioneered the practical application of stochastic techniques to model the heterogeneities of fluvial environments. Especially in cases where offshore well spacing is on the order of square kilometers, operators have had to rely on the results of fluid-flow modeling to design and manage production projects.

The initial investigations with advanced reservoir characterization of fluvial reservoirs in the North Sea were spawned from the feasibility studies of the Snorre field consisting of the Statfjord and Lunde formations. Here, the Boolean approach (marked point method) was developed and tested (Clemetsen et al., 1990; Stanley et al., 1990; Henriquez et al., 1990). The positive reception of these techniques by geologists and reservoir engineers led to further testing and developments of the Boolean methods (Tyler et al., 1992a) and integration of fiber processes (Georgsen and Omre, 1993) into these models. Boolean methods also are being used to investigate stochastic modeling of faults (Munthe et al., 1993).

During the development of the Boolean methods, other methods have been investigated to solve similar problems. The truncated Gaussian random field model (Matheron et al., 1987) was used to simulate the facies units of a progradational shoreface (MacDonald et al., 1991), and Markov random field (Tjelmeland and Omre, 1992) was used to describe and simulate the reservoir architecture of a deltaic coastal plain environment (Fælt et al., 1991; Tyler et al., 1993). In this chapter, we illustrate the application of stochastic modeling for describing the heterogeneities of fluvial environments together with an example of a North Sea reservoir.

CONVENTIONAL MODELING

Classical reservoir modeling (conventional modeling) is based on the use of horizon maps and two-dimensional maps of reservoir parameters. The maps used include top and bottom structure maps based on seismic and well data. These maps are used to generate thicknesses. Other geological and petrophysical parameters, such as sand/shale ratio, permeability, and porosity, are generated between wells using standard interpolation techniques. This type of model is referred to as layer-cake model (Weber and van Geuns, 1990) and is applicable to reservoirs with fairly laterally continuous permeability and thicknesses.

The transition from conventional layer-cake modeling to full 3-D modeling is limited by several factors:

1. Time limitations for geological evaluation at the feasibility and planning stages have encouraged the use of quick interpolation techniques between wells.
2. Geological sampling and data collection from outcrop analogs have been limited, and the data collected are only now beginning to include the parameters needed for modeling architectural flow units.
3. Computer hardware capabilities have been a limiting factor for the development of high resolution 3-D models. The algorithms and software necessary are demanding of both processing and storage capacities.

In offshore field studies, large well spacing, lack of complete core and well traces, and poor-quality seismic data between wells leave conventional modeling the tool most often used by explorationists. Fluvial reservoirs are typically complex, displaying large permeability contrasts between channel sands and the shale matrix. Sand continuities are often poor and show directional dependence, and isochores can vary rapidly due to the erosive character of the channel belts. In the offshore setting with large areas of unknown interwell data, fluvial reservoirs are not suited to the use of the interpolation and smoothing techniques of conventional modeling.

PROCESS RESPONSE MODELING

Process response modeling is a 3-D technique to simulate the sedimentological processes that have formed the reservoirs in geological time. Processes that are simulated include the discharge of clastic sediments, transport and deposition of these sediments according to physical equations, erosion of earlier deposits, compaction, and, finally, subsidence. These processes require input of the paleotopography, sediment discharge rates into the basin, compaction and subsidence rates, and sea level changes.

Bridge and Leeder (1979) used this type of modeling for a fluvial environment in two dimensions. This work gives helpful insight to other stochastic modeling because of the ability to better describe the resulting sand-body continuity (Henriquez et al., 1990) and repulsion functions. The model of Alexander (1986) includes tectonic control in a 3-D model to predict concentrations and orientations of sand channels in an alluvial setting.

SEDSIM3 (Tetzlaff and Harbaugh, 1989) is a software package that models through time the transport and deposition of clastic grains to obtain 3-D representations of sedimentary basins. Both the long-term processes, such as eustatic sea level changes, and short-term processes, such as storms and waves, are considered along with the appropriate time scales. This creates a model in 3-D that can incorporate sequence stratigraphic concepts to define stochastic elements, such as calcite layers, that tend to be associated with maximum and minimum sea levels.

Because of the complexity of the input data for this method and the difficulties in conditioning to well observations, this type of modeling still remains in the research environment; however, the work done with such models will give knowledge that can be used in the simpler descriptive models.

STOCHASTIC MODELING

The picture of the depositional environment and heterogeneities will never be complete for petroleum reservoirs, thus the goal of geological modeling is to characterize the formation using reservoir parameters together with their associated uncertainties. Stochastic modeling, conditioned to well observations, provides the reservoir engineer with several realizations of the geologic parameters on which to base reservoir production simulations.

The 3-D geological modeling of reservoir parameters and heterogeneities is becoming a popular alternative to the conventional approach. Recent literature has shown not only the theoretical development of statistical algorithms (e.g., MacDonald et al., 1991; Journel and Alabert, 1988), but also their practical application in case studies (Tyler et al., 1993; Damsleth et al., 1992; Hove et al., 1992).

The application of 3-D geological modeling requires four main stages in preparing the dynamic reservoir simulation (Figure 1).

Structural Model

Because geological parameters tend to be more uniform within individual layers in a sequence, the reservoir is divided into time zones that can be modeled independently. Each time zone is modeled as a parallelepiped. The reservoir geometry is obtained by applying erosion and compaction maps. In this case, the grid follows relatively flat features of the reservoir, which reduces errors when transferring the data to a gridded numerical simulator. These time zones are bounded by chronostratigraphic surfaces. At times, these surfaces cannot be interpreted from seismic data, and other techniques that depend on the amount of data available from the field can be used.

Thin shale layers having large lateral continuity can act as pressure barriers between overlying and underlying sandstones. In some fields where historical pressure data are available, these shales have been shown to be correlated in space. Such layers must be treated as deterministic data, and thus are not modeled within the time zones but placed at the boundaries between time zones.

In a fluvial scenario where high-resolution sequence stratigraphic interpretations of parasequences and sequence boundaries have been made, the flooding surfaces have been used as the two-dimensional maps

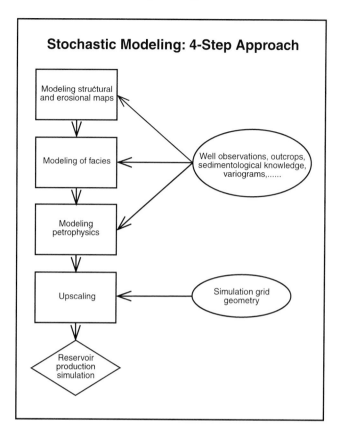

Stochastic Modeling: 4-Step Approach

Modeling structural and erosional maps

Modeling of facies

Modeling petrophysics

Upscaling

Reservoir production simulation

Well observations, outcrops, sedimentological knowledge, variograms,......

Simulation grid geometry

Figure 1. A flow chart of the steps encountered in stochastic modeling.

bounding the time zones to be simulated. This method allows for the sequence boundaries, interpreted as highly erosive and highly variable surfaces, to be modeled less explicitly. The sequence boundaries were modeled implicitly as the bottom of the channel belt systems within the overbank fines. The channel belts were the entity varying from one realization of the reservoir zone to the next.

In another deltaic coastal plain environment, sequence boundaries were used as erosive surfaces when more transgressive systems were underlying them; however, the flooding surfaces at the top of the lowstand system tracts were modeled as nonerosive boundaries where fine-grained shales and coals have been deposited (Tyler et al., 1993).

At the phase of feasibility studies of an oil field, it may not be reasonable to expect that the geological interpretation of data has included the interpretation of sequence stratigraphy. Often at this stage, simple lithostratigraphic correlations must suffice as defining zonations of the reservoir. In this case, the lack of knowledge strengthens the need for stochastic modeling with the corresponding impact on revealing the degree of uncertainty.

Sedimentary Facies Model

Geological Setting of the Fluvial Reservoir

The discontinuous sand bodies of fluvial reservoirs in the North Sea are observed in the Statfjord

Formation, the Lunde Formation, and the Ness Formation in the area of the Viking Graben, where the Ness is less marine and more continental in type. The evaluation of reserve volumes and production performance of these reservoirs has been a challenge to geologists and engineers, because it is very difficult to extract information about the geometry of the architectural elements and correlate between wells, even with relatively good well spacings of 500 or 600 m.

Thickness, width, sinuosity, and orientation of the channel sandstones are of great importance for the contact volume of producing wells, as well as the fluid flow paths between injection and production wells. Due to the dependence of production behavior on the geometry of channel sandstones, the conventional modeling technique is not suitable in a fluvial domain. Two characteristics of fluvial reservoirs, the fairly regular attributes of individual river beds and the occurrence of river beds and channels forming a labyrinth through a background matrix of nonpermeable sediments, make these reservoirs well suited for description using a marked point process (Stoyan et al., 1987).

Modeling Techniques

Detailed modeling using a marked point process (Boolean) for description of fluvial deposits in the North Sea began with the SISABOSA model (Augedal et al., 1986). SISABOSA simulates the reservoir using infinitely long parallelepipeds describing the permeable river beds that are placed randomly within the otherwise impermeable shale matrix (Figure 2). The program calculates sand/shale ratios in each grid block; the ratios are used with porosity to calculate pore volumes in the reservoir simulator. Also, correction factors for transmissibility between neighboring grid blocks are calculated with the purpose of combining with permeability from the reservoir simulator to calculate the flow between these blocks. The focus of this tool is to evaluate sand body connectedness. Due

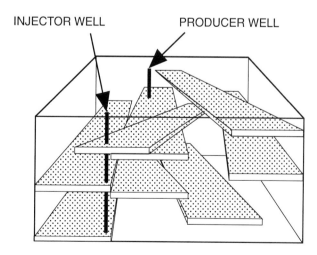

INJECTOR WELL   PRODUCER WELL

Figure 2. A schematic of the SISABOSA model, modified from Augedal et al. (1986). Shown are the parallelepipeds representing sand channels that penetrate the impermeable shale matrix.

to the random placement of the river beds, this value tends to be overestimated (Henriquez et al., 1990) as shown in Figure 3. Development of the FLUREMO model (Clemetsen et al., 1990) remedied this problem.

The assumption underlying the FLUREMO model is that each channel belt consists of a series of river bed deposits. The river beds are treated as line segments that meander with a sinuous behavior around the center of the channel belt axis. Depending on the accommodation space of the fluvial plain, a channel belt may exist for a period of time at any one area of the fluvial plain before it will avulse, creating a new channel belt deposit. With little accommodation space, rivers will seldom avulse and thick, extensive, flood-plain shale barriers are formed; however, more accommodation space will allow for avulsion of rivers with new positions of the channel belts having the possibility of stacking or eroding into older channel belts on the plain. Even though the FLUREMO model does not include the time dependence of a process model, a repulsion function allows for the interaction between channel belts (Figure 4). This function gives a greater flexibility to the system and can be used to simulate more realistic continuity ratios.

Recognizing that heterogeneities must be quantified at scales less than the river beds (Miall, 1985), these earlier models were enhanced to account for other architectural elements (Tyler et al., 1992b). In MOHERES, each river bed in the channel belt is simulated (Figure 5), followed by a simulation that randomly places internal elements, such as gravel bars, sandy bed forms, and passive infills, dependent upon their probability of size and spatial occurrence within the river beds.

Petrophysical Model

The earlier SISABOSA and FLUREMO models do not attempt to describe the petrophysical attributes of the modeled fluvial sand channels. These models calculate transmissibility factors (Augedal et al.,

A

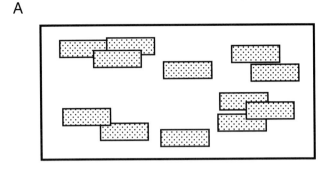

B

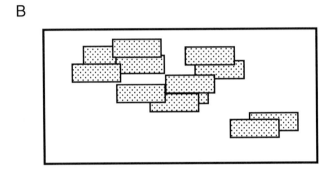

C

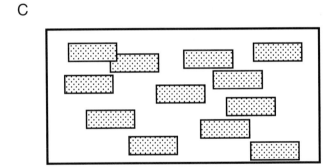

Figure 4. Example of interactions between channels: (A) no specified interaction, (B) attraction, (C) repulsion (modified from Clemetsen et al., 1990).

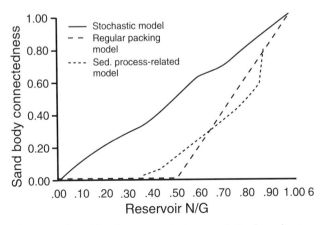

Figure 3. Sand body connectedness plotted against net/gross (N/G) ratio as calculated by different models. Modified from Henriquez et al., 1990.

1986) that are used together with the permeability values in the numerical simulator to calculate the total grid block transmissibility. The transmissibility factor was calculated from the amount of sand simulated within each block of the coarser numerical reservoir simulation grid.

The theory of Gaussian random fields (Journel and Huijbregts, 1978) is most often used for the simulation of petrophysical (continuous) parameters within each discrete facies type. This algorithm has been used in connection with the truncated random function for facies distributions in the HERESIM software (Rudkiewicz et al., 1990) and the DESIRE software (Aasen et al., 1990). The algorithm also is used in connection with the Markov random field (Fælt et al., 1991; Tyler et al., 1993), indicator simulation (Alabert and Massonnat, 1990), and the marked point process

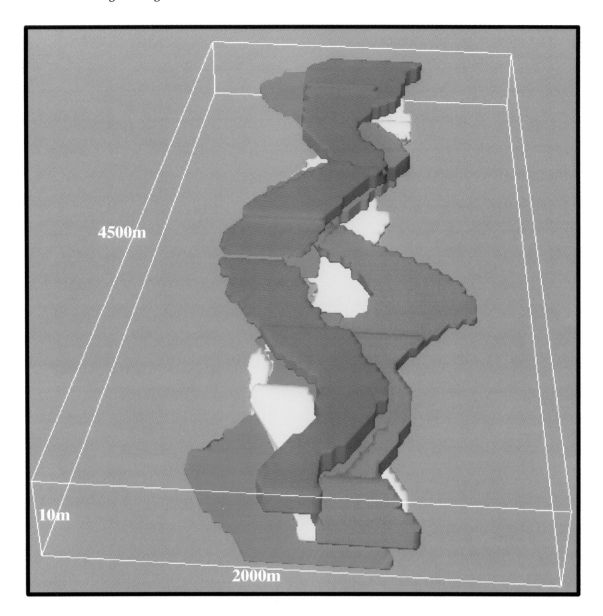

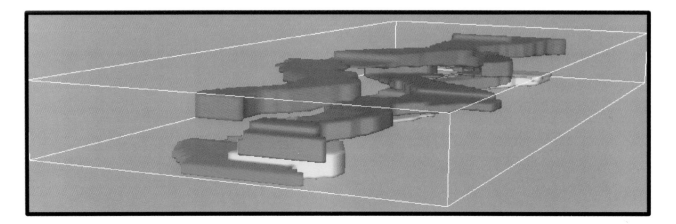

Figure 5. An example from the MOHERES software of a single channel belt consisting of four channels.

(Damsleth et al., 1992; Tyler et al., 1992c). The Gaussian random model allows simulation of continuous distributions of permeability, porosity, and initial water saturation that are spatially correlated, including vertical and horizontal trends, and conditioned on the facies architecture. In the case of fluvial reservoirs, the thick sand sequences often are represented and identified by their characteristic fining-upward permeability trend. Note that other models for modeling petrophysical parameters also are being developed. Other approaches include sequential indicator simulation (Deutsch and Journel, 1992), fractal simulations (Hewett and Behrens, 1988), and simulated annealing (Deutsch, 1992).

Upscaling

Unfortunately, even the largest computers to date cannot compute fluid flow directly on the high-resolution grids of the geological model due to data handling limitations. For the purpose of dynamic calculations, it is necessary to reduce the volume of data from the high-resolution grid by estimating effective properties for larger numerical simulation grid blocks. Additive parameters, such as porosity and water saturation, require no more than classical arithmetic averaging. Absolute and relative permeability, however, represent complex problems requiring more refined techniques.

Absolute permeability has been scaled up using several techniques. Aasen et al. (1990) used arithmetic averaging for horizontal permeability and harmonic averaging for vertical permeability. A quick, yet sophisticated, method of renormalization (King, 1989) essentially replaces four grid values at a time (eight in three dimensions) by an equivalent permeability. The process is repeated until the scale to be used in the full fluid calculations is obtained. A more mathematically formalized algorithm includes solving pressure equations over the three directions of the coarse grid (Holden and Lia, 1992). This method is much slower, but often gives very accurate estimates for effective permeabilities (McGill et al., 1993).

INFLUENCE OF HETEROGENEITIES ON PRODUCTION PROFILES: A CASE STUDY

To accurately simulate fluid-flow processes, the reservoir engineer must combine the numerical model of dynamic processes with an accurate description of the porous medium. The recent emphasis on improved oil recovery has increased our awareness of the importance of heterogeneities. The work of Aasen et al. (1990) entailed gas injection in a vertical cross section with heterogeneities produced using a Boolean process for facies, followed by a Gaussian random field for the continuous description of porosity and permeability. Their results showed a marked variability of production performance dependent upon the variability of the geological model (Figure 6). Published results (Hewett and Behrens, 1988; Damsleth et al., 1992; Wolcott and

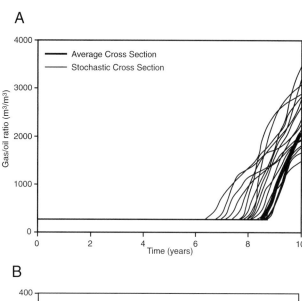

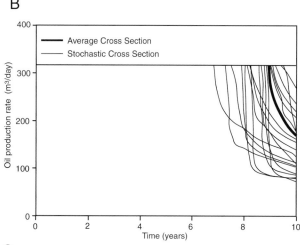

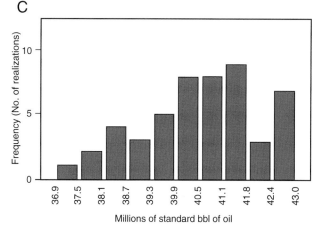

Figure 6. Production profiles showing the variability of the DESIRE model and a conventional model: (A) gas/oil ratio, (B) oil production rate, (C) oil recovery after 10 yr. Modified from Aasen et al. (1990).

Chopra, 1993; among others) show that for heterogeneous porous media, recovery efficiencies are much lower from realizations generated using geostatistical and statistical techniques than from models using kriging or other conventional modeling techniques. Figure 7 shows results from Damsleth et al.'s (1992) work.

A

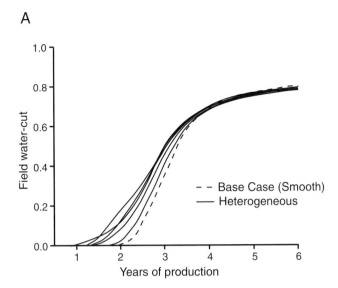

B

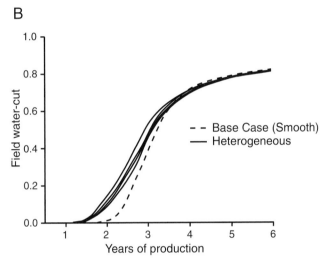

Figure 7. Field water-cut buildup showing the variability of the SESIMIRA model and a conventional model: (A) five facies realizations, (B) five permeability realizations.

Figure 7A shows the results of simulation using five realizations of facies, while Figure 7B shows the results of five realizations of the permeability model conditioned to the same facies realization. The spread in the curves results in the conclusion that the heterogeneities introduced through the facies model have a stronger impact on the water cut than the finer scale heterogeneities at a subfacies scale.

A detailed geological model also has been used for fields that have several years of production history. Tyler et al. (1993) simulated the depositions of the coastal plain environment of the Ness reservoir of a North Sea field. The modeling of a 3-D element was started after a high-resolution sequence stratigraphic study had been performed for the entire field. The sequences of the Ness are dominated by bay head deltas, lagoons, coal swamps, and incised valley fills. Fifteen realizations were created using a Markov random field for the facies simulation combined with the

Gaussian random field for the simulation of permeability, porosity, and initial water saturation. Numerical reservoir simulations for the realizations resulted in six of the realizations giving good estimates of water buildup relative to measured rates in the production wells (Figure 8A, B). The six realizations were then used to improve production forecasting (Figure 8C) and for quantifying uncertainty in the decline phase when tie-in projects are operationally feasible. Earlier studies (Hewett and Behrens, 1988) have looked at geological modeling in reference to estimating historical data; however, this study estimated water cuts on a field basis, rather than on a well basis.

Several advantages to using detailed geological modeling are mentioned in Tyler et al. (1993). Among them are the following.

1. Stochastic simulation of reservoir heterogeneities requires much less time and fewer resources to build and update the models than a conventional simulation, which requires much time to iterate history matching.
2. Several realizations are available for studies of future predictions, contrary to the traditional approach of a single simulation with the one probable model for sensitivity studies.
3. Results of the high-resolution sequence stratigraphy, together with further analysis of unswept oil in these realizations, can provide better knowledge of reservoir sweep efficiency, infill drilling potential, and Induced Oil Recovery (IOR) potentials.

Case Study from the Statfjord Formation

Structural Model

This case study of the Statfjord Formation of a North Sea field involves the quantification of uncertainty for a field being studied for its feasibility for production with an offshore platform. The study was confined to the largest fault block of the many comprising the field. In this fault block, data were limited to one exploration well; however, three other exploration wells penetrate the Statfjord Formation in other areas of the field and were used to specify input parameters. Due to the sparse data from the field and the time available to reach feasibility conclusions, simple lithostratigraphic interpretations were used to generate the five time zones making up the simulated reservoir volume.

Facies Modeling

Data from the four exploration wells that penetrate the Statfjord Formation were used to establish the input parameters for the facies model; however only the well penetrating the major fault block was used for conditioning the facies simulation. The conditioning data in the well included the thickness of the sedimentary units and the interpretation of the grouping of the observed river beds into channel belts.

Data concerning the geometry of the flow unit types were difficult to extrapolate or identify from well data. In this case with few wells and little well

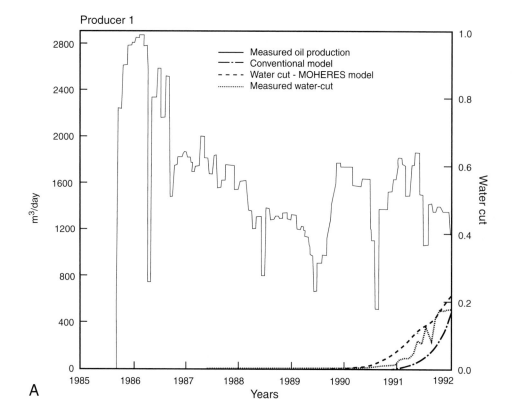

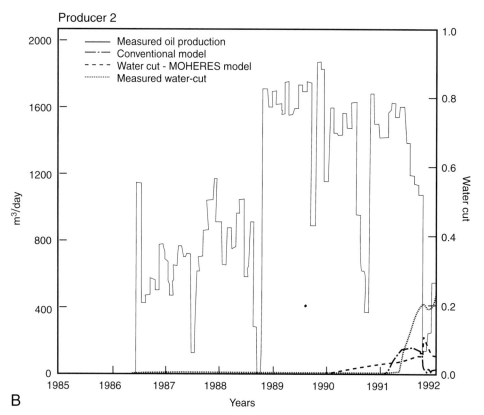

Figure 8. Production profiles comparing the results of the history matched and the MOHERES model for a coastal plain environment with over 6 yr of historical data: (A) well profile for producer 1 in the element, (B) well profile for producer 2 in the element.

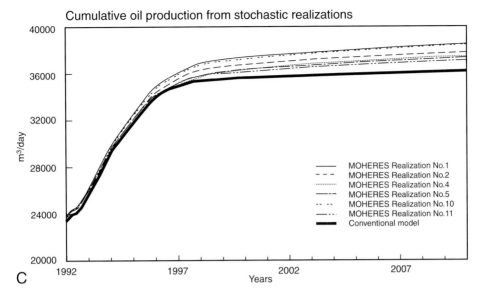

Figure 8 (continued). (C) field future oil production for the element predicted using six realizations of the MOHERES model and the history-matched conventional model.

test data, it was necessary to supplement well data with outcrop information to generate quantitative data for sand geometries. The most important field analogs used for this study include the Escanilla Formation (Dreyer et al., 1992) and Reptile Sandstone (T. Dreyer, 1991, personal communication) of the Spanish Pyrenees, the Sherwood Sandstone of southern England (Dranfield et al., 1987), and the Castlegate formation of Utah (Geehan et al., 1986). Experience and geological knowledge accumulated from outcrop studies were used to derive values for channel widths, number of channels per channel belt, and repulsion between channel belts. For modeling internal facies, width, length, frequency, and vertical placement within channels also were obtained in this manner.

The internal geometry of the fluvial system determined the continuity for fluid flow. Thus, the spatial and size distribution of individual channels within channel belts, and the interaction between channel belts, must be modeled correctly. The following parameters are specified in the model:

- Thickness and width of channels, with expected value and standard deviation
- Erosion and stacking of channels within a channel belt, simulated using specified parameters to describe vertical and horizontal distance between channels
- Avulsion of channel belts simulated with parameters that influence stacking and erosion of channel belts

Each channel belt observed in the core data taken from the four exploration wells in the field consists of the four following types of architectural elements: channel, gravel bar, sandy bed form, and passive

infill. All of these elements are described in the hierarchy of architectural elements of Miall (1985), and modeled subsequent to the channels.

Petrophysical Modeling

Permeability data were available from logs, core plugs, and minipermeameter measurements on cores from the four wells. Analysis of permeability data shows it distributed lognormally; therefore, it was handled in logarithmic form. Variograms were calculated for all four sand facies using data from logs, cores, and minipermeameter (where available).

The grid resolution used for modeling permeability was 45 m along the channel direction, 30 m across the main channel direction, and 0.5 m in the vertical direction. The resulting grid consisted of approximately 6 million grid cells. Input parameters for the facies and permeability modeling are given more specifically in Tyler et al. (1992c).

Numerical Reservoir Simulation

The simulation grid for permeability, which consisted of over 6 million grid cells, required large amounts of computer time when upscaling using the method of Holden and Lia (1992); thus, we used simpler and less-computer-intensive methods. The arithmetic mean was used in the x and y directions, and the harmonic average was used in the vertical direction. The harmonic mean weights the interspersed impermeable shales and, therefore, was chosen for the scaling in the vertical direction. The harmonic mean values compare favorably with the method of Holden and Lia (1992).

Reservoir production forecasts were obtained using 15 realizations of facies and permeability from the stochastic model. The purpose of these fluid flow simulations was to improve the quality of the production forecasts while considering the reservoir uncertainty.

The conventional model used constant permeability within each layer of the model of the fault block. Values used were the geometric average from core data. The conventional model used equal values for k_x, k_y, and k_z, and a transmissibility multiplier to decrease transmissibility between layers. When simulating with the stochastic model, the upscaled permeability from MOHERES was used and the transmissibility multiplier was removed. This multiplier was no longer needed because the facies and permeability models take into consideration the shales interspersed between sands.

The simulation model included one gas injection well and five horizontal production wells in a line drive configuration. The target oil production rate for the wells was tuned for the total reserves in the conventional model and, thus, was not altered for the stochastic model.

The 15 realizations from the stochastic model gave oil production trends similar to the conventional model (Figure 9A). The reservoir simulation models of the stochastic geological models were run with the same drilling schedules, maximum oil production rates, and maximum gas production rates as the previously simulated conventional model to have a valid comparison. The conventional model had a draining strategy designed to recover as much as possible from this heterogeneous reservoir, and the relatively high number of five horizontal wells had been planned. Thus, the variations in channel distribution in the different stochastic realizations had little impact on the oil production profiles during the first years of production. All of the simulations from the stochastic model, however, gave water production much higher than the conventional model (Figure 9B). This result is due to the high-permeability channel belts that yielded high initial oil rates, but also allowed for high production of water and gas from the underlying aquifer and the gas injection well. The spread in the water profiles of the realizations is due to the placement of the channel belts in relationship to the placement of the wells. The remaining oil saturation is high in the areas with lower permeability in interchannel areas.

Figure 9 shows production profiles for the conventional model and realizations number 8 and number 4, which gave the highest and lowest oil production, respectively, of the 15 realizations. The water production profiles show the uncertainty involved in this geological model. Platform capacity for water treatment would have been greatly underestimated if only the conventional model had been used in the planning stages of field development for this field due to water production rates nearly twice as high as the conventional model after approximately 4 yr of production, and twice the total water production after 20 yr. Gas production was quite similar for the conventional model and the stochastic models.

Due to a lack of permeability measurements from outcrop data, and the difficulties of fitting permeability data with variograms, it was difficult to arrive at a variogram model. Three tests were performed involving identical facies distributions, but changing the vari-

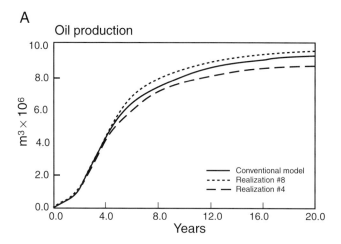

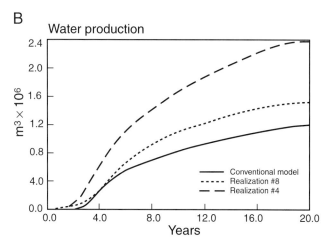

Figure 9. Field production profiles for the conventional model compared with the two most extreme realizations from the MOHERES model: (A) oil production, (B) water production.

ogram model. These three production profiles are nearly identical, indicating that production is highly dependent upon the placement of sands and shales in the model, and less dependent upon the permeability model used (Tyler et al., 1992c) for this particular case.

FURTHER WORK

In this chapter, we present methods and tools that reservoir geologists and engineers can use in managing heterogeneous reservoirs. In an effort to make these tools both more user-friendly and faster, and to include remaining sources of uncertainty in the production profiles, several areas of research are evolving.

• A quick and efficient method is needed for sorting the realizations of the geological model for selecting those for simulation with the numerical reservoir simulator. Guerillot and Morelon (1992) developed a method that involves flow calculations in time, while the pressure field is held constant. The advantage of this approach is that the computer time is reduced because the pressure is not included in the calculations at each time step. Another sorting method is being developed by two of this chapter's authors (Tyler and Henriquez). Here, a random walk method to simulate fluid flow is combined with a very quick method for calculating the pressure field. The method is superior to other approaches with respect to human and engineering effort needed to initialize the ranking model.

- The stochastic modeling approach is well suited for describing characteristics dependent on facies types. These characteristics may include initial water saturation, porosity, relative permeability, and capillary pressure curves. At present, water saturation and porosity can be modeled with MOHERES, SESIMIRA, HERESIM, and other tools, but more work needs to be done to include relative permeability and capillary pressure curves using the method described by Corbett and Jensen (1991). The approach involves beginning the description of rock properties at the lamina scale, and scaling to the facies scale in several steps using fluid-flow simulation.

- Many of the offshore oil fields in the North Sea are influenced by faults below seismic resolution. A stochastic approach is proposed by Munthe et al. (1993), for modeling these phenomena using a Boolean approach. We have used this technique to predict fluid flow and it has shown a very important impact on compartmentalization of the reservoir.

- Uncertainties from architectural elements and permeability can be quantified with the modeling technique discussed here; however, it is also important to direct efforts toward the uncertainties in total reserves that stem from poor seismic resolution and chronostratigraphic time markers.

- The limitation with modeling two-dimensional chronostratigraphic maps lies in the gridding for fluid-flow simulation. Hove et al. (1992) proposed a gridding technique to follow the main geological features of the model. A more automated approach has been developed by Farmer (1992), and has recently been extended to three dimensions. Guerillot and Swaby (1993) have also proposed a technique for defining an optimal grid in three dimensions.

- At this point, stochastic modeling techniques may condition the realizations to well observations of facies and permeability; however, well test data also must be used for conditioning realizations. Deutsch (1992) used simulated annealing as an optimization function when swapping the pixel values from a geostatistical model to obtain results comparable to well test data.

CONCLUSION

The stochastic approach to reservoir modeling has moved beyond the realm of the academic and research environment, and is a common technique in many oil companies. The large number of plausible realizations, all consistent with available well and analog data, is used to quantify the uncertainty in production performance. The production profiles have been used in reservoir management decisions. Areas of application include production feasibility, fluid handling capacities for new projects, uncertainties in plateau lengths and tie-in possibilities, evaluating IOR and infill drilling potentials, and planning injection strategies.

Realizations using stochastic modeling tools depend upon sedimentological knowledge for the depositional environments. The amount of geological information included in the modeling is considered an advantage to developing quality descriptions of the reservoir. As the volume of geological data made available is increased, the less uncertainty there will be in the resulting models.

ACKNOWLEDGMENTS

We would like to express our appreciation to Statoil management for permission to publish this work. We also thank the researchers at the Norwegian Computing Center for their cooperation in developing the MOHERES software and to the many sedimentologists and geologists in Statoil who have contributed very important information for the software development and project work.

REFERENCES

Aasen, J. O., J. K. Silseth, L. Holden, H. Omre, K. B. Halvorsen, and J. Hoiberg, 1990, A stochastic reservoir model and its use in evaluation of uncertainties in the results of recovery processes, in A. Buller et al., eds., North Sea oil and gas reservoirs II: London, Graham and Trotman, p. 425–436.

Alabert, F. G., and G. J. Massonat, 1990, Heterogeneity in a complex turbidic reservoir: stochastic modeling of facies and petrophysical variability: SPE Paper 20604, SPE Annual Technical Conference and Exhibition, Proceedings, p. 775–790.

Alexander, J., 1986, Idealized flow models to predict alluvial sandstone body distribution in the Middle Jurassic Yorkshire basin: Marine and Petroleum Geology, v. 3, p. 298–305.

Bridge, J. S., and M. R. Leeder, 1979, A simulation model of alluvial stratigraphy: Sedimentology, v. 26, p. 618–644.

Clemetsen, R., A. R. Hurst, R. Knarud, and H. Omre, 1990, A computer program for evaluation of fluvial reservoirs, in A. Buller et al., eds., North Sea oil and gas reservoirs II: London, Graham and Trotman, p. 373–386.

Corbett, P. W. M., and J. L. Jensen, 1991, An application of small scale permeability measurements—prediction of flow performance in a Rannoch facies, lower Brent Group, North Sea: Minipermeametry in Reservoir Studies Conference, Proceedings, p. 1-22.

Damsleth, E., C. B. Tjolsen, H. Omre, and H. H. Haldorsen, 1992, A two-stage stochastic model applied to a North Sea reservoir: Journal of Petroleum Technology, April, p. 402–408.

Deutsch, C., 1992, Annealing techniques applied to reservoir modeling and the integration of geological and engineering (well test) data: Ph.D. thesis, Stanford University, Stanford, California, 325 p.

Deutsch, C., and A. Journel, 1992, GSLIB: geostatistical software library and user's guide: New York, Oxford University Press, 340 p.

Dranfield, P., S. H. Begg, and R. R. Carter, 1987, Wytch garm oilfield: reservoir characterization of the Sherwood Sandstone for input to reservoir simulation studies, in J. Brooks and K. Glennie, eds., Petroleum geology of north-west Europe: London, Graham and Trotman, p. 149–160.

Dreyer, T., et al., 1992, Sedimentary architecture of field analogues for reservoir information (SAFARI): case study of the fluvial Escanilla Formation, Spanish Pyrenees, in The geological modeling of hydrocarbon reservoirs and outcrop analogues: International Association of Sedimentologists, Special Publication, v. 15, p. 57–80.

Farmer, C. L., 1992, Grid generation as a discrete optimisation problem, in M. J. Baines, ed., Proceedings from Conference on Numerical Methods for Fluid Dynamics: Oxford University Press.

Fælt, L. M., A. Henriquez, L. Holden, and H. Tjelmeland, 1991, MOHERES, a program system for simulation of reservoir architecture and properties: European Symposium on Improved Oil Recovery, Proceedings, p. 27–39.

Geehan, G., et al., 1986, Geological prediction of shale continuity in Prudhoe Bay field, reservoir characterization: London, Academic Press, p. 63–81.

Georgsen, F., and H. Omre, 1993, Combining fibre processes and Gaussian random functions for modeling fluvial reservoirs, in A. Soares, ed., Geostatistics Troia '92: Netherlands, Kluwer Academic Publishers, p. 425–440.

Guerillot, D., and I. Morelon, 1992, Sorting equiprobable geostatistical images by simplified flow calculations: SPE Paper 24891, SPE Annual Technical Conference and Exhibition, Proceedings, p. 327–342.

Guerillot, D., and P. Swaby, 1993, An interactive 3D mesh builder for fluid flow reservoir simulation: SPE Petroleum Computer Conference, Proceedings, p. 79–90.

Henriquez, A., K. Tyler, and A. Hurst, 1990, Characterization of fluvial sedimentology for reservoir simulation modeling: SPE Formation Evaluation, September, p. 211–216.

Hewett, T. A., and R. A. Behrens, 1988, Conditional simulation of reservoir heterogeneity with fractals: SPE Paper 18326, SPE Annual Technical Conference, Proceedings, p. 645–660.

Holden, L., and O. Lia, 1992, A tensor estimator for the homogenization of absolute permeability: Transport in Porous Media, v. 8, p. 37–46.

Hove, K., G. Olsen, A. Nilsson, M. Tonnesen, and A. Hatløy, 1992, From stochastic geological description to production forecasting in heterogeneous layered reservoirs: SPE Paper 24890, SPE Annual Technical Conference and Exhibition, Proceedings, Washington, D.C., October 4–7.

Journel, A. G., and F. G. Alabert, 1988, Focusing on spatial connectivity of extreme-valued attributes: stochastic indicator models of reservoir heterogeneities: SPE Paper 18324, SPE Annual Technical Conference and Exhibition, Proceedings, p. 621–632.

Journel, A. G., and C. H. Huijbregts, 1978, Mining geostatistics: London, Academic Press, 600 p.

King, P. R., 1989, The use of renormalization for calculating effective permeability: Transport in Porous Media, v. 4, p. 37–58.

MacDonald, A., T. Hoeye, P. Lowry, T. Jacobsen, J. O. Aasen, and A. Grindheim, 1991, Stochastic flow unit modelling of a North Sea coastal-deltaic reservoir: First Break, v. 10, p. 124–133.

Matheron, G., et al., 1987, Conditional simulation of the geometry of fluvio-deltaic reservoirs: SPE Paper 16753, SPE Annual Technical Conference and Exhibition, Proceedings, p. 591–599.

McGill, C., P. King, and J. Williams, 1993, Estimating effective permeability: a comparison of techniques, in B. Linville, ed., Reservoir characterization III: Tulsa, Oklahoma, PennWell Books, p. 829–834.

Miall, A. D., 1985, Architectural element analysis: a new method for facies analysis applied to fluvial deposits: Earth Science Reviews, v. 19, p. 33–81.

Munthe, K., et al., 1993, Sub-seismic faults in reservoir description and simulation: SPE Paper 26500, SPE Annual Technical Conference and Exhibition, Proceedings, p. 843–850.

Olsen, G., A. Skauge, and J. A. Stensen, 1992, Evaluation of the potential application on the WAG process in a North Sea reservoir: Revue de l'Institut Français du Petrole, v. 47, n. 1, p. 81–93.

Rudkiewicz, J. L., D. Guerillot, and A. Galli, 1990, An integrated software for stochastic modelling of reservoir lithology and property prediction with an example from the Yorkshire Middle Jurassic formation, in A. Buller et al., eds., North Sea oil and gas reservoirs II: London, Graham and Trotman, p. 399–406.

Stanley, K. O., K. Jorde, N. Raestad, and C. P. Stockbridge, 1990, Stochastic modelling of reservoir sand bodies for input to reservoir simulation, Snorre field, northern North Sea, Norway, in A. Buller et al., eds., North Sea oil and gas reservoirs II: London, Graham and Trotman, p. 91–101.

Stoyan, D., W. S. Kendall, and J. Mecke, 1987, Stochastic geometry and its applications: New York, Wiley, 345 p.

Tetzlaff, D. M., and J. W. Harbaugh, 1989, Simulating clastic sedimentation: New York, Van Nostrand Reinhold, 202 p.

Tjelmeland, H., and H. Omre, 1992, Semi-Markov random fields, in A. Soares, ed., Geostatistics Troia '92: Netherlands, Kluwer Academic Publishers, p. 493–504.

Tyler, K., A. Henriquez, F. Georgsen, L. Holden, and H. Tjelmeland, 1992a, A program for 3D modeling of heterogeneities in a fluvial reservoir: 3rd European Conference on the Mathematics of Oil Recovery, Proceedings, p. 31–40.

Tyler, K., A. Henriquez, A. MacDonald, T. Svanes, L. Holden and A. L. Hektoen, 1992b, MOHERES—a collection of stochastic models for describing heterogeneities in clastic reservoirs: 3rd International Conference on North Sea Oil and Gas Reservoirs, Proceedings, p. 213–221.

Tyler, K., T. Svanes, and A. Henriquez, 1992c, Heterogeneity modelling used for production simulation of fluvial reservoir: SPE Formation Evalua-tion, June, p. 85–92.

Tyler, K., T. Svanes, and S. Omdal, 1993, Faster history matching and uncertainty in predicted production profiles with stochastic modeling: SPE Paper 26420, SPE Annual Technical Conference and Exhibition, Proceedings, p. 31–43.

Weber, K. J., and L. C. van Geuns, 1990, Framework for constructing clastic reservoir simulation models: Journal of Petroleum Technology, v. 42, p. 1248–1297.

Wolcott, D., and A. Chopra, 1993, Incorporating reservoir heterogeneity with geostatistics to investigate waterflood recoveries: SPE Formation Evaluation, March, p. 26–32.

Chapter 9

A Prototype Procedure for Stochastic Modeling of Facies Tract Distribution in Shoreface Reservoirs

Alister C. MacDonald
Jan Ole Aasen
Statoil Research Centre
Trondheim, Norway

ABSTRACT

A stochastic model comprising a truncated Gaussian field with a linear expectation trend has been developed for modeling facies tract distribution in shoreface reservoirs. The expectation trend is described in terms of a progradation direction ϕ, which controls facies tract orientation, and an aggradation angle θ, which governs the distances over which the tracts can prograde. Interfingering between facies tracts is regulated by the spatial covariance function of the underlying Gaussian field. The estimation of ϕ and θ uses a statistical method known as Bayesian inference, which uses general geologic information and analog data, in addition to well observations.

A controlled test has been done based on two Cretaceous parasequences exposed in the Book Cliffs, near Helper, Utah. The test involved a synthetic extrapolation exercise where four measured sections were treated as appraisal wells and a fifth (blind test) well was used to evaluate the results of the simulations. Fifty realizations of facies tract distribution in each parasequence were simulated. The stochastic model generated geologically realistic facies tract distributions and the thicknesses of shoreface facies in the blind test well were predicted satisfactorily. The main limitations of our model are that it cannot capture abrupt updip pinch-outs or systematic distal thickening of facies tracts. Modifications are necessary to include such effects.

Despite the limitations, our modeling procedure has potential to assist in the management of shoreface reservoirs where the facies tract distribution is uncertain. The procedure also has potential to help identify and estimate risk for exploration prospects.

INTRODUCTION

Although shoreface reservoirs are relatively homogeneous when compared, for example, with fluvial reservoirs, they still comprise a hierarchy of different heterogeneities that influence the appraisal and development of hydrocarbon resources. Where the distribution of heterogeneities is associated with significant uncertainty, it is necessary to use stochastic methods to capture this uncertainty and anticipated complexity in a reservoir model. In this chapter, we are concerned with the design and testing of a prototype stochastic model for describing one particular type of heterogeneity in shoreface reservoirs: the distribution of so-called facies tracts within parasequences (Cross et al., 1993). The geologic aspects of the modeling procedure are presented here, whereas the statistical aspects will be published in more detail elsewhere.

The distribution of facies tracts, such as foreshore and upper shoreface, define the spatial distribution of high- and low-quality reservoir facies. Facies tract distribution therefore is an important variable to model correctly. Many shorefaces are associated with rather laterally continuous facies tracts. Thus, in densely drilled fields, the position of the facies tracts can be mapped using conventional deterministic procedures based on well data, whereas in less densely drilled fields (for example during an appraisal phase) significant uncertainty can be associated with facies tract distribution. Uncertainty is also particularly acute where there is a need to extrapolate to the peripheral parts of a field or to a possible satellite. Smaller scale lagoonal shorefaces, for example those that form

important reservoirs in the Middle Jurassic Ness Formation of the Brent Group in the North Sea, are associated with rather less continuous facies tracts that are difficult to map in a deterministic manner using well data.

A crucial consideration is that the design of a stochastic model should reflect as realistically as possible the underlying conceptual geologic model; therefore, we outline a conceptual model for shoreface reservoirs before we describe the stochastic model. A controlled test of the modeling procedure has been done using data from the Book Cliffs in Utah to illustrate a potential application of the stochastic model for exploration and appraisal problems, an area that, so far, has received little attention in the literature.

CONCEPTUAL MODEL

Hierarchies implicit in sequence stratigraphy provide a framework for subdividing and modeling shoreface reservoirs. Although a number of different hierarchies can be used, a simple conceptual model of a shoreface reservoir (Figure 1) should include parasequences bounded by marine flooding surfaces as a fundamental subdivision (Van Wagoner et al., 1990). The flooding surfaces are laterally continuous, approximate time lines, and provide a high-resolution chronostratigraphic framework for reservoir description and modeling. They can be associated with thin shale beds and can be preferentially cemented. Each parasequence is internally composed of facies tracts, such as foreshore and upper shoreface, which typically interfinger and

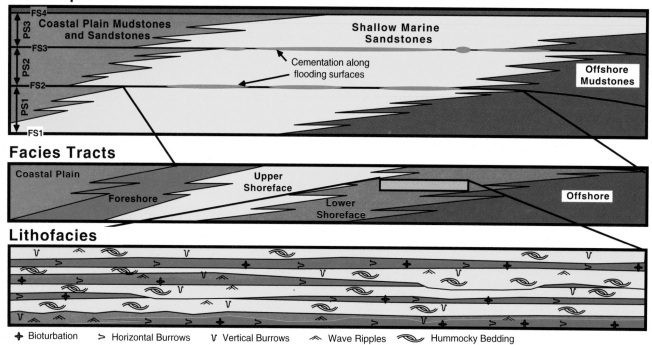

Figure 1. Conceptual model of prograding shoreface reservoirs illustrating three hierarchical levels of heterogeneity: (1) parasequence level, (2) facies tract level, and (3) lithofacies level.

are ordered in a systematic manner. The interfaces between facies tracts typically crosscut time lines, and each facies tract, in turn, is composed of individual lithofacies, such as bioturbated beds that typically are aligned parallel to time lines. Different types and proportions of lithofacies are associated with the different facies tracts. For example, shale beds within a shoreface sequence generally are re-stricted to the lower shoreface and offshore transition zone. Also, the proportion and dimensions of lithofacies within a particular facies tract varies between successive parasequences due to different preservation potential under varying accommodation rates. For example, under high rates of sea level rise, the proportion and continuity of shales beds within a lower shoreface can be expected to be greater than under lower rates of sea level rise. This effect is termed volumetric partitioning (Cross et al., 1993).

This conceptual model (Figure 1) provides the basis for a systematic modeling procedure involving:

- Subdivision of the shoreface reservoir into parasequences based on the correlation of flooding surfaces
- Simulation of vertical connectivity between parasequences
- Simulation of facies tract distribution
- Simulation of facies distribution within facies tracts

MacDonald et al. (1992) outlined a modeling procedure for coastal-deltaic reservoirs based on a similar conceptual model. This present paper presents the stochastic model for step 3.

FORMULATING THE STOCHASTIC MODEL

The present literature on stochastic reservoir modeling tends to focus on the modeling algorithms, and the geologic and reservoir technical problems present often appear to be of secondary importance. This, in our opinion, is unfortunate. The design of stochastic models should be tailored to a particular reservoir management problem; the model should reflect as realistically as possible the underlying conceptual geologic model and capture the uncertainty associated with the geometric variables. The formulation of the conceptual geologic model as a mathematical (stochastic) model is a particularly crucial step (Figure 2). The stochastic model also should be designed to take into account the

available field specific data and to maximize the use of these data. In other words, the choice and design of a stochastic modeling procedure should be application and data driven rather than algorithm driven.

A model, by definition, is an approximation and simplification of a real phenomenon, and cannot represent all aspects of the particular phenomenon. Thus, when you are formulating a stochastic model, you must be explicit with regard to the assumptions and simplifications implicit to the model. Likewise, you must be aware of the limitations when applying the model.

The three most important geometric characteristics to be considered when formulating a stochastic model for facies tract distribution are as follows.

- A systematic ordering of facies tracts within a single parasequence. For example, offshore facies cannot be juxtaposed against coastal plain facies. Instead, there is a systematic ordering of facies tracts from offshore through lower and upper shoreface and foreshore to coastal plain.
- Most shoreface parasequences are progradational, i.e., the positions of facies tracts migrate gradationally or stepwise in a seaward direction with time. Most parasequences are also aggradational to a certain extent (i.e., the sediments are deposited and build upward), and the interfaces between facies tracts tend to migrate upward during deposition.
- Facies tracts tend to interfinger as a result of smaller scale cyclicity within the parasequences. The cyclicity may reflect high-frequency relative sea level fluctuations or autocyclic processes.

We have adopted a method based on a truncated Gaussian field with a linear trend in the expectation value to accommodate these geologic characteristics in the stochastic model. The use of a truncated Gaussian field ensures a systematic ordering of the facies tracts. The expectation trend describes the direction of progradation and the degree of contemporaneous aggradation, and the underlying covariance function can be used to regulate the local continuity and interfingering between the different facies tracts.

TRUNCATED GAUSSIAN MODEL

Truncated Gaussian fields are used to describe the distribution of discrete or categorical variables, such as facies or facies tract type. The method requires

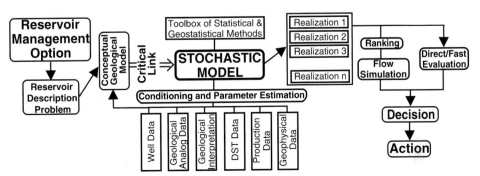

Figure 2. Flow diagram illustrating the steps involved in a stochastic modeling study aimed at a specific, well-defined reservoir management problem. The efficient formulation of the conceptual geologic model as a stochastic model is often the critical link in the procedure.

truncation of a continuous Gaussian function at one or more threshold values ($t_1,..., t_n$), where the number of categorical variables to be simulated is equal to n + 1 (Figure 3). Category k occurs wherever the Gaussian function assumes a value between t_{k-1} and t_k (with t_0 and t_{n+1} set to $-\infty$ and $+\infty$). The method generally involves:

1. Coding categories as intervals of a Gaussian function bounded by thresholds
2. Calculating Gaussian values in the wells corresponding to the observed categories
3. Simulating the Gaussian values between wells
4. Transforming the simulated Gaussian values back to categories to define the spatial distribution of categories

The main parameters that describe the model are the threshold values ($t_1,..., t_n$) and the parameters of the underlying Gaussian field (expectation E, variance σ^2, and spatial correlation function ρ). For a stationary field, the threshold values control the relative proportions of simulated facies. For example, method 1 in Figure 3 illustrates a one-dimensional Gaussian function with a constant mean value E of 0 and a variance σ^2 of 1. The function is truncated by two thresholds (t_1 and t_2) to produce three different facies types (green, red, and yellow). The relative proportions of simulated facies types can be changed by varying the threshold values. For example, if t_2 were increased from 0.6 to 1.0, the proportion of red facies would increase, and the yellow facies would decrease correspondingly.

The variance σ^2 and spatial correlation function ρ control the spatial continuity of the simulated facies.

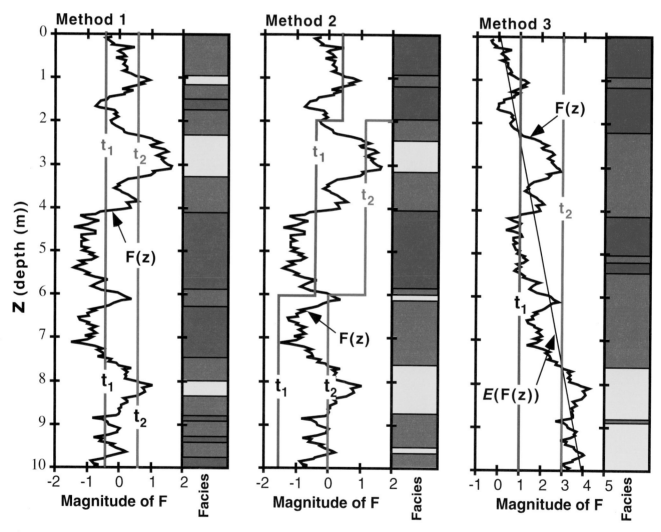

Figure 3. Three methods of truncating the continuous one-dimensional Gaussian function [F(z)]. Two thresholds t_1 and t_2 are used to define the distribution of three categories/facies (green, red, and yellow). Method 1 involves a stationary Gaussian function and stationary threshold values. Method 2 involves a stationary Gaussian function, but nonstationary thresholds. This method is the HERESIM method where the nonstationary thresholds are termed proportion curves (Rudkiewicz et al., 1990). Method 3 involves stationary thresholds and a nonstationary function with a linear expectation trend {E[F(z)]}, which is the method used in this paper. The function [F(z)] in method 3 is the sum of {E[F(z)]} and the stationary function [F(z)] in methods 1 and 2.

The correlation lengths of the correlation function are of particular importance. The Gaussian function [F(z)] in Figure 3 is simulated using a spherical variogram model (Isaaks and Srivastava, 1989) and a correlation length of 2 m. An increase in the correlation length would lead to the simulation of a more smoothly changing function, which, in turn, would produce thicker facies units.

A truncated Gaussian field is a relatively simple model for describing the spatial distribution of a discrete variable, and the strengths of such a model lie in its simplicity; the model is easy to specify, simulation is fast, and it is relatively simple to condition the model on well data. The main disadvantages also lie in its simplicity. A single correlation function ρ and variance σ^2 is used and the different simulated categories thus are associated with similar spatial continuity. Also, where more than two categories are to be simulated, the truncation of a continuous variable naturally produces a preferred ordering, i.e., category 2 always occurs between categories 1 and 3, whereas 3 can never occur adjacent to 1 (in Figure 3 the yellow facies can never occur adjacent to the green facies). In some situations, this property is undesirable, but it is desirable for modeling facies tracts that are typically highly ordered (Figure 1).

The use of truncated Gaussian fields for modeling lithofacies distributions has been popularized by the HERESIM program package (e.g., Rudkiewicz et al., 1990). This method emphasizes the definition of the threshold values $(t_1 - t_n)$ using so-called proportion curves, which allow the simulation of vertical or horizontal variation in the proportions of different facies types where a stationary Gaussian function and nonstationary thresholds are used (method 2, Figure 3).

The model developed here (method 3, Figure 3) for simulating facies tracts differs from the model used in HERESIM in that proportion curves are not used (i.e., the thresholds are constant), but the underlying Gaussian field is nonstationary. The expectation function of the three-dimensional Gaussian field F is a linear expression in the spatial coordinates x,y,z:

$$E\{F(x,y,z)\} = a + b\text{x} + c\text{y} + d\text{z} \qquad (1)$$

An expectation function defined by a vector with an arbitrary direction was adopted to capture the desired large-scale ordering of facies tracts (Figure 1). Note that the HERESIM software allows only for vertical or horizontal trends in threshold values (proportion curves) and cannot be applied directly for modeling the systematic ordering of facies tracts (Figure 1), which requires a vector with an arbitrary direction in space.

The simulated interfaces between facies tracts occur normal to the defined trend, i.e., normal to the vector (b,c,d) (Figure 4). The adoption of a *linear* trend implies that the interfaces between facies tracts are more or less parallel. It also assumes that the shoreline was more or less linear and maintained an approximately constant orientation during the progradation of the parasequence (Figure 4). In Figure 4, note that the distribution of facies tracts is the same as would be generated if the underlying Gaussian function had a variance of 0. The spatial covariance function $(\sigma^2 \times \rho)$ of the Gaussian field therefore will produce variations from this distribution. Although the use of a linear trend is a reasonable first approximation, the model outlined here is inappropriate if facies tracts within a shoreface reservoir are associated with large and complex lateral thickness variations. In such cases, the development and application of more complex models are necessary.

MODELING PROCEDURE

The truncated Gaussian field model described in this chapter forms the basis for a modeling procedure for facies tract distribution illustrated in Figure 5. The

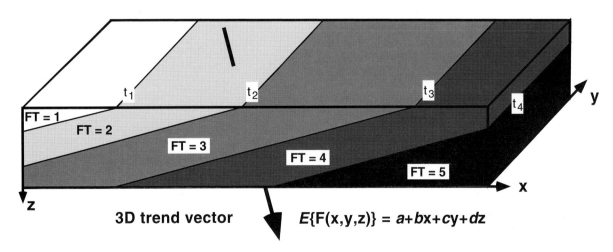

Figure 4. The linear trend vector produces a systematic ordering of facies tracts illustrated by different gray tones. The use of a linear trend results in parallel interfaces between facies tracts that are normal to the trend vector. FT = facies tract, $t_1 - t_4$ = thresholds.

different stages in the procedure are outlined in the following paragraphs.

Defining the Simulation Grid

Stochastic heterogeneity modeling is generally done using a regular, orthogonal simulation grid. It is expedient to align the grid with the x-y plane parallel to interpreted time lines (MacDonald et al., 1992). The definition of the simulation grid thus depends on the interpretation of time lines within the modeling volume. Flooding surfaces that bound parasequences (Figure 1) provide the most reliable time lines within shoreface reservoirs, and the simulation grid should generally be aligned parallel to flooding surfaces that have been correlated from well to well.

The creation of a regular grid framework is simply achieved by transforming well data depth values to produce standard thicknesses for each parasequence in each well (MacDonald et al., 1992). Each parasequence then can be assigned a constant number of grid block layers. Note that estimates of input parameters must be based on the transformed thicknesses and a regular coordinate system, and not on the true thicknesses and Cartesian coordinates.

Interpreting the Well Data

Within each parasequence you must interpret the different facies tracts observed in the wells and order them beginning with the most landward (proximal) and working toward the most distal (seaward). The most proximal facies tract is assigned to category 1 and the most distal to category n + 1, producing n thresholds. It is then necessary to interpret the position of the various facies tracts in the wells (using the transformed coordinate system) and identify the depths where interfaces between facies tracts cross the well positions (Figure 6). The identification of the interfaces is important because they define the thresholds that are to be used in the truncation of the continuous Gaussian function (Figure 3). The first threshold (t_1) occurs between facies tract 1 and 2, the second threshold (t_2) occurs between facies tract 2 and 3, and so on (Figure 6). Note that you do not need to define the absolute values of the thresholds; the defining of their ordering and occurrence at the well locations is sufficient.

Estimating the Expectation Trend

Estimating the expectation trend (equation 1) is fundamental to the modeling procedure because it strongly controls the simulated distribution of facies tracts. Because geologists cannot intuitively estimate the constants in the trend (equation 1), a geometric relationship between the constants and two geologically meaningful angles [the progradation direction ϕ (phi) and the aggradation angle θ (theta)] has been set up as follows (Figure 7) where

$$a = 0; b = \sin\theta \sin\phi;$$
$$c = \sin\theta \cos\phi; \text{ and } d = \cos\theta \qquad (2)$$

The parameter a can be set to 0 and the trend direction vector (b,c,d) can be given a length of unity without loss of generality (Figure 7). Both angles typically will be associated with varying degrees of uncertainty and capturing this uncertainty in the stochastic model

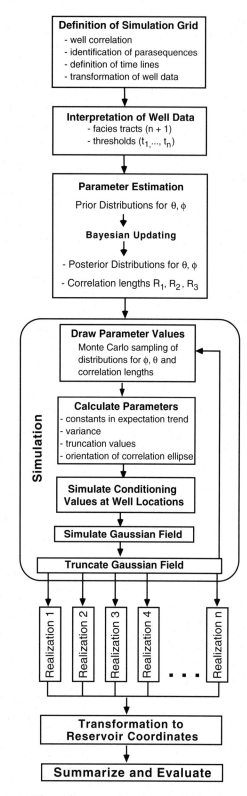

Figure 5. Flow diagram for the modeling procedure.

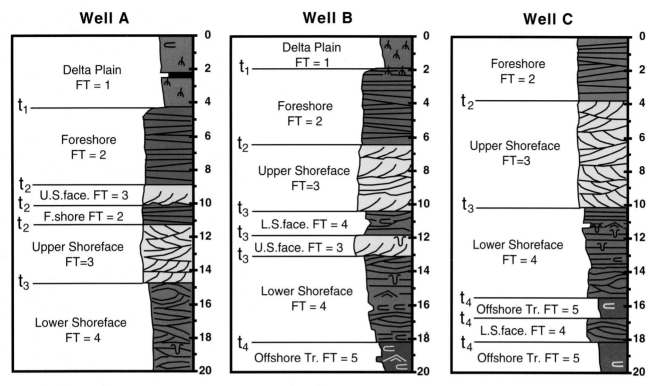

Figure 6. Interpretation of well data. The identified facies tracts in three wells are ordered from proximal to distal and assigned integer values [e.g., delta plain facies tract (FT) is assigned to category 1]. Thresholds (t_1–t_4) are then assigned to the facies tract interfaces observed in the wells. U.S.face. = upper shoreface, F.shore = foreshore, Offshore Tr. = offshore transition, L.S.face. = lower shoreface.

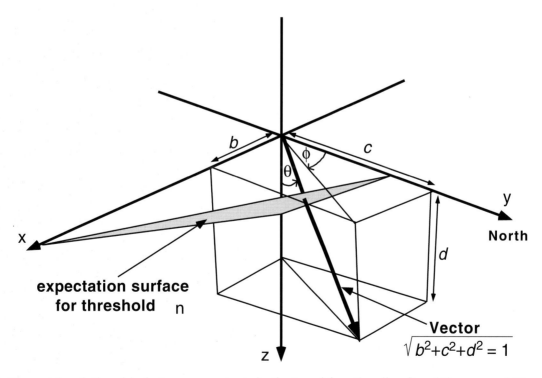

Figure 7. Geometric relationship between constants in the trend function (b,c,d) and the progradation direction, ϕ, and the aggradation angle, θ, where the aggradation angle is defined as the average angle of climb of facies tract interfaces within a parasequence (Figure 4). Note that because the trend vector is orthogonal to the parallel interfaces (Figure 4), the angle between the trend vector and the vertical axis (θ) is identical to the aggradation angle.

is fundamentally important. Accordingly, you should not use constant values. You must estimate probability distributions for φ and θ as input to capture the uncertainty associated with a particular reservoir management option. This uncertainty often will be more significant than the inherent uncertainty in a stochastic model with constant input parameters (i.e., the uncertainty that results from the use of different random seeds); nevertheless, many stochastic modeling studies aimed at quantifying uncertainty use constant input parameters that cannot be known precisely, and therefore ignore one of the most important real sources of uncertainty.

Classical statistical inference techniques rely on observations of a phenomenon to estimate the parameters that describe the phenomenon. Although the observed positions of the thresholds in the wells provide information about appropriate angles for φ and θ, the observations often will be insufficient to define the parameters precisely; thus, you should draw on general geologic information and analog data to assist in estimating the parameter.

To account for general geologic information, analog data, and the available well data, we adopted a Bayesian inference approach (Box and Tiao, 1992; Omre and Halvorsen, 1989). Bayesian inference differs from classical statistical inference in that it supposes that general information about the phenomenon is useful for estimating parameters in addition to observing the phenomenon. General geologic information and analog data define "prior" distributions that are modified using well data to produce "posterior" distributions that are then used as input to the stochastic simulation. The definition of the prior distributions using geologic information is outlined in the following sections, as is a brief description of the Bayesian method for estimating φ and θ.

Prior Distributions

The progradation direction is an important parameter to estimate correctly because it governs the orientation and direction of elongation of the facies tracts. The intention is to choose values and build a prior probability model that reflects a geologist's interpretation of the uncertainty and spread in φ. The estimate generally will draw heavily on regional or local paleogeographic information rather than on analog data.

We have generally used a von Mises distribution (Mardia et al., 1979) to define φ. A von Mises distribution is a circular equivalent of the normal distribution and is conveniently defined in terms of only two parameters: a mean direction a and a concentration parameter κ (kappa) (Davis, 1986). The concentration parameter κ in the von Mises distribution describes the possible dispersion in progradation directions. The parameter is analogous to the inverse of variance in a standard normal distribution: low κ values produce a wide dispersion of directions and imply that little is known about φ, whereas high κ values result in little dispersion and imply that φ is tightly constrained.

The aggradation angle is also an important parameter because it strongly controls the distances over

which the facies tracts can prograde; high aggradation angles lead to the simulation of laterally restricted facies tracts, whereas low angles lead to the simulation of continuous, sheetlike facies tracts. Analog data can assist in estimating aggradation angles. Based on a few calculations from well-exposed shoreface parasequences, angles apparently range from 0.5° for highly aggradational parasequences to 0.01° for strongly progradational parasequences; however, many more outcrop measurements from shorefaces deposited in a variety of settings are required to improve our understanding of this parameter.

The application of sequence stratigraphy concepts is of particular importance here because the aggradation angles are strongly influenced by the balance between sediment supply and rates of relative sea level rise. If the rate of sediment supply were constant, low rates of relative sea level rise would tend to produce low aggradation angles, whereas accelerated sea level rise would lead to higher aggradation angles. Sequence stratigraphy supposes that relative sea level varies in a systematic manner. Aggradation angles between consecutive parasequences therefore should vary in a systematic manner depending on their position within different systems tracts (Posamentier and Vail, 1988). For example, in the early part of a highstand systems tract (where the rate of relative sea level rise is gradually decreasing) the aggradation angle between successive parasequences can be expected to decrease upward (Figure 8).

We have generally used a truncated normal distribution defined by a mean value, a standard deviation, and maximum and minimum values to describe the prior distribution of θ.

Bayesian Updating of Distributions

A thorough Bayesian method requires that prior distributions are defined for all of the input parameters [φ, θ, σ^2, ρ, t_1,..., t_n], and that calculation of the posterior distributions be based on the exact stochastic model. This is a complex problem and, for the prototype modeling procedure, we adopted a simplified approach to define the input distributions just for φ and θ. We assume that the variance σ^2 and the correlation function ρ have little influence on the estimate of φ and θ, and that we have no information about the magnitude of the thresholds (t_1,..., t_n).

To formulate the method, let (x_1, y_1, z_1),...., (x_N, y_N, z_N) be a set of observations where an interface between two facies tracts has been observed. The stochastic model implies, for given values of the model parameters, a posterior distribution for the depths to the interfaces at (x_1, y_1),..., (x_N, y_N), with a density

$$f[z_1,..., z_N \mid (\sigma^2, \rho, \phi, \theta, t_1,..., t_n)] \qquad (3)$$

We have made the simplifying assumption that f is multivariate normal. If we also assume that the variance σ^2 and the correlation function ρ are known, the set of observations (x_1, y_1, z_1),..., (x_N, y_N, z_N) may then be used, through Bayes's theorem, to transform a

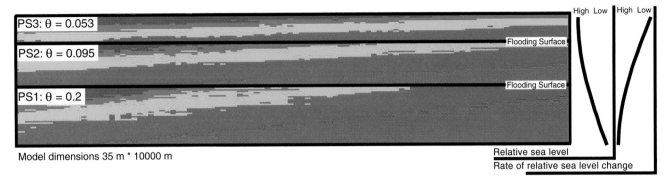

Figure 8. Aggradation angles θ and rate of relative sea level change. The three parasequences (PS1–PS3) are simulated with decreasing θ to reflect deposition in the early part of a highstand systems tract where the rate of relative sea level change is decreasing (Posamentier and Vail, 1988).

prior distribution for the other parameters into a posterior distribution with a density

$$
\begin{aligned}
&P_{post}(\phi, \theta, t_1,..., t_n) \\
&= C \times f[z_1...z_N \mid (\sigma^2, \rho, \phi, \theta, t_1,..., t_n] \\
&\times P_{prior}(\phi, \theta, t_1,..., t_n)
\end{aligned}
\tag{4}
$$

where C is a normalization factor. In our calculations we have used a prior of the form

$$
P_{prior}(\phi, \theta, t_1,..., t_n) = P_\Phi(\phi) P_\Theta(\theta)
\tag{5}
$$

where $P_\Phi(\phi)$ and $P_\Theta(\theta)$ are the defined prior distributions. This implies that we assume that nothing is known beforehand about the values of the thresholds. This assumption allows the calculation of the marginal posterior probability density for ϕ and θ having the form

$$
P_{post\Phi\Theta}(\phi, \theta) = c \times \exp\left\{-\frac{1}{2\sigma^2} \overset{\text{r}}{u} \times M\overset{\text{r}}{u}\right\}
$$
$$
\times P_\Phi(\phi) P_\Theta(\theta)
\tag{6}
$$

where c is a normalization factor, M is a matrix dependent only on the data $(x_1, y_1, z_1),..., (x_N, y_N, z_N)$, and $\overset{\text{r}}{u}$ is the trend direction vector with components $\sin\theta\sin\phi$, $\sin\theta\cos\phi$, and $\cos\theta$ (Figure 7). Bayes's rule therefore gives the posterior probability density of the model parameters ϕ and θ as a product of the prior density and the likelihood function, which in this case is given by

$$
\exp\left\{-\frac{1}{2\sigma^2} \overset{\text{r}}{u} \times M\overset{\text{r}}{u}\right\}
\tag{7}
$$

An example of a Bayesian modification of prior to posterior distributions is illustrated in Figure 9. The prior distribution for ϕ has been defined using a von Mises distribution with a mean direction of 090° and a κ of 5. The likelihood function is calculated using nine observations of thresholds in five wells and is associated with a maximum likelihood direction of 059°. Because the resultant posterior distribution is a prod-

uct of the likelihood function and prior distribution (equation 6), it lies between the likelihood and prior models with a modal value of 077°. The relative importance of the prior distribution as opposed to the well observations depends on the number of well observations available. If only a few observations of thresholds are available, the posterior distribution will be very close to the prior distribution; however, if 100 wells were available, the prior distribution would be unimportant and the posterior distribution would be similar to the likelihood function. The influence of the prior distribution also depends on a number of other factors, including the assumed variance σ^2 of the underlying Gaussian function; the higher the assumed variance, the stronger the influence of the prior distribution.

Estimating the Correlation Function

The correlation function ρ of the underlying Gaussian field controls the local continuity of the different facies tracts in different directions and the degree of interfingering between them. The function is defined by a model type (e.g., exponential, spherical, Gaussian) and orthogonal correlation lengths ($r_1 - r_3$). In the present procedure, a spherical variogram model is used as default, whereas the magnitude of the correlation lengths ($r_1 - r_3$), which define a correlation ellipsoid (Figure 10), have to be estimated. Vertical correlation lengths reflect bed thicknesses, whereas horizontal correlation lengths reflect continuity of the interfingering in the horizontal plane. Correlation lengths can be expected to be significantly larger when they are normal rather than parallel to the progradation direction, and it is necessary to orient the ellipse so that the longest axis (r_1) is parallel to the shoreline strike and normal to the progradation direction (Figure 10).

In most situations well data are too few to calculate the horizontal correlation lengths experimentally, and estimating the parameters again relies on the use of analog outcrop statistics or general geologic knowledge. Calculating the experimental variograms using outcrop data has not been done as part

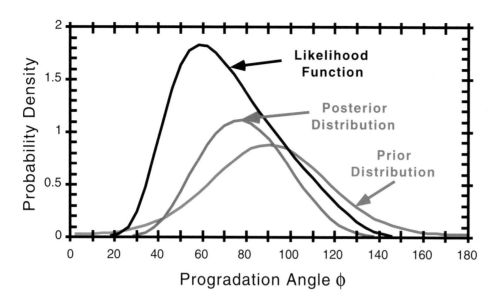

Figure 9. Updating of prior distribution using likelihood function of well observations to produce a posterior distribution. The prior distribution for ϕ has been defined using a von Mises distribution with a mean direction of 090° and a κ of 5. The likelihood function is calculated using nine observations of thresholds in five wells. The posterior distribution is simply the product of the likelihood function and prior distributions (equation 6) and, therefore, lies between the likelihood and prior models.

Parallel to Progradation Direction (x-z section)

Normal to Progradation Direction (y-z section)

Parallel to time lines (x-y sections)

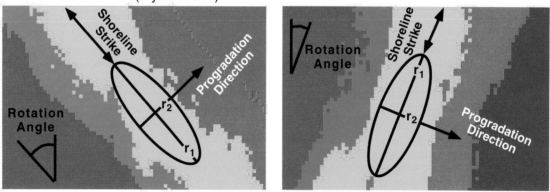

Figure 10. Orthogonal correlation lengths (r_1–r_3), which define the dimensions of a correlation ellipsoid, control the continuity of the interfingering between facies tracts. The major horizontal axis r_1 is aligned parallel to the shoreline strike, the minor horizontal axis r_2 is aligned parallel to the progradation direction ϕ, and the vertical axis r_3 is parallel to the z-axis of the simulation grid. The scale is schematic.

of this study, and outcrop data should be analyzed to help estimate covariance functions.

Because the horizontal correlation lengths (r_1 and r_2) cannot be estimated using well data alone, their

exact magnitudes cannot be known precisely. Accordingly, you must include this uncertainty in the stochastic model by defining distributions for the correlation lengths instead of using constant values for all simulations. We have generally used normal distributions defined by mean value and standard deviations to describe the magnitudes of $r_1 - r_3$.

Simulation

Defining the simulation grid, interpreting well data, and estimating the parameters are the most important "geologic" input to the modeling procedure. After the geologic input has been defined, the simulation process, which involves a variety of statistical and mathematical techniques, can begin (Figure 5).

The first stage involves Monte Carlo sampling from the distributions for ϕ, θ and ($r_1 - r_3$). The drawn values are then used to estimate a number of other parameters that are required as input to the simulator as outlined below.

1. Constants in expectation trend (equation 1). The drawn angles of ϕ and θ define the constants in the expectation trend through the simple relationships outlined in equation 2.
2. Absolute values of thresholds $t_1,..., t_n$. The chosen trend function is used to define the threshold values to be used in a simulation by optimizing P_{post} (ϕ, θ, $t_1,..., t_n$) with respect to $t_1,..., t_n$ (in the maximum likelihood sense).
3. Variance of Gaussian field. The distances of the threshold depths seen in the wells from the expected position of the threshold surfaces (using the chosen trend function) provides an indication of the variance associated with the underlying Gaussian field.
4. Orientation and magnitude of correlation function. The correlation function to be used in the simulation is calculated having a correlation ellipsoid where the horizontal axes are aligned with respect to the drawn progradation direction ϕ, and correlation lengths drawn from the specified normal distributions.

To achieve proper conditioning, you must define the values of the underlying Gaussian field at the well locations. After calculating the optimal threshold values for a given vector, the absolute values of the underlying Gaussian field at the well locations are known at interfaces between facies tracts (Figure 11); however, you also need to assign conditioning values between the observed thresholds by using a simple one-dimensional sequential simulation conditioned to the observed threshold values (Figure 11). The method uses the specified variance σ^2 and correlation function ρ, and ensures that each simulated value lies between the correct thresholds (Figure 11).

Conditional simulation of the three-dimensional Gaussian field between the conditioning wells is done using a standard technique (Fournel and Huifbregts, 1978). The Gaussian field is then truncated using the thresholds ($t_1,..., t_n$) to define the spatial distribution of facies tracts.

The final step in the modeling procedure involves transforming the orthogonal simulation grid to the real reservoir coordinates. This requires that the

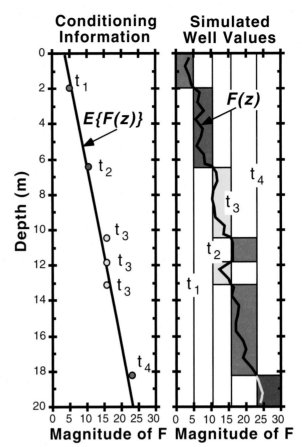

Figure 11. Generation of well conditioning values. The example is based on well 2, Figure 6. The 1D sequential simulation is conditioned on: (1) the depths of the thresholds at the well location; (2) the absolute values of the thresholds; and (3) the expectation trend {E[F(z)]}. The illustrated function [F(z)] is simulated with a correlation length of 1.5 m and a variance of 2.25. The simulated facies tracts at the well location are illustrated using the same color coding as in Figure 6.

geometry of the bounding flooding surfaces be defined, and the transformation involves a simple linear interpolation between the two bounding surfaces (e.g., Gómez-Hernández and Srivastava, 1990).

TEST EXAMPLE

A controlled test based on Cretaceous shorefaces exposed in the Book Cliffs, near Helper, Utah, illustrates the procedure for modeling facies tract distribution. The Book Cliffs outcrops are well known to many geologists and have been instrumental in the development of sequence stratigraphy concepts (Van Wagoner et al., 1990). The test is based on two parasequences within the Spring Canyon Member of the Campanian Blackhawk Formation (Figure 12). The Spring Canyon Member is the lowermost member of the Blackhawk Formation and forms a pro-

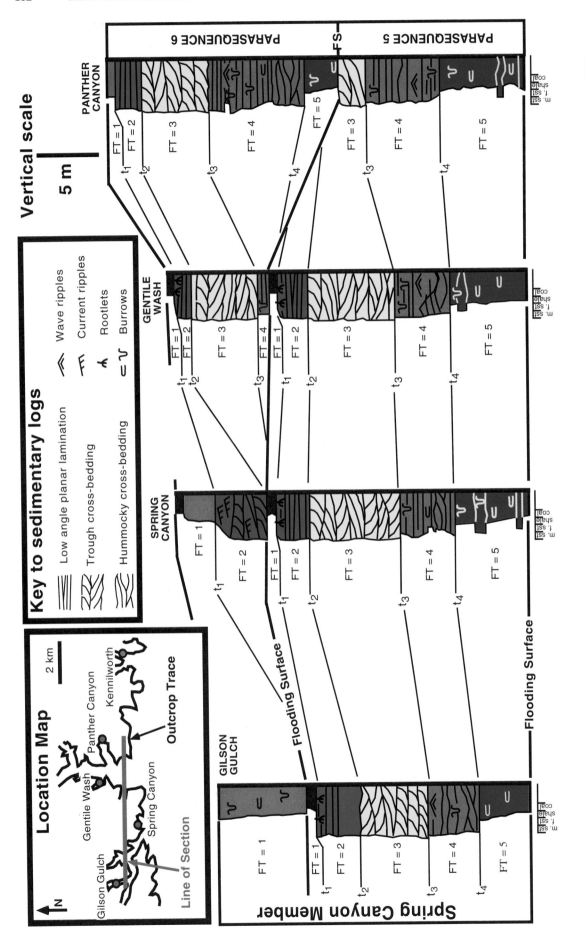

Figure 12. Sedimentary logs from the four conditioning wells used in the test study. Five facies tracts are identified in two parasequences within the Spring Canyon Member of the Blackhawk Formation and are ordered from proximal to distal (coastal plain FT = 1, backshore/foreshore FT = 2, upper shoreface FT = 3, lower shoreface FT = 4, offshore transition FT = 5). The interfaces between facies tracts define four thresholds ($t_1 - t_4$). The two parasequences are exposed along the Book Cliffs outcrop directly to the north of Helper, Utah. The location map illustrates the location of the measured sections and the trace of the Book Cliffs outcrops. The Kennilworth location is used as the blind test well.

gradational parasequence set comprising at least eight shoreface parasequences. Parasequences 5 and 6 were used for the test. Facies tract geometry within these two parasequences has been interpreted and mapped between outcrops along the Book Cliffs for a lateral distance of over 30 km (see Van Wagoner et al., 1990). As the facies tract geometries are more or less known, they provide an excellent basis for testing the modeling procedure where logged sections from the outcrop can be treated as synthetic "wells."

Our test involved a synthetic extrapolation exercise where four measured sections were treated as appraisal wells and a fifth (blind test) well (Kennilworth) was used to evaluate the results of the simulations (Figure 12). The reservoir description challenge was to define the thickness of reservoir facies at the blind test well to guide further appraisal of the "field." Fifty realizations of facies tract distribution in each parasequence were simulated.

Defining the Simulation Grid

Parasequences 5 and 6 (PS5 and PS6) range in thickness from 16 to 24 m and from 8 to 20 m, respectively. The well data have been transformed to produce a constant thickness of 20 m for PS5 and 15 m for PS6. A simulation volume of 15,000 × 10,000 m in x- and y-directions with a grid-block size of 200 m in the horizontal plane was established. The x-axis was aligned parallel to the assumed (prior) progradation direction (090°). The selected vertical resolution of 0.5 m for both parasequences requires 40 grid layers in the PS5 model and 30 in the PS6 model. The two parasequences were simulated separately using 150,000 and 112,500 grid nodes.

Interpreting the Well Data

In the four wells, we interpreted five facies tracts: coastal plain, backshore/foreshore, upper shoreface, lower shoreface, and offshore transition (Figure 12). The most proximal facies tract (coastal plain) was assigned to category 1 and the most distal (offshore transition) to category 5. The interfaces between these five facies tracts defined four thresholds $(t_1 - t_4)$ for the truncated Gaussian field. The foreshore and upper and lower shoreface were considered reservoir facies tracts, whereas the coastal plain and offshore transition zone deposits were considered nonreservoir.

Estimating the Parameters

The general paleogeographic setting for the Blackhawk Formation involves a source area to the west and a depositional basin to the east, and the deposition of the Spring Canyon Member was assumed to be related to shorelines that prograded in an easterly direction. Approximately north–south–oriented shorelines and easterly progradation have been interpreted for other members within the Blackhawk Formation (Taylor and Lovell, 1992; O'Byrne and Flint, 1993).

At this large scale, however, the Book Cliffs outcrops are rather one-dimensional, and it is difficult to map the paleoshoreline orientations precisely. A degree of uncertainty can therefore be expected for the progradation direction. A prior distribution with a mean direction of 090° and a von Mises concentration parameter κ of 5 has been used for both parasequences (Figure 13).

As previously outlined, estimating the aggradation angles depends on the sequence stratigraphic interpretation of the parasequences. Taylor and Lovell (1992) proposed that the entire Blackhawk Formation represents a highstand sequence set (Mitchum and Van Wagoner, 1991). Aggradation angles within the Blackhawk Formation as a whole can be expected to decrease upward as a result of gradually decreasing rates of sea level rise in an manner analogous to that illustrated in Figure 8. The aggradation angles within the Spring Canyon Member thus should be (on average) the highest within the Blackhawk Formation. Interpreting changes in the rates of relative sea level rise within the Spring Canyon Member is difficult, and we have a poor empirical basis for estimating aggradation angles. Based on a few measurements of aggradation angles from other parasequences, we used a mean value of 0.1° and a standard deviation of 0.025° to define the prior distributions of θ for both parasequences (Figure 13).

Likelihood functions (equation 7) were calculated based on the observations of the thresholds [$(t_1 - t_4)$ in Figure 12] and used to update the prior distributions, resulting in the posterior distributions illustrated in Figure 13. Probably, PS5 prograded in an easterly–southeasterly direction, whereas the overlying PS6 prograded in a slightly more northerly direction. PS5 is associated with significantly lower aggradation angles than PS6 (modal values of 0.042 and 0.093, respectively). The low aggradation angles estimated for PS5 are rather surprising considering its sequence stratigraphic context. These angles are lower than, for example, those associated with the Grassy Member at a higher level within the Blackhawk Formation, where angles of 0.15 and 0.25° have been estimated on the basis of data published by O'Byrne and Flint (1993). Note that φ and θ are dependent variables in the posterior distributions (Figure 13). Thus, the posterior bivariate distributions, rather than the marginal distributions, must be used as the basis for the Monte Carlo sampling.

Long horizontal and short vertical correlation lengths have been assigned to the Gaussian field. A correlation length of 7000 m with a standard deviation of 2000 m has been assigned to the major horizontal axis (r_1), reflecting an expected high level of continuity parallel to the paleoshoreline (Figure 10). A correlation length of 2000 m with a standard deviation of 1000 m was used parallel to the progradation direction and normal to the paleoshoreline (r_2). A vertical range (r_3) of only 1 m with a standard deviation of 0.5 m was used to reflect small-scale interfingering between facies tracts.

PARASEQUENCE 6

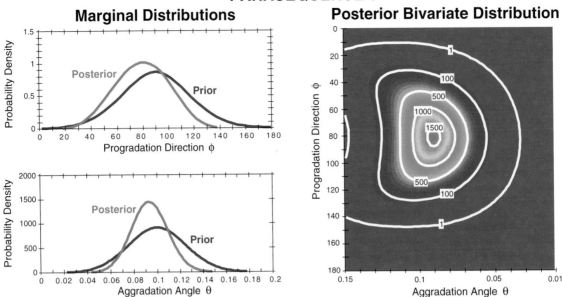

PARASEQUENCE 5

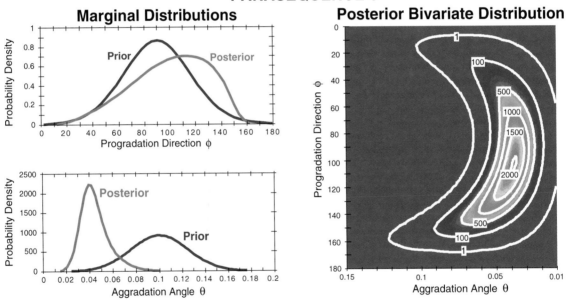

Figure 13. Probability distributions for ϕ and θ used in the simulation of the two Spring Canyon Member parasequences. The same prior distributions have been used for both parasequences. A von Mises distribution with mean = 090° and $\kappa = 5$ was used for the progradation direction, and a normal distribution with mean = 0.1 and $\sigma = 0.025$ was used for the aggradation angle. The contours of the bivariate posterior distributions refer to probability densities that define the relative probability of different ϕ, θ vectors. Note the inverted scale on the x-axis of the bivariate plots. An inverted scale is used because the distance of progradation is inversely proportional to the aggradation angle. The x-axis in this case becomes a pseudodistance axis (i.e., PS5 prograded relatively farther east than PS6).

Evaluation of Results

The range of ϕ and θ values used in the 50 simulations (Figure 14) provides a satisfactory representation of the posterior distributions (Figure 13). The simulations of PS5 are associated with generally lower aggradation angles and more variable progradation directions than those of PS6.

Two realizations of each parasequence are used to illustrate the simulated facies tract geometries. Three-dimensional models are presented in Figure 15 and two cross-sections through both parasequences are

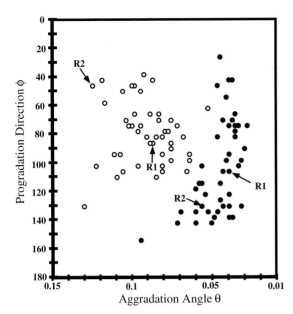

STATISTICAL SUMMARY

PROGRADATION DIRECTIONS, ϕ

PS5		**PS6**	
Mean:	105°	Mean:	078°
St. dev.:	32°	St.dev.:	21°
Minimum:	026°	Minimum:	038°
Maximum:	154°	Maximum:	132°

AGGRADATION ANGLES, θ

PS5		**PS6**	
Mean:	0.046°	Mean:	0.091°
St. dev.:	0.012°	St.dev.:	0.017°
Minimum:	0.028°	Mininum:	0.053°
Maximum:	0.095°	Maximum:	0.131°

Figure 14. ϕ,θ vectors used in the 100 test simulations. Closed circles = PS5, open circles = PS6. Compare this figure with the bivariate posterior distributions of Figure 13. R_1 and R_2 refer to the realizations illustrated in Figures 15 and 16.

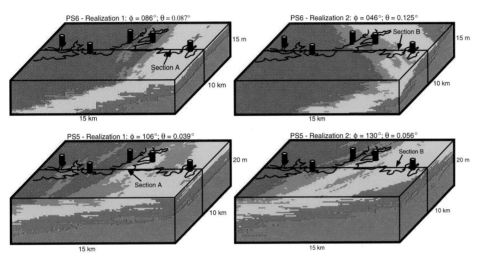

Figure 15. Two realizations of facies tract distribution in PS5 and PS6 simulated using different ϕ and θ values. The location of the four conditioning wells, the blind test well (Kennilworth), and the trace of the Book Cliffs outcrop (Figure 12) are illustrated on the realizations. The color code for the different facies tracts is the same as that used in Figure 12.

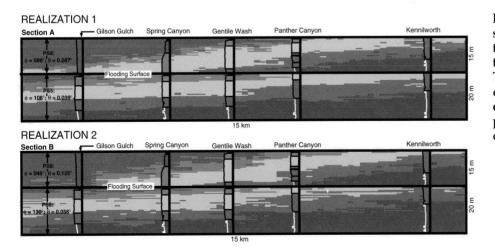

Figure 16. West-east cross sections through the realizations of facies tract distribution illustrated in Figure 15. The well data from the conditioning wells are projected onto the cross sections. The projection is normal to the cross section.

presented in Figure 16. Realization 1 for both parasequences is associated with an approximately easterly progradation; PS5 realization 2 prograded to the southwest and PS6 realization 2 to the northwest. Thus, the cross section through realization 2 in Figure

16 comprises two parasequences with approximately orthogonal progradation directions. Aggradation angles range from 0.039° for PS5 realization 1 to 0.125° for PS6 realization 2. Note the influence of θ on facies tract geometry: low angles lead to laterally con-

tinuous facies tracts, whereas higher angles lead to narrower facies tracts. The different values of φ and θ produce realistic facies tract geometries, demonstrating the considerable uncertainty associated with these parameters using data from just four wells.

The influence of the correlation length normal to the shoreline strike (r_2) is also illustrated in Figures 15 and 16. The PS5 realizations are associated with slightly longer r_2 correlation lengths, leading to the simulation of more continuous interfingering between facies tracts. The interfingering between lower shoreface (red) and offshore transition (blue) is particularly illustrative.

A large number of different maps can be generated in order to summarize the results of a series of stochastic simulations. In our synthetic example we have assumed that the option of drilling an appraisal well at the Kennilworth location is strongly dependent on the presence of good reservoir facies (foreshore, upper shoreface, and lower shoreface), and we have produced a series of maps illustrating probability of finding reservoir intervals (PS5 and PS6 combined) with net:gross ratios over particular threshold values (Figure 17). The first map illustrates the probability for a net:gross ratio greater than 0.48 which is the actual net:gross at the Kennilworth location. Net:gross ratios greater than 0.48 were predicted in 40% of the simulations, and less than 0.48 in the other 60%. If the economic success of an appraisal well was dependent on a net:gross greater than 0.48, the stochastic model predicts that a well at the Kennilworth location has a 40% chance of success. If the economics of the prospect require higher net:gross ratios, the Kennilworth location becomes unattractive (Figure 17B). Note that the probability for encountering net:gross values greater than 0.65 at Kennilworth is very low despite the fact that the two nearest wells (Panther Canyon and Gentile Wash) are associated with net:gross ratios greater than 0.65.

The simulated facies tract distribution at Kennilworth location is illustrated in Figure 18. In general, the stochastic model has been a good predictor of the depth to the base of the lower shoreface, and has produced reasonable estimates of the reservoir thicknesses. The simulated thickness of reservoir facies tracts (foreshore, upper shoreface, and lower shoreface) for PS5 has a mean value of 8.3 m and compares favorably with the measured value of 7.1 m.

The model for PS6, however, simulates slightly too thin reservoirs with an average value of 7.0 m, compared with a measured value of 10.1 m.

Although the overall thickness of reservoir facies is reasonably well predicted, the stochastic model has not been a good predictor of the relative thicknesses of lower and upper shoreface facies. Upper shoreface facies in both PS5 and PS6 are observed to pinch out a few hundred meters west of the Kennilworth location on the Book Cliffs outcrop, whereas the stochastic model predicts upper shoreface facies up to several meters thick at the Kennilworth location (Figure 18). This difference is a consequence of two factors. The first is that, for PS5, the aggradation angles used in the simulations are generally too low and, as a result, the upper shoreface facies prograde farther out into the basin. The second factor is related to thickness variations of the different facies tracts. The relative thicknesses of facies tracts (in the transformed coordinate system) changes from proximal to distal along the Book Cliffs outcrop from west to east. In particular, the thicknesses of the lower shoreface facies in both PS5 and PS6 increase distally. The use of a *linear* expectation trend implies that lateral thickness variations (between two thresholds) are limited (Figure 4), and the model cannot be applied to describe parasequences associated with systematic lateral thickness variation.

A similar limitation is also illustrated in the cross-sections (Figure 16), particularly at the projected Spring Canyon location in PS6. At the Spring Canyon outcrop, upper shoreface and foreshore facies in PS6 pass rapidly updip into flood tidal delta sands then into coastal plain mudstones over a distance of approximately 1 km (Van Wagoner et al., 1990, Figure 13). This rapid pinch-out is associated with a rather high-angle interface between facies tracts, which differs significantly from the aggradation angle observed elsewhere within PS6. The rapid pinch-out is difficult to capture using the stochastic model with a linear expectation trend. As a consequence, upper shoreface (yellow) and foreshore (green) facies tracts are simulated west of the Spring Canyon where they are not observed on the Book Cliffs outcrop. Modification to the stochastic model to account for systematic distal thickening and abrupt updip pinch-out is evidently necessary.

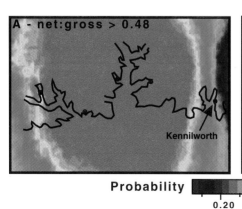
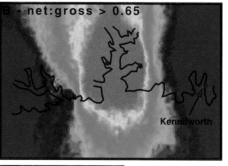

Figure 17. Net:gross probability maps. The trace of the Book Cliffs outcrop and the synthetic well locations (Figure 12) are superimposed on the maps which cover an area of 15 × 10 km.

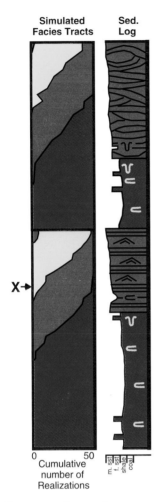

**Simulated
Facies Tracts   Sed.
Log**

X→

0   50
Cumulative
number of
Realizations

Figure 18. Sedimentary log and summary of simulated facies tract distribution at the Kennilworth location. The simulated facies tract log illustrates the proportion of different facies tracts simulated at different depths. For example, at depth x, the upper shoreface (yellow) was simulated in eight of the realizations, the lower shoreface (red) in 40 realizations, and the offshore transition (blue) in two realizations.

CONCLUSIONS

- A truncated Gaussian field with an expectation function defined by an arbitrary linear trend is a relatively simple and easy to use model for describing facies tract distribution in shoreface reservoirs. The model allows simulation of facies tract geometries that appear geologically reasonable.
- The most important parameters controlling simulated facies tract distributions are the progradation direction ϕ, aggradation angle θ, and spatial covariance function ($\sigma^2 \times \rho$). They govern the orientation of facies tracts, the distances over which the facies tracts can prograde, and the degree of interfingering.
- The magnitudes of these parameters cannot be known precisely and they typically are associated with significant uncertainty. It is fundamental to include this uncertainty in the modeling by defining

probability distributions for the parameters instead of using constant values.

- Classical statistical inference techniques often are inadequate for parameter estimation due to insufficient well observations.
- Bayesian inference provides an attractive approach for estimating parameters because it uses all available information: geologic interpretations, analog data, and well observations.
- The realistic estimation of aggradation angles θ and spatial covariance functions ($\sigma^2 \times \rho$) relies on an improved understanding of the geologic controls and the collection of empirical data from geologic analogs (e.g., outcrops).
- Although the prototype stochastic model can generate facies tract geometries that appear geologically reasonable, it cannot capture effects such as abrupt updip pinch-outs and systematic distal thickening of facies tracts. Modifications to the model are necessary to include such effects.
- Continual development and modification of stochastic models is to be expected because any model can only approximate and simplify a real phenomenon, and cannot represent all aspects of a particular phenomenon.
- Despite minor limitations, the modeling procedure is considered to have potential to assist the management of shoreface reservoirs where the distribution of facies tracts is associated with significant uncertainty.
- As has been illustrated by the Book Cliffs test example, the procedure also has potential application within exploration where it can provide a more rigorous basis for describing the probability of finding reservoirs within different structures, and thus improve the basis for identifying and risking prospects (e.g., White, 1993).
- A modeling procedure similar to that described in this paper may be applied to other depositional settings characterized by contemporaneous progradation and aggradation, such as alluvial fans, submarine fans, and tidal sand waves.

ACKNOWLEDGMENTS

The stochastic model that forms the basis for the procedure described in this paper was programmed by the Norwegian Computing Centre during 1990 as part of the Norwegian Government SPOR program. The model was originally a module within the Desiré package and is now an integral part of Statoil´s MOHERES stochastic modeling package. We thank Alistair Jones and Jim Doyle of BP and Anthony T. Buller of Statoil for comments on earlier drafts of the manuscript, and we also thank Statoil for permission to publish the paper.

REFERENCES

Box, G. E. P., and G. C. Tiao, 1992, Bayesian inference in statistical analysis: John Wiley and Sons, New York, 588 p.

Cross, T. A., M. R. Baker, M. A. Chapin, M. S. Clark, M. H. Gardner, M. S. Hanson, M. A. Lessenger, L. D. Little, K-J. McDonough, M. D. Sonnefeld, D. W. Valasek, M. R. Williams, and D. N. Witter, 1993, Application of high-resolution sequence stratigraphy to reservoir analysis, *in* R. Eschard and B. Doligez, eds., Subsurface reservoir characterization from Outcrop Observations: Editions Technip, Paris, p. 11–33.

Davis, J. C., 1986, Statistics and data analysis in geology: John Wiley and Sons, New York, 646 p.

Gómez-Hernández, J. J., and R. M. Srivastava, 1990, ISIM3D: an ANSI-C three-dimensional multiple indicator conditional simulation program: Computers and Geosciences, v. 16, p. 395–440.

Isaaks, E. H., and R. M. Srivastava, 1989, An introduction to applied geostatistics: Oxford University Press, New York, 561 p.

Journel, A. G., and C. J. Huifbregts, 1978, Mining geostatistics: Academic Press, London, 600 p.

MacDonald, A. C., T. H. Høye, P. Lowry, T. Jacobsen, F. O. Aasen, and A. O. Grindheim, 1992, Stochastic flow unit modeling of a North Sea coastal-deltaic reservoir: First Break, v. 10, p. 124–133.

Mardia, K. V., J. T. Kent, and J. M. Bibby, 1979, Multivariate analysis: Academic Press, London, 521 p.

Mitchum, R. M., Jr., and J. C. Van Wagoner, 1991, High-frequency sequences and their stacking patterns: sequence stratigraphic evidence of high-frequency eustatic cycles: Sedimentary Geology, v. 70, p. 131–160.

Omre, H., and K. B. Halvorsen, 1989, The Bayesian bridge between simple and universal kriging: Mathematical Geology, v. 21, p. 767–786.

O'Byrne, C. J., and S. Flint, 1993, High-resolution sequence stratigraphy of Cretaceous shallow marine sandstones, Book Cliffs outcrops, Utah, USA—application to reservoir modelling: First Break, v. 11, p. 445–459.

Posamentier, H. W., and P. R. Vail, 1988, Eustatic control on clastic deposition II—sequence and systems tract models, *in* K. Wilgus et al., eds., Sea level changes—an integrated approach: SEPM Special Publication no. 42, p. 125–154.

Rudkiewicz, J. L., D. Guérillot, A. Galli, and Group Heresi, 1990, An integrated software for stochastic modeling of reservoir lithology and property with an example from the Yorkshire Middle Jurassic, *in* A. T. Buller et al., eds., North Sea oil and gas reservoirs II: Graham and Trotman, London, p. 399–406.

Taylor, D. R., and R. W. W. Lovell, 1992, Recognition of high-frequency sequences in the Kennilworth Member, Blackhawk Formation, Book Cliffs, Utah, *in* J. C. Van Wagoner et al., eds., Sequence stratigraphy applications to shelf sandstone reservoirs: AAPG Field Conference Guide, p. 1–9.

Van Wagoner, J. C., R. M. Mitchum, K. M. Campion, and V. D. Rahmanian, 1990, Siliciclastic sequence stratigraphy in well logs, cores, and outcrops: AAPG Methods in Exploration Series, No. 7, 55 p.

White, D. A., 1993, Geologic risking guide for prospects and plays: AAPG Bulletin, v. 77, p. 2048-2061.

Chapter 10

Numerical Facies Modeling Combining Deterministic and Stochastic Methods

Andres S. Hatløy
Geomatic
Oslo, Norway

ABSTRACT

In the North Sea only a small number of wells can be economically justified before important and often irreversible reservoir development decisions are made. The incorrect production forecasts the oil industry experienced during the 1970s and 1980s to a large extent likely were due to the use of oversimplified (i.e., too homogeneous) reservoir models. It became evident that detailed reservoir descriptions incorporating different types of heterogeneities were needed to quantify the uncertainty and, possibly, reduce the economic risk.

Heterogeneities exist on all scales from cementation of pores to displacement of fault blocks and must be accounted for, in one way or another, when generating a detailed reservoir description. Reservoir modeling is often divided into four steps to reflect the different scales of heterogeneity. The first step is to build the structural framework of the reservoir. The second step is to model the facies architecture and the distribution of the different facies types. The third step is to describe the petrophysical properties, and the fourth step is homogenization and scale-up to generate a model that can be applied in fluid flow simulations.

This paper focuses on numerical facies modeling and describes the SESIMI-RA concept, which is an object-based method; however, the SESIMIRA approach is general and enables the user to combine deterministic and stochastic parameters. A case study demonstrates how a fluvial sequence has been modeled using a combination of empirical relations, probability distributions, and (geo)logical rules. The result is a set of realizations of the model in a computer format that can be used as input to petrophysical property modeling, reservoir communication studies, and fluid-flow simulations.

SESIMIRA has, during the last six years, been used with success to model different reservoir types. These models have been used to quantify the uncertainty in the facies architecture, as a basis for detailed models of petrophysical parameters, and to quality check sedimentological interpretations.

INTRODUCTION

During the last 10 years we have seen a tremendous increase in the performance/price ratio of computers in terms of both calculating power and speed of graphics. Development of advanced graphics and user-friendly, cheaper software has run in parallel with this evolution.

The oil business, at the same time, has experienced a decreasing oil price and unexpected reservoir performances. This has triggered a search for methods that can provide detailed reservoir descriptions and quantify the inevitable inherent uncertainty of such descriptions.

Use of oversimplified geological models based on the data from a limited number of widely spaced wells has been regarded as the main cause for failures in predicting field performances. A variety of methods for detailed stochastic modeling has thus been developed and gained popularity.

This chapter presents a concept that combines stochastic and deterministic methods in modeling the facies body distribution. A case study illustrates how this method has been used to model a fluvial reservoir.

NUMERICAL FACIES MODELING

Motivation

Numerical facies modeling is here defined as the process of obtaining, in a computer format, a detailed model of the distribution of the different facies types in a reservoir. The motivation for this type of modeling is that research and production history have shown that the flow-unit distribution strongly influences the fluid-flow behavior and, therefore, the need to account for the uncertainty in reservoir architecture when making reservoir management decisions.

The term "facies" is perhaps the most frequently abused geologic term, thus making it difficult to define precisely. In this chapter we use the term "facies" to mean any subdivision and grouping of geometric units believed to be present in a reservoir.

Bear in mind that the objective of this type of facies modeling is not to model how and when the different facies bodies were deposited, but rather where the different bodies are present today, their sizes, and how they relate to each other.

Studies and measurements of outcrops and cores show that the variability in the petrophysical properties is greater between bodies of different facies types than between facies bodies of the same type. The way the petrophysical properties varies (i.e., the correlation structure) is a function of the facies type. Thus, a proper facies model is needed to obtain a sound and detailed petrophysical model for reservoirs containing multiple facies types.

The process of generating a numerical facies model is as valuable as the numerical model itself because this process forces the geoscientists working on a project to quantify and formalize their knowledge, experience, and ideas. It is easier to draw some paleogeographic maps illustrating different depositional models than to define width, length, thickness ratios, orientation ranges, and volume fractions of the different facies types. The hardest task is to define a quantitative formalized model that honors the well data and is coherent with sedimentological principles.

The main objectives for establishing a numerical facies model can be summarized as follows:

- Generate a detailed facies body distribution model for input to rock property modeling and flow simulations
- Study sensitivities and the uncertainty in the reservoir architecture
- Perform detailed volume calculations (bulk volume per facies types, etc.)
- Investigate and analyze reservoir communication
- Generate a model that can quickly and conveniently be edited, altered, and updated as more information is collected
- Create several different models in a computer format, not only in the minds of geoscientists.

SESIMIRA Modeling Concept

Several facies modeling methods have emerged during the last 10 years as stochastic methods have gained popularity. An overview and introduction to stochastic modeling is given by Haldorsen and Damsleth (1990).

This paper describes the SESIMIRA modeling concept (SEdimentological SIMulation in IRAp; IRAP is a mapping system developed by Geomatic). This method is best classified as an object-modeling method or a marked-point process. A detailed description is given by Gundesø and Egeland (1990). Damsleth et al. (1992) described a two-stage method using SESIMIRA as the first stage to generate a facies model, and conditional simulation to model the petrophysical properties.

The SESIMIRA software system includes a high-level geological macro language enabling the user to combine or use either a deterministic or stochastic approach. The stochastic part of the SESIMIRA system is also often called a Boolean method and belongs to the class of discrete stochastic models. The basic and reasonable assumption is that the facies body distribution in a reservoir is discrete. This means that just one facies type is present at any location in the reservoir. The SESIMIRA concept also implies that mathematical and logical relations can be used to model how bodies of equal or different types are interrelated and located.

Object modeling may be regarded as an automatic three-dimensional drawing method. The investigator must draw (or insert) bodies of different shapes, sizes, and properties at different locations inside a volume.

For a fluvial reservoir, the problem, simply put, is to fill a volume (the reservoir) with shoestrings (the channels) of different size and orientation, and let some of them penetrate some lines (the well paths) at certain known locations (the well intervals where channels are found); however, this method requires that you know the location, size, and shape of every body in the reservoir.

This is difficult today due to large well spacing, limited seismic resolution, and the ambiguous nature of the available data. There exists more than one possible model of the reservoir. Accordingly, one must turn to stochastic numerical modeling and generate a family of plausible and possible models of the reservoir. Analyzing those will quantify the uncertainty in the flow unit architecture.

One must honor all nonambiguous deterministic information whenever that information is present. Stochastic modeling is not a substitute for conventional (deterministic) modeling, but rather a helpful tool for modeling the uncertain aspects of a reservoir. SESIMIRA has the possibility to combine deterministic and stochastic approaches. This ability enables the user to generate a stochastic reservoir model during the exploration stage, further refine the model, and include an increasing proportion of deterministic elements when moving from the appraisal to the production phase.

Numerical facies modeling does not replace old conventional methods. Instead, it requires additional and more detailed studies. A modeling project using the SESIMIRA concept requires contributions from an interdisciplinary team and is often divided into five stages:

1. Collecting and analyzing input data
2. Defining a quantitative model
3. Generating realizations of the numerical model
4. Analyzing results
5. Transferring realizations to flow simulation

Collecting and Analyzing Input Data

A prerequisite when defining a quantitative facies model is a thorough analysis of both hard and soft facies information. Hard information is measurable (deterministic) data regarded as unambiguous, whereas soft information is concepts, experience, and interpretations. The sources of hard information are mainly well data, whereas soft data commonly stem from studies of outcrops, seismic data, and well data from other reservoirs believed to be similar to the reservoir under investigation. For example, interpretation of cores and well logs gives you a picture of what kind of facies bodies are present along the well path, but gives you only a clue (a range) as to what may be the thickness and size of the bodies (Figure 1).

SESIMIRA requires four different types of input: structural data, well data, geometric data, and volume (length, thickness, width) data.

Structural data are depth maps of the top and bottom of the different reservoir zones; zonation is necessary only if the variation in the facies body distribution cannot be described using simple depth-related rules (described in a following section). Well data are the facies distribution in the wells simply as a facies type interval log per well. Geometric data can be further divided into two groups: geometric parameters and geometric relationships. Geometric parameters are parameters defining the size and shape of the different facies types. The parameters defining the length, thickness, and width can be given as distributions (histograms or predefined distribution types), mathematical functions and expressions, or constant numbers.

A generic shape for a facies type can be generated in several ways: by using a contour map, or a geometric

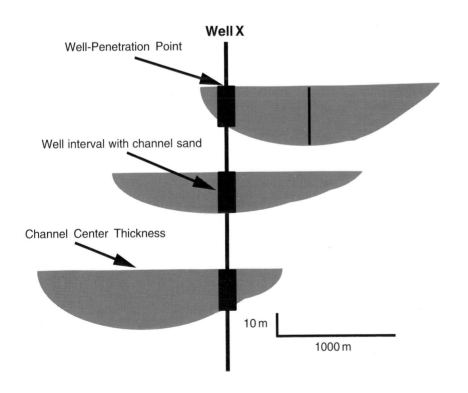

Well X

Well-Penetration Point

Well interval with channel sand

Channel Center Thickness

10 m

1000 m

Figure 1. Well-penetration point versus channel thickness. (See text for discussion.)

primitive (box, ellipsoid, half-ellipsoid, half-cylinder, etc.), or by digitizing the desired facies shape.

Boxes and ellipsoids may be looked upon as nongeological shapes, but bear in mind that the goal is to model what the reservoir looks like after deposition, burial, and postdepositional faulting. A degree of degeneration is inevitable when representing detailed geology in a three-dimensional grid cell model (also often referred to as a voxel model).

Relations and rules describing how different facies types are located and interrelated are harder to define. Locations of facies bodies are either independent of other facies bodies (e.g., calcite stringers) or are dependent on or connected to other facies bodies of the same or a different type (e.g., crevasse splays must be positioned along the banks of channels).

The location of independent bodies may be governed by distributions for the x, y, and z coordinates of the body's center point or by simple mathematical expressions. Random location within a zone is accomplished simply by finding a point from three draws from three uniform distributions defined by the minimum and maximum x, y, and z coordinates, respectively. Skewed distributions make it possible to place, for example, all calcite stringers at the bottom of the southern part of a zone. The powerful, high-level macro language of SESIMIRA enables the user to mix the distributions as required.

The distributions may be dependent or independent and one or two dimensional. Location of connected bodies is determined using logical and mathematical expressions such as, for example, connecting a mouth bar at the downstream end of a channel; connecting crevasse splays at drawn locations along the channel bank given that the number of splays does not exceed more than two per kilometer of channel length; connecting no splays if the channel thickness is less than 2.5 m.

Interrelationships also fall into two groups: relations between bodies of the same type and relations between bodies of different types. For both groups, the user must answer the same questions regarding whether the bodies avoid (repulse) each other, melt together, or erode other bodies. These questions are normally answered by generating a facies-types hierarchy defining which facies types can erode other types. For relations between bodies of the same type, rules saying, for instance, that thinner channels cannot erode thicker channels may be applied. If the channels branch, you have to define the probability of branching or the number of branches as a function of the channel length.

Volume data are the volume of each facies type as a fraction of the total reservoir bulk volume. These fractions can be determined explicitly for each facies type by calculating the average facies-to-gross ratio seen in the wells. Alternatively, these ratios can be determined implicitly, especially for fluvial models. A typical example is the number of crevasse splays that often are determined implicitly by defining the average number of splays per kilometer of channel length.

Defining a Quantitative Model

The result of the input data analysis is a set of surface grids defining the structural framework of the model; facies-type interval logs showing what kind of facies bodies the wells have penetrated; and probability distributions, mathematical functions, and rules describing the abundance, location, size, shape, and spatial relation of the different facies types.

The next step is to define a logical sequence (the facies model) that, by using the input data, produces the desired number of realizations of this facies model. This facies model must honor both the deterministic (hard) and the stochastic (soft) information so that all realizations will be equal in the well intervals, but vary in accordance with the applied stochastic model outside the wells.

The SESIMIRA macro language provides a flexible and efficient way of setting up a numerical facies modeling procedure. The procedure is frequently split into two parts as shown in the flow charts in Figures 2 and 3. The first part of the procedure is the modeling of the bodies found in the wells, the second part is the modeling of bodies believed to be present, but not encountered in the wells.

The volume fraction of each facies type is calculated after the first step as a quality assurance. If, at this stage, the volume fraction of a facies type exceeds the defined fraction, two possibilities exist: (1) the parameters regulating the size of the bodies have values that are too high or (2) the assumed volume fraction is too low. Advanced visualization features can help the user to quality check the modeling outcome (Figures 4–13).

Generating Realizations and the Numerical Model

A realization can be generated basically in two ways, either by executing the model procedure interactively or executing the model procedure in batch mode. The interactive mode displays the bodies as they are placed into the reservoir zone. In batch mode, the quality check of the realizations must be performed after they have been generated. The SESIMIRA macro language has possibilities for generating displays and panels that can be composed of sliders, scroll bars, pull-down menus, input fields, and selection buttons. This enables the expert user to set up standard model procedures for different depositional environments, and makes it easy for novice users.

Analyzing Results

The result of a SESIMIRA modeling is a set of three-dimensional (3-D) integer matrices of cells called realizations. Each cell has a number code telling which facies type occupies the volume represented by that cell. It is possible to analyze the realizations both quantitatively and qualitatively. Examining cross sections and horizontal slices is an efficient method used to compare and rank realizations. Three-dimensional displays give an instant feeling for the facies body distribution, but two-dimensional views must be used if one wants to study details (Figures 4–13).

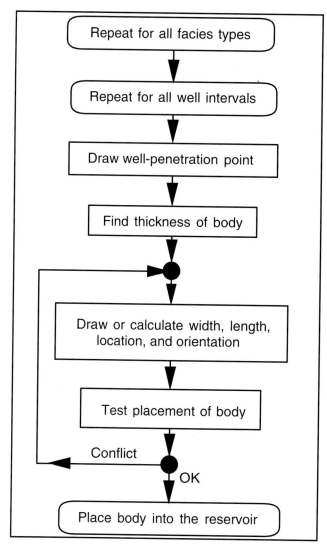

Figure 2. Modeling procedure for facies bodies found in wells.

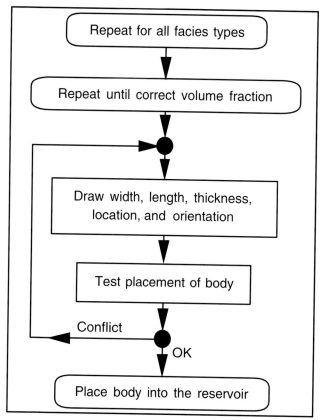

Figure 3. Modeling procedure for facies bodies believed to be present, but not seen in the wells.

that can be investigated are

- Connected volumes as function of the number of wells
- Connected volumes as function of the number of well patterns (e.g., by applying different sets of production well locations)
- Variation in the connected volumes in the different realizations

Transferring Realizations to Flow Simulation

A detailed numerical model cannot be transferred directly to a flow simulator due to the present computer technology and cost of using computers; therefore, models of the petrophysical properties and the facies distribution must be scaled up (i.e., decrease the number of cells and calculate average cell values). The problem is that the regridding and homogenizing techniques often neglect the information in the facies model.

A reservoir engineer is interested in a 3-D model of the porosity, permeability, and fluid saturation in the reservoir, but not the facies body distribution, although the variation in petrophysical properties depends on this distribution. The result is that the permeability and porosity contrasts due to different facies types may be lost completely or in the best case smeared out. Inherent in SESIMIRA is a regridding technique that generates a coarser simulation grid dependent on a dense grid representing a real-

Time and budget constraints often limit the number of realizations that can be used in flow simulation studies; therefore, a method for ranking realizations prior to petrophysical modeling often is needed. Volume calculations may be used to study the different realizations and quantify the differences, but also to find, for example, the connected sand volumes as a function of the number of wells, well patterns, or well drainage radius.

Reservoir Communication

Connected volume is the volume of those cells in contact within a given radius from a line representing a well. The criterion for horizontal connection is that at least one cell wall must be common. The criterion for vertical connection may be set by the user. The well positions may be given in a file, interactively, or drawn from distributions. Examples of sensitivities

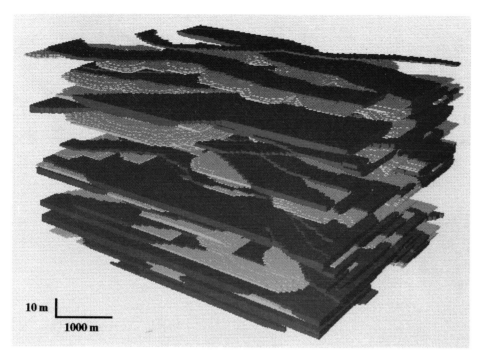

Figure 4. Three-dimensional view of one realization of the model of the channel belt and crevasse splay distribution. Red = channel belts, yellow = crevasse splays, and the background facies is not displayed.

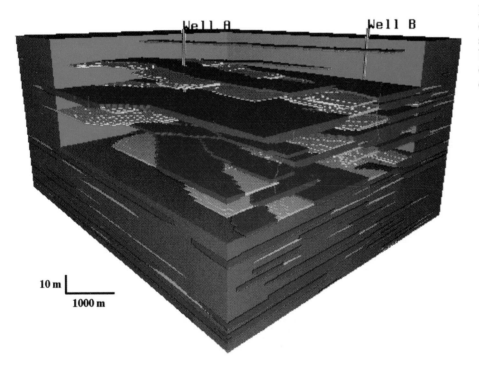

Figure 5. Chair mode display of the same realization as shown in Figure 4. Green = background facies, red = channel belts, and yellow = crevasse splays.

ization of a facies model. This new gridding technique is described in detail by Hove et al. (1992).

The technique is a two-step procedure and the input is two, 3-D grids, one defining the zone to be scaled up and one representing the flow unit distribution in that zone. The first step is to search through the zone and find how the different bodies are distributed. The result is a frequency diagram showing the distribution of flow units at different levels within the zones. The user must then, based on this information

and the desired vertical resolution in the coarse grid, select the location and the number of layer boundaries in the coarse grid.

The next step generates a set of layer boundaries at the chosen locations. The constraints are that no flow units are intersected by a boundary and that a boundary conforms to either the top or the bottom of the flow units at that particular stratigraphic level. The result is a unique flow simulation grid for each realization of the flow unit model.

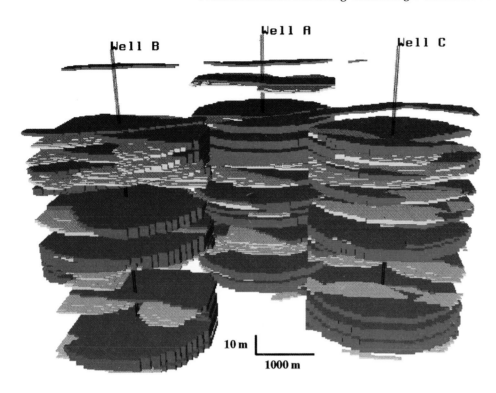

Figure 6. Three-dimensional view of the sand volumes within a radius of 1000 m from the three wells. Red = channel belts and yellow = crevasse splays.

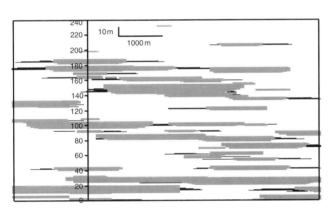

Figure 7. East-west cross section through well A. Grey = channel belts, black = crevasse splays, and white = background facies.

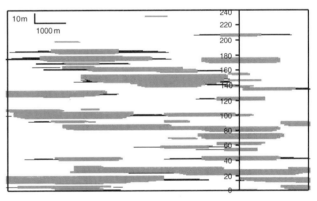

Figure 9. East-west cross section through well C. Grey = channel belts, black = crevasse splays, and white = background facies.

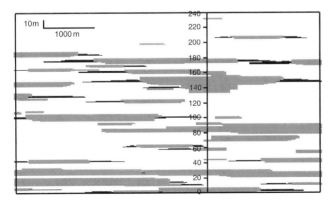

Figure 8. East-west cross section through well B. Grey = channel belts, black = crevasse splays, and white = background facies.

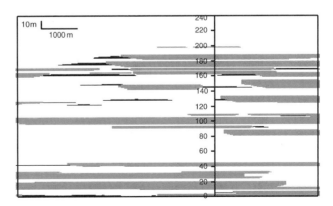

Figure 10. South-north cross section through well A. Grey = channel belts, black = crevasse splays, and white = background facies.

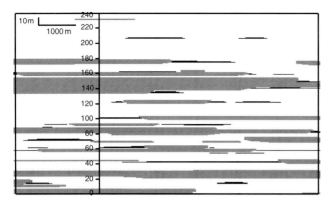

Figure 11. South-north cross section through well B. Grey = channel belts, black = crevasse splays, and white = background facies.

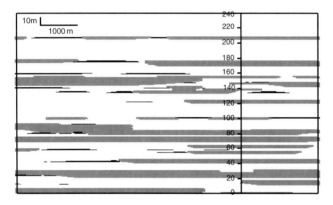

Figure 12. South-north cross section through well C. Grey = channel belts, black = crevasse splays, and white = background facies.

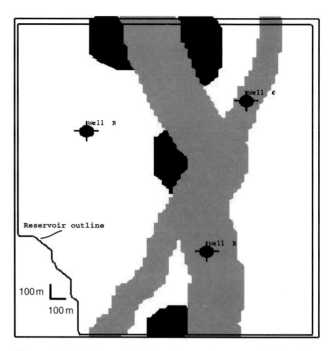

Figure 13. Horizontal slice through one model realization 140 m above the base of the model. Grey = channel belts, black = crevasse splays, and white = background facies.

CASE STUDY

The goal of this study was to establish a numerical facies model of part of a fluvial reservoir, and by generating several realizations of this model, investigate the reservoir communication and the influence different facies architectures have on flow simulations.

Structural Setting and Depositional Environment

The model area covers one fault compartment and is 4 km wide and 5 km long. The reservoir is made up of a series of fault blocks considered to be independent flow regimes. The unit under study is an approximately 200-m-thick sequence of fluvial sediments. The sequence was divided into three flow-unit types based on interpretation of well logs and studies of cores and analog outcrops. The three types of flow units are channel belts, crevasse splays, and a no-flow background matrix of silty mudstones, shales, and coal. The channel belts are believed to represent multistory channels of low to moderate sinuosity. Data from three wells inside the model area were available at the time of this study.

Model Size and Resolution

The 3-D outline of the numerical model was defined using the isochore map of the zone as the model boundary. This method gives an almost rectangular box that is 4 km wide, 5 km long, and 240 m thick; a horizontal view of the model area is shown in Figure 13. A cell width of 50 m, a length of 50 m, and a thickness of 0.5 m were used to achieve the resolution needed to model properly the smallest facies bodies believed to be present in the zone. This scale gives a total of approximately 4 million cells.

Facies Geometry and Distribution

Figures 14 and 15 summarize the most important facies geometry parameters used for the channel belts and the crevasse splays. The channel belts were modeled as half-cylinders of low to moderate sinuosity having a cross section with a flat base and top (Figure 14D). The belt length was regarded as infinite (i.e., none of the channel belts terminate inside the model area). The channel belt width was modeled as dependent on the thickness and depth of the belt. The applied formulas stem from studies of analogs (both in the field and in the literature), well correlations, and production tests. The channel belt orientation was drawn from a uniform distribution between northwest and northeast (Figure 14B).

A smoothed version of the frequency diagram of the channel belt thicknesses found in the wells (Figure 14C) was used as the thickness distribution for the channel belts not found in the wells. The channel belts are not

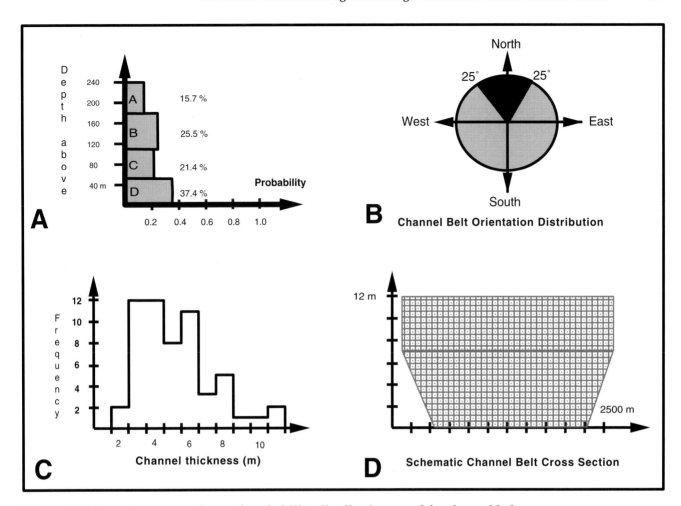

Figure 14. Schematic cross section and probability distributions used for channel belts.

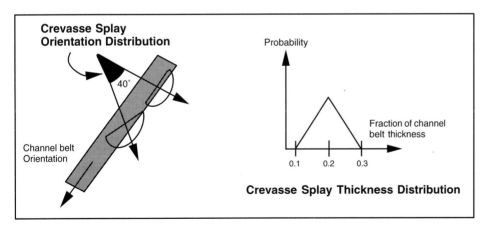

Figure 15. Probability distributions used for crevasse splays.

evenly distributed throughout the zone. Four depth intervals (labeled A–D in Figure 14A) were defined according to core descriptions and well logs from the three wells inside the compartment and other adjacent wells. The distribution given in Figure 14A was derived from the well data and used to draw the z coordinate (i.e., the depth) of the stochastic channel belts. A volume fraction of 0.36 was applied for the channel belts.

The crevasse splays were modeled as quarter-ellipsoids having the flat base nearest the channel belt bottom and the shortest plane side along the channel belt flank (Figure 15). The formulas for the length and width of the crevasse splays are given in Figure 16. The splay thickness was drawn from a triangular distribution given by 10%, 20%, and 30% of the channel belt thickness.

The orientation of the splays was assigned using a uniform distribution ranging from perpendicular to the channel belt orientation, to perpendicular –40° in the downstream direction (Figure 15). The splays

1. Minimum channel belt thickness: 2m

2. Channel belt width = $A \times$ Thickness[a]

3. Channel belts may erode all other facies types

4. The channel belt length is infinite

5. Channel belts thinner than 3m have no associated crevasse splays

6. Minimum crevasse splay width : 200m

7. Minimum crevasse splay thickness : 0.5m

8. Crevasse splay width = $B \times$ Thickness + 160

9. Crevasse splay length = $1.5 \times$ Width

10. Draw other geometrical parameters from distributions given in Figure 3 and 4

Figure 16. Applied modeling rules.

were distributed unevenly along the channel belt flanks until the volume fraction reached 0.08 of the total volume; however, no splays were connected to channel belts thinner than 3 m.

Applied Modeling Rules and Procedure

A two-step procedure, similar to that shown in Figures 2 and 3, was designed to generate realizations of the facies body distribution. The first step consists of modeling the channel belts found in the wells and their associated crevasse splays. The second step is generating the so-called stochastic channel belts, that is, channel belts believed to be present based on volume fraction distributions seen in the wells and in analog reservoirs and outcrops. This last step continues until the facies model is filled with the user-defined channel belt and splay volume fractions. The most important modeling rules are given in Figure 16.

In considering with the bodies in the wells, it is natural to start with what you know: the location. Although the well data show the facies body distribution along a line in reservoir space, they do not tell where the well penetrates the different facies bodies. Figure 1 illustrates the relationship between the channel belt thickness and the point on the channel belt at which well penetrates. This penetration point must be drawn, and from this point, channel belt center thickness calculated. Then, the width can be found by applying the formulas defined in the previous section, and the orientation by a draw from the distribution in Figure 14B.

At this stage, the channel-belt geometry is defined and the procedure must now test if the channel belt can be placed into the reservoir without coming into conflict with the well information. If the given set of drawn or calculated geometric attributes has created a channel belt that intersects any interval in any of the wells that is not made up of channel sand, the channel

belt is rejected or some of the attributes are modified. This sequence is repeated until all channel belts found in the wells have been modeled. The same principles are used for the crevasse splays (see the schematic overview of the modeling procedure in Figure 2).

For the stochastic channel belts, neither the thickness nor the location is known; therefore the thickness and the location must be drawn from the predefined probability distribution. Other than these two items, the procedure is the same as for the first step (Figure 2). The channel-belt volume fraction is calculated after each channel belt has been placed into the numerical model to test whether the modeling shall terminate or continue to generate additional channel belts.

Results

Facies Body Distribution

Figures 4–6 show three different 3-D views of one of the realizations. The channel belts are gray (red in 3-D displays), the crevasse splays are black (yellow in 3-D displays), and the background facies is pale gray (green in Figure 5). The background facies is not displayed in Figures 4–6 to better visualize the 3-D distribution of the sand bodies. Figures 7–9 show three east-west cross sections through the three wells. Figures 10–12 are three south-north cross sections through the same three wells.

The white areas inside the frames in those figures represent the nonpay facies. As expected and desired, Figures 4–13 show a model of a fluvial sequence of north-south–trending channel belts of low to moderate sinuosity with some associated crevasse splays. The figures also show that the majority of the channel belts is concentrated at certain depths as specified in the input.

Figure 13 is a horizontal slice through one realization and shows two intersecting channel belts, each intersecting one of the wells. Most deterministic modeling methods would have correlated a single channel-belt intersecting well A and well B. Figure 13 is a good example of how stochastic methods are capable of generating a range of possible reservoir descriptions, whereas deterministic approaches give just one outcome.

Reservoir Communication

Figure 6 illustrates how connected volumes are calculated; the red and yellow cells represent those parts of the channel belts and the crevasse splays that are within the drainage radius of wells A, B, and C. Such calculations can be used to analyze how interconnected the sand bodies are, and to indicate how many wells are needed to drain a certain percentage of the pay sands.

Table 1 gives the result of a set of calculations of the connected sand volume. Five different realizations were investigated and the analysis consists of six cases, each applying a different set of locations for four vertical production wells. Each well position was drawn randomly within the model area and a

Table 1. Results from an analysis of the connected sand volumes in five different realizations for six different well patterns*

	Realization number				
	1	2	3	4	5
Well pattern A	26.2	33.1	29.6	27.3	30.0
Well pattern B	20.0	30.8	30.1	25.8	21.6
Well pattern C	33.3	32.3	36.6	32.6	33.6
Well pattern D	33.5	41.4	37.2	36.2	35.5
Well pattern E	33.6	28.2	30.6	28.6	33.0
Well pattern F	27.7	29.6	32.2	33.6	31.5

*The numbers show the percentage of the total sand volume (channel belts + crevasse splays) that theoretically can be drained from four vertical wells all having a drainage radius of 1000 m. The four well locations were chosen randomly within the reservoir border and the volumes were calculated for six outcomes of the 4-well pattern.

drainage radius of 1000 m was used for all wells. The sand volume is the volume of the channel belts and the crevasse splays. The percentage of the sand volume in contact with the four wells ranges from 20.0 to 41.4% for the 30 different calculations (Table 1). The average connected volume is 32.6%.

Use of Realizations in Fluid-Flow Simulations

Fluid-flow simulations using some of the realizations generated during this study are planned, but unfortunately were not completed at the time of writing. A detailed numerical facies model (made of several million cells) cannot readily be transferred to a flow simulator. Hove et al. (1992) described how a SESIMIRA model had been applied in a flow simulation study and how one simulation grid was generated for each facies realization to honor the varying facies body distribution.

DISCUSSION

Facies Geometry and Distribution

The facies body distribution in a SESIMIRA model is the computerized version of a geoscientist's formalized and quantitative sedimentological description. This realization implies that "garbage in" gives "garbage out," but also that a geoscientist is able to study and test different interpretations.

It is very important to test and compare different interpretations because many interpretations are possible because you often have few observations and, thus, large uncertainty. Use of 3-D modeling methods like SESIMIRA will quickly rule out some interpretations that do not give realistic 3-D models, but that are not in conflict with core descriptions and well log interpretations.

Reservoir Communication

The results from the analysis of the connected sand volumes in the different realizations for the dif-

ferent well patterns should not be considered as absolute; however, it is apparent that the connectivity in this reservoir is good. The figures are purely theoretical estimates and are included only to demonstrate this feature of the SESIMIRA concept. This feature enables the user to analyze and quantify the variability between different realizations of a SESIMIRA facies model.

Detailed reservoir communication studies must include a model of the petrophysical properties, heterogeneities within the channel belts, fault distribution, degree of fault sealing, fluid contacts, and recovery factors.

CONCLUSIONS

A numerical facies modeling concept called SESIMIRA has been presented. This concept enables the user to apply either a deterministic or stochastic approach, or combine them to model the distribution of the facies bodies in a reservoir. The result is a set of 3-D matrices of cells each having a code representing the facies type. These realizations can be used as input to petrophysical parameter modeling systems, reservoir communication analyses, and fluid-flow simulations.

A case study of a fluvial reservoir demonstrates SESIMIRA's ability to model complex spatial relationships between facies bodies of different facies types as seen in Figures 4–13. The case model consists of three flow-unit types: (1) channel belts and (2) crevasse splays embedded in a (3) no-flow background matrix of silt, shale, and coal. A combination of empirical equations, (geo)logical rules, and probability distributions was used to obtain the desired spatial distribution of the facies bodies. The variability between different realizations of the model was studied by visually inspecting and calculating the difference in connected sand volumes.

Stochastic modeling methods do not reduce the uncertainty in reservoir architecture models, but in contrast to deterministic methods, they give the geoscientists the possibility to quantify this uncertainty.

ACKNOWLEDGMENTS

I thank the management of Geomatic a.s. for permission to publish this paper, to my colleague Olav Egeland (the inventor of the SESIMIRA concept in cooperation with Norsk Hydro) for valuable suggestions and comments, and to the geoscientists at Saga Petroleum, Norsk Hydro, and Mobil MEPTEC in Dallas, Texas, with whom I have cooperated on numerical facies modeling projects, and who provided the background for this chapter.

REFERENCES

Damsleth, E., et al., 1992, A two-stage stochastic model applied to a North Sea reservoir: Journal of Petroleum Technology, April 1992, v. 44, p. 402-408.

Gundesø, R., and O. Egeland, 1990, SESIMIRA—a new geological tool for 3-D modeling of heterogeneous reservoirs, in A. T. Buller and P. R. King, eds., North Sea oil and gas reservoirs II: Oxford University Press.

Haldorsen, H. H., and E. Damsleth, 1990, Stochastic modeling: Journal of Petroleum Technology, April 1990, v. 42, p. 404-412.

Hove, K., G. Olsen, S. Nilsson, M. Tønnesen, and A. Hatløy, 1992, From stochastic geological description to production forecasting in heterogeneous layered reservoirs: SPE paper 24890, presented at the 67th Annual SPE Technical Conference and Exhibition, Washington, D.C., Oct. 1992.

Chapter 11

Geostatistical Analysis of Oil Production and Potential Using Indicator Kriging

Michael Edward Hohn
Ronald R. McDowell
West Virginia Geological and Economic Survey
Morgantown, West Virginia, U.S.A.

ABSTRACT

Heterogeneous reservoirs lead to serious discontinuities in production from well to well, whether caused by sedimentological or structural factors. Geostatistical methods, such as ordinary kriging, are useful both for assessing degree of heterogeneity and for drawing maps of production and initial potential. Because of small-scale spatial variability in production from heterogeneous reservoirs, conventional contour maps suffer from numerous closed contours and ill-defined trends, whereas maps of kriged estimates suffer from a lack of fine details because of the smoothing property of kriging. Indicator kriging uses multiple transforms based on selected cutoffs to model spatial variability. For each indicator variable, the practitioner computes variograms, fits a model, and computes kriged estimates representing the probability of exceeding the cutoff. Applied to cumulative production and initial potential of oil from wells in a Lower Mississippian sandstone in West Virginia, indicator kriging with three cutoffs showed the presence of distinct north-south trends and a major east-west discontinuity in production and potential. These trends could be related to similar anisotropies in the presence of interbedded shales and high-density zones within the reservoir, north-south oriented basement faults, and an east-west structural discontinuity.

INTRODUCTION

Uppermost sandstones in the Price Formation (Lower Mississippian) (Figure 1) have been prolific sources of oil and gas production in the Appalachian basin since the 1880s (Ruley, 1970). Granny Creek field is located in central West Virginia (Figure 2), and produced an estimated 6.5 million bbl of oil between 1924, its discovery date, and the 1960s (Whieldon and Eckard, 1963). Drillers' experiences in this and other fields producing from these sandstones have led to the conclusion that production comes from high-

porosity zones of limited stratigraphic and geographic extent (Haught, 1959); this heterogeneity means that offsetting highly productive wells often results in wells having mediocre to low oil production.

In this chapter, we describe geostatistical methods for analyzing initial potential and cumulative production data with the primary objective of assessing the degree and nature of heterogeneity in production. A secondary goal is identifying geologic factors that account for observed trends in oil production within this field. Geostatistical tools were used for illustrating continuity and discontinuity in oil production at

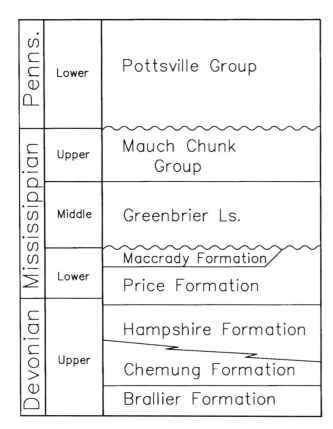

Figure 1. Generalized geologic column for the study area in central West Virginia.

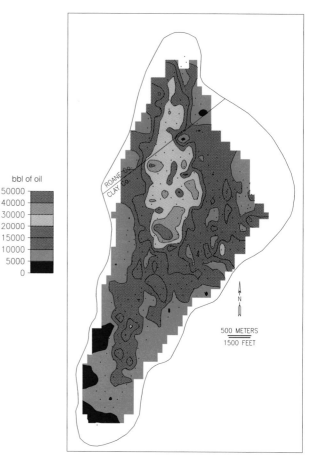

Figure 3. Contour map of cumulative oil production from uppermost Price Formation sandstone during the first 10 yr of each well in Granny Creek field. Units are barrels of oil.

Figure 2. Location of Granny Creek field in West Virginia.

scales up to field size and for mapping initial potentials and cumulative production. Indicator kriging was used specifically because data were not distributed normally, and conventional contour maps tended to have numerous closed contours around single wells (Figure 3); ordinary kriging produced smoothed

maps lacking small-scale features and discontinuities. Maps of indicator variables exhibited fewer closed contours, yet preserved small-scale discontinuities.

For regionalized variables, such as cumulative production, the indicator approach provides a suite of maps, each representing the probability that observed cumulative production exceeds a given cutoff. Maps of probabilities at low cutoffs, such as the 25th percentile or the median, tend to exhibit broad regional patterns in production, whereas maps corresponding to higher cutoffs tend to connect the most productive wells into one or more hot spots, and spatial discontinuities in oil production become clear. Thus, indicator kriging with multiple cutoffs satisfies the geologists' need for maps defining both regional trends and local hot spots.

Indicator kriging is a useful method for handling data that are not distributed normally, contain outliers, or might be a mixture of populations. Linear estimation methods, such as kriging, tend to smooth data in the process of generating point estimates, particularly when nugget effects are substantial. In fact, one can correctly describe kriging as a smoothing or moving average technique. This property of kriging makes it optimal in a least-squares sense, and makes linear estimation an excellent method for estimating economic analysis; however, maps of kriged estimates actually may have fine details smoothed away, and the geologist wishing to use such maps can interpret only the broader trends. If the target of the analysis is reservoir heterogeneity, maps of ordinary kriged estimates have little usefulness.

When data are nonnormal or outliers are present, linear estimation either creates a number of isolated highs if the spatial continuity of the highs values is low, or, where these values lie in poorly sampled areas, tends to produce large areas of overly optimistic estimates. This is a problem with almost any interpolation method, such as the variants of inverse distance. In some cases, the upper tail of the observed distribution represents a population different from the bulk of observations, which represent a background to the hot spots. Spatial continuity can differ between the background values and these hot spots. Consider an oil field: field limits may be well defined by the trapping mechanism, and so, at a low level, oil volume exhibits a high degree of spatial continuity, i.e., there is a clear distinction between field and non-field areas. Within the field, at average values of oil volume, spatial continuity might be low because of almost random variations in porosity and permeability across the field. Hot spots, resulting from discrete porosity trends, yield high spatial correlations that should be evident on variograms. Kriged estimates for a suite of cutoffs can be used both for observing trends at a number of levels and spatial scales, and for generating estimates that are robust to nonnormality and presence of outliers (Journel, 1983).

The next section describes the sources of production and potential data; how these data were modified before analysis; the source of porosity, thickness, and water saturation information; and how original oil in place was calculated. The methods section describes indicator kriging as we used it. Following the results section, we briefly describe possible geologic controls.

DATA

Initial potentials for oil were obtained from public completion records and from records provided by companies operating within Granny Creek field. Analysis was limited to primary production obtained before the period of secondary recovery by water flooding in the 1960s; commingled oil volumes were excluded, leaving data on 301 wells.

Annual and monthly production data were obtained from operators. Missing annual volumes were estimated from available data for some wells. Another group of wells was reported by year on a lease basis; annual production on a per-well basis was estimated from the date production began for each well, and from monthly per-well production data available for 1940–1949. Most of these wells were drilled in the 1920s and 1930s. No commingled production was used, and wells were limited to those

completed before 1960. Final maps are based on 313 wells.

Porosity was calculated from gamma-ray and density logs digitized every 0.25 ft (0.8 m), using the following formula for reservoir porosity (ϕ):

$$(\phi) = (\rho ma - \rho_{\log}) / (\rho ma - \rho_{H_2O})$$

where ρ_{ma} = "typical" rock density for reservoir (2.68), $\rho_{\log}$ = bulk density log value, and ρ_{H_2O} = density of formation water (1.00). Calculated porosities were computed over the lower, fine-grained part of the sandstone, the portion from which oil shows were most frequently observed by drillers; most wells were completed only in this interval.

Reservoir thickness was determined from gamma-ray and density logs. Water saturation was computed for the productive interval of 20 wells, using the equation for the formation factor (F):

$$(F) = 0.81 / \phi^2$$

and the equation for water saturation (S_w):

$$(S_w) = \sqrt{(F \times R_w) / R_{\log}} = \sqrt{(0.81 \times R_w) / (R_{\log} \times \phi^2)}$$

where ϕ = porosity, R_w = formation water resistivity (0.05), and $R_{\log}$ = deep induction log resistivity. Water saturations were averaged over the productive interval and estimates were computed by ordinary kriging over the extent of the field. The resulting grid of estimates was then sampled at each of the 698 well locations in the field.

At each of these locations, original oil in place (OOIP) was estimated using

$$(OOIP) = [A \times H \times \phi \times (1 - S_w)] / B_o$$

where A = effective area drained by well(s), H = reservoir thickness, ϕ = porosity, S_w = water saturation, and B_o = formation volume factor at reservoir pressure and temperature (1.113).

Because the drainage radius was not known, a value was chosen such that recovery efficiency would average about 18%, which is the overall value for this field when computed from data given by Whieldon and Eckard (1963).

Generally, wells having both gamma-ray and density logs and those represented by production data were different sets; therefore, to compute recovery efficiency, original oil in place was interpolated to a grid and then sampled at locations of wells for which production data were available. Full production histories were not available for many wells. Recovery efficiency was estimated for each well as the ratio of 10-yr cumulative production to original oil in place, multiplied by 100%. Different values for drainage radius affect the scale of values obtained for original oil in place and efficiency of recovery, but leave geographic patterns in these parameters unchanged.

METHODS

The term $z(x)$ is the value of a regionalized variable at location x, z_c is a cutoff value of the regionalized variable, and $i(x;z_c)$ is the indicator variable calculated through the indicator transform:

$$i(x;z_c) = 1 \text{ if } z(x) < z_c$$
$$\text{else} = 0.$$

The indicator transformation converts the original, continuous variable into one that takes on the values of 0 and 1. For each indicator variable created, directional variograms were calculated and modeled with anisotropic variograms.

Estimated probability that the regionalized variable exceeds a given cutoff at a location equals:

$$1 - P^*(x;z_c) = 1 - \Sigma\lambda_k \, i(x_k;z_c)$$

where k is the number of wells in the neighborhood of location x. Weights λ_k were calculated by ordinary kriging. For a subset of locations on a grid, problems in order relations occurred, that is

$$P^*(x;z_i) < P^*(x;z_{i+1})$$

where cutoff z_i is less than cutoff z_{i+1}. This relationship is inconsistent with the definition of the indicator variables because the probability of exceeding the higher cutoff can never exceed that of exceeding the lower cutoff. Order relations were checked for all grid cells for pairs of cutoffs, and resolved using the procedure of Journel (1987) as implemented in the GSLIB package of geostatistical software (Deutsch and Journel, 1992).

Further details regarding indicator kriging are presented in Journel (1983) and Hohn (1988), including use of the method for generating robust estimates of regionalized variables. Isaaks and Srivastava (1989) described a streamlined variant of indicator kriging for robust estimation, which they called median indicator kriging.

RESULTS

Cutoffs chosen in computing indicator variables were 12,335, 18,244, and 25,000 bbl of oil, corresponding to the 50th, 75th, and 90th percentiles, respectively. Directional variograms reflect the spatial anisotropy at all cutoffs and, in particular, at the highest percentile (Figure 4). Using variograms models illustrated in Figure 4, kriged maps of the three indicator variables show a number of north-south trends. At the lowest cutoff (Figure 5), three trends are apparent: a broad swath of elevated probabilities in the north-western part of the field (point A, Figure 5); point B (Figure 5) to the east; and a narrow band of elevated probabilities in the southern part of the field (point C, Figure 5). Localized highs (point D, Figure 5) are separated from the main north-south trend by a west–southwest to east–northeast band of depressed probabilities.

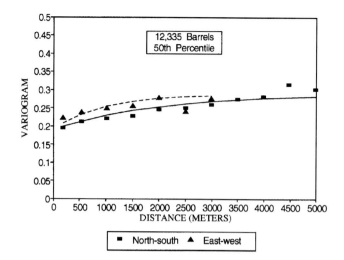

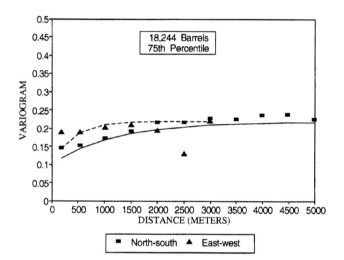

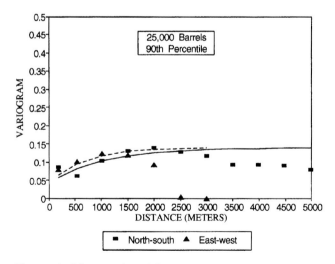

Figure 4. Observed and fitted variograms for three indicator variables calculated from oil cumulative production data.

At the 75th percentile, the broad swath of high probabilities remains (Figure 6), but all other trends appear discontinuous. The contrast between the

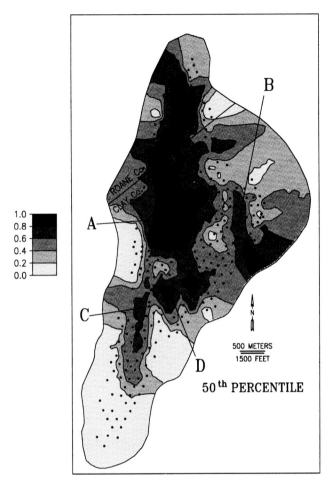

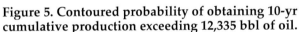

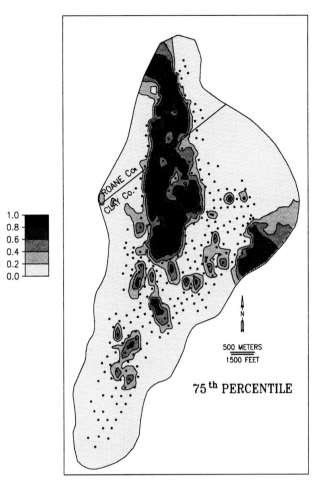

Figure 5. Contoured probability of obtaining 10-yr cumulative production exceeding 12,335 bbl of oil.

Figure 6. Contoured probability of obtaining 10-yr cumulative production exceeding 18,244 bbl of oil.

northern and southern halves of the field becomes most apparent at higher cutoffs; no wells produced more than 25,000 bbl of oil in the southern portion of the field (Figure 7).

Three cutoffs were useful in the study of spatial variation in initial potential: 10, 15, and 25 bbl of oil per day (BOPD), corresponding to the 25th, 50th, and 75th percentiles, respectively. Variograms for all three cutoffs reflected the strong north-south anisotropy at lags beyond 2 km (Figure 8). Isotropic variograms were used because average well spacing is small relative to the 2-km lag at which consistent anisotropy is observed.

The map of the 25th percentile (Figure 9) highlights areas with the poorest initial well performance, including the northwestern part of the field where the reservoir gradually thins to nothing, and north-south bands that outline the narrow fairway in the southern part of the field. Probabilities associated with the median initial potential (Figure 10) and the 75th percentile (Figure 11) follow trends very similar to oil in place (Figure 12) for the northern half. Compared with the corresponding maps for cumulative production (Figures 5, 6), initial potentials vary considerably from well to well, even on probability

maps. Initial potentials are not measured consistently among companies or over time, and because flow is measured during a short interval of time, measurement error can be high. Nevertheless, general trends of initial potential resemble trends in cumulative production, with some exceptions. For instance, feature B on Figure 5 does not show up at all on the maps of initial potential. There appear to be fewer north-south features on the map of initial potential, largely because the probability low between features A and B on Figure 5 is not obvious. The support for maps of cumulative production and initial potential are not the same, so some differences between the maps is expected.

Calculated recovery efficiency (Figure 13) averages 18%, but ranges between less than 10% to over 60%. Based on maps of recovery efficiency, oil in place, and cumulative production, the proportion of oil in place recovered correlates more with 10-yr cumulative production than with calculated oil in place. Some trends in recovery efficiency show up very poorly if at all on the map of cumulative production (Figure 3), but appear on the map showing the probability of exceeding the median value of 12,335 bbl of oil (Figure 5), namely, features C and D. In other

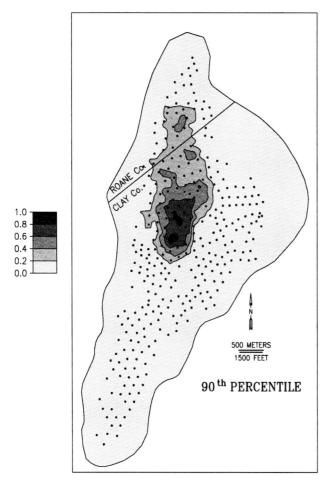

Figure 7. Contoured probability of obtaining 10-yr cumulative production exceeding 25,000 bbl of oil.

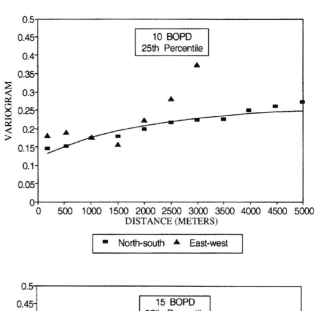

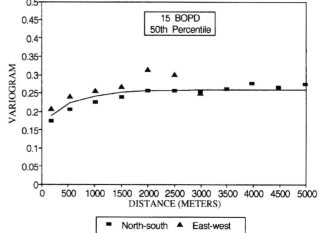

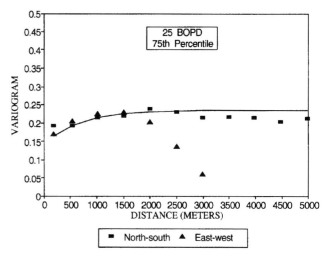

Figure 8. Variograms and fitted models of three indicator variables calculated from oil initial potential.

places, recovery efficiency is low, yet the probability of above-median wells is high, such as in the area of feature B (Figure 5).

GEOLOGIC CONTROLS ON PRODUCTION

Granny Creek field lies within a northeast-plunging syncline surrounded to the west, south, and southeast by gas fields producing from the same interval (Cardwell and Avary, 1982). Although structure contours run north-south in the northern part of the field, trends in cumulative production and initial potential cut across contours in the southern portion. Trends within the field would be difficult to explain purely in terms of structure; nevertheless, the western extent of the field might be limited by a gas cap, and the northeast extent is definitely limited by erosional thinning of the reservoir in that direction.

The uppermost Price Formation comprises two sandstones representing different depositional environments within a fluvially-dominated prograding delta: an upper, coarse sandstone deposited as channels, and a fine-grained sandstone deposited as dis-

tributary mouth bars that form the reservoir. These sandstones are overlain unconformably by limestone of the Greenbrier Group (Middle Mississippian). Pre-Greenbrier erosion completely removed the

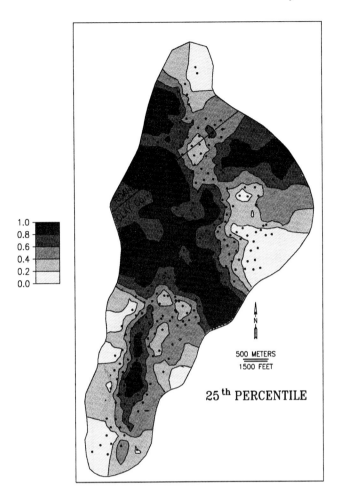

Figure 9. Contoured probability that initial potential of oil exceeds 10 bbl of oil per day.

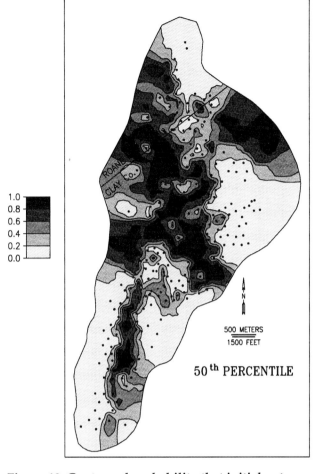

Figure 10. Contoured probability that initial potential of oil exceeds 15 bbl of oil per day.

upper unit in the northeastern part of Granny Creek field, and parts of the lower, fine-grained sandstone. Thinning of the reservoir rock accounts for a decrease in calculated original oil in place to the northeast (Figure 12). A north-south trend in oil in place exceeding 80,000 bbl occurs on the west side of the field. In general, the northern one-half of the field has the largest areas with relatively high expected quantities of oil; thickness of the reservoir sandstone here ranges between 25 and 30 ft (8 and 9 m) except where it thins to the northeast. In the southern one-half of the field, the lower sandstone is locally thickened up to 50 ft (15 m), but higher water saturations and lower average porosity lead to lower oil volumes. These trends generally correspond to trends in cumulative production, but details in the spatial variation in cumulative production are hard to relate to calculated oil in place.

The main trend of oil production is only partially related to reservoir thickness, porosity, and water saturation, and therefore to inferred oil in place. The north-south trend of high production corresponds to a region of high amounts of oil in place. Adjacent areas have lower calculated volumes of oil in place with few trends, in contrast with probability maps of

cumulative production that show distinct, narrow north-south trends.

Lithologic and diagenetic features within the reservoir sandstone may account for much of the small-scale variability in oil production. Shale, siltstones, and high-density zones determined from geophysical logs (Figure 14) to be within the reservoir sandstone tend to occur in a north-south band coinciding with the region of highest observed oil accumulation. Based on log and core data, these features represent lenticular zones up to 12 ft (4 m) thick of extensive carbonate cement (A. Vargo, 1993, personal communication) generally oriented along bedding, but in some cases inclined relative to the base of the reservoir. In the eastern part of the field, siltstone or shale interbeds were observed near the base of the sandstone; the locations of these breaks describe a north-south trend coinciding with the area of reduced probabilities separating features A and B on Figure 5. Another group of shales were observed near the base of the reservoir sandstone in the southern part of the field, again roughly describing a trend similar to that in production.

How could these spatial patterns in shales, siltstones, and diagenetic features affect production?

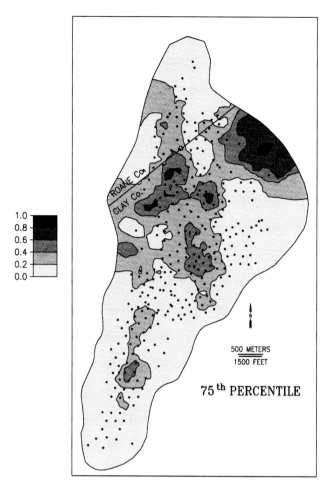

Figure 11. Contoured probability that initial potential of oil exceeds 25 bbl of oil per day.

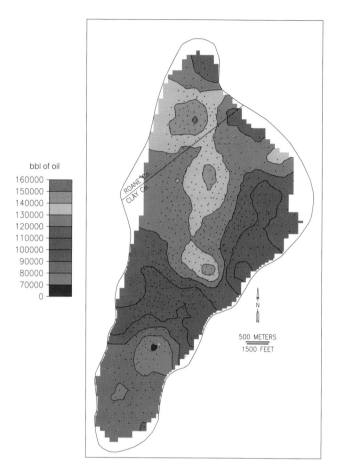

Figure 12. Calculated per-well oil in place for uppermost Price Formation sandstones within Granny Creek field. Units are bbl of oil.

Together, they support the conceptual model of the reservoir as a prograding deltaic sequence comprised of several lenticular units oriented north-south, offlapping from east to west, the source of sediments being to the east or northeast (Zou and Donaldson, 1992). Maps of average reservoir porosity describe north-south trends, although with many small-scale east-west trends. Permeability is statistically correlated with porosity and might be expected to vary along north-south trends; therefore, a feature-for-feature connection between oil production and sedimentology may be difficult to make. Clearly, depositional or diagenetic features play a major role in controlling accumulations of oil in the field.

Within the field, a contribution from structure is not ruled out. Sedimentological features do not explain the sharp contrast between the northern and southern halves of the field, particularly the discontinuity in production separating areas C and D from A to the north (Figure 5). Zheng et al. (1993) described several north-south basement faults in both the northern and southern halves of the field that coincide or lie adjacent to the areas of highest production. These faults are offset in the northern one-half to the east by a zone of east-west faults occurring in this area of discontinuity in oil production and potential.

SUMMARY OF INDICATOR KRIGING IN PRACTICE

Indicator kriging is useful for displaying production data from heterogeneous reservoirs. A recommended procedure for using the method is to begin by examining a histogram of the raw data and determining at least the median, 25th, 75th, and 90th or 95th percentiles for use as cutoffs. The median can be considered a "typical" value, such that one can speak of a typical well with respect to production. The high percentiles highlight the unusual, presumably very profitable wells. At high cutoffs, the small number of wells exceeding the value means that variograms become irregular and model fitting becomes difficult.

For each cutoff selected, perform the indicator transform and compute directional variograms. Here, one can learn about the presence and degree of directional anisotropies as reflected in the different ranges, relative degree of discontinuity at small scales as reflected in the apparent Y intercept (nugget effect), and changes in spatial continuity with increased cutoff value.

The final step is to use the variogram models and kriging to calculate the probability surfaces, check and correct order relations, and plot maps. Products may

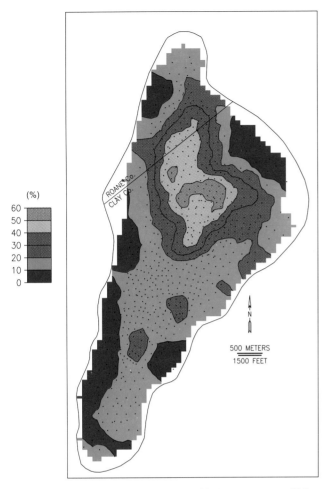

Figure 13. Contoured recovery efficiency per well in Granny Creek field. Units are percentages.

be used for economic assessments, for siting new wells in favorable areas, and for comparison with geologic maps. Probabilities calculated for low cutoff values can be subtracted from one to give maps of the probability of falling below a cutoff; such a map highlights areas of problem wells to be avoided in future drilling.

ACKNOWLEDGMENTS

We thank D. Matchen and A. Vargo for providing data and maps, and for critical review of the first draft of this chapter. J. Donovan and R. L. Chambers also deserve our gratitude for their reviews. Research was supported by the U. S. Department of Energy under contract DE-AC22-90BC14657.

REFERENCES

Cardwell, D., and K. L. Avary, 1982, Oil and gas fields of West Virginia: West Virginia Geological and Economic Survey, Mineral Research Series, MRS-7B, 119 p.
Deutsch, C., and A. G. Journel, 1992, GSLIB: geostatistical software library and user's guide: New York, Oxford University Press, 340 p.

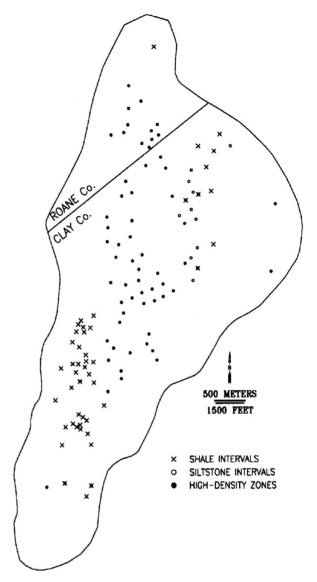

Figure 14. Locations of wells in which high-density zones and shale breaks are observed from geophysical logs.

Haught, O., 1959, Oil and gas report and map of Doddridge and Harrison counties, West Virginia: West Virginia Geological and Economic Survey, Bulletin 16, 39 p.
Hohn, M. E., 1988, Geostatistics and petroleum geology: New York, Van Nostrand Reinhold, 264 p.
Isaaks, E., and M. Srivastava, 1989, An introduction to applied geostatistics: New York, Oxford University Press, 561 p.
Journel, A. G., 1983, Nonparametric estimation of spatial distributions: Mathematical Geology, v. 15, no. 3, p. 445–468.
Journel, A. G., 1987, Geostatistics for the environmental sciences: U.S. Environmental Protection Agency, Project No. CR 811893, 135 p.
Ruley, E., 1970, "Big Injun" oil and gas production in north-central West Virginia: AAPG Bulletin, v. 54, no. 5, p. 758–782.

Whieldon, C. E., and W. E. Eckard, 1963, West Virginia oilfields discovered before 1940: U.S. Bureau of Mines, Bulletin 607, 187 p.

Zheng, L., R. Shumaker, and T. Wilson, 1993, A seismic interpretation of the structural development of the Granny Creek oil field: 24th Annual Appalachian Petroleum Geology Symposium, Program and Abstracts, p. 125–126.

Zou, X., and A. Donaldson, 1992, Regional Big Injun (Price/Pocono) subsurface stratigraphy of West Virginia: Geological Society of America Abstract, v. 27, no. 7, p. 309.

Chapter 12

Integrating Well Test–Derived Effective Absolute Permeabilities in Geostatistical Reservoir Modeling

Clayton V. Deutsch
Exxon Production Research Company
Houston, Texas, U.S.A.

André G. Journel
Department of Petroleum Engineering
Stanford University
Stanford, California, U.S.A.

ABSTRACT

In many cases, an estimate of effective absolute permeability may be derived from a pressure-transient well test. This effective permeability does not resolve local details of the permeability distribution; however, it does constrain the average permeability in the vicinity of the well. This paper presents an approach, based on simulated annealing, that integrates well test–derived effective permeabilities in stochastic reservoir models.

The volume and type of averaging informed by the well test must first be calibrated by forward simulating the well test on stochastic reservoir models that are consistent with the geological interpretation, core, well log, and seismic data. Stochastic reservoir models are then constructed with simulated annealing to additionally honor the well test–derived average permeabilities.

We present an example that illustrates how the methodology is implemented in practice. The improvement in the stochastic reservoir models is demonstrated by more accurate and precise prediction of future reservoir performance.

INTRODUCTION

The concept underlying stochastic reservoir modeling is to construct numerical models of the reservoir properties that are consistent with all relevant data. History matching is easier and forward predictions are more reliable when the numerical geological models honor more information. Applying a flow simulator to multiple numerical models allows an appreciation for the uncertainty in the reservoir response. A maximum amount of relevant prior information, e.g., core measurements, log data, geophysical data, geological interpretations, and well test data, must be integrated to ensure realistic output from the simulator. This chapter is concerned with integrating effective permeabilities derived from pressure-transient well tests into stochastic models of absolute permeability.

Conventional conditional simulation techniques such as Gaussian (Matheron et al., 1987; Deutsch and Journel, 1992), fractal (Hewett, 1986), or indicator (Journel and Alabert, 1990) models have the ability to account for local conditioning data (core and well log data), a global histogram, and varying amounts of spatial information in the form of variogram models. Variants of these techniques based on cokriging (Doyen et al., 1988) or some type of trend model (Marechal, 1984) allow geophysical information to be integrated into the resulting realizations. None of these techniques, however, allow the integration of well test–derived effective permeabilities.

Simulation techniques based on marked point processes (Haldorsen and Damsleth, 1990) are well suited to simulating spatial phenomena characterized by a repetition of easily depicted shapes. The resulting spatial structure is implicitly controlled by the placement of the digitized or analytically defined objects. Conditioning to local data and a global histogram is achieved by disallowing inconsistencies at data locations and controlling the number of objects placed in the realization. The integration of geophysical interpretations and well test–derived effective permeabilities is difficult with these latter techniques.

A well test–derived effective permeability does not directly resolve the smaller scale permeability values near the well bore; however, it does account for a complex nonlinear average of the small-scale values. Flow simulation results are directly influenced by the average flow characteristics as indicated by well test–derived effective permeabilities. Historically, well test–derived permeabilities were used in homogeneous reservoir models for reservoir performance forecasting (Ramey, 1990). Two inadequacies of such homogeneous and deterministic models are that they do not allow one to assess uncertainty and they do not allow for the important influence of small-scale permeability heterogeneities. Stochastic reservoir models allow the small-scale permeability heterogeneities to be accounted for; however, as mentioned, current stochastic modeling techniques do not account for the information carried by average properties measured by well tests.

Another possible method to account for well test–derived permeabilities is to rescale the core or well log–derived permeability values such that the permeability/thickness product of the core data is the same as the well test result. This method may be appropriate when the well test is measuring a different component of the permeability, such as fractures. We do not address this issue in this chapter. The assumption made here is that the core and well log data are representative and their spatial distribution may be altered to reproduce the well test result.

The problem of conditioning stochastic models to well test–derived effective permeability is difficult because the average is nonlinear. Conventional simulation techniques allow data of different volumetric supports only when the averaging is linear (Journel and Huijbregts, 1978).

One way to achieve this conditioning would be to discard all models that do not yield a forward-simulated well test response close enough to the actual measured pressure response. Such selection procedures may be practical when building models with a single well test; it is not practical, however, in the presence of multiple well test interpretations, some based on more advanced multirate tests, that give a number of different permeability averages near the well. In general, a prohibitively large number of realizations would be required to find a few that simultaneously match all well test data.

The algorithm proposed in this paper generates realizations with simulated annealing (Kirkpatrick et al., 1983) that honor the measured well test effective permeabilities in addition to conventional data, such as the local log-derived permeability values data, a histogram, and a variogram. The application of simulated annealing requires that the generation of a stochastic realization be posed as an optimization problem. The two-part objective function in this optimization problem consists of the deviation from the model variogram plus the deviation from the well test effective permeabilities. The deviation from the measured well test results could be known through forward simulation of the well test on all candidate realizations. Again, it would be impractical to forward simulate the well test after each perturbation as demanded by simulated annealing; therefore, the forward simulation must be replaced by a more easily calculated numerical approximation. A nonlinear power average (Korvin, 1981; Deutsch, 1987; Alabert, 1989) of the small-scale permeability values has been adopted for this purpose.

More precisely, the well test is first interpreted to provide the effective permeability-thickness product near the well bore. The effective permeability is computed knowing the reservoir thickness. Then, the type of averaging, as quantified by an averaging power, and the volume of averaging must be calibrated. The averaging power depends essentially on the connectivity of the extreme permeability values. The volume of averaging depends on the duration of the well test. Finally, the initial stochastic realizations are altered by systematically changing the elementary grid block permeability values so that the previously calculated effective permeability and the variogram model are honored. The effective permeability can be calculated very fast after each perturbation using the previously calibrated power average.

The following algorithm will clarify the implementation details and acknowledge a number of limitations. One limitation is that the method imposes the well test results without accounting for uncertainty in the underlying well test interpretation. A second limitation is that the method requires that the full well test response be summarized by a single nonlinear weighted power average. The severity of these limitations may be judged through numerical experimentation.

QUANTIFYING WELL TEST–DERIVED EFFECTIVE PERMEABILITY

Pressure transient well tests are performed by generating a flow rate impulse in the reservoir and measuring the pressure response. Well test interpretation consists of interpreting the pressure response by using an appropriate mathematical model to relate the pressure response (output) to flow rate history (input) (Horne, 1990). Provided that the mathematical model is appropriate, the model parameters can be associated to certain reservoir parameters. Of particular interest is the effective absolute permeability k_e.

To apply the envisaged optimization technique it is necessary to translate this well test–derived k_e into a more easily calculated property while retaining the flexibility to differentiate a wide variety of heterogeneous systems encountered in practice. The power averaging formalism is used to model the nonlinear averaging of absolute permeabilities (equation 1). The assumption is that the elementary permeability values average linearly after a nonlinear power transformation, that is

$$\bar{k}(\omega) = \left[\frac{1}{N} \sum_{u_i \in V} k(u_i)^{\omega} \right]^{\frac{1}{\omega}} \qquad (1)$$

where $\bar{k}(\omega)$ is the ω-power average permeability of the N permeability values $k(u_i)$, $i = 1,\ldots,N$, at locations u_i within the volume of interest V. The power ω ranges between the bounding values of -1 and 1 corresponding to the harmonic and arithmetic averages, respectively (the geometric average is obtained for $\omega = 0.0$). The idea is to use relation (1) and to calibrate the averaging volume V and averaging power ω for each particular well test.

To define the appropriate averaging volume it is necessary to consider the portion of the pressure response used to derive the well test effective permeability k_e. In practice, k_e is obtained by interpreting the pressure response during the time at which the response resembles infinite-acting radial flow. Early

time effects, such as well-bore storage, and late time boundary effects are not considered in the interpretation. It is possible to define an inner radius r_{min} and an outer radius r_{max} that correspond to the time limits of the interpretation because the pressure response, at any time t, may be related to block permeabilities within a time-dependent radius of drainage $r(t)$. Consider a typical pressure response shown on the Miller-Dyes-Hutchinson (MDH) plot in Figure 1. The pressure response between 1 and 10 hr is used to derive an estimate of the effective permeability $k_{welltest}$. The inner and outer limits of the shaded region (on the schematic illustration of the reservoir) correspond to the radius of drainage at 1 and 10 hr, respectively. Actually, these limits will not be circular due to local heterogeneities. The impact of this assumption will be revealed in the experimental calibration described in a following section.

The time interval during which the pressure response resembles infinite-acting radial flow is easily determined by standard interpretation techniques. Evaluating the radius of drainage $r(t)$ at the time limits is not as straightforward; depending on the arbitrary definition chosen for $r(t)$, the radius can change by as much as a factor of 4. It will be necessary to calibrate the radius of drainage $r(t)$ by repeated flow simulations. The block permeabilities contributing to the pressure response measured up to time t are approximately enclosed by a circular volume centered at the well and defined by a time-dependent radius $r(t)$ written as (van Poollen, 1964; Johnson, 1988; Alabert, 1989)

$$r(t) = A \sqrt{\frac{k_e t}{\phi \mu c_t}} \qquad (2)$$

where A is a constant, k_e is the reservoir permeability around the well, ϕ is the porosity, μ is the fluid viscosity, and c_t is the total compressibility. Depending on the definition chosen for the radius of drainage, the value of A ranges from 0.023 to 0.07 (for oil field units). Alabert (1989), in evaluating the averaging volume of a well test and for specified levels of discretization and test durations, found an optimal A_{opt} value of 0.010 in oil field units.

The averaging power ω describes the type of averaging within the volume. In many cases, this averaging power is close to the geometric average ($\omega = 0$). For

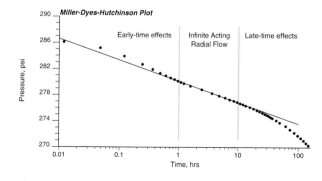

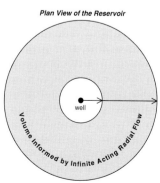

Figure 1. A schematic illustration of the volume measured by a given well test interpretation. The inner and outer limits of the shaded region, on the plan view of the reservoir, correspond to the radius-of-drainage at 1 and 10 hr, respectively (the limits of infinite acting radial flow).

practical test durations and for complex heterogeneous permeability distributions, the type of averaging can differ significantly from the geometric average.

There are no reliable universal values for the constant A and the averaging power ω; they must be calibrated using the following procedure (see also Alabert, 1989; Deutsch, 1992).

1. Generate n_s (20–100) multiple realizations of the permeability field with relevant statistical properties.
2. Forward simulate a well test, with conditions as close as possible to those used in the field to arrive at k_e, on each realization to obtain n_s pressure response curves.
3. Deduce an effective permeability $\bar{k}_i$, $i = 1,\ldots,n_s$ from each pressure curve using established well test interpretation techniques (Horne, 1990).
4. Compute average permeabilities $\bar{k}(A,\omega)_i$, $i = 1,\ldots,n_s$ for A values between the practical bounding limits of 0.001 and 0.020 and for ω values between practical bounding limits of −0.5 and 0.5.
5. Choose the pair (A_{opt},ω_{opt}) that yields the closest agreement between the reference $\bar{k}_i$, $i = 1, \ldots, n_s$ values and the approximate $\bar{k}(A_{opt},\omega_{opt})_i$, $i = 1,\ldots,n_s$ values.

After establishing appropriate A_{opt} and ω_{opt} values the validity of the power average approximation can be checked by creating a scatterplot of the $\bar{k}(A_{opt},\omega_{opt})_i$ values versus the well test–derived $\bar{k}_i$ values.

In summary, the weighted nonlinear power average (equation 1) is proposed as a computationally simple replacement for the full well test response. In practice, a well test response is interpreted to yield an estimate of the effective permeability k_e and the averaging volume parameter A_{opt} and power ω_{opt} are calibrated for the particular geological setting. The next step is to impose the well test–derived effective permeability k_e, i.e., the appropriate ω average $\bar{k}(A_{opt},\omega_{opt})$, on stochastic models. We use the technique of simulated annealing.

SIMULATED ANNEALING

The "annealing" approach to stochastic simulation has no explicit random function model, rather the creation of a simulated realization is formulated as an optimization problem. The first requirement of this class of methods is an objective (or energy) function, which is some measure of difference between the desired spatial characteristics and those of a candidate realization. The essential feature of annealing methods is to iteratively perturb (relax) the candidate realization and then accept or reject the perturbation with some decision rule. The decision rule is based on how much the perturbation has brought the candidate image closer to having the desired properties. One possible decision rule is based on an analogy with the metallurgical process of annealing, hence the name simulated annealing. Technically the

name "simulated annealing" applies only to those stochastic relaxation methods based strictly on simulated annealing (Aarts and Korts, 1989; Kirkpatrick et al., 1983); however, through common usage the name "annealing" is used to describe the entire family of methods that are based on the principle of stochastic relaxation.

Annealing is the process where a metallic alloy is heated so that molecules may move positions relative to one another and reorder themselves into a low-energy crystal (or grain) structure. The probability that any two molecules will move relative to one another is known to follow the Boltzmann probability distribution. Simulated annealing is the application of the annealing mechanism of perturbation (swap the attribute values assigned to two different grid node locations) with the Boltzmann probability distribution for accepting perturbations.

At first glance this approach appears terribly inefficient. For example, millions of perturbations may be required to arrive at an image that has the desired spatial structure. However, these methods are more efficient than they might seem as long as few arithmetic operations are used to update the objective function after a perturbation; virtually all conventional global spatial statistics (e.g., a variogram) may be updated locally rather than globally recalculated after a local perturbation. Also, the power average representation of the well test k_e developed earlier is easily updated.

The objective function is defined as some measure of difference between a set of reference properties and the corresponding properties of a candidate realization. The reference properties could consist of any quantified geological, statistical, or engineering property. In the context of this paper, the reference properties consist of traditional variogram functions and the well test–derived effective permeability. Thus, the objective function could be written as

$$O = \lambda_1 \times \sum_{i=1}^{n_h}\left[\gamma^{ref}(\mathbf{h}_i) - \gamma^{real}(\mathbf{h}_i)\right]^2$$
$$+ \lambda_2 \times \sum_{j=1}^{n_K}\left[k_{e_j}^{ref} - k_{e_j}^{real}\right]^2 \tag{3}$$

where O is the objective function, n_h is the number of variogram lags, $\mathbf{h}_i$ considered important, $\gamma^{ref}(\mathbf{h}_i)$ is the reference variogram value for lag $\mathbf{h}_i$, $\gamma^{real}(\mathbf{h}_i)$ is the variogram value taken from the candidate realization, n_K is the number of well test–derived effective permeabilities to be reproduced, $k_{e_j}^{ref}$ is the jth well test–derived effective permeability, $k_{e_j}^{real}$ is the jth effective permeability calculated from the candidate realization using power average, and λ_1 and λ_2 are relative weights to ensure that the variogram contribution has the same importance as the well test contribution.

The starting image is a three-dimensional array, $z(\mathbf{u}_i)$, $i = 1,\ldots,N$, of permeability values. The annealing methodology to achieve a realization l, $z^{(l)}(\mathbf{u}_i)$, $i = 1,\ldots,N$, with a low objective function (see equation 3) is as follows.

1. Generate an easily constructed initial realization: the initial realization is either the output of a more conventional stochastic simulation algorithm or the initial realization could be generated by assigning each nodal value at random from the stationary univariate distribution $F(z)$.

2. Establish the reference components in the objective function: the reference variogram values $\gamma^{ref}(\boldsymbol{h}_i)$, $i = 1,...,n_h$ are based on experimental data or on an analytical model. The reference effective permeabilities k_{ej}^{ref}, $j = 1,...,n_K$, are the direct result of applying standard interpretation techniques to the pressure transient responses measured in the field.

3. Compute the realization components in the objective function: the variogram values $\gamma^{real}(\boldsymbol{h}_i)$, $i = 1,...,n_h$, and the effective permeabilities k_{ej}^{real}, $j = 1,...,n_K$, are calculated from the candidate realization.

4. Compute the objective function O based on the reference and the realization statistics (see equation 3).

5. Perturb the realization to generate a new realization by swapping the permeability values at any two locations.

6. Update each component in the objective function and recompute the objective function O_{new} with the perturbation. The change to the objective function is $\Delta O = O_{new} - O_{old}$.

7. The perturbation is accepted or rejected on the basis of a specified decision rule. One approach would be to accept all helpful perturbations $\Delta O \leq 0.0$ and to reject all disruptive perturbations $\Delta O > 0.0$. This choice, which corresponds to a steepest descent approach, can lead to a local minimum. The essential contribution of simulated annealing is a prescription for when to accept or reject a given perturbation. The acceptance probability distribution is given by

$$P\{accept\} = \begin{cases} 1, & \text{if } \Delta O \leq 0.0 \\ e^{\frac{-\Delta O}{t}}, & \text{otherwise} \end{cases} \tag{4}$$

All favorable perturbations ($\Delta O \leq 0.0$) are accepted and some unfavorable perturbations are accepted with an exponential probability distribution. The parameter t of the exponential distribution is analogous to the "temperature" in annealing. The higher the temperature, the more likely an unfavorable perturbation will be accepted. Accepting the perturbation causes the image $z(\boldsymbol{u}_i)$, $i = 1,...,N$ and the objective function O to be updated.

8. When the objective function O gets close to zero then the realization is considered finished because it now honors both the reference variogram and the well test data; otherwise, return to step 5 and continue the perturbation process.

The idea is to start with an initially high temperature parameter t and lower it by some multiplicative factor λ (say 0.1) when enough perturbations have been accepted ($K_{accept} = 10$ times the number N of grid nodes in the system) or too many have been tried ($K_{max} = 100 \times N$). The algorithm is stopped when efforts to lower the objective function become sufficiently discouraging (Press et al., 1986).

One remaining issue is to establish the weights λ_1 and λ_2 applied to each component in the objective function. The purpose behind these weights is to have each component play an equally important role in the global objective function. Without any weighting, the component with the largest units would dominate the objective function. The weights λ_1 and λ_2 are established so that, on average, each component contributes equally to a change in the objective function ΔO. That is, each weight λ_c is inversely proportional to the average change of that component objective function:

$$\lambda_c = \frac{1}{|\overline{\Delta O_c}|}, \quad c = 1,...,2 \tag{5}$$

In practice, the average change of each component $|\overline{\Delta O_c}|$ cannot be computed analytically; however, it can be numerically approximated by evaluating the average change of M (say 1000) independent perturbations:

$$|\overline{\Delta O_c}| = \frac{1}{M}\sum_{m=1}^{M}|O_c^{(m)} - O_c|, \quad c = 1,...,2 \tag{6}$$

where $|\overline{\Delta O_c}|$ is the average change for component c, $O_c^{(m)}$ is the perturbed objective value, and O_c is the initial objective value. Each of the M perturbations $m = 1,...,M$ arises from the swapping mechanism employed for the annealing simulation.

All of the elements needed for integrating well test data are now in place. The resulting realizations obtained after going through the simulated annealing procedure with the objective function (see equation 3) are conditional to both the initial geological/statistical description and the well test–derived permeabilities. The following example illustrates how the methodology is implemented in practice.

AN EXAMPLE APPLICATION

Consider an example where the block horizontal absolute permeabilities are known to follow a distribution $F(z)$, which results from the sum of a constant 2.5 md and a lognormal component with a mean of 10.0 md and a variance of 225 md^2. This two-part distribution reflects the simplifying assumption of representing a three-dimensional permeability distribution by a two-dimensional areal model. The two-dimensional spatial distribution of permeability is characterized by a spherical normal scores variogram model $\gamma_Y(\boldsymbol{h})$ with a range of 25 grid block units. A 101 × 101 unit realization of this lognormal permeability field was generated to serve as a reference distribution (Figure 2). A core (or well log–derived) permeability and a drawdown well test–derived permeability are known at the five

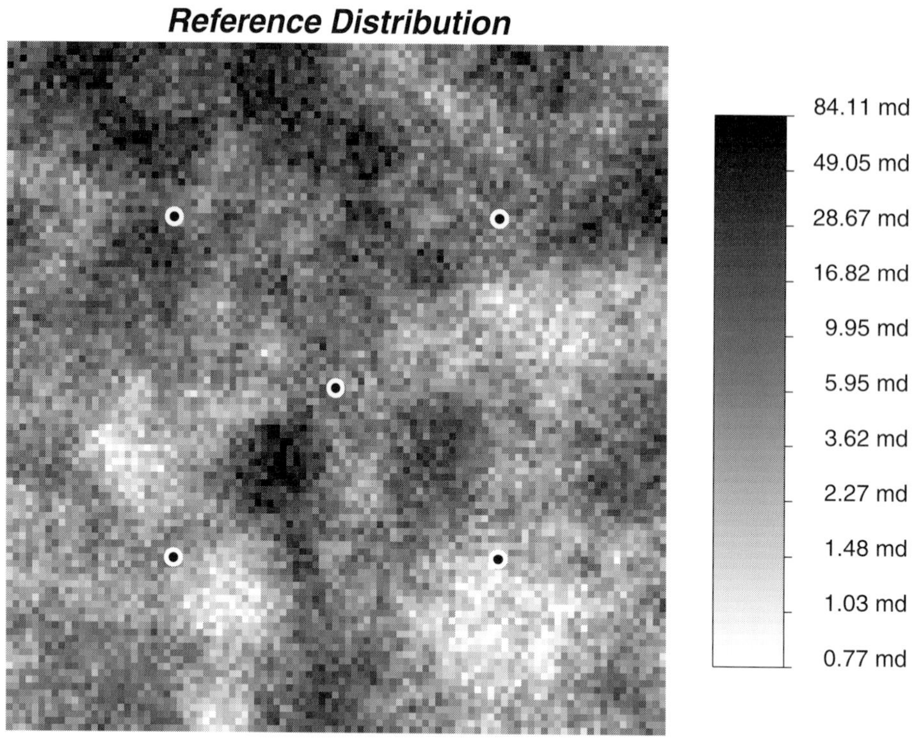

Reference Distribution

Figure 2. The reference distribution of permeability considered as the true reservoir for the purposes of obtaining the well test responses and interpreted effective permeabilities. The conditioning data, falling on a five-spot pattern, are highlighted by the black dots.

well locations (see Table 1). The five drawdown pressure response curves are shown on Figure 3.

Simulated annealing was used to generate 100 realizations conditional to the distribution model $F(z)$, the variogram model $\gamma_Y(h)$, and the five core data; the following objective function was considered:

$$O = \sum_{i=1}^{n_h} \left[\gamma^{ref}(\mathbf{h}_i) - \gamma^{real}(\mathbf{h}_i) \right]^2$$

where n_h corresponds to the most compact arrangement of 200 lags (Figure 4). Annealing allows this objective function to be reduced close to zero, i.e., the resulting realizations provide an excellent reproduction of the variogram. The first four realizations are shown on Figure 5.

These 100 initial realizations reproduce the core permeability at the five well locations. The well test values, however, are not reproduced. Figure 6 shows five histograms of the well test–derived permeability values obtained from the 100 simulated realizations. The vertical line in each histogram is the result of the reference case. Ideally, the realizations would be conditional to the reference well test values.

To check the validity of the numerical well test simulation, the five well tests were simulated using uniform permeability fields. The results for all five wells with three different uniform permeability fields (5.0, 10.0, and 50.0 md) are shown on Figure 7. The time

limits for infinite acting radial flow are shown by the vertical lines. The well test–derived effective permeabilities are shown on Table 2. The well test–derived values appear consistently lower than the input permeability field; all of the wells show the same bias. This bias was not considered significant for permeabilities of less than 50 md. Additional validation runs would be warranted if well test–derived permeability values were observed outside of this range.

These 100 effective permeability values, derived from forward simulating the well test on 100 unconditional realizations, were used to calibrate the parame-

Table 1. Core- and well test–derived effective absolute permeability from five wells in the study area

Well	Core Permeability (md)	Well Test Permeability (md)
1 (top left*)	7.25	14.16
2 (top right)	10.95	9.63
3 (center)	8.89	9.41
4 (bottom left)	5.04	5.59
5 (bottom right)	4.03	4.54

*Position in parentheses refers to well positions shown in Figure 2.

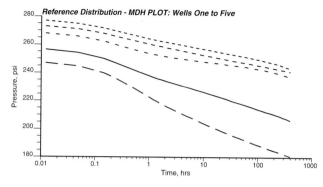

Figure 3. The pressure response for the five wells is shown on this Miller-Dyes-Hutchinson (MDH) plot.

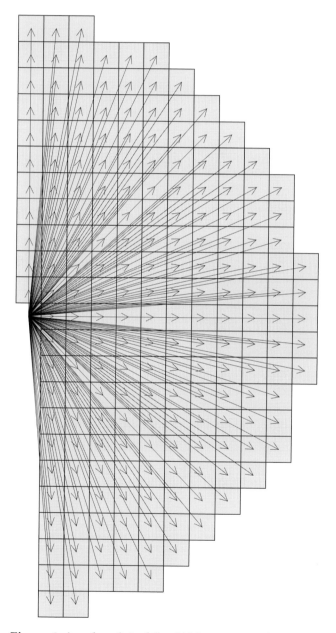

Figure 4. A polar plot of the 200 lag vectors that enter the objective function of the annealing simulation.

Table 2. Well test–derived effective absolute permeabilities from uniform permeability fields*

Well	Uniform Permeability Fields (md)		
	5.00	10.00	50.00
1 (top left**)	4.96	9.99	48.75
2 (top right)	4.96	9.99	48.77
3 (center)	4.96	10.00	48.80
4 (bottom left)	4.96	9.98	48.73
5 (bottom right)	4.96	9.99	48.75
Difference (%)	–0.1	–0.1	–2.5

*Simulated values appear slightly lower than those of the uniform permeability field.
**Position in parentheses refers to well positions shown in Figure 2.

ters A_{opt} and ω_{opt} that provide the best numerical approximation to the well test result. The average permeabilities $k\,(A,\omega)_i$, $i = 1,...,n_s$ for A values between the practical bounding limits of 0.001 and 0.020 and for ω values between practical bounding limits of –0.5 and 0.5 were computed using the 100 initial realizations. The criteria for an optimal pair (A_{opt},ω_{opt}) was to simultaneously maximize the correlation and minimize the bias between the power average approximation (equation 1) and the true well test values. The optimal pair A_{opt} = 0.003 and ω_{opt} = 0.0 was found to both maximize the correlation and minimize the bias.

Figure 8 shows a scatterplot of the power average numerical approximation and the true well test–derived effective permeability. Note the lack of any bias (the average effective permeabilities are 9.3 md in both cases) and the excellent correlation of 0.83. The physical volume informed by the well test can be determined by the constant A_{opt} and knowledge of the time limits of the infinite acting portion of the pressure response: r_{min} = 1.0 grid units (70.1 ft) and r_{max} = 4.6 grid units (320.0 ft).

The annealing simulation now uses a two-part objective function (equation 3). Annealing is able to generate realizations that lower this objective function to zero. The first four realizations are shown on Figure 9.

To verify that the calibrated power average is a fair approximation of the actual well test result, a full well test was forward simulated on the final 100 realizations. The distributions of effective permeabilities, after processing, is extremely close to the reference values.

This illustrates that well test effective permeabilities can be imposed on stochastic realizations to a fair degree of approximation. What has not been shown yet is that accounting for well test–derived effective permeability actually helps predict future reservoir performance. To illustrate the improvement in future prediction, all 201 realizations (the reference, the 100 initial, and the 100 well test–conditioned realizations) have been associated to the five-spot injection/production pattern shown on Figure 2 with an injector at all corner locations and a central producer. All variables except the block absolute permeabilities have been held constant and the performance of each realization

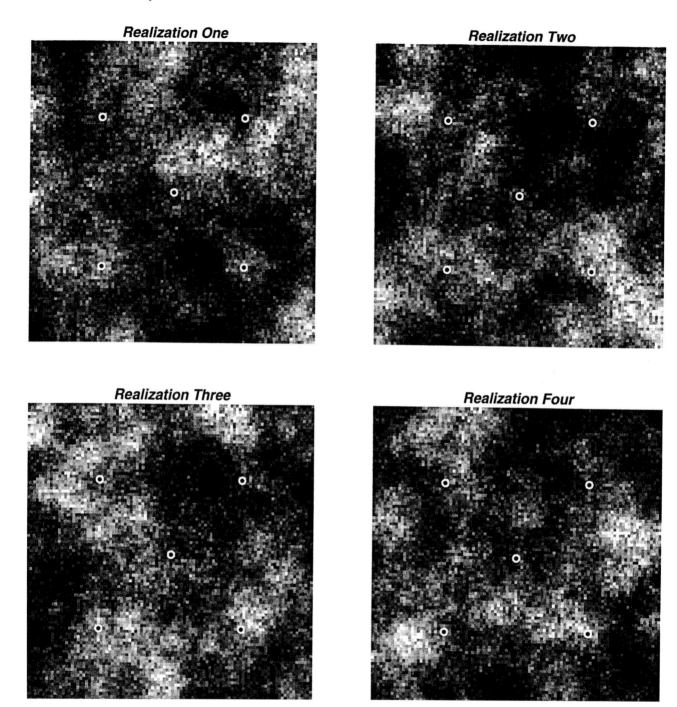

Figure 5. Four realizations conditional to the five data (represented by the black dots), the global histogram, and the global variogram.

has been simulated with ECLIPSE (Exploration Consultants Limited, 1984).

Three response variables are presented here: (1) the time to reach 90% water cut, (2) the time of first water arrival at the producer, and (3) the time to reach 50% water cut (the units are important only in a relative sense). The reference image yielded response variables of 86.8, 8.93, and 20.8, respectively. The histograms of values obtained before and after conditioning to the

well test data are shown on Figure 10. Note that the well test–conditioned distributions show less uncertainty as measured by the 90% probability interval.

REMARKS AND CONCLUSIONS

Our methodology integrates well test–derived properties into stochastic reservoir models. The

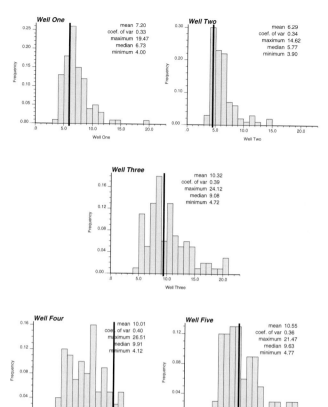

Figure 6. The well test–derived effective absolute permeabilities for each of the five wells. One hundred realizations were generated and five well test simulations were performed on each realization to obtain these histograms.

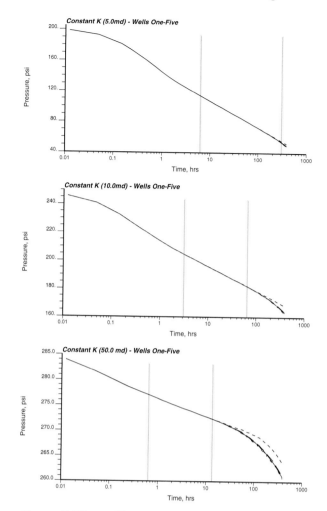

Figure 7. The well test response for three different uniform permeability cases. The response and interpreted results are shown for all five wells in the five-spot pattern.

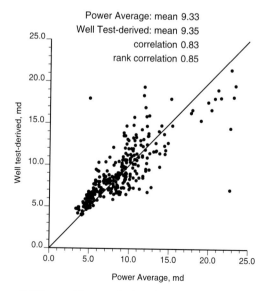

Figure 8. The well test–derived effective permeability versus the power average approximation for the 500 (5 wells × 100 realizations) well tests.

methodology consists of generating realizations with simulated annealing. The objective is to minimize deviations from the initial variogram model and yet honor a numerical approximation of the well test–derived effective permeability. The power average numerical approximation is useful because it would not be feasible to rerun a flow simulation after each perturbation called for by the annealing technique.

An example shows how the methodology could be implemented in practice. The results are encouraging even though the example may not fully reflect the heterogeneity encountered in practice.

The annealing methodology presented to integrate well test data is quite general. The method has the potential to integrate many disparate data, as long as these data can be quantified to enter a global objective function. For example, multiple-point statistics could be used to input complex curvilinear geological structures and seismic data could be incorporated by adding another component to the global objective function (Deutsch, 1992; Doyen et al., 1989). Other sources of data that could be incorporated are the results of multiple-rate and tracer tests.

Realization One

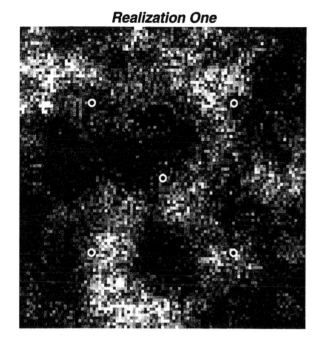

Realization Two

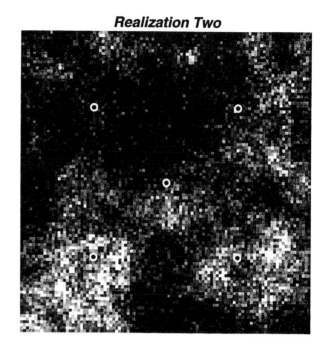

Realization Three

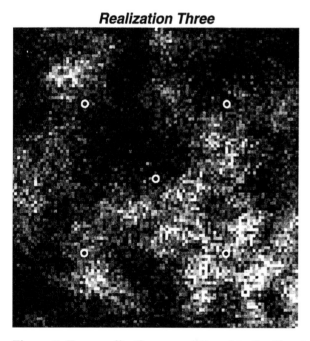

Realization Four

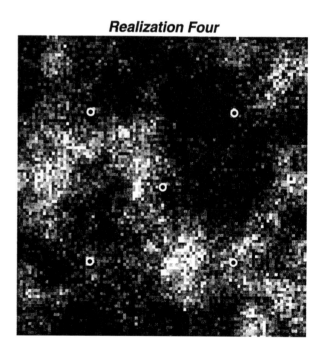

Figure 9. Four realizations conditional to the five data (represented by the black dots), the global histogram, the global variogram, and the five well test–conditioned results.

ACKNOWLEDGMENTS

We thank the management of Exxon Production Research Company for allowing publication of this paper.

REFERENCES

Aarts, E., and J. Korst, 1989, Simulated annealing and Boltzmann machines: New York, John Wiley.

Alabert, F., 1989, Constraining description of randomly heterogeneous reservoirs to pressure test data: a Monte Carlo study: SPE Paper 19600, SPE Annual Conference and Exhibition, Proceedings.

Deutsch, C., 1989, Calculating effective absolute permeability in sandstone/shale sequences: SPE Formation Evaluation, p. 343–348.

Deutsch, C., 1992, Annealing techniques applied to reservoir modeling and the integration of geological and engineering (well test) data: Ph.D. thesis, Stanford University, Stanford, California.

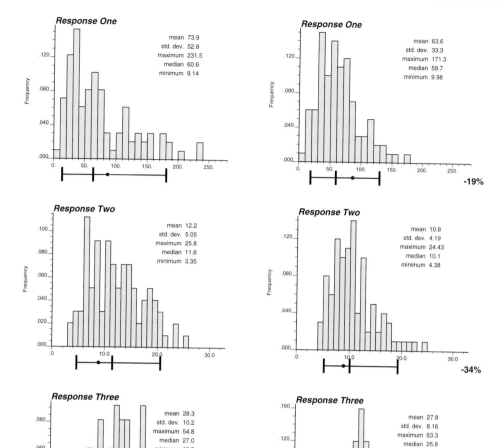

Figure 10. The histograms of the breakthrough time and the final oil in place obtained by running flow simulations on the realizations before (left) and after (right) integrating the well test data. Below the horizontal axis, the black dot is the reference value, the thicker vertical line is the median, and the outermost vertical line represents the 95% probability interval.

Deutsch, C., and A. Journel, 1992, GLSLIB geostatistical software library and user's guide: New York, Oxford University Press.

Doyen, P., 1988, Porosity from seismic data: a geostatistical approach: Geophysics, v. 53, n. 10, p. 1263–1275.

Doyen, P., T. Guidish, and M. de Buyl, 1989, Seismic discrimination of lithology in sand/shale reservoirs: a Bayesian approach: SEG 59th Annual Meeting Extended Abstracts.

Exploration Consultants Limited, 1984, ECLIPSE reference manual: Henley-on-Thames, England, Exploration Consultants Limited.

Haldorsen, H., and E. Damsleth, 1990, Stochastic modeling: Journal of Petroleum Technology, April, p. 404–412.

Hewett, T., 1986, Fractal distributions of reservoir heterogeneity and their influence on fluid transport: SPE Paper 15386.

Horne, R., 1990, Modern well test analysis: Palo Alto, California, Petroway, Inc.

Johnson, P., 1988, The relationship between radius of drainage and cumulative production: SPE Formation Evaluation, March, p. 267–270.

Journel, A., and F. Alabert, 1990, New method for reservoir mapping: Journal of Petroleum Technology, February, p. 212–218.

Journel, A., and C. J. Huijbregts, 1978, Mining geostatistics: New York, Academic Press.

Kirkpatrick, S., C. Gelatt, Jr., and M. Vecchi, 1983, Optimization by simulated annealing: Science, v. 220, n. 4598, p. 671–680.

Korvin, G., 1981, Axiomatic characterization of the general mixture rule: Geoexploration, v. 19, p. 267–276.

Marechal, A., 1984, Kriging seismic data in presence of faults, in G. Verly et al., eds., Geostatistics for natural resources characterization: Dordrecht, D. Reidel.

Matheron, G., H. Beucher, H. de Fouquet, A. Galli, D. Guerillot, and C. Ravenne, 1987, Conditional simulation of the geometry of fluvio-deltaic reservoirs: SPE Paper 16753.

Press, W., B. Flannery, S. Teukolsky, and W. Vetterling, 1986, Numerical recipes: New York, Cambridge University Press.

Ramey, H., Jr., 1990, Advances in practical well test analysis: SPE 65th Annual Technical Conference and Exhibition, Proceedings, p. 665–676.

van Poollen, H., 1964, A hard look at radius of drainage and stabilization-time equations: Oil & Gas Journal, September, p. 139–147.

Chapter 13

◆

Constraining Geostatistical Reservoir Descriptions with 3-D Seismic Data to Reduce Uncertainty

Richard L. Chambers
Amoco Production Research
Tulsa, Oklahoma, U.S.A.

Michael A. Zinger
Amoco Production Company
Cairo, Egypt

Michael C. Kelly
Amoco Production Research
Tulsa, Oklahoma, U.S.A.

◆

ABSTRACT

The geostatistical external drift method is used to integrate three-dimensional (3-D) seismic data into a reservoir description and is illustrated with an application on a west Texas Permian basin interbedded carbonate-clastic reservoir. Seismic reflection amplitude, inverted to acoustic impedance, supplements sparse well control to estimate interwell porosity.

The North Cowden unit is a mature field and serves as a laboratory for many reservoir characterization experiments. The extensive wireline and core database available for the area covered by a high-resolution 3-D seismic survey was resampled to mimic scenarios similar to three stages in a reservoir's life: (1) a late exploration/appraisal phase; (2) a development phase; and (3) a mature production phase, typical of many Permian basin fields.

Spatial interpolation by kriging porosity with and without seismic data are compared. Stochastic (Monte Carlo) simulations are used to evaluate interpolation uncertainty (standard error). Interpolation uncertainty is greatly reduced when seismic data are integrated into the reservoir description.

INTRODUCTION

Because the detail needed to develop reservoirs far exceeds the detail required to find them, industry is seeing an increased use of three-dimensional (3-D) seismic data in reservoir management (Robertson, 1991). Some recent examples include defining complex fault patterns, segmenting some reservoirs into many small compartments, and the ability to book larger reserves (Nestvold and Nelson, 1992; Nestvold, 1987, 1991).

Most recently, seismic data have been used to predict spatial variations in lithology and porosity using a geostatistical approach (Doyen, 1988; Doyen et al., 1988; Doyen and Guidish, 1992; Journel and Alabert, 1990). Inference of rock properties from seismic attributes, such as reflection amplitude and acoustic impedance, is not a unique problem because more than one geologic variable influences the magnitude of these values; however, porosity, lithology, and gas occurrence contribute most to changes in acoustic properties (Doyen and Guidish, 1992). Bashore et al. (this volume) illustrate the use of seismic data to constrain the geological and petrophysical models used for flow simulation. Seismic reflection amplitude was used as an external drift to estimate hydrocarbon pore volume and conditional simulations to generate risk maps in an exploitation study (Wolf et al., this volume).

This case study illustrates the quantitative integration of 3-D seismic data into a reservoir description during three phases of a reservoir's life. An extensive wireline and core database available for an area covered by a high-resolution 3-D seismic survey was resampled to mimic scenarios similar to (1) a late exploration/appraisal phase; (2) a development phase; and (3) a mature production phase, typical of many Permian basin fields. Seismic reflection amplitude, inverted to acoustic impedance, supplements sparse well control in the estimation of interwell porosity.

Spatial interpolation by kriging porosity with and without seismic data are compared. Stochastic (Monte Carlo) simulations are used to evaluate interpolation uncertainty (standard error). Interpolation uncertainty can be significantly reduced with integration of "soft" seismic data into the reservoir description.

STUDY AREA

The North Cowden unit (NCU) is located on the eastern edge of the Central Basin platform in the west Texas Permian basin (Figure 1). This field serves as a laboratory for many reservoir characterization experiments because it is a very mature, well characterized reservoir. In the study area, production is from the Guadalupian Grayburg Formation (Permian), which is transitional between the previously more open-marine conditions of the San Andres Formation and the more arid sabkha and siliciclastic eolian dune field environment of the younger Queen Formation.

Lithologically, the Grayburg is composed of alternating dolomite and siltstone for a total thickness of about 140 m. Dolomites range from anhydritic skeletal wackestones through mudstones. Porosity is moldic or vuggy and can be extensively plugged by anhydrite. The siltstones are dominantly angular to subrounded quartz grains with angular feldspathic grains, which commonly alter to clay, partially plugging pore throats. Siltstone porosity is intergranular. This formation has a characteristic shoaling-upward, prograding sedimentary motif, ranging from shallow open-marine to tidal flat/sabkha sediments. The silt is believed to be of eolian origin, reworked by strandline processes into a series of thin, offlapping shoals.

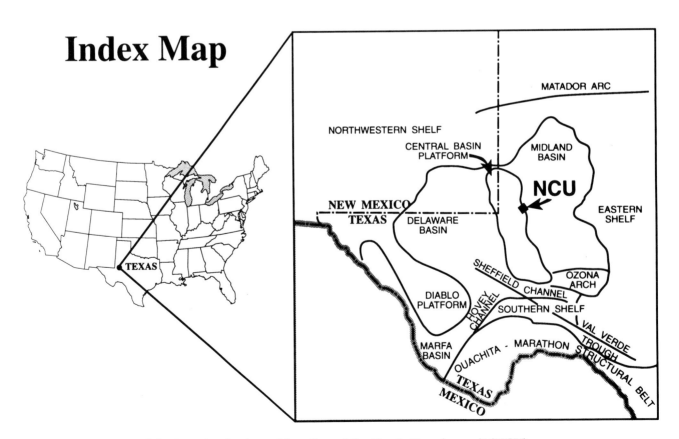

Figure 1. Index map of the Permian basin and location of the North Cowden unit (NCU).

Progradation of the carbonate shelf was approximately from west to east in the NCU.

The Grayburg Formation is divided into five dolomite intervals (D1–D5), each separated by a siltstone interval (S1–S3), with one exception. The D2/D3 dolomite intervals are separated by a very diagnostic "hot" gamma-ray layer (shale) 1–3 m thick. Within the study area the second siltstone, S2, is the most productive interval. The dolomites tend to be nonproductive and form seals. The S1 and S3 siltstone intervals have only a minor contribution to the total production in this part of the field.

Structurally, the reservoir is a north-south–trending asymmetrical anticline, dipping gently eastward into the Midland basin.

This study focuses on the S2, or second siltstone interval, because it is one of the most productive Grayburg Formation intervals; locally, however, the S2 acts as a thief zone to injected water. Thus, one must understand the variability in this zone to better manage a large portion of the NCU.

DATA SETS

Well Data and Description of the S2 Interval

The typical wireline suite in the NCU consists of gamma-ray and acoustic logs, with an occasional neutron-density log. Normalized gamma-ray logs are used as a lithologic discriminator in Permian basin fields. Previous NCU field studies have used 54 API as the siltstone/dolomite cutoff, with siltstones correlating to larger API values (Cairns and Feldkamp, 1992). The motif for the S2 is similar to the Grayburg Formation with thin dolomite beds alternating with thicker siltstone beds, the dominant facies.

The entire Grayburg Formation was cored at more than 80 locations. Whole-core porosity and permeability were measured for each 1-ft segment. From acoustic transit time and whole-core porosity crossplots, 12 regression equations were calculated; one for each interval, D1–D5 and S1–S3, and each facies within an interval based on the API cutoff. At uncored well locations, porosity was computed using the appropriate equation. The data set was made consistent by also calculating porosity at cored well locations. The correlation coefficient for whole-core porosity and acoustic transit time for only the S2 interval is –0.82 with a 0.56 mean error variance (residual). This coefficient takes into account both facies types.

Prior to selecting a site for the seismic survey, average porosity and average acoustic transit time were calculated over the S2 interval at 261 well locations. These data and S2 thickness were mapped over a 73 km^2 central area of the NCU. The purpose of this study was to evaluate the spatial distribution of these S2 properties over a large area of the NCU.

The S2 interval is 30 m thick in the western, or landward, side of the study area, thinning to 10 m basinward to the east. Although gross thickness decreases basinward, net siltstone thickness remains almost constant. Reservoir quality increases from west to east, eventually diminishing in quality as the

interval thins and becomes more dolomitic basinward, outside of the present study area.

Based on gamma-ray logs, the S2 can be described as a package of separate, distinct beds that individually have a limited areal extent, but as a group appear to be laterally continuous across the study area. On the more landward side, the S2 consists of three distinct, porous siltstone beds (acoustically slow), separated by tight dolomite (acoustically fast). Moving basinward about 1 km, there is a distinct change in the acoustic log response with little change in the character of the gamma-ray logs. Within this transitional zone, about 500–800 m wide, the upper S2 bed has acoustic traveltimes similar to the overlying D3 dolomite. Thin sections show that pore spaces are filled with anhydrite. Within the central and eastern portions of the study area, interbedded dolomite decreases significantly. The S2 interval appears as a more singular body of siltstone, with both gamma-ray and acoustic logs showing a similar pattern. Porosity patterns track the acoustic log response and do not always follow the gamma-ray response. Porosity ranges from 3–13%.

On the basis of these results, an area of 4 km (west to east) by 3.2 km (north to south) was selected as the site for a high-resolution 3-D seismic survey. Within this area, the S2 interval has an average thickness of 16 m, ranging from 12–21 m. Porosity ranges from 6–8% in the northwestern part of the seismic grid, to a low of 5–6% in the transitional zone, and to a high of 10–13% within the sweet-spot of the field. Fifty-five of the 261 wells used in the evaluation study are located within the seismic grid.

Seismic models for two five-well cross sections of the Grayburg Formation show that a 12-m S2 interval should be nearly resolved with seismic data containing frequencies greater than 90 Hz; however, some thin-bed tuning can be expected in the western to northwestern part of the study area where there are three distinct siltstone beds. Significant lateral changes in S2 reflection amplitude are seen in the modeling results. By convention, negative amplitudes or low acoustic impedance implies slower interval velocity and correlates to less-dense rock. In this study, such a response correlates most closely with more porous rock, with a secondary contribution from facies type. There is no gas cap in this reservoir to produce an amplitude anomaly.

In this study, porosity, not pore volume, was inferred from acoustic impedance because the correlation between porosity and thickness is about 0.2.

3-D Seismic Data

The 3-D seismic survey used a swath acquisition technique, with four active receiver lines for each source line, with the second line in the swath serving as the source line. One hundred thirty-two (132) parallel lines were laid out, east to west, with no crosslines. This configuration was chosen to take advantage of the stack array, designed to work most effectively inline. Refraction static corrections are more easily handled with two-dimensional lines extracted from a swath acquisition scheme.

Thirty-three swaths were acquired. Each line was configured as a split spread, with a 1402-m tail spread and 100 source points at 24.5-m intervals, for a total inline length of 3840 m. Crosslines have a 49-m separation, extending 3200 m north to south. Approximately 42,000 CDP (common depth points) were acquired over a 13 km^2 area. Maximum inline fold is 58, with one fold crossline (times 58 fold from the inline direction). Frequency content at a reservoir depth of 1220–1370 m is 7–110 Hz.

The processing sequence was designed to preserve "true" reflection amplitude. Inversion to acoustic impedance followed the method of Lindseth (1979). The low-frequency (0–7 Hz) contribution to the impedance, due to density, was derived from conversion of seismic stacking velocities to interval velocities, using a Gardner's relationship. The high-frequency (7–110 Hz) contribution to the impedance was derived by first scaling the true amplitude data to reflectivity, then computing an exponentiated running sum. Care was taken to ensure that the high- and low-frequency components were correctly balanced. The final impedance is the product of the low- and high-frequency components. These results compared favorably with impedance curves computed from wireline acoustic and neutron-density data.

The S2 interval was identified on stacked and migrated seismic sections using synthetic seismograms. Although the S2 interval was interpreted on amplitude images, the overall stratigraphy is better illustrated on an instantaneous phase section using a reverse image (Figure 2A). The S2 horizon is the peak located immediately above the San Andres/Grayburg unconformity. Once the S2 interval was interpreted and the horizon flattened, reflection amplitude and acoustic impedance values were extracted at each CDP location. Acoustic impedance for the S2 interval is shown in Figure 2B. Impedance values were rescaled, with a change in sign, to range between –1 and 0, with the darker shades correlating to smaller impedance values (lower density and slower interval velocities). The impedance data were resampled to an equivalent CDP spacing of about 24.5 m. The 16,900 impedance values were used as "soft" or secondary data in estimating interwell porosity.

DATA INTEGRATION METHOD

Geostatistical data integration methods are analogous to a regression problem. The objective is to make a prediction at an unsampled location for variable A based on its relationship (linear combination) with one or more secondary variables. Variable A is sparsely sampled compared to the secondary data. Variable A in this study is average porosity at seven, 14, or 55 well locations, with 16,900 acoustic impedance (AI) values comprising the secondary data set.

If a correlation can be established between the primary variable, porosity, and the secondary variable, AI, then it would be desirable to use AI to predict porosity at interwell locations. Such an integration takes advantage of the locally more precise, but sparse well data and the densely sampled, but locally "soft" seismic data.

Several geostatistical data integration methods are available. Generally, the choice depends somewhat on the type and amount of data available and the purpose of the study. The external drift method is used in this study.

In nonstationary geostatistics, the drift (or local trend) can be modeled using a set of polynomials, which are functions of the x and y coordinates. These drift functions are polynomials of degree 0, 1, or 2. Degree 0 is no drift; therefore, the data are stationary and should be modeled with a variogram. A linear drift is degree 1, and a quadratic drift is degree 2. In matrix form, n lines and columns are added to the matrix of covariances and n dimensions added to the vector of kriging weights (Galli and Meunier, 1987; Moinard, 1987).

Once a meaningful correlation is established, in this case between porosity and AI, functions of AI are used as a drift function, rather than the functions of the x and y coordinates. Thus, AI becomes an "external drift" or a correlated shaping function that introduces local trends into the kriging system and accounts for spatial variability not sampled by sparse well data (Galli and Meunier, 1987; Moinard, 1987; Araktingi, et al., 1993). This approach requires that the external drift be known at all estimation points (grid nodes) and at all data points (well locations).

Note that the external drift method has an intrinsic self validation property. In the case of low correlations, kriging with a drift function filters out the effect of the drift (Marbeau, 1993, personal communication).

CALIBRATION OF THE SEISMIC DATA

Acoustic impedance was estimated by point kriging at each well location using a linear variogram model and an eight CDP neighborhood from the 42,000 impedances, not from the resampled data. The AI values were simply transferred to the grid nodes using a linear variogram. There was no interpolation in the last step because the grid node dimensions are congruent with the 16,900 CDP locations in the resampled seismic data.

The correlation coefficient for AI and porosity is 0.56 for all 55 data values; however, after removing anomalous impedance values that are probably due to the effect of thin-bed tuning, the new correlation coefficient for 47 data points is 0.76 (Figure 3). This is close to the –0.82 correlation found between averaged whole core porosity and acoustic log values for the same interval. Point correlation measures, like the product-moment coefficient, are not always the best indication of correlation, especially in data sets that exhibit spatial correlation; however, some reasonable degree of correlation must exist among the data for meaningful results. Data integration procedures, such as external drift, cokriging, and collocated cokriging, may be beneficial even with a point correlation as low as 0.4 (Araktingi, 1991, personal communication).

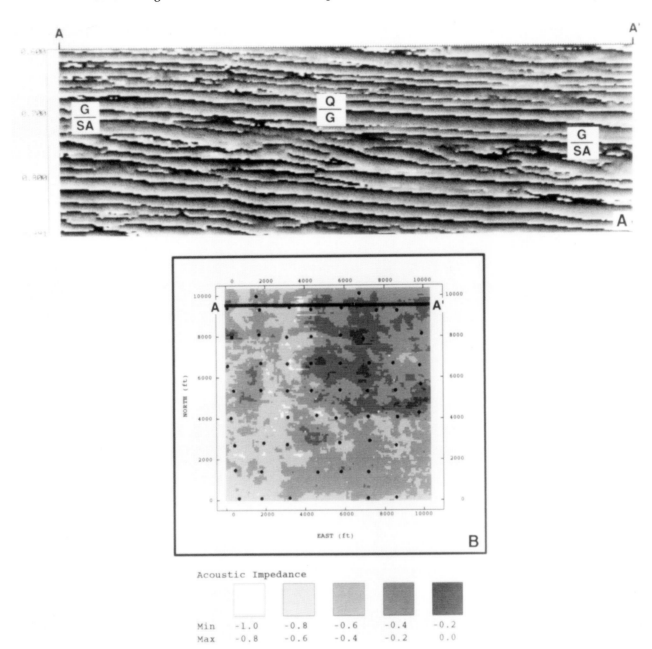

Figure 2. (A) Instantaneous phase display (reverse image) of a seismic line AA′ in the northern portion of the NCU. The Grayburg Formation (G) is between the Queen (Q) and San Andres (SA) Formations. The S2 interval is the peak indicated at the G/SA boundary. (B) A normalized acoustic impedance map of the S2 horizon, and seismic line AA′.

THREE CASE STUDY SCENARIOS

Because the purpose of this study is to illustrate the effect of integrating seismic data as secondary or soft constraining data into a reservoir description, three case studies were designed to mimic different stages in a reservoir's life: (1) late exploration/appraisal; (2) development; and (3) maturity.

Generalized covariance models were computed for average porosity for the three cases with and without an external drift. During the modeling of the generalized covariance, the nugget term was forbidden as

one of the possible sets of coefficients. The residual variance (mean error variance) for the regression of whole-core porosity against acoustic traveltime is 0.56 and 1.1 for porosity against AI. These values were used as the nugget terms for kriging and simulations with and without an external drift, respectively. This method of estimating a nugget term for generalized covariance modeling was suggested by Marbeau (1989, personal communication).

The grid cell size for kriging without an external drift is 122 × 122 m (26 × 26 cells). For simulation without external drift the grid cell size is 61 × 61 m (52 × 52

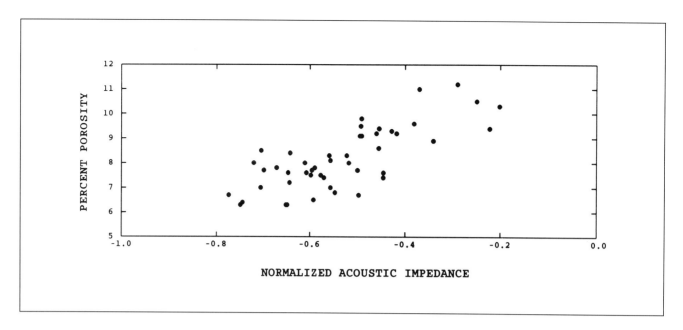

Figure 3. Cross plot of average S2 porosity and normalized acoustic impedance. Correlation coefficient is 0.76.

cells). For kriging and simulation with an external drift, the grid cell size is 24 × 24 m (130 × 130 cells); equivalent to the decimated seismic data set of 16,900 CDP. The x and y dimension for all grids is 3170 m.

55 Well Data Case

This data case represents a typical Permian basin well spacing with a drainage area of 40-ac. A map of kriged average S2 porosity is shown in Figure 4A. This result serves as a baseline against which the seven- and 14-well data cases are compared. To illustrate the spatial relationship between average S2 porosity and AI, kriged results of the 55 AI values are shown in Figure 4B. The correlation coefficient for these maps is 0.7, a result lower than expected, considering the apparent good visual correlation. You may recall that the S2 consists of three distinct siltstone beds separated by dolomite in northwestern portion of the study area. The low AI in that area are probably due to the effect of thin-bed tuning, as previously predicted in the seismic models, lowering the overall correlation coefficient between the maps in Figure 4. To make this study more accurate, the local impedance anomalies should have been corrected prior to data integration.

Kriging with external drift (KED) is shown in Figure 5. The smoothing effect of kriging is apparent if the overall texture of Figure 5 is compared to Figure 4B, the AI map; however, there is more local texture in the KED results than by simply kriging only the well data. Notice that the local AI anomaly in the northwest corner of Figure 4B does not appear as a local porosity high in the KED results. Because the local correlation among the data is low, the well control is honored more than the seismic data. The porosity high in the southwestern corner of Figure 4A is not as apparent in the KED map, although a higher poros-

ity trend is seen in the KED results. In this case, the well control suggests high porosity, but seismic data suggest low porosity. In the northeast quadrant there is good agreement between the KED results and kriging. KED results tend to show more area in the 9–12% porosity range, with some locally lower porosity areas not apparent on Figure 4B.

Seven Well Data Case

A seven-well subset of the 55 well locations was selected using the following criteria: wells were sampled with an even spatial distribution of data points to create a strong linear increase in porosity from west to east, so that the highest porosity appears to be in the southeast quadrant of the map. The number of wells in this case study is representative of a late exploration or appraisal phase. A kriged solution of the seven porosity data is shown in Figure 6A and should be compared to Figure 4A, the 55 well case.

Compare the seven-well KED (Figure 6B) with the 55–well KED (Figure 5). The seismic data have a much stronger conditioning influence in a data set with only seven wells. The high porosity zone in the northeast quadrant is reproduced, but a calculated pore volume for the seven-well KED is much greater than in the 55–well KED. Although the seven-well KED results are optimistic, porosity trends closer to "reality" are reproduced. Further field development using the seven-well KED map, rather than a map based on just the seven-well data, could avoid drilling potentially dry or only marginally productive wells in the southeast quadrant.

One could argue a case for simply converting the impedance map to porosity through a regression procedure. Granted, a similar map would be produced, but regression techniques have two main drawbacks: (1) the methods do not honor the data unless local

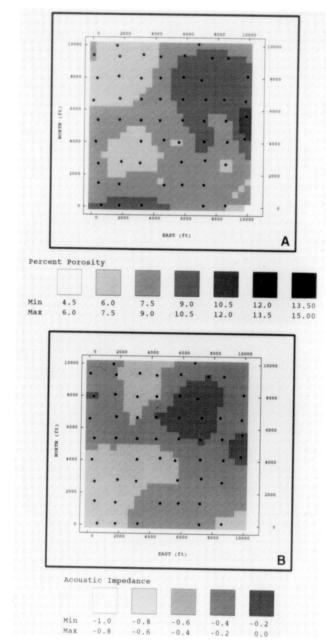

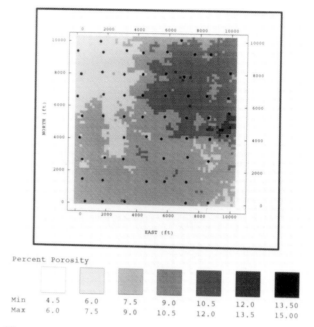

Figure 5. Kriging of 55 porosity data with normalized acoustic impedance as an external drift.

Figure 4. (A) Kriging of 55 porosity data and (B) 55 normalized acoustic impedance data estimated at the well location.

(1) sample the northeast quadrant; (2) to test reservoir limits; and (3) to sample the total range of porosity known from the 55–well data case. This scenario is representative of a development phase.

The 14-well kriged map and fourteen-well KED map are shown in Figure 7. The conditioning effect of the seismic data is intermediate to the seven- and 55–well data cases. A calculated pore volume for this scenario is less optimistic than the seven-well case, but still overestimated when compared to the 55 case.

STOCHASTIC SIMULATIONS

The smoothing effect characteristic of all gridding algorithms, kriging included, can bias results aimed at understanding reservoir heterogeneity. To reproduce local variability in rock properties in the inter-well regions, one must use stochastic simulations. KED can add interwell variability, but the kriging process also smoothes the drift function. Stochastic simulation, coupled with external drift, adds interwell variability that is likely present in the reservoir. The simulations are said to be "conditional" because they honor the sample data.

During the simulation process, the kriged estimate and kriging variance are computed at each grid node, then a random number is drawn from a normal distribution, with mean and variance equal to the sample data mean and variance, and placed at the grid node. Each simulated value is then treated as real data in the simulation of subsequent grid values. The simulation process preserves not only the spatial information, but also the sample data variance. In this study, simulated porosity values were allowed to range between 4.5 and 15% porosity, a range slightly

regressions are made and (2) to honor the data, residuals should be gridded and added to the regression, but using what method? This step is analogous to universal kriging, but intrinsic random functions of order k always beat a universal kriging in a cross validation. A map of estimated porosity is not the only desired result, we also want some assessment of uncertainty in the mapped results, a topic addressed in the section on stochastic simulations.

14 Well Data Case

An additional seven wells were added to the data integration process. Well locations were selected to

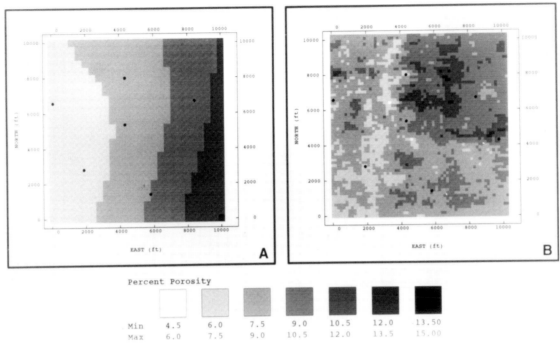

Figure 6. (A) Kriging of seven porosity data and (B) kriging of seven porosity data with normalized acoustic impedance as an external drift.

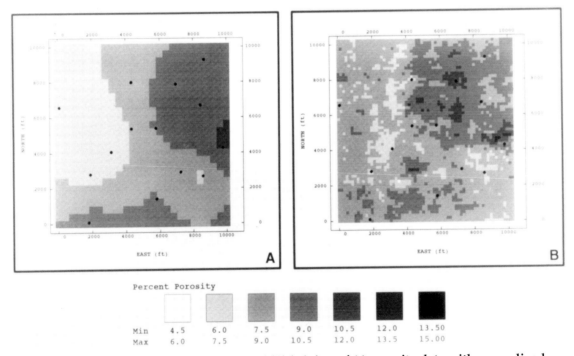

Figure 7. (A) Kriging of 14 porosity data and (B) kriging of 14 porosity data with normalized acoustic impedance as an external drift.

larger than the actual range for the 55 porosity values, and was similar to the range measured in whole core for the S2 siltstone.

Besides honoring the sample data, or a user-imposed distribution, stochastic simulations generate multiple equally probable realizations of a data set. Analyzing the variability in numerous simulations is useful for assessing uncertainty in the kriged results or for assessing the reduction in uncertainty through integration of secondary data. The mean value of n simulations is the kriged solution. Uncertainty is assessed by computing a standard deviation map from the n simulations. Histograms, probability plots, and measures of central tendency and spread can be computed for any

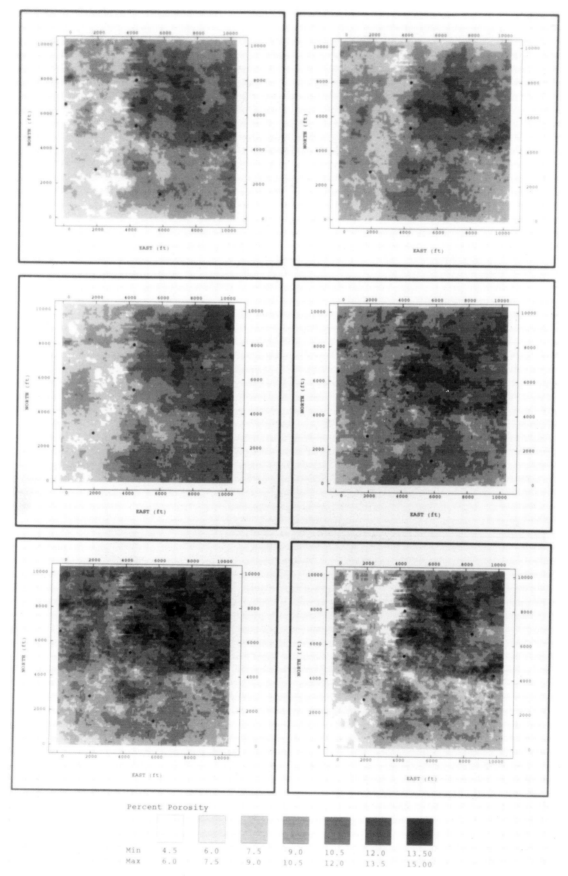

Percent Porosity

| | Min | 4.5 | 6.0 | 7.5 | 9.0 | 10.5 | 12.0 | 13.50 |
| Max | 6.0 | 7.5 | 9.0 | 10.5 | 12.0 | 13.5 | 15.00 |

Figure 8. Six conditional simulations of seven porosity data with normalized acoustic impedance as an external drift.

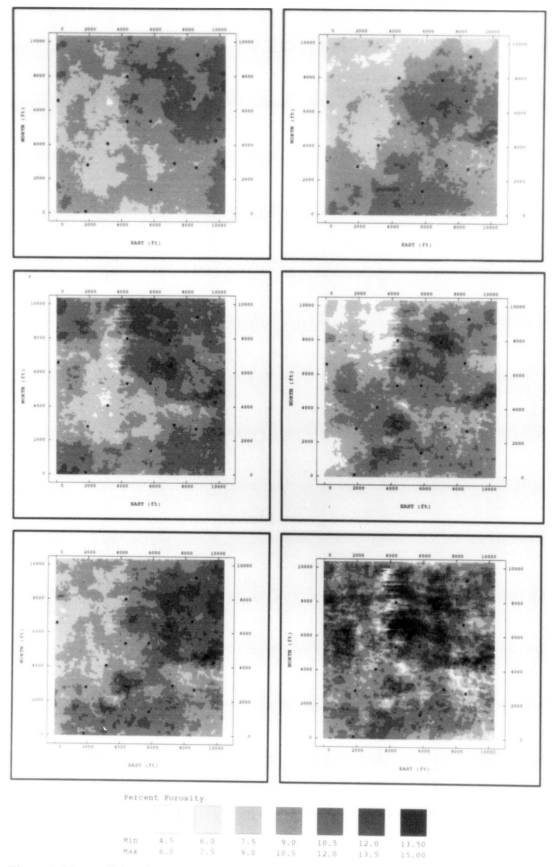

Figure 9. Six conditional simulations of 14 porosity data with normalized acoustic impedance as an external drift.

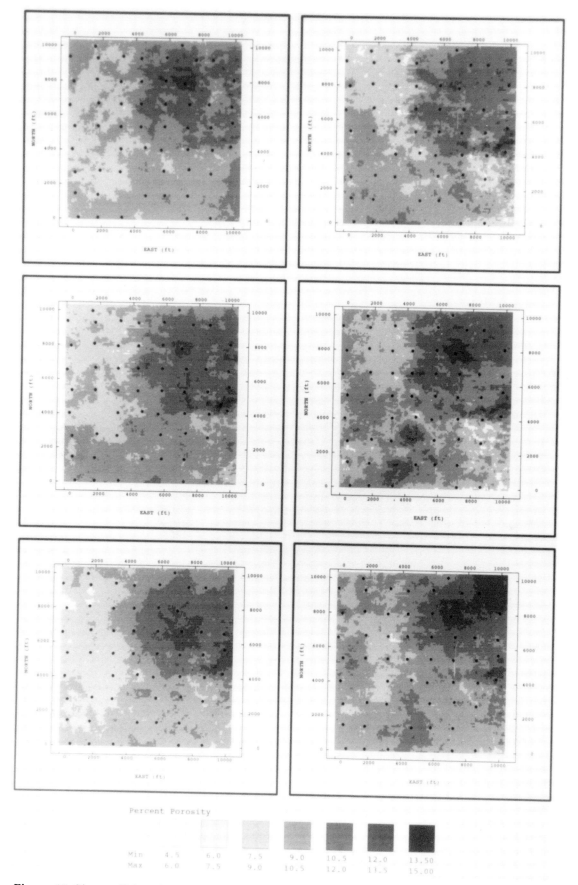

Figure 10. Six conditional simulations of 55 porosity data with normalized acoustic impedance as an external drift.

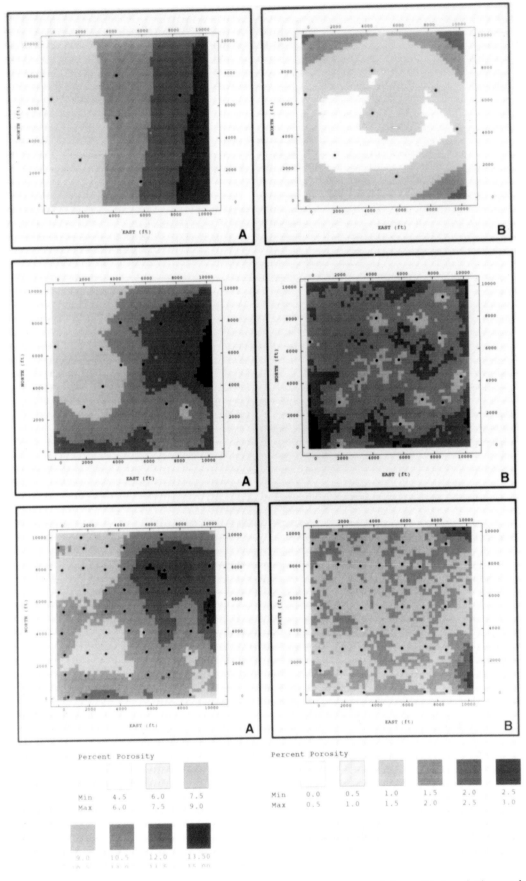

Figure 11. (A) Mean and (B) standard deviation maps computed from 30 simulations of seven, 14, and 55 porosity data, respectively, from top to bottom.

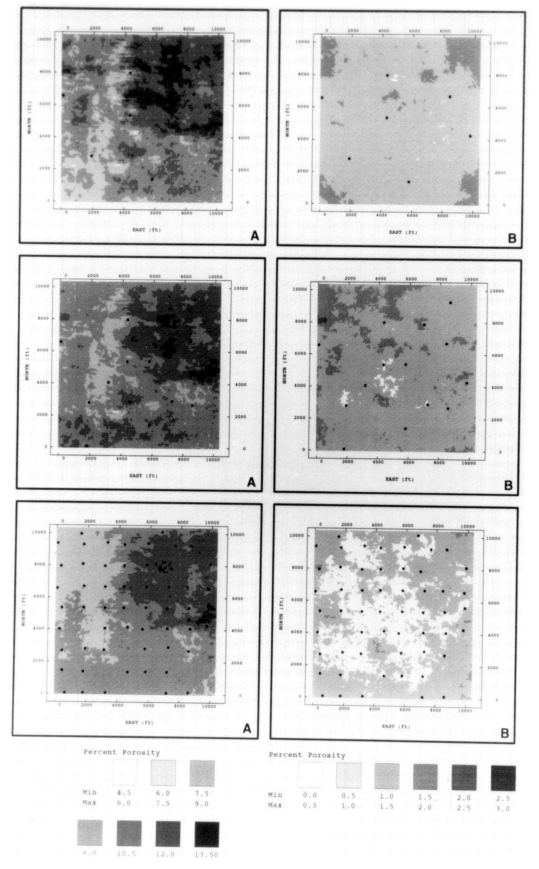

Figure 12. (A) Mean and (B) standard deviation maps computed from 30 simulations of seven, 14, and 55 porosity data, respectively, from top to bottom, with normalized acoustic impedance as an external drift.

grid node. Risk or probability of exceedance maps also can be generated for any porosity cutoff or probability value. For an example of the last two possibilities see the paper by Wolf et al. (this volume).

Thirty conditional simulations, with external drift, were generated for the seven-, 14-, and 55–well data cases. Six simulations were selected from each set of 30 to illustrate the range of outcomes and are shown in Figures 8–10. Simulations without external drift are not shown. The mean of the 30 simulations (without and with external drift) and their standard deviations are shown in Figures 11 and 12. Thirty realizations probably do not capture the total variance possible from a simulation process. Notice, however, the similarity of the means of the 30 simulations (Figure 12A) to their KED counterpart (Figures 5, 6B, 7B). As the number of simulations increases, their average is closer to the kriged solution.

Figure 11 illustrates the mean and standard deviation maps for 30 simulations without external drift. The large area of low standard deviation in the seven-well case (Figure 11B, upper right) is not a surprise. The data area centrally located and the generalized covariance model used a linear drift with cubic coefficients. The standard deviation map for 55 wells (Figure 11B, lower right) is about one-half the error for 14 wells (Figure 11B, center right). The results simply reflect the amount of conditioning data.

Maps of the mean and standard deviation for 30 simulations with external drift (SED) are shown in Figure 12. The amount of data for each case is almost the same, the only difference is in the amount of hard conditioning data and seven, 14, or 55 porosity values. Making a case for a reduction in error for a seven-well simulation (Figure 11B) versus a seven-well SED (Figure 12B) is difficult. In the seven-well SED example, part of the increased error can be partially explained by the magnitude of the nugget value. Recall that the nugget value is 1.1 for KED and SED, almost twice the value for the nonexternal drift examples. The error for 14- and 55–well SEDs (Figure 12B) was reduced to almost one-half the error in the nonexternal drift examples (Figure 11B).

DISCUSSION AND CONCLUSIONS

Three reservoir scenarios representative of late exploration/appraisal, development, and mature-field development phases were designed to illustrate the impact of using seismic data as a secondary variable to constrain a reservoir description.

The general motivation behind data integration is to combine data in an effort to obtain more accurate results having reduced uncertainty than would be achieved by using only a portion of the data. In theory, this is great; however, in practice we often find it difficult to combine data measured by different tools and at different scales. Several geostatistical methods are available that provide the fundamental framework for quantitative data integration.

The integration of secondary data sets, such as acoustic impedance data from the inversion of 3-D seismic data, can significantly reduce interwell estimation uncertainty. The soft data also add small-scale variability not sampled from well data alone. Extra care is required in the data processing and inversions steps if seismic data are to be quantitatively integrated into a reservoir description, especially if the final product is a fluid flow model. To realize the full potential of the secondary data set, data types must be properly correlated and scaled.

The external drift method was used in this study to avoid some of the problems encountered when the primary variable is grossly under-sampled with respect to the secondary variable. In this situation, the cokriging covariance model can be very underdetermined, producing a covariance model not admissible (positive definite). In general, cokriging is best used when the variables are measured at the same location. Other difficulties encountered with cokriging include the calculation and modeling of cross variograms and finding a suitable search neighborhood. Calculating and modeling cross variograms are not done directly and finding a suitable search neighborhood is more straightforward with the external drift approach.

One of the most important benefits of geostatistical methods is the ability to assess uncertainty associated with kriging estimates using stochastic methods. These methods also allow us to assess the contribution of a secondary data set by assessing its ability to reduce overall variance, thereby reducing uncertainty. Figures 11 and 12 compare the standard deviation maps with and without, respectively, the constraints imposed by a secondary variable. The industry as a whole is becoming more risk conscious. We need to ask more often how good our estimates are, what the uncertainties are, and what risk factor is involved.

Although geostatistical studies can involve more resources in terms of time and computer costs, updating previous results as additional data are acquired is fairly easy. The initial investment spent learning the intimate details about a data set, using classical statistical and geostatistical methods in tandem, is well worth the effort. We end up with a product that integrates all available data and the ability to assess the uncertainty in the results.

ACKNOWLEDGMENTS

We thank Amoco Production Company for permission to publish this paper. Special thanks go to the seismic processing team: Joey Hammond, Duane Ehrle, and Nikki Cutburth, from Amoco Production Research. We also extend thanks to John Meyers, our seismic data acquisition field foreman. Al Frisillo, the group supervisor, provided valuable advice in designing the seismic data acquisition scheme and later during the data processing. Patty Barron assisted in the preparing the figures. Finally, our appreciation goes to Jean-Paul Marbeau, Larry Lines, and Mike Ross for reviewing this paper.

REFERENCES

Araktingi, U. G., W. M. Bashore, T. T. B. Tran, and T. A. Hewett, 1993, Integration of seismic and well log data in reservoir modeling, *in* B. Linville, ed., Reservoir characterization III: Tulsa, Oklahoma, PennWell Publishing, p. 515–554.

Cairns, J. L., and L. D. Feldkamp, 1992, Three-dimensional visualization for improved reservoir characterization: SPE Paper 24269, SPE European Petroleum Computer Conference, Proceedings.

Doyen, P. M., 1988, Porosity from seismic data: a geostatistical approach: Geophysics, v. 53, p. 1263–1295.

Doyen, P. M., and T. M. Guidish, 1992, Seismic discrimination of lithology and porosity, a Monte Carlo approach, *in* R. E. Sheriff, ed., Reservoir geophysics: investigations in geophysics, v. 7: Tulsa, Oklahoma, Society of Exploration Geophysicists, p. 243–250.

Doyen, P. M., T. M. Guidish, and M. H. de Buyl, 1988, Lithology prediction from seismic data, a Monte Carlo approach: 58th Annual International Meeting of the Society of Exploration Geophysicists.

Galli, A., and G. Meunier, 1987, Study of a gas reservoir using the external drift method, *in* G. Matheron and M. Armstrong, eds., Geostatistical case studies: Dordrecht, D. Reidel, p. 105–109.

Journel, A. G., and F. G. Alabert, 1990, New method for reservoir mapping: Journal of Petroleum Technology, v. 42, no. 2, p. 212–218.

Lindseth, R. O., 1979, Synthetic sonic logs—a process for stratigraphic interpretation: Geophysics, v. 44, p. 3–26.

Moinard, L., 1987, Application of kriging to the mapping of a reef from wireline logs and seismic data: a case history, *in* G. Matheron and M. Armstrong, eds., Geostatistical case studies: Dordrecht, D. Reidel, p. 93–103.

Nestvold, E. O., 1987, The use of 3D seismic in exploration, appraisal and field development: Proceedings of the Twelfth World Petroleum Congress, v. 2, p. 125–132.

Nestvold, E. O., 1991, 3-D seismic: is the promise fulfilled?: Expanded Abstracts, v. 1, 61st Society of Exploration Geophysicists Annual Meeting, p. 717–720.

Nestvold, E. O., and P. H. H. Nelson, 1992, Explorers still hope to improve on 3-D seismic's wealth of data: Oil & Gas Journal, March 16, p. 55–61.

Robertson, J. D., 1991, Reservoir management using 3-D seismic data, *in* L. W. Lake, H. B. Carroll, Jr., and T. C. Wesson, eds., Reservoir characterization II: San Diego, Academic Press, p. 340–354.

Chapter 14

Importance of a Geological Framework and Seismic Data Integration for Reservoir Modeling and Subsequent Fluid-Flow Predictions

William M. Bashore
Udo G. Araktingi
Reservoir Characterization Research and Consulting, Inc.
Fullerton, California, U.S.A.

Marjorie Levy
William J. Schweller
Chevron Petroleum Technology Company
La Habra, California, U.S.A.

ABSTRACT

Fluid-flow simulation results are used extensively as reservoir performance predictions upon which to base economics for reservoir management decisions. The generation of numerical models for simulation purposes may be easily facilitated by computer-aided algorithms, regardless of the quality of data or input parameters. This study contrasts two different geological interpretation styles: lithostratigraphic and chronostratigraphic. Specifically, comparisons are made as to properly integrating seismic-based information and to potentially erroneous conclusions deduced from simulator predictions if the simulation models are built without a sound geological framework. Models derived from the two different correlation strategies using only well logs are compared. Seismic inversions are included within the chronostratigraphic framework as a third model type. Multiple realizations of each model type are input to a fluid-flow simulator. Selecting the appropriate simulation results is closely tied to the reservoir management objective in question. Histograms of breakthrough times indicate little difference between the well-only models of either correlation strategy, whereas water displacement patterns are significantly different. Models that have been conditioned by the seismic pseudologs show substantially different results for both breakthrough and displacement distributions, and the spread or uncertainty in breakthrough estimates is greatly reduced compared to the well-only model results.

A cloud transform method with correlated probability fields is introduced for stochastically estimating one model parameter (porosity) from another (impedance). This method allows incorporating the scatter in the relationship between the two parameters (crossplot).

INTRODUCTION

Generating highly detailed, numerical "geological" models is becoming common practice in the absence of close collaboration between reservoir engineers and development geologists. This situation has been reinforced by the advent and use of newly available geostatistical workstation applications (e.g., Araktingi et al., 1992). The minimum input data required to generate a reservoir simulation model are some well logs and a few correlated markers at each well location. Fluid-flow simulations are then performed on these models to provide performance predictions needed for reservoir management decisions. Given the financial significance of the results obtained from reservoir models, two questions are raised: how critical is the geological interpretation as a guiding framework for estimating interwell reservoir properties, and what are the consequences for fluid-flow predictions? The intent of this study is to explore these questions and to demonstrate the importance of geoscientists and reservoir engineers working together to build reservoir simulation models and the ultimate impact on economic decisions.

Two specific objectives are proposed for this study:

- Investigate the effects of the geological correlation strategy selected (either lithostratigraphic or chronostratigraphic) on reservoir model building, especially when incorporating seismic data, and on the flow predictions simulated using these models.
- Investigate the subsequent differences in fluid-flow predictions based upon these models.

The estimation and conditional simulation techniques employed are based on using models of spatial continuity and the kriging algorithm (e.g., Isaaks and Srivastava, 1989). Specifically, three sets of cross sectional models are generated from (1) impedance logs correlated by prominent lithologic packages; (2) impedance logs correlated chronostratigraphically through an integrated sequence interpretation of well-log, seismic, biostratigraphic, and geochemical data; and (3) the same chronostratigraphic correlation scheme, but incorporating seismic pseudologs (traces created from seismic inversion) as an additional constraint. Conversion to porosity is accomplished by applying a "cloud transform" technique to capture the uncertainty associated with the impedance-poros-

ity scattergram.[1] Permeabilities are calculated as an exponential function of porosity.

Each set of model types contains about 100 equiprobable cross sections that are waterflooded in a reservoir simulator. Histograms of breakthrough times, a flow parameter commonly analyzed, are displayed for comparative analysis of flow predictions using each of the three model building approaches. Fluid-displacement cross sections are also displayed.

STUDY AREA, STRATIGRAPHIC INTERVAL, AND DATA SETS

The study area is a continental shelf and upper slope setting along a north-south passive margin. The Upper Cretaceous stratigraphic interval used in this study consists of predominantly siliciclastic sedimentary rocks that thin from about 150 m on the paleoshelf to less than 50 m on the paleoslope. Subsidence and sedimentation have buried the interval to a depth of over 1300 m and present-day regional dip is about 2° to the east. The 200-km² area contains twelve wells and eight seismic lines (Figure 1). The log suites used for lithology determination primarily consist of gamma-ray, sonic, density, and neutron logs. Palynological and micropaleontological data were extracted and interpreted from sidewall cores, where available, and from cuttings. Pyrolysis information, albeit sparse, was also collected from several wells.

The seismic lines, originally acquired at a 25-m subsurface interval, were decimated to every fourth trace (100-m spacing) prior to loading onto the workstation for interpretation and prior to inverting for acoustic impedance. These traces were further decimated to one every 400 m for use in estimation and conditional simulation. Unfortunately, until smarter search and random data access routines are implemented, this decimation is required owing to memory limitations in computer software. This eventual 16:1 decimation has eliminated significant information, but the variogram analysis indicates that the spatial variance in the seismic data is well modeled by the theoretical variogram (see Figure 5). This model of spatial dependency implies that the eliminated traces can be predicted with small error from the remaining traces as long as the intertrace distances are small

[1]A stochastic approach for transforming one variable to another to preserve the scatter or uncertainty in their cross-relationship (i.e., crossplot or scattergram). See the porosity and permeability models following for discussion.

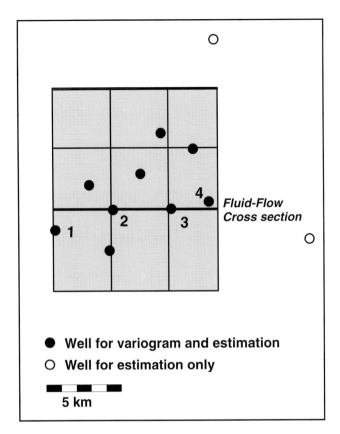

Figure 1. Base map showing relative positions of seismic lines, well control, and the cross section used in fluid-flow simulations.

enough to capture the range of variation (i.e., at least two per correlation range). The minimum range of correlation defined from the seismic data is 1200 m, so the 400-m sampling equals three traces for this range.

GEOLOGICAL CORRELATION STRATEGIES

The two types of geological correlation strategies applied in this study are lithostratigraphic and chronostratigraphic (van Wagoner et al., 1990). The lithostratigraphic approach assumes that packages of similar lithologies are laterally connected. A lithostratigraphic correlation would suggest that as a typical delta builds across a continental shelf, the sandy nearshore and shallow-marine sediments build outward across muddy prodelta deposits (Figure 2). This strategy predicts that common lithofacies correlate over long distances. For example, the upper part of the delta forms a laterally continuous sandy layer. Thus, lithostratigraphic correlation lines in a deltaic system parallel regional stratigraphic markers.

In contrast, the chronostratigraphic model recognizes that a deltaic depositional system progrades through time and that at any given time, facies grade laterally from one to another. As a result, minor lateral changes in depositional environments through time can create abrupt vertical facies changes, as well as lateral changes in lithology. Chronostratigraphic correlation lines highlight changes in depositional geometry (Figure 2). This correlation scheme does not correlate based upon prominent lithologic changes, unless there are additional data, such as seismic, biostratigraphic, or geochemical data, indicating it is an isochronous layer.

LITHOSTRATIGRAPHIC VERSUS CHRONOSTRATIGRAPHIC INTERPRETATION

The comparative cross section consists of four wells that closely tie one of the dip-oriented seismic sections (Figure 3). These four wells were correlated

Figure 2. Diagram illustrating differences between lithostratigraphic and chronostratigraphic interpretations of two well logs in a dip-oriented cross section through a prograding deltaic margin. Lithostratigraphic correlation assumes that similar lithologies connect to form a single reservoir layer. Chronostratigraphic correlation recognizes the importance of depositional processes in determining the geometry and connectivity of reservoir sands.

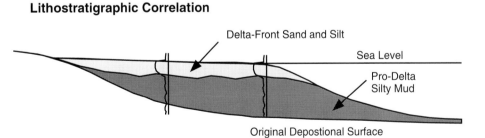

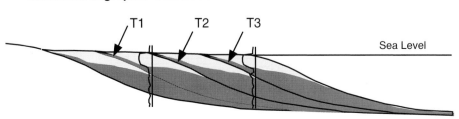

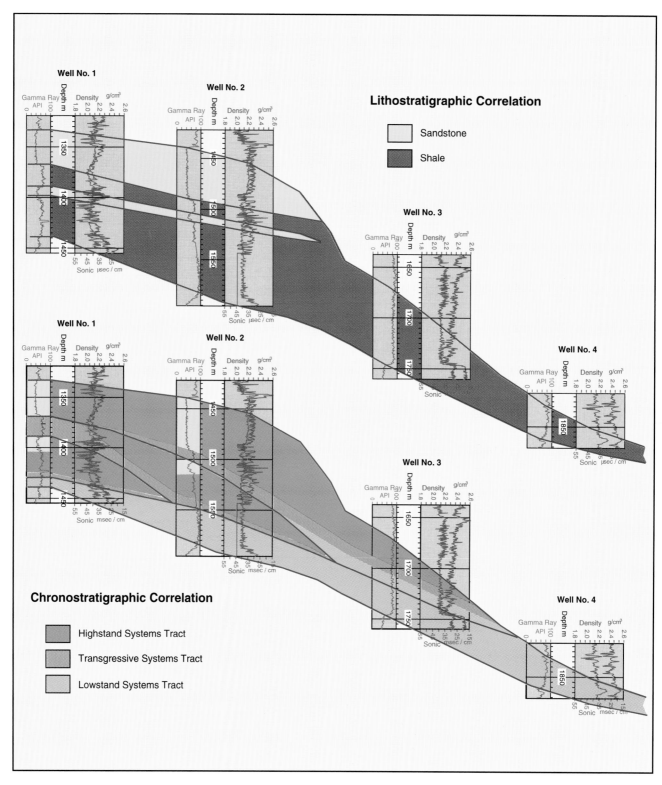

Figure 3. (A) The upper diagram shows the interpreted lithostratigraphic correlation. In this interpretation, two layer-cake sand bodies are present separated by an intervening shale. The lower diagram shows the chronostratigraphic correlation and the systems tracts present in each prograding package (Brown et al., 1977; Vail et al, 1977; Vail, 1987; van Wagoner et al., 1990). In this interpretation, the upper sand body is still connected; however, the thinner sand units in wells 1 and 2 are separated by an intervening, inclined shale layer and are not connected. Cross sections are not referenced to a specific datum.

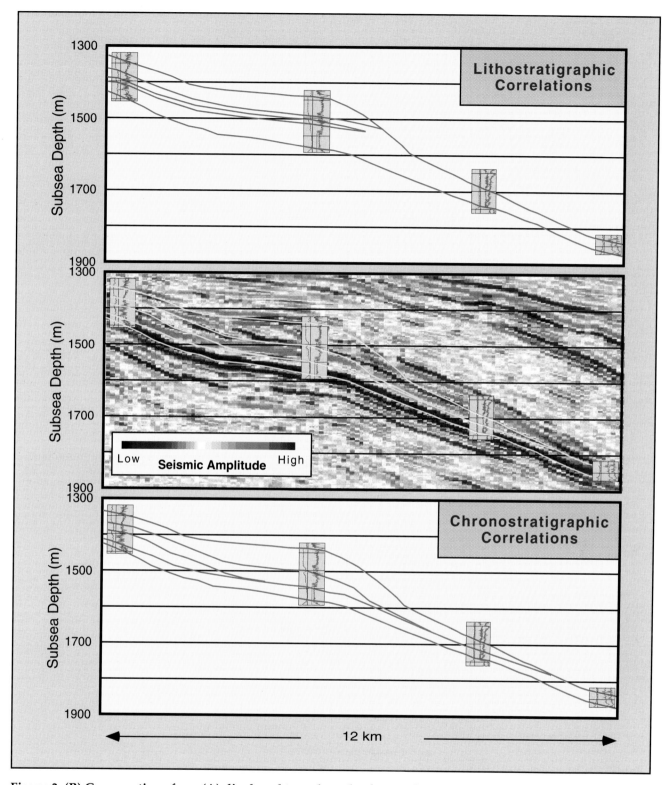

Figure 3. (B) Cross sections from (A) displayed to scale and referenced to sea level. The middle cross section contains the seismic line, whose reflectors clearly shows a clinoform geometry. In conjunction with biostratigraphic data, the seismic and well-log data were reconciled to produce a consistent chronostratigraphic interpretation [lower cross section in (A)].

lithostratigraphically across the shelf and upper slope by interpreting lithology from the available log suite. Sparse biostratigraphic data were used to locate major boundaries. The chronostratigraphic approach involved multiple iterations of interpreting the well-log, seismic, and biostratigraphic data so as to achieve a single interpretation consistent with all available information.

The lithostratigraphic interpretation recognizes the presence of two major sand bodies, each of which builds outward across the shelf and pinches out or is truncated before it reaches the upper slope (Figure 3). These sand bodies are separated by an intervening shale, which either merges with the lower shale on the outer shelf or is similarly truncated. Either lithostratigraphic interpretation predicts layer-cake geometry for the sand and shale units, and suggests high connectivity within the thick upper sand bodies in well 1 and well 2, as well as within the thinner sand bodies.

The chronostratigraphic interpretation presumes that individual deltaic sand packages thin and grade into shale basinward along any chronostratigraphic surface, although they coarsen upward in a vertical profile. The interpretation shows three prograding packages overlying a basal fine-grained unit (Figure 3). The oldest package is present only in well 1 and consists of a lower transgressive shale portion and an upper prograding sandstone unit that thins and becomes finer basinward. The second package overlies the prograding sandstone unit of the oldest package, and consists of a basal transgressive shale across the entire study area and a progradational sandstone package that pinches out updip east of well 1 and downdip west of well 3. The youngest prograding package caps the entire sequence. This package consists of a fine-grained basinal deposit in wells 3 and 4, overlain by a transgressive shale that pinches out updip east of well 1, which, in turn, is overlain by a thick prograding sandstone unit that thins and grades into shale east of well 2. This interpretation predicts a prograding clinoform geometry, and suggests that the thin sandstone bodies in wells 1 and 2 are isolated from each other and form discrete sand bodies.

When comparing the results of the two interpretation strategies, it is important to understand the integrated chronostratigraphic interpretation philosophy. The interpretation of no one data set is allowed to drive the interpretation of another. For example, a common variation on the seismic stratigraphy interpretation procedure espoused by Vail (1987) for obtaining an "integrated" sequence interpretation is to transfer the biostratigraphic interpretation (e.g., tops, paleobathymetry, etc.) onto the well logs. The well log interpretation is made for lithologies and system tracts based upon log character. Finally, the interpreted sequence markers are transferred to the seismic data via a synthetic seismogram. The sequence and system tract boundaries are then correlated along continuous seismic reflectors. Maps are made from these horizons; unfortunately, any errors made in any interpretation step are propagated and, possibly, magnified in the final interpretation.

The creation of precise time-depth functions was crucial to making the chronostratigraphic correlations. The match filter scheme employed to generate these functions is driven purely by statistical cross-correlation between the well synthetic seismogram and the associated seismic trace. As stretches and shifts were applied to the sonic log values, we paid careful attention to the alignment of well-log depth markers and their seismic time picks. If differences were found, several things were checked: (1) the match filter process for unusual changes in the sonic log values, unreasonable time shifts, and estimated waveforms with poorly behaved characteristics, (2) changes in the log interpretation adjusted to the seismic interpretation, (3) changes in the seismic interpretation to fit the log interpretation, or (4) changes in the interpretation of biostratigraphic data to fit the correlation of the seismic from one well to the next. For example, the age assigned to a specific sequence boundary greatly changed as correlation misties along a prominent seismic sequence boundary were reconciled at several well locations. A check of the micropaleontological data found that older fossils, originally thought to date the sequence boundary, were reworked and the younger fossils, originally thought to be from hole caving, were in fact in situ and provided a consistent age for the sequence boundary.

The final chronostratigraphic interpretation through the four wells would not have been possible without the other wells and seismic lines in the data set. Also, without interpreting the sequence above the interval of interest, a different interpretation might have been reached. A process-oriented interpretation must be couched in terms of the processes not only locally, but in a basinal context as well.

SEISMIC DATA INVERSION

Seismic amplitudes are not measurements of point properties in the subsurface, but aggregate responses indirectly attributed to point properties. Much has been written and many case histories published demonstrating the utility of seismic amplitudes for predicting such things as the likelihood of gas accumulations, porosity build-ups, or channel sands. Although analyzing and interpreting seismic amplitudes have great merit, it is not possible to integrate them directly into reservoir models for engineering purposes, such as flow simulations or volumetric calculations. Thus, a method for converting these aggregate responses back to their underlying point properties is required; this method is seismic inversion (Lavergne and Willm, 1977; Lindseth, 1979; Bamberger et al., 1982).

The first step in the inversion process is estimating the imbedded waveform and removing it from the seismic traces, i.e., wavelet processing. A match-filter scheme between the well reflectivities and the seismic traces, where they are coincident, is used not only for wavelet extraction, but also for generating time-depth functions. One assumes that seismic data are better estimators of travel time than integrated sonic logs, and that well logs contain better estimates of depth than depth-converted seismic data. Log velocity values are adjusted to tie the well synthetic seismogram to the corresponding seismic trace. This processing must be carefully monitored in terms of the magnitude of changes made to the sonic log and the behavior of the waveform estimated.

The resulting depth-time functions produced from the match filter approach are kriged (or cokriged if other correlated velocity information from seismic processing is available) throughout the study area. In this way, one can convert all inverted seismic traces to depth and maintain the high level of sample-to-sample cross-correlations between well logs and seismic pseudologs.

Wavelets were extracted for eight well locations, and their constant-phase components have a range of about 32°. Modeling has shown flow results are insensitive to phase misestimation on the order of ±20° (Araktingi and Bashore, 1992); therefore, given the 32° phase range in this study, applying a single inverse wavelet to all lines does not alter results of fluid-flow simulations. The wavelet-processed data are then converted to sparse reflectivity sections by a spectral extrapolation algorithm (Walker and Ulrych, 1982; Oldenburg et al., 1983). This additional step will produce pseudologs with more blocky forms instead of pseudologs with unrealistic modulations created by the waveform's dominant period.

Two specific problems arise in the inversion of seismic reflectivities to impedance profiles: the lack of low-frequency or background information upon which to place the high-frequency information contained in the seismic bandwidth; and the need for dynamic scaling, both in space and time, of the seismic reflectivities to earth reflectivities prior to inversion. The information required to solve these problems is available at the well locations, but not elsewhere. Fortunately, the same statistical methods used for building the reservoir models may be used to estimate these seismic processing models (i.e., background impedance and reflectivity balancing models) at every seismic trace location.

One must maximize the sample-by-sample correlation between the inversion results and the log data if one intends to use both during the estimation process. To obtain absolute values from the inversion, it is extremely important to include the background model in the inversion. Araktingi and Bashore (1992) have demonstrated that the correlation between hard and soft data is critical to how much influence or conditioning the soft data will have on estimation and conditional simulation. Well-log spectral density plots, especially for acoustic logs, indicate a tremendous power contribution from the low frequencies, i.e., those not normally recorded in seismic data.

For this study, the background model is obtained by generating the low-frequency response of the impedance well logs (Araktingi et al., 1990), and then kriging these to the seismic trace locations. The kriging is constrained by marker surfaces picked from the seismic data. In this way, the background model and the high-frequency seismic reflectivities are coincident in time. Because the actual addition of the low- and high-frequency components in the inversion is performed in the frequency domain, they must be aligned in time prior to inversion (Santosa et al., 1986). This technique ensures this condition. Because the background model is low frequency in nature, and therefore has low vertical resolution (100s of meters), only the top and bottom markers of the geological interpretations were used; the question of lithostratigraphic versus chronostratigraphic correlation for this model is not germane.

The reflectivity balancing model is created in a similar manner to the background model. Reflectivity envelopes from synthetic seismograms are calculated for each impedance log, and these are kriged to each seismic trace location. Envelopes for the seismic traces are also generated, and the scaling function is computed as the desired envelope model (kriged solution) divided by the existing envelope model (trace envelope). The scaling function is applied to the sparse reflectivity estimated from the seismic traces. Univariate statistics determined for the balanced seismic reflectivity and for the reflectivity series of each well log are found to be quite similar. No further balancing is deemed necessary.

Figure 4 is a crossplot of the pseudo–impedance logs derived from the seismic data at the well locations with the actual impedance log measured in the well. The correlation coefficient between pseudologs

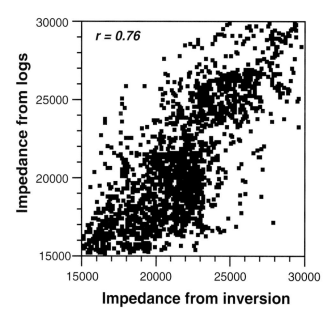

Figure 4. Crossplot of the impedance estimates, from the seismic inversion, with the impedance log derived from the sonic and density logs. The strong correlation is a result of carefully constraining the inversion with well-based background and reflectivity-balancing models.

Figure 5. Diagram illustrating the differences among the three spatial continuity models (variograms) used in generating the impedance models for each model type. The longer correlation ranges for the variograms in which only well logs were used for determination are due to longer interwell spacings versus seismic trace spacing. As a result, features estimated using these variograms will be more continuous than features estimated using the seismic-based variogram.

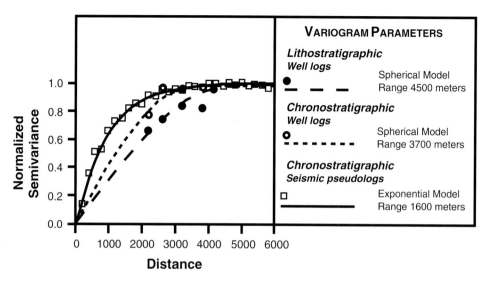

and impedance well logs is 0.76 as a result of the well-log constraints imposed when building the background and balancing models. The high degree of correlation indicates that seismic data can be used to predict interwell impedance better than using the sparse well control alone.

ACOUSTIC IMPEDANCE MODELS

To perform reservoir simulations one must construct models with porosity and permeability properties defined at each simulation grid node. Unfortunately, porosity and permeability logs do not exist for each well or for the complete interval of interest. To overcome the inability to estimate these interwell properties directly from logs, we used the relationship between porosity and acoustic impedance. Enough core porosity and sonic-density log measurements exist from this area and other areas of similar depositional style and history to establish the transform of porosity from impedance. The cloud transform technique used to obtain porosity from impedance is discussed in the next section.

Figure 5 shows the variogram models used for generating the kriging solutions for acoustic impedance along the cross section used for flow studies. The long correlation ranges (4.5 km and 3.7 km, lithostratigraphic and chronostratigraphic, respectively) for the variograms in which only well logs were used are consistent with the average well spacing (≈3 km). It is also not surprising that the chronostratigraphic correlation scheme indicates slightly less lateral continuity because it does not necessarily try to align common lithologies from well to well. The correlation length predicted from the seismic pseudologs is substantially shorter (1.6 km); moreover, this range is better sampled with a 400-m trace spacing than the well-only situation.

Comparison of estimated cross sections using these variograms may be seen in Figure 6. Interpolation of the upper model is guided by the lithostratigraphic correlations. As expected, the prominent trends in impedance are subparallel to the top and bottom mark-

ers. Although this cross section appears geologically plausible, it does not reflect the progradational nature of sediment deposition typically associated with outbuilding of the basin margin during periods of high sediment influx. The second model in Figure 6 does show a more progradational geometry of high- and low-impedance beds mimicking the clinoform patterns in the seismic section. Because of the similar correlation ranges for the two variograms defined by the two different correlation schemes (wells only), the lateral extent of high- and low-impedance layers are roughly equivalent. This similarity in correlation distance may induce similar breakthrough predictions, although the patterns of fluid displacement would be different.

The lowermost cross section in Figure 6 is the result of including the interwell impedance variations from the seismic inversions as a function of the confidence level defined by the cross-correlation of well impedance logs and pseudoimpedance logs (Figure 4). The technique used to perform the joint estimation is called collocated cokriging (Xu et al., 1992). The shorter range of correlation modeled from the seismic information suggests that the impedance distributions are not as laterally continuous as those modeled from the well data only. This reduction in lateral continuity is seen by comparing the two lower cross sections in Figure 6. Because of the complexity of deltaic depositional systems, it is likely that individual sand and shale deposits (high and low impedance, respectively) are not laterally extensive. Flow patterns from models with the second data set included are expected to be more disrupted than those without the second data set. Water displacements in the models without the seismic information should be more restricted to the continuous high-impedance zones.

POROSITY AND PERMEABILITY MODELS

Ten conditional simulations were created for each of the three model types using the Gaussian sequential simulation (GSS) approach (Deutsch and Journel, 1991).

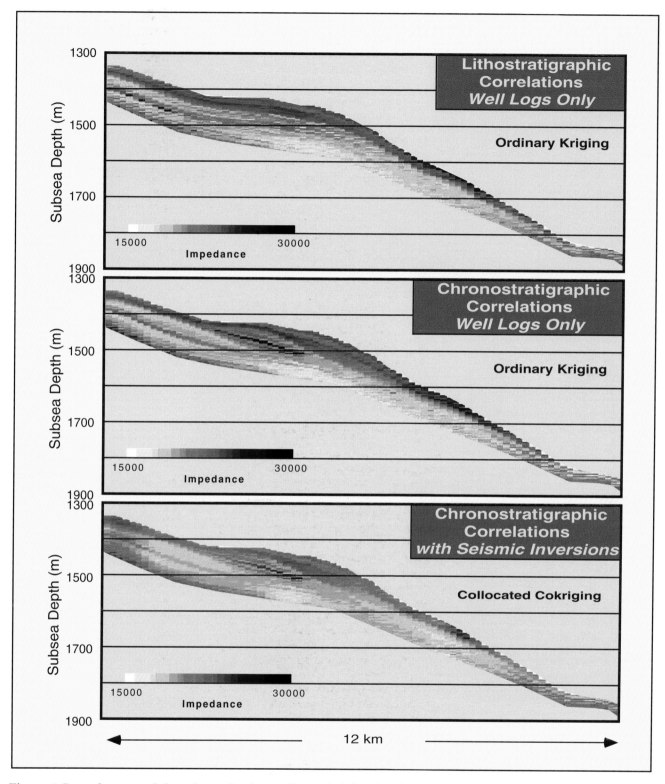

Figure 6. Impedance models estimated using ordinary kriging for the upper two cross sections (wells only) and collocated cokriging for the lowermost cross section (wells and seismic inversions).

By running ten equiprobable impedance simulations, one assumes that a large degree of the possible local variations will be captured. Similarly, for each impedance model, ten porosity realizations were created resulting in 100 porosity models associated with each type of model (lithostratigraphic, chronostratigraphic, and chronostratigraphic with seismic).

A stochastic approach, referred to here as a cloud transform, was used to convert impedance to porosity. Figure 7 contains the fundamental elements of this technique. Impedance (independent variable) is divided into bins or classes to describe the distribution of porosity (dependent variable) for a given class of impedance. Bin widths are adjusted so that a significant number

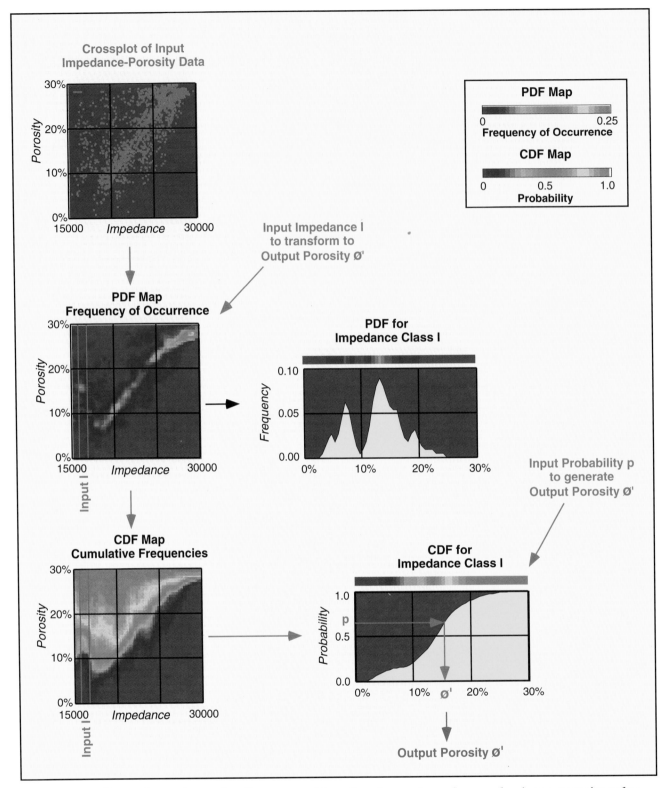

Figure 7. This figure shows the stochastic process of transforming an impedance value into a porosity value given an input of impedance-porosity pairs. The impedance axis is divided into classes for which the histogram (PDF) of porosity values is computed. For each impedance PDF, a cumulative frequency plot (CDF) is generated. The appropriate CDF for the impedance value, I, is sampled by an input probability value, p, to determine a porosity value, $\emptyset$. If a different probability is input for the same impedance value, then a different porosity value will be output.

(about 30) of porosity values are contained in each impedance class to adequately describe their porosity distribution. These histograms or probability density functions (PDF) are accumulated to form a cumulative frequency plot or cumulative density functions (CDF), one for each impedance class. The CDF provides the probability that a random porosity value is less than or equal to a particular porosity value.

As with any statistical technique, it is important to define the input data as well as possible into appropriate statistical populations. These populations should be based upon related geological factors, whether genetic or diagenetic. For example, had core descriptions been available in the study area, the crossplot shown in Figure 7 could have been divided into separate crossplots for refined rock groupings within this clastic system. The bimodal distribution for the impedance class delineated in the figure may then have been reconciled into two separate distributions (i.e., multiple cloud transforms).

To transform a specific impedance value into a porosity value using the cloud transform, take the following steps:

- Determine to which impedance class the impedance value belongs;
- Randomly select a probability value from the uniform distribution between 0 and 1; and
- "Look up" the porosity value corresponding to this probability on the appropriate CDF for the impedance class determined.

An additional step is included to preserve the spatial correlation structure in the output porosity data set. If every impedance value were transformed using a purely random probability value, then a highly random component would be manifested in the output porosity field that may not be present in reality. To control the spatial variability in the output, a two-dimensional correlated probability field is created instead of a purely random one. This field is generated by kriging a grid of random numbers with a specified two-dimensional variogram. The grid is established as two random values per correlation range in each direction.

Figure 8 contains the probability field that was used to "look up" the porosity values in the appropriate CDF in Figure 7 to transform the impedance realization into the porosity realization, also shown in Figure 8. Although the cloud transform would normally be applied to compute permeability from porosity and to capture the uncertainty associated with this relationship, it was not used. The intent of this study is to analyze the effect of geological framework and the uncertainty that exists in the 100 porosity models. Thus, a lognormal equation is sufficient to convert porosity to permeability.

FLUID-FLOW SIMULATIONS

Three hundred cross-sectional models of porosity and permeability were built, as previously described,

upon which to simulate a waterflood displacement (see Figure 1 for cross section location). Figures 9 and 10 show one example of porosity and permeability, respectively, for each of the three model types. A water injector was placed at the downdip end of the cross section and an oil producer at the other end. The grid dimensions of the cross sections are 99 nodes in the vertical direction and 81 nodes in the horizontal direction. The maximum interval thickness is about 200 m with a cross-sectional length of 12 km, resulting in cells approximately 2 m thick and 150 m wide. A black oil simulator (Chien et al., 1989) was used for the fluid-flow simulations (mobility ratio = 2). Upscaling was not performed on the property grids, and identical relative permeability curves were used for each simulation.

The effects on flow behavior as a result of the different correlation strategies and data used in generating the reservoir models are quantified by determining water breakthrough time at the producer. Comparative analysis is aided by plotting the breakthrough times as histograms for each of the three model types (Figure 11). For the lithostratigraphic case, the average breakthrough time occurred at 0.178 pore volumes injected (PVI), whereas the average breakthrough time for the chronostratigraphic case (logs only) occurred at 0.184 PVI. The standard deviations of the two distributions are 0.015 and 0.022 PVI, respectively, indicating that the magnitude of the sweeps in both cases are very similar. However, fluid-flow visualizations for both cases, as shown in Figure 12, demonstrate that the use of the different correlations causes different connectivity patterns of porosity and permeability. The differences are apparent in the water saturation distributions for each model type and for each time step.

The models generated with the addition of the seismic-based information produce waterfloods with an average breakthrough time of 0.242 PVI, which is 31% later than the chronostratigraphic case without the seismic inversions and 35% later than the lithostratigraphic case. In the third case, water saturation snapshots demonstrate better sweep efficiency and later breakthrough. The high-permeability channels that contributed to earlier breakthrough times and less sweep efficiency in the well-only cases are not prevalent in the realizations that include the seismic inversions. Furthermore, the standard deviation of the breakthrough estimates for the third case is 0.0056 PVI, which is substantially smaller than for the first two cases (0.015 and 0.022 PVI, respectively).

DISCUSSION

The fluid-flow results seem to indicate that the chosen geological model does not have a significant impact on predicting breakthrough time or the ensuing oil recovery. This is problematic because the intent of this chapter is to show that the geological framework is important; however, another major element besides the geological framework contributes to spatial continuity (which governs flow): the variogram

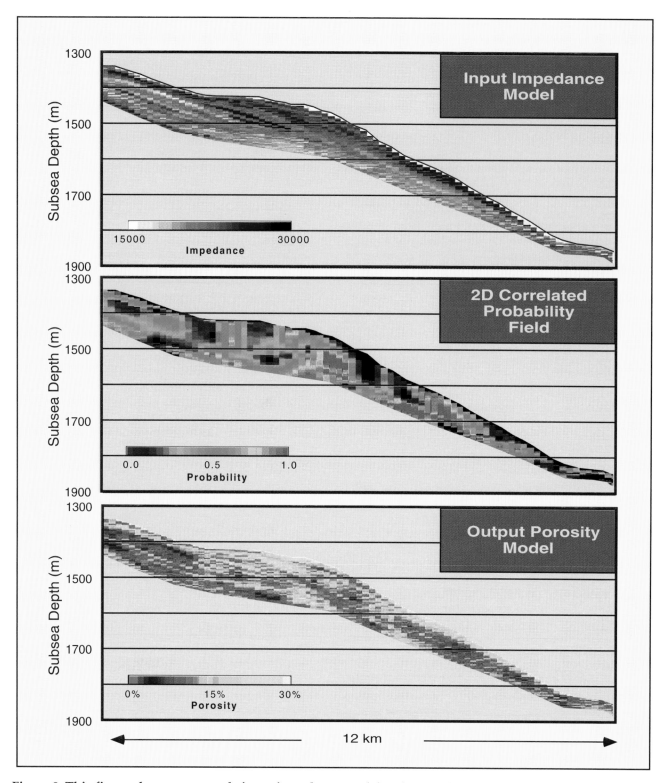

Figure 8. This figure shows an example input impedance model and a correlated two-dimensional probability field that were used with the cloud transform in Figure 7 to generate the porosity model. To create a new realization of porosity for the same input impedance model, a new probability field is generated and used.

model used to estimate the interwell heterogeneity (Figure 5). In both cases where only the well-log data were used to calculate spatial variances, the interwell distances (average 4 km) are roughly the same magnitude as the correlation range modeled. Because of this similarity, the same high-permeability zones near the top of the interval, both for lithostratigraphic and chronostratigraphic correlations, are the primary cause

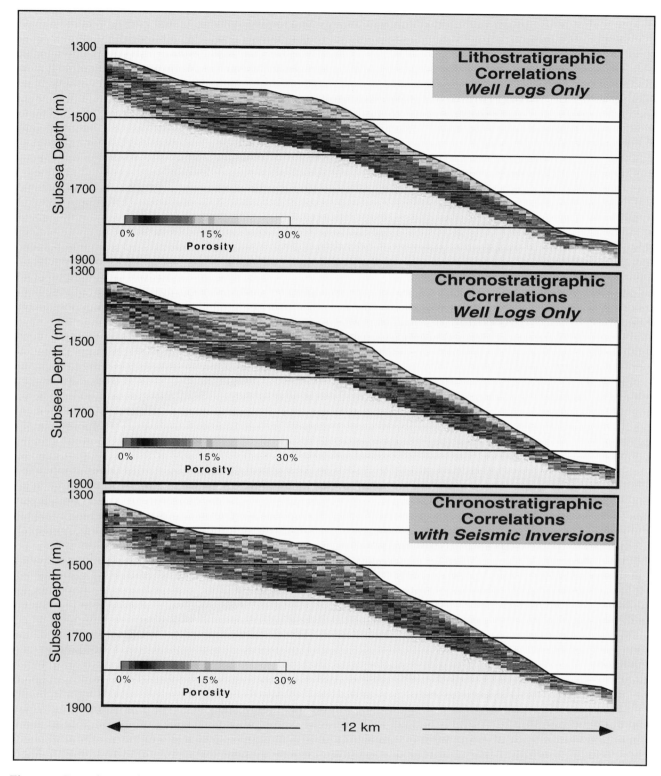

Figure 9. Porosity models generated using the cloud transform in Figure 7. Each one is from a conditional simulation of impedance for the three model types.

for similar breakthrough times. Although the breakthrough times are similar, the overall sweep patterns are different.

Nonetheless, the additional work of deriving chronostratigraphic markers through integrated inter-

pretation of well-log, seismic, and biostratigraphic data sets is warranted not only for geologically sound reasons, but also for reservoir management concerns. For example, it may prove crucial for solving reservoir problems where predicting swept and unswept

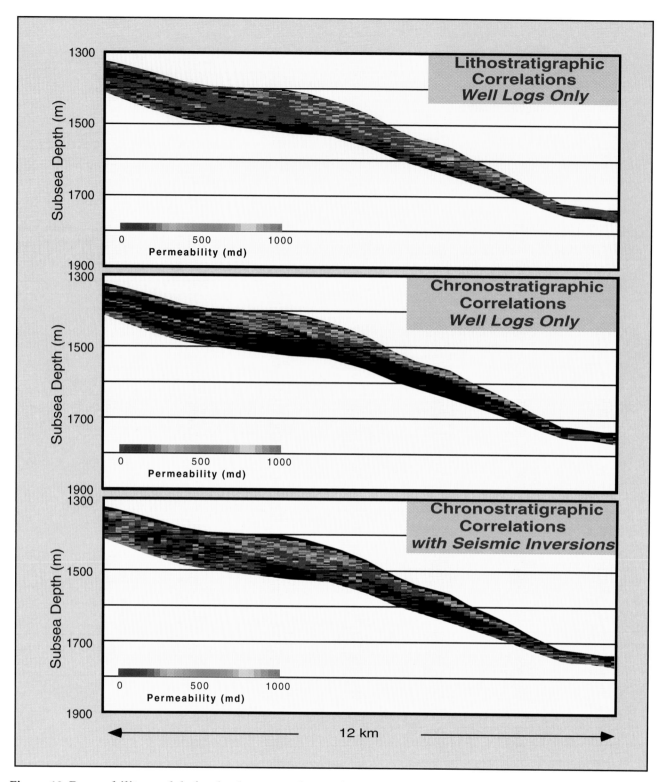

Figure 10. Permeability models for the three porosity models shown in Figure 9. Permeability is computed as an exponential function of porosity.

areas is important, as in developing infill drilling programs, conformance control, and designing sophisticated tertiary recovery processes. Moreover, as is discussed in a following section, a chronostratigraph-

ic model provides the proper geological framework for including seismically derived log data.

Integrating seismic pseudologs with well logs to generate reservoir models produced very different

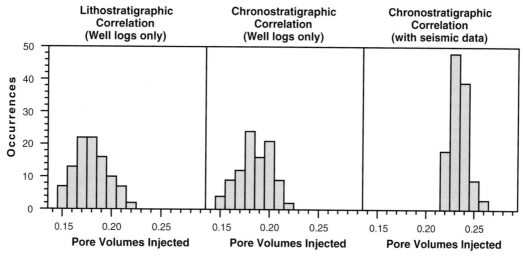

Figure 11. Histograms of breakthrough times in PVI from waterflood simulations for 100 realizations of each model type.

flow results (breakthrough times) and fluid displacement patterns. These results indicate that the reservoir is not as well connected as predicted by the well-only models. The vertical fluid front moves more uniformly through the seismically influenced reservoir model as though the reservoir was more homogeneous[2] as compared to the well-only models. The variogram model for the seismic pseudologs shows a smaller range of correlation than the well-based variogram. Modeling the shorter range was possible only because of the smaller separation distances between seismic traces. Furthermore, the additional information provided by the seismic data greatly reduced the overall uncertainty in breakthrough predictions and improved the confidence in those estimates.

One could argue that similar results could have been obtained by not incorporating the seismic data and simply using a shorter variogram range. This probably would have yielded comparable breakthrough results, although with wider variance; however, these models would be inappropriate for decisions on delineation drilling or bypass oil determination because the variations observed among the wells would be minimally constrained, if at all. Geostatistical techniques do not create data nor do they replace the need for additional data. In fact, their greatest strength is providing methods for integrating diverse data sets tempered by the relative degree of confidence or predictability in these data sets.

CONCLUSIONS

The choice of geological framework for guiding geostatistical estimations will have a significant effect on the characteristics of the reservoir models generated;

however, the effect on predictions of fluid-flow behavior may not necessarily be as significant and is controlled by at least two factors:

- The flow parameter selected, such as breakthrough time, may not necessarily indicate differences in flow behavior. Other parameters, such as fluid displacement patterns, may be substantially different. One must be careful in selecting an appropriate indicator to address the reservoir management problems in question.
- The choice of the spatial continuity model (i.e., variogram required for the estimation algorithms) may have a larger effect than the choice of correlation strategy. The incorporation of secondary data sets, such as seismic inversions, can provide information on small-scale variations in interwell heterogeneity, which may have a profound effect on fluid-flow estimates and estimate uncertainty; however, to realize the potential contribution of the secondary data set, the framework relating this data to the primary data set must be carefully defined. The soft data will be ignored or misapplied by the geostatistical algorithms if the data sets are not properly correlated. Again, the underlying geological model is crucial. The extra time and care necessary to integrate the additional seismic information about interwell heterogeneity are well invested for producing better flow predictions.

ACKNOWLEDGMENTS

We thank Chevron Petroleum Technology Company for the permission to publish this material. We also thank the following colleagues at Chevron: K. L. Finger and S. R. Jacobson for supplying the biostratigraphic and geochemical interpretations used in this study; N. J. Hancock for aiding the interpretation of lithologies

---

[2]A randomly heterogeneous permeability field will behave similarly to a perfectly homogeneous permeability field in terms of fluid flow and saturation front (Araktingi and Orr, 1988).

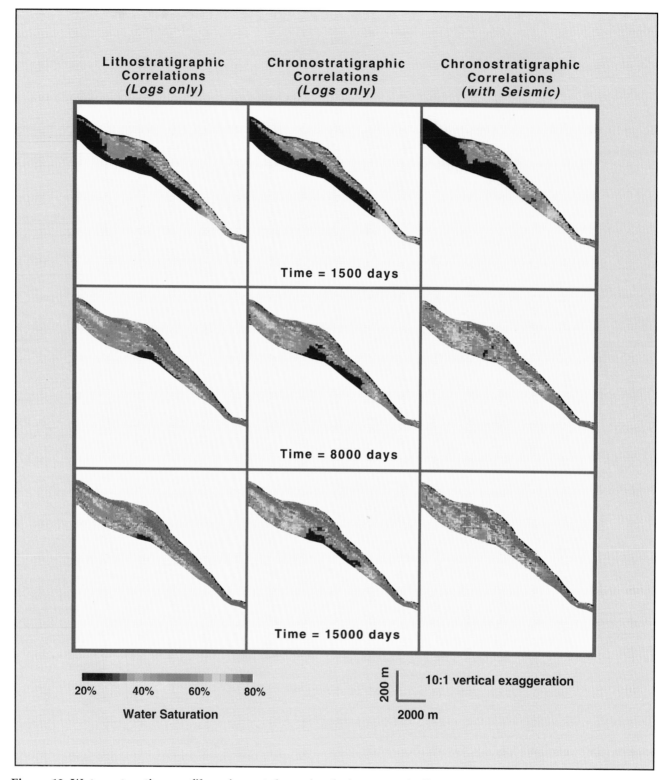

Figure 12. Water saturation profiles, given at three simulation steps, indicate different displacement patterns for the three model types.

from well logs; and D. S. McCormick for his critical review of and comments on this chapter.

REFERENCES

Araktingi, U. G., and W. M. Bashore, 1992, Effects of properties in seismic data on reservoir characteri-

zation and consequent fluid-flow predictions when integrated with well logs: Proceedings of the Society of Petroleum Engineers Technical Conference, p. 913–926.

Araktingi, U. G., and F. M. Orr, 1988, Viscous fingering in heterogeneous porous media: Proceedings of the Society of Petroleum Technical Conference.

Araktingi, U. G., W. M. Bashore, T. A. Hewett, and T. T. Tran, 1990, Integration of seismic and well-log data in reservoir modelling: Proceedings from the Third Annual NIPER Conference on Reservoir Characterization, in press.

Araktingi, U. G., T. A. Hewett, and T. T. Tran, 1992, GEOLITH: an interactive geostatistical modeling program: Proceedings of the Society of Petroleum Engineers Computer Conference.

Bamberger, A., G. Chavent, C. Hemon, and P. Lailly, 1982, Inversion of normal incident seismograms: Geophysics, v. 47, p. 757–770.

Brown, L. F., Jr., and W. L. Fisher, 1977, Seismic-stratigraphic interpretation of depositional systems: examples from Brazilian rift and pull-apart basins, in C. E. Payton, ed., Seismic stratigraphy—applications to hydrocarbon exploration: AAPG Memoir 26, p. 213–248.

Chien, M. C., H. Yardumian, E. Y. Chung, and W. W. Todd, 1989, The formulation of a thermal simulation model in a vectorized, general purpose reservoir simulator: Proceedings of Society of Petroleum Engineers Reservoir Simulation Symposium.

Deutsch, C. V., and A. G. Journel, 1991, GSLIB: geostatistical software library user's guide: Stanford Center for Reservoir Forecasting, Stanford University, p. 55–59.

Isaaks, E. H., and R. M. Srivastava, 1989, An introduction to applied geostatistics: New York, Oxford University Press, 561 p.

Lavergne, M., and C. Willm, 1977, Inversion of seismograms and pseudo-velocity logs: Geophysical Prospecting, v. 11, p. 231–250.

Lindseth, R. O., 1979, Synthetic sonic logs—a process for stratigraphic interpretation: Geophysics, v. 44, p. 3–26.

Oldenburg, D. W., T. Scheuer, and S. Levy, 1983, Recovery of the acoustic impedance from reflection seismograms: Geophysics, v. 48, p. 1318–1337.

Santosa, F., and W. W. Symes, and G. Raggio, 1986, Inversion of band-limited reflection seismograms using stacking velocities as constraints: Inverse Problems, v. 3, p. 477–499.

Vail, P. R., 1987, Seismic stratigraphy interpretation using sequence stratigraphy. Part I: Seismic stratigraphy interpretation procedure, in A. W. Bally, ed., Atlas of seismic stratigraphy: AAPG Studies in Geology no. 27, v. 1, p. 1–10.

Vail, P. R., R. M. Mitchum, Jr., R. G. Todd, J. M. Widmier, S. Thompson III, J. B. Sangree, J. N. Bubb, and W. G. Hatlelid, 1977, Seismic stratigraphy and global changes of sea level, in C. E. Payton, ed., Seismic stratigraphy—applications to hydrocarbon exploration: AAPG Memoir 26, p. 49–212.

van Wagoner, J. C., R. M. Mitchum, K. M. Campion, and V. D. Rahmanian, 1990, Siliciclastic sequence stratigraphy in well logs, cores, and outcrops: AAPG Methods in Exploration Series, no. 7, 55 p.

Walker, C., and T. J. Ulrych, 1982, Autoregressive recovery of the acoustic impedance: Geophysics, v. 47, p. 1160–1173.

Xu, W., T. Tran, R. M. Srivastava, and A. G. Journel, 1992, Integrating seismic data in reservoir modeling: the collocated cokriging alternative: Proceedings of Society of Petroleum Engineers Technical Conference, p. 833–842.

Chapter 15

Integration of Well and Seismic Data Using Geostatistics

D. J. Wolf
Providian Bancorp
San Francisco, California, U.S.A.

K. D. Withers
M. D. Burnaman
Mobil Exploration and Producing Technical Center
Dallas, Texas, U.S.A.

ABSTRACT

The geostatistical method can be applied to quantitatively relate well and seismic data, assess the quality of the resulting map, and estimate the probability of success directly from the available data. To illustrate the steps of the geostatistical method, we present a case study in which the thickness (in feet) of an oil-bearing sand was estimated using seismic amplitude from a three-dimensional seismic survey as a guide.

The geostatistical method is a four-step procedure that calls on several statistical tools. The first step is to quantify the spatial continuity of the well data using variogram analysis. The second step is to find and quantify a relationship between the well and seismic data. The third step is to use what has been learned to grid the well data using the seismic as a guide via kriging with external drift. The last step is to assess the accuracy of the map just made. Traditionally, a geoscientist creates a map that is assumed to be correct until additional information becomes available. Only rarely is an estimate made of the map's accuracy. A geostatistician creates an expected value or average map and has a quantitative estimate of its accuracy. Conditional simulation is a geostatistical tool that yields a quantitative measure of the error in a map.

In the case study, the geostatistical method was used to estimate the expected value and error in a sand isochore. Cumulative probability distributions and risk maps are two tools used to display the range of error expected in the sand thickness map. These tools, along with economic thresholds, were used to directly derive the probability of success from the data available.

Two wells drilled after geostatistical analysis confirmed that using the geostatistical method is more accurate than traditional, nongeostatistical methods.

The application of geostatistics presented in the second part of this paper has been masked to protect proprietary information. Although the data have been distorted, the application of geostatistics has in no way been affected.

INTRODUCTION

The geostatistical method can be applied to quantitatively integrate well and seismic data and quantitatively assess map accuracy. The advantage of this method is that it makes geoscientists and engineers think about the data before making a map. Not only does the method yield maps with certain statistically desirable properties, but also gives the user a quantitative measure of the reliability of the map.

In traditional computer mapping packages, the user starts with an x,y,z listing of data and produces a contoured map. The process is fairly straightforward: data in, contours out. There are few parameters to adjust and those that do exist do not relate to geology. Geostatistics is very different. The user starts with the same x,y,z data and then follows a logical, ordered set of procedures before producing a map. Geology and common sense are infused in the process in several places. Often, the parameters needed can be adjusted to directly relate to geology. The data that go into the geostatistical method are studied and evaluated to extract geologically useful information that is then used to make the final map. Geostatistics usually takes more time than traditional mapmaking methods, but it is well worth the extra effort.

OVERVIEW OF GEOSTATISTICS

Geostatistics is the application of several statistical tools that are used to determine the spatial distribution of geologic variables.

Variograms, kriging, kriging with external drift, cokriging, and conditional simulation are some of the geostatistical tools commonly used. These tools have been tied together in a rational methodology that is known as the geostatistical method. The geostatistical method is a new way to solve mapping problems. It elevates the solutions of mapping problems to a new plane. At present, geoscientists collect data, post the data on a map, and contour the data. The geostatistical method gives the user a methodology for quantitatively determining the spatial characteristics of geologic variables prior to contouring. The spatial information is then used to make the maps and then assess, in a reasonably objective way, the accuracy of the resulting maps.

GEOSTATISTICAL METHOD

Briefly, the geostatistical method is a four-step procedure that calls on several geostatistical tools. These steps are as follows.

1. Learn from the data through simple statistical data analysis (such as calculating means, variances, minimum and maximum values, and histogram plots) and variogram analysis.
2. Find relationships between data sets (if possible) through crossplots, usually done in geophysics by trying to find relationships between sparse well data and relatively dense seismic data.
3. Use what has been learned and found in the data to determine the spatial distribution of control points (i.e., grid for subsequent contouring). The tools most often used in this step are kriging, kriging with external drift, and cokriging.
4. Assess the accuracy/error/risk of the map made in step 3 using conditional simulation. The assessment of risk is perhaps the greatest leap forward that geostatistics provides in solving mapping problems.

Finding a relationship between well and seismic data is not absolutely necessary; however, if a relationship is found, it greatly reduces uncertainty in the results.

Learning from the Data: Variogram Analysis

A variogram is a mathematical technique to quantify the spatial correlation or continuity of a geologic variable. A variogram is a plot of the average squared difference in value between data points against the distance between data points (Figures 1, 2). Control points close together, on average, do not differ in value as much as control points farther apart. The calculated variogram is modeled by fitting certain mathematical functions through the data. Variogram modeling is a mathematical necessity that ensures that kriging will work.

The key parameters that describe the variogram model (Figure 3) are

- Nugget effect, or the value of the model at zero distance
- Sill, or the variance of the data
- Range, or the breakover point from the correlated to uncorrelated zone of the variogram

The nugget is a gauge of measurement uncertainty. If it were zero, then the data would be honored exactly, otherwise the grid values would not honor the well data. Variogram analysis can be used to identify and quantify the fact that spatial continuity can be longer in one direction than another (anisotropic) in the control points via directional variograms. The direction and magnitude of the anisotropy can be used in subsequent steps of the geostatistical method.

Finding Relationships

The second step in the geostatistical method is to find relationships between seismic and well data. It is the most difficult and least developed (in terms of tools available) of the four steps. Crossplotting of the well variable and the seismic variable at the well locations will sometimes lead to finding usable relationships for cokriging. The user must physically justify any relationship that is found.

Using What Has Been Learned and Found

Kriging is a gridding algorithm that estimates a grid node value such that for the parameter of interest, the squared difference between the grid node

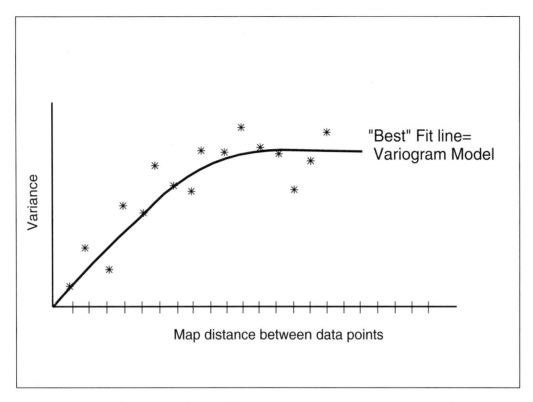

Figure 1. Learning from your data with variograms. A variogram is a graph that is used to express the spatial continuity of a regionalized (mappable) variable. It is a crossplot of the average squared difference of the variable of interest between all data pairs a given distance apart (variance) versus the distance apart.

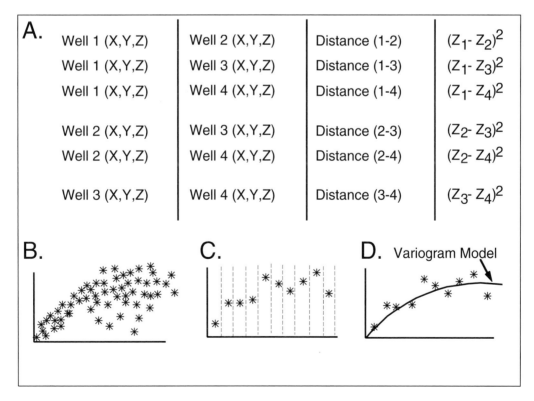

Figure 2. Calculating a variogram. (A) List all of the possible data pairs and compute the distance apart and $(Z_i - Z_j)^2$, where Z is the variable of interest. (B) Plot the distance apart against $(Z_i - Z_j)^2$. (C) Set up a bin system and average all of the data in a bin. (D) Fit a variogram model.

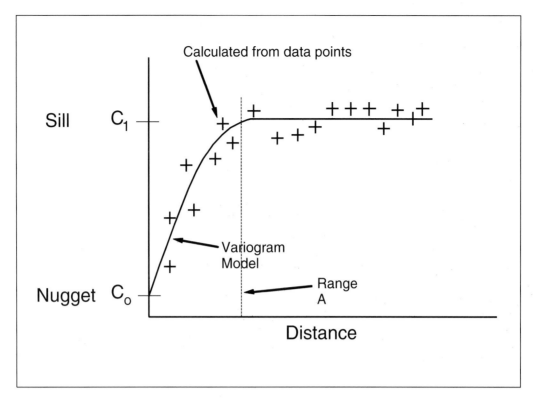

Figure 3. Parts of a variogram. The nugget quantifies measurement inconsistency and the range is the break point between correlated and uncorrelated data.

value and the surrounding control points is consistent (in a least-squares sense) with the variogram model (Figure 4). Kriging answers the following question: "What value should the grid point be assigned so that it best fits the variogram model?" Spatial relationships that have been "learned" by the variogram from the well control are applied to the areas between the control points in the kriging process.

Kriging is mathematically achieved by calculating the grid value as a weighted average of the surrounding control points. Kriging takes into account the distance between control points and the grid node to be calculated, and how close the control points are to each other (automatic declustering), and maintains the spatial relationship given by the variogram model. The weights are assigned in such a way as to minimize the variance in the least-squares sense, thus eliminating systematic overestimation or underestimation error.

Cokriging (Figure 5) looks at not only spatial relationships in the data to be gridded (usually a well parameter, such as porosity or depth), but also spatial relationships in a second, denser data set (usually a seismic parameter) that correlates with the data to be gridded. The second data set is referred to as the guide data. In addition to using variogram models for both the control data and the guide data, cokriging also uses a cross-variogram that quantifies the spatial cross correlation between the two data sets.

Cokriging calculates a grid value as a weighted average of control and guide points. Cokriging takes into account how far the control and guide points are from

the grid point to be computed, and how close control and guide data are (declustering); cokriging also honors spatial relationships found in the variogram for the control data and guide data, and in the cross-variogram between the control data and guide data.

There is no requirement in cokriging for the units (e.g., feet, grams/cubic centimeter, feet/second) of the control points and guide data to be identical. Cokriging works when the units of the control points and guide data are vastly different.

Unfortunately, cokriging is computer intensive and cross-variograms are difficult to model. Instead, kriging with external drift (Figure 6) can be used as a first-order approximation of cokriging. Kriging with external drift (KED) calculates a grid value as a weighted average of control points using the guide data directly. Although cokriging is more desirable from a theoretical point of view, KED is more desirable from a practical point of view. Our experience has found little difference between the methods.

Assessing Map Accuracy: Conditional Simulation

Conditional simulation (Figure 7) is a mapping process that quantifies contouring uncertainty (Figure 8). When contouring data, a geoscientist always has the ability to move (or "swing") a contour line. Figure 8 is a seven-well structure problem contoured four different ways. If the triangle represents a proposed well location, then the estimate of the well parameter (in this case depth to a horizon in thousands of feet) at

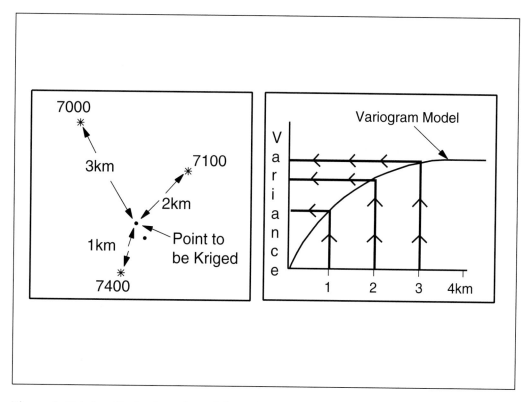

Figure 4. Kriging finds the value of the point to be kriged so that it best fits the variogram model.

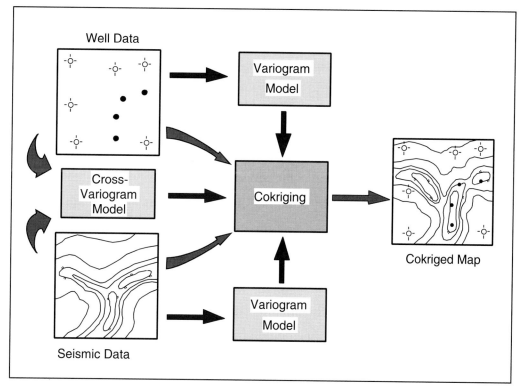

Figure 5. Cokriging using well and seismic data. Note the cross-variogram model in addition to variogram models.

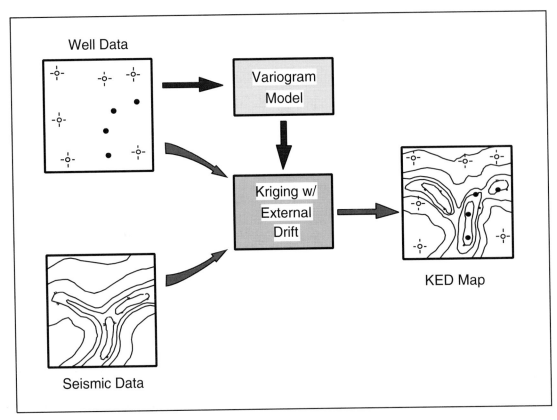

Figure 6. Kriging with external drift (KED) uses well data and its variogram with seismic data.

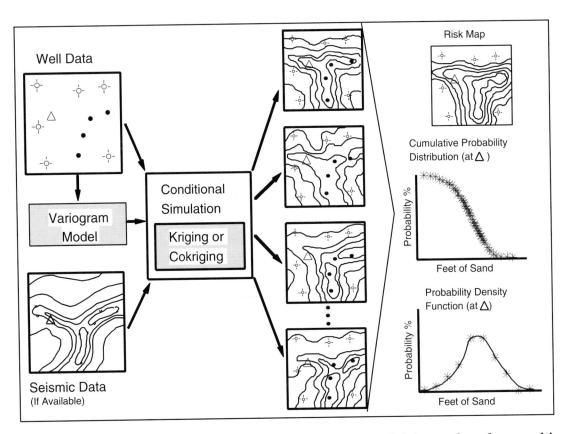

Figure 7. Conditional simulation can be used with kriging or cokriging, and produces multiple realizations of the same map.

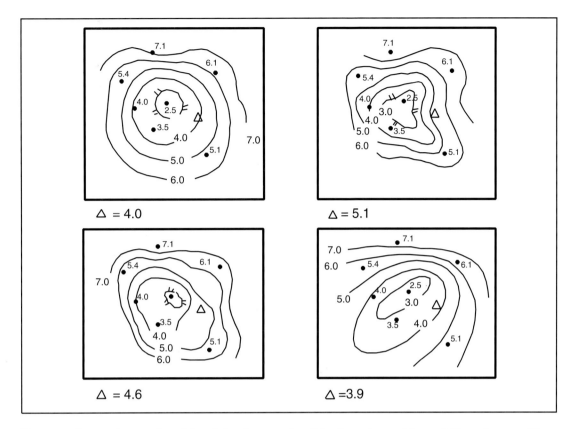

Figure 8. Seven data points (depth in thousands of feet) contoured four different ways. The value at the triangle ranges from 3.9 to 5.1, illustrating contouring uncertainty.

the proposed location is different depending on how the data is contoured. The value at the proposed well ranges from 3900 to 5100 ft. Which estimate is correct? No one will know for sure until the well is drilled. Until then, perhaps the mean of the four estimates, 4400 ft, would be a good estimate.

Conditional simulation is a mathematical way to contour a set of data (with or without a guide data set) several hundred different ways. The conditional simulation maps, sometimes referred to as images, honor what is known about the true surface. Each conditional simulation image

- Has approximately the same mean and variance as the control points
- Has approximately the same variogram of the control data
- Approximately honors the control point values

A guide data set (such as seismic data) can be used if so desired. Each image can be considered as a slightly different and reasonable way to contour the data. Honoring the control data and the mean, variance, and variogram model of the control data ensures that each map is geologically reasonable. The average of the several hundred images is (by mathematical definition) the kriged solution. If a guide data set is used, then the average of the several hundred conditional simulation images is the KED solution.

Another way to explain conditional simulation is to say that it is a set of pseudorandom numbers in the form of a grid that is organized to

- Have the same spatial continuity or variogram model as the control points
- Have the same mean and variance as the input data
- Honor the control data

In general, conditional simulation maps contain much more detail than the map produced by kriging or KED. Conditional simulation maintains detail in a map, but kriging or KED demands that the map be unbiased and minimum variance (or be smooth) at the expense of detail.

The resulting several hundred maps can be analyzed statistically and the contouring uncertainty can be displayed in several forms to assess accuracy in the map. Techniques such as probability density functions, cumulative probability distributions, and risk maps are commonly used for analysis. Uncertainty or accuracy for a specific point, such as a proposed well location on a map, can be quantified by sampling the map value at the proposed well location on all of the several hundred conditional simulation images. This gives the reasonable range of possible values given the surrounding data. The closer the proposed location is to existing control, the less the uncertainty;

conversely, the farther the location is from existing control, the greater the uncertainty. At a control point location there is no uncertainty because the value is exactly known, as long as the nugget effect of the variogram model is zero.

A risk map is a map of the probability of being greater than or less than a threshold level. A risk map is produced by first generating many conditional simulation images and then counting, at each grid node, how many of the conditional simulation images exceed a threshold value. The number of images that exceed the threshold value is divided by the total number of images, resulting in a probability. Repeating this at each grid node results in a map of the probability of exceeding a particular threshold value. For example, assume we create 1000 conditional simulation images of depth to the top of a producing formation and count, at each grid node, the number of images that are above the oil/water contact. A map of the probability of encountering the formation above the oil/water contact at each grid node can be made. The probability map so generated is called a risk map.

APPLICATION OF THE GEOSTATISTICAL METHOD

The following example illustrates how the geostatistical method can be applied to determine the thickness (in feet) of an oil-bearing sand between well control points using 3-D (three-dimensional) seismic amplitude as a guide. Conditional simulation will be used to estimate the uncertainty in the final sand thickness map.

PROJECT GOAL AND PROBLEM

The goal of this project is to quantitatively integrate seismic and well data to estimate sand thickness at a proposed well location and evaluate quantitatively the quality of the resulting sand isochore map. Secondary goals are to evaluate the value of an open lease by estimating the volume of oil-bearing sand and to give a quantitative estimate of upside and downside potential of the volume of oil sand based on the data. This estimate is currently provided either by a geoscientist's past experience or by optimistic/pessimistic contouring of the data set.

The interval of interest in this study is an oil-bearing sand of relatively constant porosity. No formation water has been encountered to date. The sand is secondary pay under an existing field. Sand thickness does not correlate with overlying production. Sand thickness has been difficult to predict from well control. To date, sands from 2 to 80 ft thick have been encountered. Wells are completed if there is 40 ft of oil sand or more. The 40 ft represents the economic cutoff. Figure 9 is a cross section of the basic

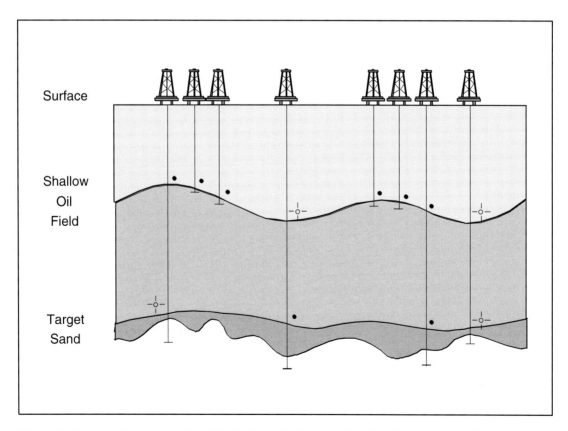

Figure 9. Schematic cross section illustrating shallow production interval and deeper target.

problem: the dry holes at the target sand level are uneconomic because they are thin. Figure 10 is a base map showing the area of good 3-D seismic coverage and well penetrations to the target oil sand. This study will concern itself with the shaded area shown in Figure 10.

The geostatistical method will be used to solve the problem of estimating sand thickness. We will first learn from the existing data via variogram analysis, then try to find relationships between the seismic and well data using crossplots. Then we will use what we have learned to quantitatively determine the spatial distribution of the well data. Finally, we will assess the accuracy of the resulting sand thickness map using conditional simulation and cross validation.

LEARNING FROM THE DATA

In Figure 11, the asterisks represent a variogram plot derived from 14 wells on the western one-half of the area. The solid line represents a variogram model of the sand thickness derived from the variogram plot. This may not look like a good fit, but geostatistically it is not bad considering there are only 14 input points. From a practical standpoint, kriging, KED, and cokriging are all fairly robust to the choice of the variogram model. In other words, small changes in variogram

parameters lead to small changes on the resulting maps. Because the nugget effect is zero, the well data will be honored in both kriging and KED.

The variogram has quantified an important geologic factor: the spatial continuity of the variable. The variogram model is used in kriging, KED, and conditional simulation to control the spatial continuity of the resulting maps.

FINDING RELATIONSHIPS

Usually, the most difficult step in the geostatistical method is finding a relationship between the well parameter of interest (in this case, sand thickness) and some aspect of the seismic data. In this case study, finding the relationship was difficult. To summarize, we noticed a relationship between the peak amplitude of the sand reflector on seismic data and the thickness of the oil sand in the wells. Figure 12 shows a typical seismic section with the sand reflector highlighted. Figure 13 is a map of the peak amplitude for that reflector. Note how wells with thick sands are in areas of high amplitudes and wells with thin sands are in areas of low amplitudes. Figure 14 shows a cross plot of sand thickness in the wells versus the peak seismic amplitude value at the well location.

The physical justification for the peak amplitude/sand thickness correlation is shown in Figure 15.

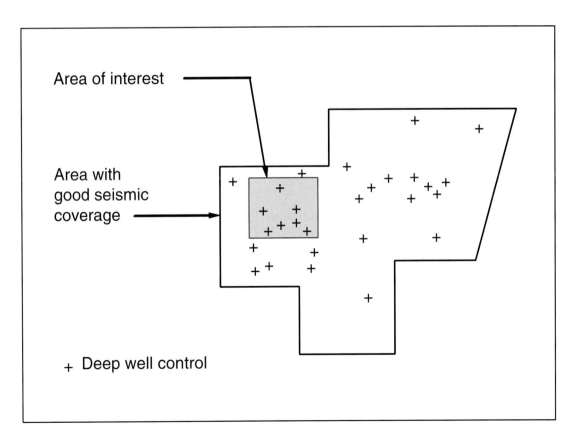

Figure 10. Base map showing three-dimensional seismic coverage, deep well penetrations (+ symbols), and study area (shaded).

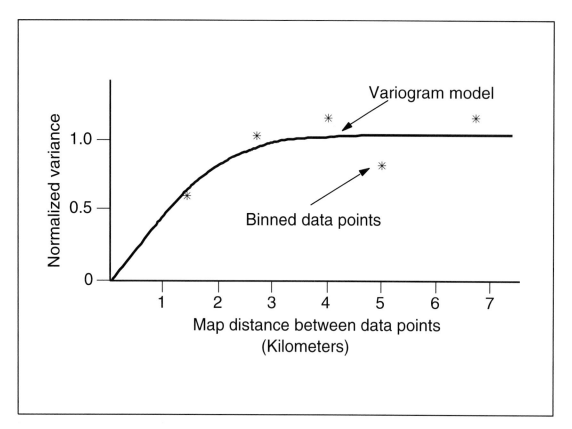

Figure 11. Variogram of sand thickness from the fourteen westernmost wells.

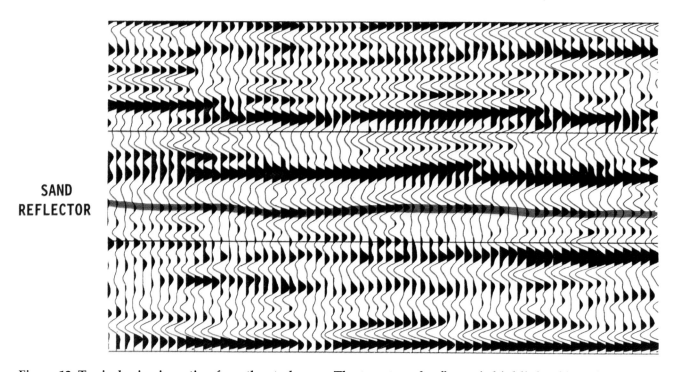

SAND
REFLECTOR

Figure 12. Typical seismic section from the study area. The target sand reflector is highlighted in red.

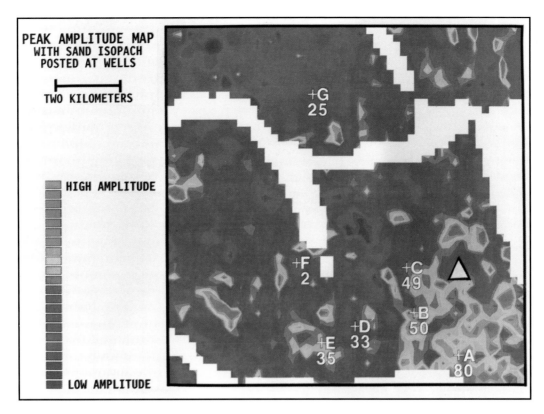

Figure 13. Peak amplitude map of seismic reflector with sand isopach values posted at well locations A–G. Thicker sands tend to occur in the higher amplitude zones (warm colors). Proposed well location shown with triangle.

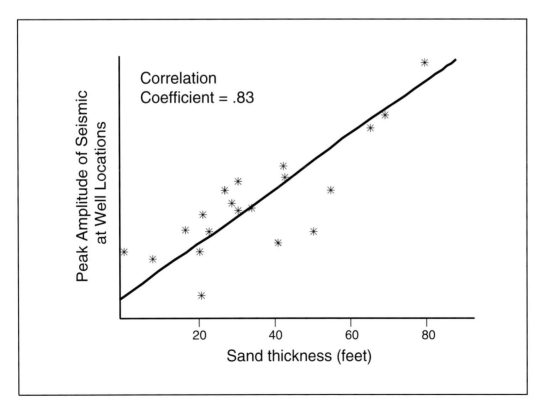

Figure 14. Crossplot of amplitude value versus sand thickness for twenty wells in the seismic coverage area. Wells in bad seismic areas (e.g., low fold) were eliminated. The correlation coefficient is 0.83.

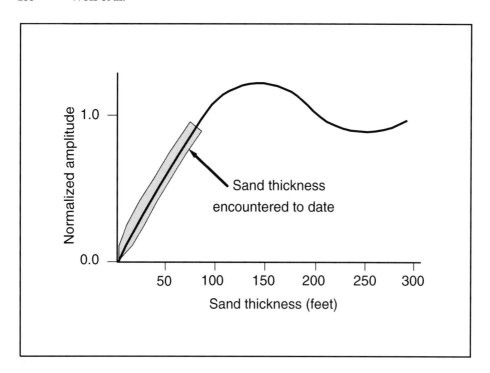

Figure 15. A simple tuning chart illustrates the physical justification of the peak amplitude/sand thickness correlation. The chart was constructed using seismic frequencies, sand velocity, and encasing material velocity.

This is a simple tuning chart constructed using seismic frequencies, sand velocity, and encasing material velocity to show that the thickness of sand encountered to date is in the linear part of the tuning chart. Finding the relationship, and physically justifying it, commonly is the most important and most difficult part of the geostatistical method. Does that mean geostatistics cannot be used if one does not find a relationship? Absolutely not! Geostatistics can be applied without seismic information; however, seismic data will greatly reduce the uncertainty and increase confidence in the results.

INTEGRATING SEISMIC AND WELL DATA

So how can we take advantage of what has been learned? KED will use the spatial continuity information in the form of a variogram model and use the peak amplitude map that was found to be a good predictor of sand thickness.

Figure 16 is a hand-contoured solution to the problem. Note that this map yields an estimate of 72 ft of sand for the proposed location.

Figure 17 shows the kriging solution to the problem. This solution is very similar to the hand-contoured solution. This solution is typical of kriging and other computer gridding algorithms. Kriging, by itself, is usually not worth the extra effort of making a variogram. The kriging solution estimate is approximately 65 ft of sand for the proposed location.

The KED solution is shown in Figure 18. KED uses the variogram model derived from the well data and the peak amplitude map as a guide. KED ties the wells to within a grid cell. Compare Figure 18 to the peak amplitude map in Figure 13. Note that Figure 18 is measured in feet and that Figure 13 is measured in units of

amplitude. The KED map using seismic amplitude indicates 45 ft of sand at the proposed location. This is 5 ft greater than the economic cutoff of 40 ft. Should the well be drilled? We will determine this answer by asking how good the estimate is. Is it 45 ft ± 45 ft?

ASSESSING ACCURACY

Conditional simulation was used to address the question of accuracy. Two hundred maps of sand thickness, using the seismic peak amplitude as a guide, were prepared. Each map

- Honors the well control to within a grid node
- Honors the variogram model of the well data to control spatial continuity
- Has the same mean and variance as the input well data
- Uses the seismic as an imperfect guide
- Is considered equiprobable

Figures 19 and 20 show twelve of the 200 maps. Note how they are basically similar (for example, all have thick sand in the southeastern corner), but differ in detail. Each map is a reasonable way to contour the well data using the seismic as a guide.

How can 200 maps be analyzed efficiently? They can be analyzed through the use of cumulative probability distributions, probability density functions, and risk maps. We can produce a cumulative probability distribution (Figure 21) if each of the 200 maps is sampled at the proposed location and the sand thicknesses are compiled, arranged in ascending order, and plotted on a graph. This graph relates the probability of encountering sand greater than or equal to a particular thickness. For example, there is a 95% chance of encountering

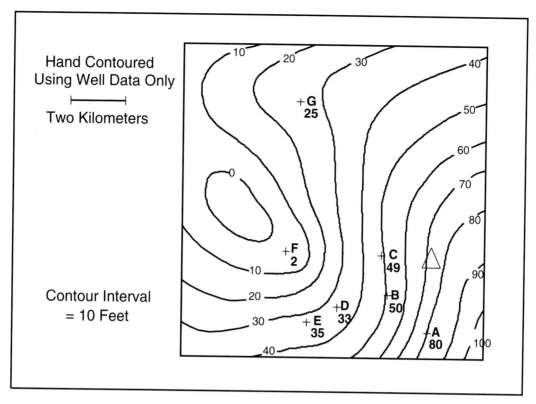

Figure 16. Hand-contoured sand isopach map. Proposed well location (triangle) indicates 72 ft of sand.

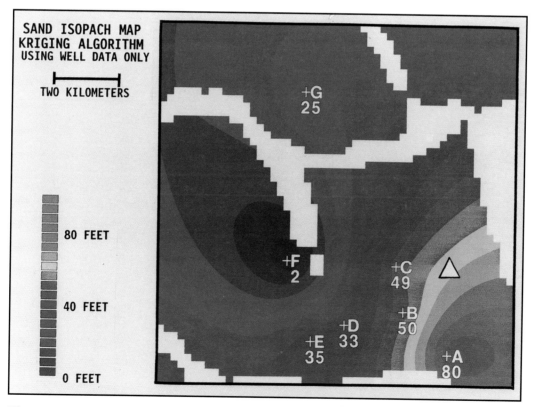

Figure 17. Kriging solution sand isopach map. Proposed well location (triangle) indicates 65 ft of sand.

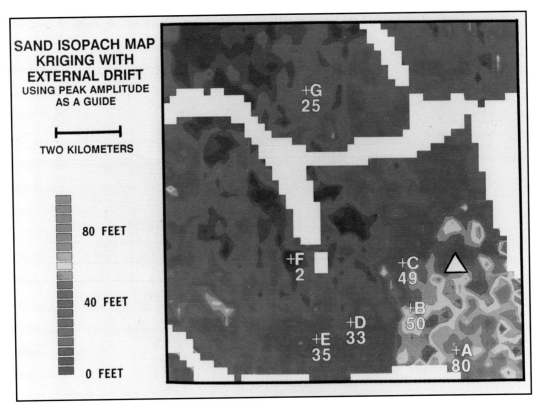

Figure 18. KED solution sand isopach map is similar to Figure 13. Proposed well location (triangle) indicates 45 ft of sand.

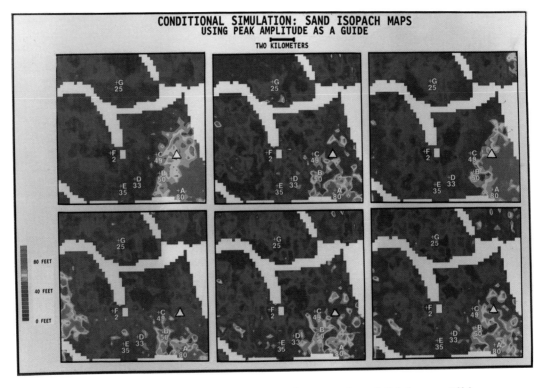

Figure 19. Six of the 200 conditional simulation images of sand thickness. All images honor the well penetrations (A–G), and are generally similar but differ in detail. Proposed well location (triangle) indicates from approximately 20 to nearly 80 ft of sand on the images shown.

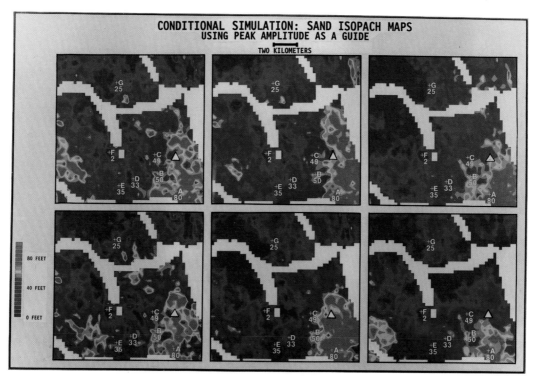

Figure 20. Six of the 200 conditional simulation images of sand thickness. All images honor the well penetrations (A–G), and are generally similar but differ in detail. Proposed well location (triangle) indicates from approximately 20 to nearly 80 ft of sand on the images shown.

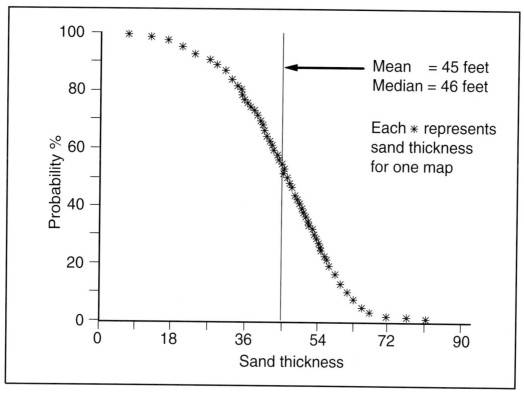

Figure 21. Cumulative probability distribution of sand thickness at the proposed location. Each asterisk represents sand thickness for one map. Vertical line intersects the distribution at the mean (45 ft) which equals the KED solution (Figure 18).

more than 18 ft of sand at the proposed location. Likewise, there is an 80% chance of encountering more than 32 ft of sand. Note that the mean for this curve is 45 ft. This is the same value as the kriging with external drift solution at the proposed location. In fact, geostatistical theory forces the mean of all conditional simulations to be the solution given by KED.

Most of the maps indicated sand thicknesses between 36 and 54 ft; however, it is also reasonable to expect as much as 80 ft and as little as 8 ft of sand. How does conditional simulation know what is geologically reasonable? Geological reasonableness is assured by the variogram model (remembering that the variogram model is derived from control points) and the fact that the conditional simulation images have the same mean and variance and also honor the control point values. In this case, the geologist for the area thought that all the maps were geologically reasonable and that none of the maps could be thrown out for being unreasonable. The images fit the geologic model for the area.

Of course this analysis can be done for any location on the map. If it is done at the location of an existing well, the cumulative probability function is a sharp stair-step at the sand thickness value of the well. By considering all 200 maps, we have quantified the upside and downside possibilities.

Another way to consider the 200 maps is to compute a probability density function, which is simply the derivative of the cumulative probability distribution. Figure 22 is a probability density function for the proposed location. The solid curve represents the geostatistical estimate. The solid vertical line at 40 ft is the economic cutoff, and the dashed vertical line is the expected value or mean of the distribution. The unshaded area represents maps that were contoured at the proposed location in which there is less than 40 ft of sand. Of the 200 conditional simulation maps, 70 were contoured with less than 40 ft of sand (70/200 = 35%) at the proposed location. The shaded area represents maps that were contoured such that there is greater than 40 ft of sand at the proposed location. A total of 130 conditional simulation maps were contoured with more than 40 ft of sand (130/200 = 65%) at the proposed location; therefore, the probability of success is 65%, given that economic success is defined as encountering greater than 40 ft of oil sand.

So far we have analyzed only one well location. What if we want to know the probability of encountering less than 40 ft of sand everywhere on the map? This kind of map is called a risk map (Figure 23). For this analysis, 200 maps were created. Of these, at the proposed well location, 70 maps had less than 40 ft of sand. This is a 35% chance of encountering less than 40 ft of sand at the grid node closest to the well location. Note that well "A" had already encountered 80 ft of sand, so the probability of encountering less than 40 ft of sand at the well location is 0%, as shown by each map. Similarly, because well "F" had already encountered 2 ft of sand, the probability of encountering less than 40 ft of sand at well "F" is 100%.

The probability density function suggests that there is a lot of error in the sand thickness map. If this is unacceptable, there are several ways to reduce the chances of error. One way is to find a better seismic data guide parameter. For example, if one could find a relationship between well and seismic data with a correlation coefficient of 1.0, then the seismic would be a perfect predictor of the well parameter. Another way is to add control points because more data would certainly help. Risk maps can be useful for optimal placement of the next well.

The narrower the probability density function, the less error or uncertainty. Primary factors that control the narrowness of the probability density function or geostatistical estimate are

- Correlation coefficient between the seismic and well data, where the higher the correlation coefficient, the narrower the probability density function (PDF)
- Variogram model range of the well data, where the longer the range, the narrower the PDF
- Distance of the proposed location to the nearest well control, where the lesser the distance, the narrower the PDF (if it is at an existing well, the PDF is a spike)

Geostatistical theory forces the average of the conditional simulations to be the KED solution. If no seismic data are used as a guide, then it becomes the kriging solution. Figure 24 shows the average of the 200 conditional simulation maps and is virtually identical to the solution using KED (Figure 18).

CROSS VALIDATION

Cross validation was used as a tool to compare the accuracy of the kriging and KED algorithms. Each well was removed, one at a time, and the remaining wells were then used to estimate the value at the location of the removed well. In other words, during each cross validation, we pretended that one specific well had never been drilled. We then used all of the other wells to estimate that missing well. Kriging and KED using the seismic data as a guide were used to estimate the missing well value. The average absolute value of the difference between the estimate of the sand thickness at the missing well locations using kriging and the actual value of the sand thickness at the well location was 22 ft. Using KED with seismic as a guide reduced the average error to 7 ft. Thus, incorporating the seismic data reduced the error by a factor of three.

At first glance, this method is very appealing; however, in practice it is not a rigorous analytical tool. The results of cross validation should not be taken too literally.

VOLUME ESTIMATES

The geoscientists involved in the project wanted to know the value of an open lease within the area. One of the parameters used to determine the amount of recoverable oil is reservoir rock volume. The reservoir

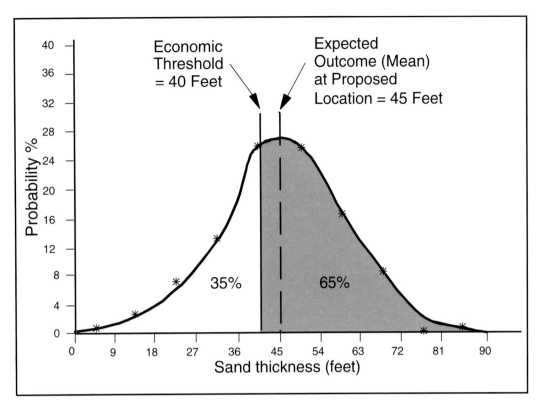

Figure 22. Probability density function (PDF) of sand thickness at the proposed location. Heavy solid curve represents the PDF, solid vertical line equals the economic threshold (40 ft), and dashed vertical line equals the mean (45 ft). The unshaded area represents the probability of economic failure (35%) and shaded area equals the probability of economic success (65%).

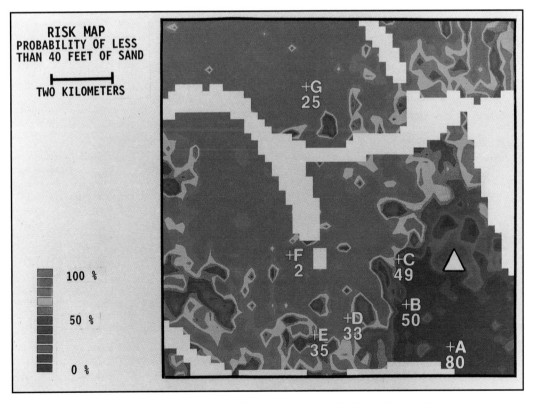

Figure 23. Risk map shows the probability of encountering less than 40 ft of sand.

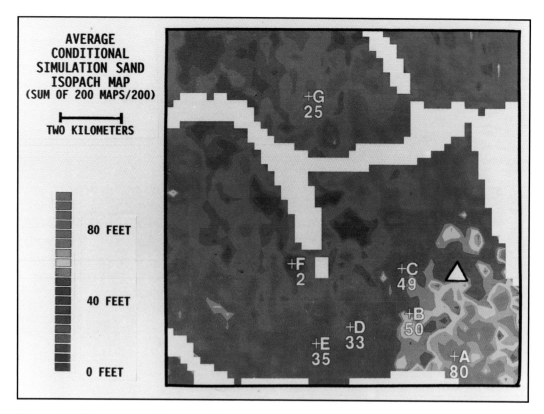

Figure 24. The average of 200 conditional simulations of sand thickness. Note similarity to KED solution (Figure 18).

rock volume of the sand over the area of the open lease using the KED map (Figure 18) was 1 million ac-ft. To get at the upside and downside of this estimate of reservoir rock volume, we turned again to the conditional simulation images.

The reservoir rock volume was calculated for each of the 200 conditional simulations. Note that the reservoir rock volume is the volume of sand only and does not take into consideration porosity, formation volume factor, or water saturation. The average reservoir rock volume of all 200 maps was 1.10 million ac-ft. Contrast this result with the 1 million ac-ft estimate determined by KED. Note that the average of the 200 conditional simulation maps is the KED map:

$$\text{KED map} = \frac{\sum_{i=1}^{200} (\text{Conditional simulation map})_i}{200}$$

However, this is not true if we calculate the volume of each map over a given area:

$$\text{Volume (KED map)} \neq$$

$$\frac{\sum_{i=1}^{200} \text{Volume (conditional simulation map)}_i}{200}$$

Replacing the left side of this equation with the quantity of 1 million ac-ft (the rock volume of the KED map) and the right side of this equation with 1.10 million ac-ft (the average rock volume of the 200

conditional simulation maps) yields
   1.00 million ac-ft ≠ 1.10 million ac-ft

The difference can be explained by the fact that truncation (isopach process) is a nonlinear operator. This concept is further explained in Appendix 1.

Which of the volume estimates (1.00 or 1.10 million ac-ft) should be used? The average volume of the 200 conditional simulations should be used (1.1 million ac-ft) instead of the volume of the average of the 200 conditional simulations (the KED solution) for the reasons explained in Appendix 1.

Figure 25 is a cumulative probability distribution of the reservoir rock volume for this example. Note that the upside, downside, and expected value of the volume are specified completely by the well and seismic data. The upside (5% level) is 1.9 million ac-ft, the downside (95% level) is 0.5 million ac-ft.

DRILLING RESULTS

The proposed location has not been drilled at this time, but two other wells in the area have been drilled. Figures 26 and 27 show the results of these wells. The heavy solid curve is the geostatistical estimate of sand thickness, and the thin, vertical solid line is the expected value of the geostatistical distribution. The vertical dotted line is the traditional predrill nongeostatistical prognosis, and the dashed vertical line is the actual drilling results. Note that the actual drilling results fall within the

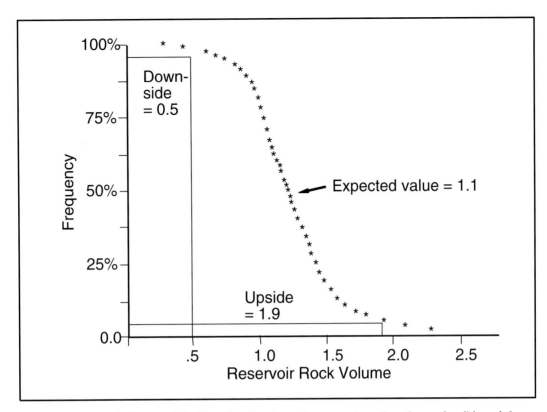

Figure 25. Cumulative probability distribution of reservoir rock volume for fifty of the conditional simulations. Each asterisk represents one volume. The downside (95% value) volume is 0.5 million ac-ft, the upside (5%) is 1.9 million ac-ft, and the expected value (mean) is 1.1 million ac-ft.

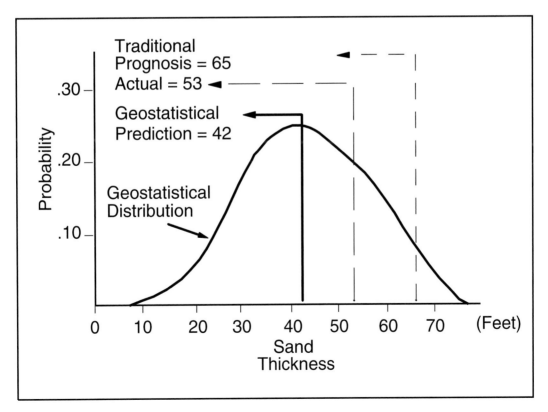

Figure 26. Drilling results of well 1; the geostatistical prediction of 42 ft was slightly closer to actual results than the traditional prediction.

196   Wolf et al.

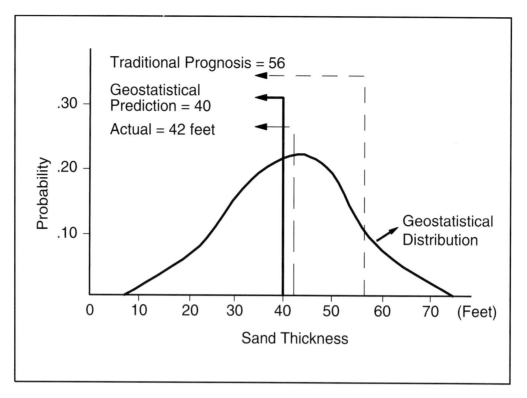

Figure 27. Drilling results of well 2; the geostatistical estimate was much closer to actual results than the traditional prediction.

geostatistical distribution and that the geostatistical estimate is closer to the actual results than is the traditional estimate.

CONCLUSION

The geostatistical method was used to first learn about the data, find relationships between seismic and well data, use what had been learned to grid the well data using the seismic data as a guide, and, most important, assess the accuracy of the results. The assessment of accuracy led to the direct estimation of the probability of success, the estimation of the upside and downside potential, and the expected value for reservoir rock volume for a given open lease.

In this project, geostatistics was used to predict oil sand thickness using seismic amplitude as a guide; however, the potential scope of applications is much broader. Velocity fields can be modeled for depth conversion, or various reservoir parameters, such as porosity, water saturation, and net-to-gross ratios, can be modeled and mapped.

Although the geostatistical method is more expensive in terms of computer and geoscientist re-sources, it is well worth the investment. We believe that the geostatistical method has led to much greater understanding of the data than is provided by any other traditional computer gridding techniques.

REFERENCES

Armstrong, M., ed., 1989, Geostatistics: Proceedings of the Third International Geostatistics Congress: Dordrecht, D. Reidel Pub. Co.

Davis, J. C., 1986, Statistics and data analysis in geology, 2d ed.: New York, John Wiley & Sons.

Doyen, P. M., Porosity from seismic data: a geostatistical approach: Geophysics, v. 53, p. 1263–1295.

Hohn, M. E., 1988, Geostatistics and petroleum geology: New York, Van Nostrand Reinhold.

Isaaks, E. H., and R. M. Srivastava, 1989, An introduction to applied geostatistics: New York, Oxford University Press, 561 p.

Journel, A. G., 1989, Fundamentals of geostatistics in five lessons, v. 8: American Geophysical Union.

Journel, A. G., and C. J. Huijbregts, 1978, Mining geostatistics: New York, Academic Press.

Matheron, G., and M. Armstrong, 1987, Geostatistical case studies: Dordrecht, D. Reidel Pub. Co.

Verly, G., ed., 1984, Geostatistics for natural resources characterization: Dordrecht, D. Reidel Pub. Co.

APPENDIX: A SYNTHETIC EXAMPLE OF CONDITIONAL SIMULATION TO TEST THE SENSITIVITY OF VOLUMETRIC CALCULATIONS

Introduction

A simple synthetic example will be used to illustrate how and why conditional simulation can be applied to volumetric problems and to show that it works better at estimating rock volume than tradition-

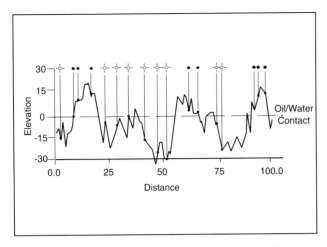

Figure 28. Synthetic example showing a depth profile with seventeen well penetrations, designated "the truth."

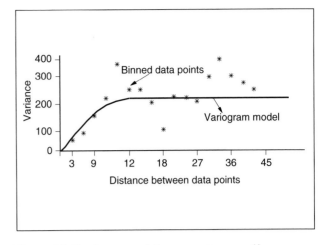

Figure 30. Variogram of the seventeen wells.

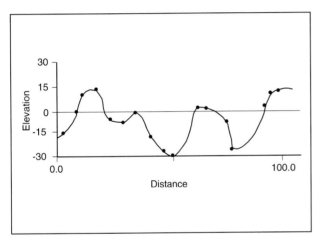

Figure 29. Hand-contoured depth profile of the seventeen wells.

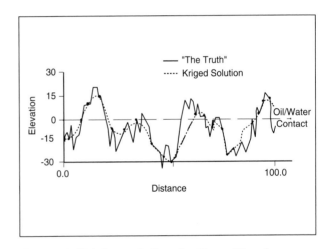

Figure 31. Kriging solution depth profile of seventeen wells is shown by the dashed line, and "the truth" profile is shown by the solid line.

al methods. A 100-point one-dimensional profile (Figure 28) was created and hereafter is referred to as "the truth." For the sake of argument, suppose that this represents a depth profile across several structurally controlled oil fields. The true depth profile was then randomly sampled seventeen times to represent seventeen wells penetrating the top of a horizon of interest. Three techniques, hand contouring, kriging, and conditional simulation, will be used to reconstruct the truth profile from the seventeen wells and estimate the reservoir rock volume (actually, the area will be calculated because this is a one-dimensional profile). These techniques will be tested and compared.

Figure 29 shows a hand-contoured solution to the problem. This is a reasonable estimate of the truth, given that there are only seventeen wells.

Kriging

A variogram and subsequent variogram model for the seventeen wells are shown in Figure 30. The seven-

teen wells were kriged (Figure 31) using the seventeen wells and the variogram in Figure 30 to reconstruct the truth. Kriging did a reasonable job of reconstruction and is similar to the hand-contoured solution. In fact, most computer gridding algorithms will give a result similar to the kriging solution.

Conditional Simulation

Conditional simulation was then used to reconstruct 100 profiles using the variogram model in Figure 30 and the seventeen wells. Figure 32 shows the first six simulations. Note how the conditional simulations look more like the truth profile. In fact, they have the same mean and variance as the well data, they all honor the variogram model of the well data, and they honor the seventeen control points. The conditional simulation images are more geologically reasonable because they contain the correct amount of spatial variation (independent of well spacing and well location) as manifested by the variogram model. They are

198   Wolf et al.

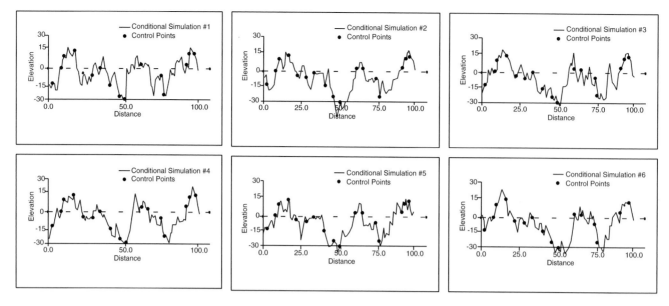

Figure 32. Six conditional simulations of the depth profile.

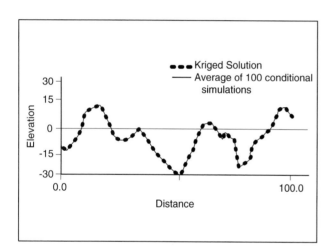

Figure 33. Average depth profile generated from all 100 conditional simulations, with the kriged solution overlain.

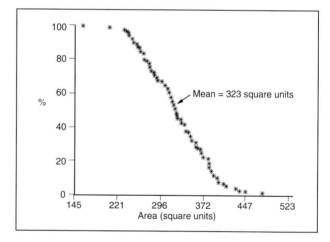

Figure 34. Cumulative probability distribution of the area under the depth profile and above the oil/water contact from 100 conditional simulations. The mean value, 323 square units, is closer to the truth profile's area than are other methods.

not an overly smooth estimate of the true profile as shown in the kriging or hand-contoured solution.

Figure 33 shows the average of all 100 conditional simulations calculated by summation and then dividing by 100. The result is very close to the kriging solution, which is not surprising. This demonstrates the mathematical property of conditional simulation: the kriged profile (or map if this were a two-dimensional problem) is the average of all of the conditional simulation images. Hence, a kriged map is sometimes referred to as the expected value map.

Area Calculations

The area (or volume if the problem were two-dimensional) was calculated between a threshold (oil/water contact) and the following curves: (1) the truth, (2) the hand-contoured solution, (3) the kriged solution, and (4) each of the conditional simulations. The results are shown in Table 1.

We conclude that conditional simulation does a much better job of estimating the truth. This occurs because it is true that the average of the 100 conditional simulation profiles is the kriging profile

$$\text{Kriged profile} = \frac{\sum_{i=1}^{100} (\text{Conditional simulation profile})_i}{100}$$

However, this is not true if we calculate the area (or volume if the problem were two-dimensional) of both

Table 1. Area Under the Profile and Above the Oil/Water Contact

Truth	339 square units
Hand Contoured	250 square units
Kriged	240 square units
Avg. of 100 Conditional Simulation Profiles	323 square units

sides of the above equation,

$$\text{Area (kriged profile)} \neq \frac{\sum_{i=1}^{100} \text{Area (Conditional simulation profile)}_i}{100}$$

Replacing the left side of this equation with 240 square units (the area of the kriging solution) and the right side of the equation with 323 square units (the average area of the 100 conditional simulation profiles) yields

240 square units ≠ 323 square units.

These values are not equal because truncation, imposing an oil/water contact, is a nonlinear operator.

Which area (or volume if the problem were two-dimensional) estimate should be used, the kriging solution (240 square units) or the average area of the 100 conditional simulations (323 square units)? The answer is the average area estimate of 323 square units based on the conditional simulations because it is closer to the actual area under the truth profile than is the area of the kriged or the hand-contoured estimate. It is also more reasonable to take the average of areas (or volumes if

the problem were two-dimensional) under the depth curve and above the oil/water contact of each of the 100 conditional simulations curve because these profiles are more geologically realistic than the kriged solution profile. It is not reasonable to estimate the area (or volume if the problem were two-dimensional) using an overly smooth representation (kriged or hand-contoured solution) of the true profile when more realistic representations (conditional simulations) could be used.

The 100 conditional simulations can also be studied to derive upside and downside estimates of area (or volume if the problem were two-dimensional). Figure 34 shows a cumulative probability function of the areas under the depth curve calculated from all 100 conditional simulations. Note that the upside (5%) and downside (95%) is read directly from the cumulative probability distribution, which is derived directly from the available well data. This graph shows the geologically reasonable range of the area calculation given that only seventeen wells exist. Calculating these graphs provides a quantitative procedure to assess uncertainty in the face of sparse data.

Conclusion

This simple synthetic example shows that using overly smooth (kriged or hand-contoured) maps to estimate rock volume leads to inaccurate results. Using a conditional simulation process leads to more geologically realistic images of depth, more accurate area (or volume for two-dimensional problems) estimates, and a fully described cumulative probability distribution for input to risk analysis programs based on the data available.

Chapter 16

◆

3-D Implementation of Geostatistical Analyses—The Amoco Case Study

Jinchi Chu
Wenlong Xu
Andre G. Journel
Stanford Center for Reservoir Forecasting
Stanford, California, U.S.A.

◆

ABSTRACT

Reservoir parameters, as with most other spatially distributed earth science variables, are distributed in three dimensions. Although most geostatistical theories are established in nD, their applications in three dimensions, as opposed to two dimensions, call for specific implementations and computer coding. In this chapter we present a three-dimensional (3-D) case study using real reservoir data, highlight problems typically associated with 3-D applications, and present possible solutions.

AMOCO DATA SET

The Amoco data originate from a producing west Texas carbonate field. The target layer, ST, is of Permian age and is one of the major producing intervals in the field. This interval represents a shallowing-upward, prograding, carbonate shelf sequence formed under moderately low-energy conditions (Chambers and Zinger, 1990). Lithologically, the formation is composed of primarily alternating dolomites and siltstones. Within the dolomites, the main factor controlling reservoir performance is anhydrite that plugs the pore space. The siltstones, which generally lack cross-bedding, present a reasonably sharp transition with the overlying and underlying dolomites. The depositional environment is thought to represent very fine grained eolian sands reworked by littoral and middle neritic processes.

The ST layer is penetrated by 90 wells evenly distributed in the field (Figure 1). ST's thickness ranges from 34 to 80 ft, as shown in the histogram of Figure 2A. A total of 4967 log-derived porosity data are available over ST and were deduced by regression from sonic logs recorded over 1-ft intervals. A porosity histogram is shown in Figure 2B.

The available seismic data are two-dimensional (2-D) seismic amplitude, seismic energy, low-frequency (*Lf*) component, and high-frequency (*Hf*) component, all associated with the ST reflector that provided only a single seismic reflection. [Acoustic impedance = $Lf \times Hf$, where *Lf* is the low-frequency information from seismic stacking velocity, and *Hf* is the high-frequency component by recursive iteration from amplitude to impedance from the seismic data (Lindseth, 1979).] These seismic data are located on a 260×130 regular grid with common-depth-point (CDP) spacing of 40 ft and line spacing of 80 ft, covering the area shaded in Figure 1. The four seismic variables are well related to each other and their maps are all visually similar to the *Lf* map shown in Figure 3.

DEFINING ST BOUNDARIES

The upper and lower surfaces of layer ST are accurately recorded at the 90 well locations. Although seismic data are available over the entire area and provide the general structure of the reservoir, their resolution is not enough to pick the ST top and bottom. The ST bounding surfaces from the 90-well data

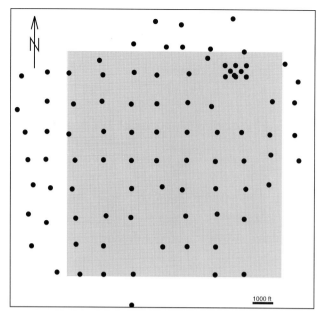

Figure 1. Well location map. The shaded area is covered with seismic lines.

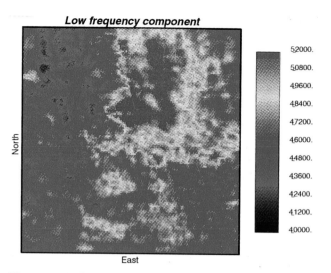

Figure 3. Color-scale map of the low-frequency component (*Lf*) associated with ST reflector. The map corresponds to the shaded area in Figure 1.

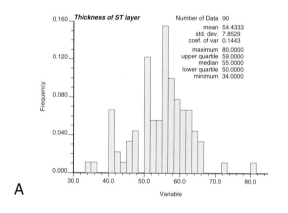

A

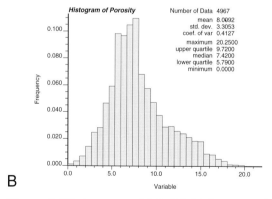

B

Figure 2. Histograms of (A) ST thickness and (B) porosity.

appear regular enough to be estimated by kriging using a continuous variogram. In a more difficult case where these surfaces might be discontinuous, faulted, or folded, complex surface modeling soft-

ware, such as GoCAD™, may be used (Mallet et al., 1991). The problem with such deterministic modeling is that the associated uncertainty is not measured and, hence, cannot be transferred into, say, uncertainty in estimating total pore volume. This study is limited to the 3-D mapping of porosity within the volume determined by kriging of the two bounding surfaces of layer ST. The uncertainty of the ST surfaces was ignored in this study.

Having decided to use kriging to estimate the top and bottom, we have two different options. One option is to krige the top and bottom separately; however, this option does not ensure that the top is always above the bottom, particularly if the variability of the ST depths is larger than that of its thickness. A better approach is to evaluate by kriging the top and thickness separately and then add them to deduce the estimate for the bottom.

The 2-D variogram map and various directional variograms of the ST top and thickness were calculated to determine the principal directions of anisotropy. The 2-D variogram map and 3-D variogram volume were calculated by laying the center of a 2-D or 3-D grid over data locations (one at a time) and, for each cell in which data existed, accumulate the square difference of the values between the center and the cell. The average of the accumulated differences in each cell is the variogram value for that cell. This procedure is much faster than the traditional algorithms for variogram calculation with tolerance distance and angles. Because it provides variograms for all directions, this method is best used for finding anisotropy from 3-D scattered data (Chu, 1993).

As shown in Figure 4A, the most continuous direction for the ST top is N30°W and the least continuous direction is N60°E.

The following anisotropic model fits well the

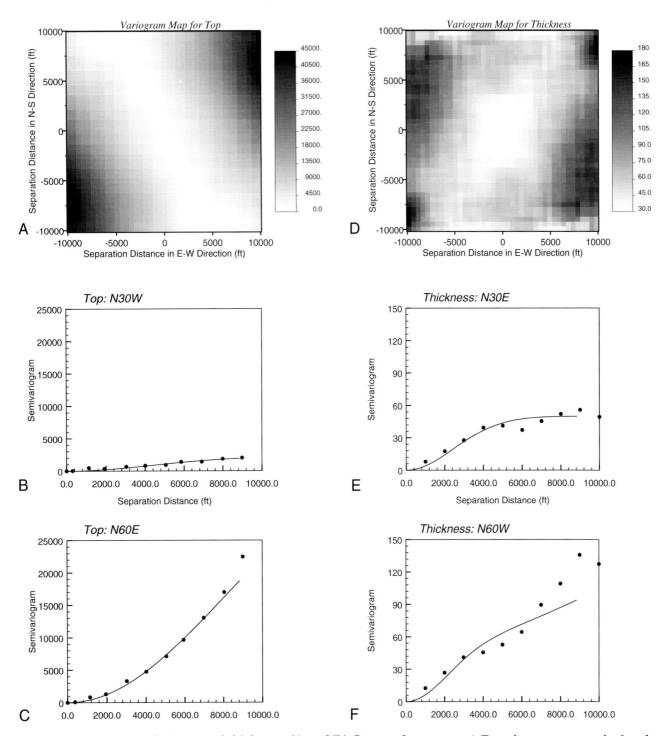

Figure 4. Variograms of ST top and thickness. (A and D) Gray-scale maps are 2-D variogram maps calculated directly from the scattered data with a grid oriented north-south–east-west; the number of lags is 20, with a lag interval of 500 ft.

experimental variogram (see Figure 4B, C):

$$\gamma_{top}(\mathbf{h}) = \epsilon + 4200\ \text{Gauss}\left[\sqrt{\left(\frac{h_x}{10680}\right)^2 + \left(\frac{h_y}{10680}\right)^2}\right]$$
$$+ 36800\ \text{Gauss}\left[\sqrt{\left(\frac{h_x}{\infty}\right)^2 + \left(\frac{h_y}{10680}\right)^2}\right] \quad (1)$$

where Gauss $(h/a) = 1 - \exp(-h^2/a^2)$ is a Gaussian model with unit sill and parameter a, that is, effective range $a\sqrt{3}$; h_x is the coordinate in the N30°W direction and h_y is the coordinate in the N60°E direction, with $\mathbf{h} = (h_x, h_y)$. Systematically a small nugget constant ϵ, set at 1% of the total sill, should be added to any Gaussian model to ensure stability of the resulting kriging matrices.

Note the consequences of the previous 2-D model

204   Chu et al.

in the two principal directions of continuity:

$$\gamma_{top}(h_x, 0) = \in + 4200 \text{ Gauss } (h_x/10680), \text{ along N30°W}$$

$$\gamma_{top}(h_x, 0) = \in + 4100 \text{ Gauss } (h_y/10680), \text{ along N60°E}$$

These are the continuous curves shown in Figure 4b and c.

As opposed to ST top, the most continuous direction for ST thickness is N30°E (Figure 4d). The model fitted is

$$\gamma_{th}(\mathbf{h}) = \in +50 \text{ Gauss } \left(|\mathbf{h}|/9000\right)^2$$
$$+ 140 \text{ Gauss } \left[\sqrt{\left(\frac{h'_x}{\infty}\right)^2 + \left(\frac{h'_y}{10680}\right)^2} \right] \quad (2)$$

with $\mathbf{h} = (h_{x'}, h_{y'})$, $h_{x'}$ being the coordinate in the N30°E direction and $h_{y'}$ being the coordinate in the N60°E direction. Note that these directions are different from those for the ST top model (equation 1).

The kriging grid comprises 88 by 88 nodes. A large search neighborhood of radius 8000 ft was considered to minimize neighborhood edge effects on the esti-mated maps for top and thickness. The two kriging maps are shown in Figure 5A and B. Their addition provides the estimated map for ST bottom shown in Figure 5C.

STRATIGRAPHIC COORDINATES SYSTEM

A common and important decision to be made in most 3-D applications relates to the coordinates system. The reservoir being studied may have nonuniform thickness, or may have been folded or even faulted, hence its anisotropy directions may vary throughout the area with the local dips. Yet, a stationary variogram model is required by most kriging and simulation algorithms.

One solution consists of splitting the global study area into several subareas, each homogeneous and flat enough to be modeled with stationary variograms defined from the traditional Cartesian coordinates system. The problems with such solution are (1) the solution requires a lot of data, enough to do the subdivision and reevaluate locally the directions of anisotropy or, in some cases, the entire variogram

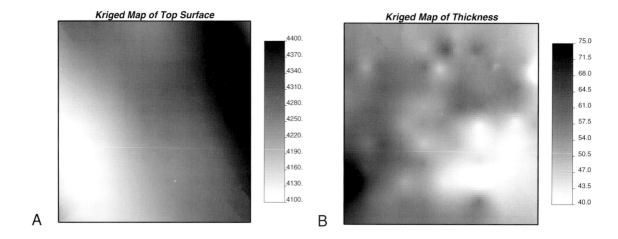

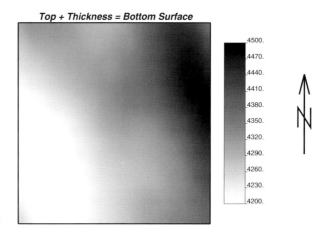

Figure 5. Kriging maps for (A) ST top, (B) thickness, and (C) bottom.

model and (2) artifact discontinuities at the boundaries between subareas may occur. An elegant, yet efficient, solution to this latter problem of discontinuity has been proposed by Soares (1990). In cases where stratigraphic conformity to either the top or the bottom of the layer can be assumed and the variability of the thickness is small (here, the coefficient of variation for thickness is 0.14, as read from Figure 2A), one could consider the coordinates transform proposed by Journel and Gomez-Hernandez (1993).

$$x' = x, y' = y, z'$$
$$= \frac{z(x,y) - \text{top}(x,y)}{\text{Thickness}(x,y)} \in [0,1] \quad (3)$$

where (x, y, z) are the original Cartesian coordinates, (x', y', z') are the transformed stratigraphic coordinates with the elevation (z') being defined relative to the local thickness of the layer. In a clastic reservoir, this transform would imply that the local sedimentation rate is proportional to the total thickness with no pinchout or erosion. This transform was adopted for this study. All variogram analyses, estimates, and simulations are performed in the stratigraphic system (x', y', z'), then the results are relocated in the original Cartesian system.

ESTIMATING POROSITY FROM WELL DATA

In the stratigraphic coordinates system, the kriging of porosity in 3-D is not theoretically different from that of 2-D. The difficulty lies in visualizing both the 3-D scattered data and the kriging results. A flexible 3-D visualization graphics software is essential for this purpose.

To define a 3-D variogram model, we need first to find the three major anisotropy directions. This is often difficult if there is little prior geological information beyond the data themselves. It is easy to miss the most continuous direction when the anisotropy ratio is high, as it is in most sedimentary environments. For the Amoco data, fortunately, we know that the most continuous direction lies in the horizontal plane (after coordinate transformation) and the least continuous direction lies in the vertical direction. In more difficult cases, for example, with nonsedimentary-type reservoirs, a 3-D variogram volume calculated from 3-D scattered data has proven very helpful in identifying the three major axes of anisotropy.

The porosity 3-D variogram volume presented in Figure 6 provides a visual appreciation of the anisotropy directions and ratios. After viewing several cross sections (an interactive process), the most continuous direction was deemed close to horizontal north (y) and the least continuous direction was vertical (z'). Directional variograms are then calculated along the principal directions, and a model can be fitted (Figure 7). The semivariogram model adopted is

$$\gamma(\mathbf{h}) = 5.5 \text{ Sph}\left(|\mathbf{h}|/10\right)$$
$$+ 8.0 \text{ Sph}\left(\sqrt{\frac{h_{x'}^2}{14000^2} + \frac{h_{y'}^2}{4000^2} + \frac{h_{z'}^2}{0.9^2}}\right) \quad (4)$$

with Sph(h/a) being the traditional spherical model with unit sill and range a; $\mathbf{h} = (h_{x'}, h_{y'}, h_{z'})$ is the separation vector in the stratigraphic coordinates system, with x' being east and y' being north. Note that the 3-D model (equation 4) results in an implicit anisotropic nugget effect (see also Figure 7). Kriging was performed on a regular grid of $88 \times 88 \times 40$, comprising 309,760 nodes.

Search Neighborhood

A proper definition of the search neighborhood and strategy is critical for 3-D applications where data are aligned along vertical wells, holes, or lines instead of being fully scattered in 3-D. Because the data density along wells is much greater than in the horizontal space, the data nearest to any single estimation node are likely to come from one or very few wells. The consequence is that artifact patches of similar estimated values may appear around each well. If all neighborhood data originate from a single well, artifact checkerboard-looking estimates may be generated due to larger kriging weights given to the data at the two extremities of the well.

One solution is to consider a very flat, ellipsoidal search neighborhood with a very small vertical radius or, more effectively, restrict the number of data retained from any single well. The GSLIB code was modified to ensure that no more than three porosity data from any single well are retained for kriging at any node. The horizontal search radius was set large (3000 ft) to use data originating from many different wells in any single kriging system. The resulting kriging of porosity is depicted in Figure 8 (after restoration of the original Cartesian coordinates). Note the stratification with strong horizontal continuity. This grid of estimated porosity values can be used directly, say, for calculating total pore volume.

SIMULATING POROSITY FROM WELL DATA

To evaluate uncertainty about the estimated porosity of Figure 8, we proceed to its simulation (stochastic imaging). The sequential Gaussian simulation algorithm was chosen here for its simplicity. To justify this choice, one should check for bigaussianity of the porosity distribution (Deutsch and Journel, 1992); however, implementation of such a check with anisotropic 3-D data can be quite tedious. Recall that if one suspects high or low porosities to be well correlated (spatially connected), the Gaussian model should not be used. Figure 9 shows the sample porosity indicator variograms in the east direction for the 0.1, 0.3, 0.5, 0.7, and 0.9 quantiles. Two important

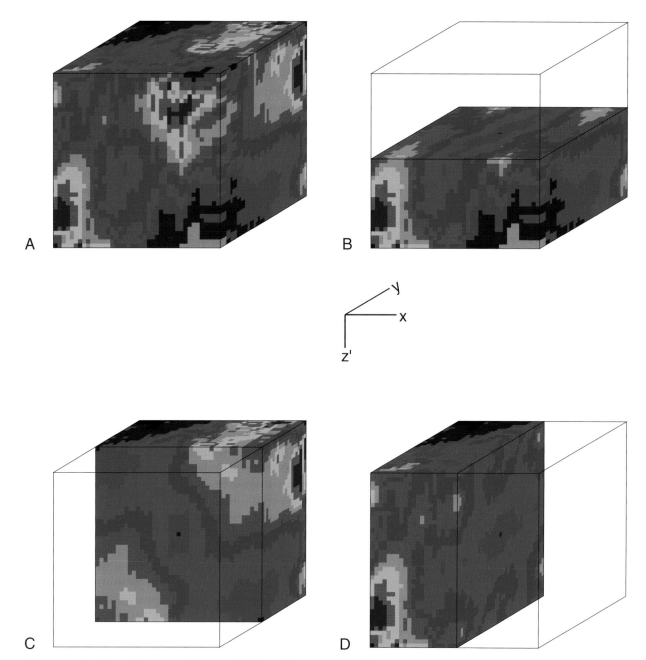

Figure 6. A 3-D variogram volume calculated from scattered 3-D porosity data. The maximum lag distance in the horizontal direction is 15,000 ft, and the horizontal to vertical exaggeration is 1:50. The darker cells have higher variogram values. (A) The whole cube; (B) the cube cut to show an (x, y) horizontal plane, indicating that the most continuous direction is roughly north (y); (C) a vertical (x, z) plane, revealing a short vertical structure and a larger eastward-dipping structure; and (D) a vertical (y, z) plane.

observations can be made from these variograms: (1) the 0.1 and 0.9 quantile indicator variograms are quasi–pure nugget effect, and (2) the variograms are reasonably symmetric around the median indicator variogram. These are features consistent with the assumption of bigaussianity of the porosity normal score transforms, and thus they do not invalidate the Gaussian model for porosity simulation.

Similar to the model (equation 4) considered for the untransformed porosity data, the semivariogram model retained for the porosity normal score transforms is as follows (see also Figure 10).

$$
\begin{aligned}
\gamma_{N\,scores}(\mathbf{h}) = 0.45\,\mathrm{Sph}\left(\sqrt{\frac{h_{x'}^2 + h_{y'}^2}{800^2} + \frac{h_{z'}^2}{0.15^2}}\right) \\
+ 0.55\,\mathrm{Sph}\left(\sqrt{\frac{h_{x'}^2}{14000^2} + \frac{h_{y'}^2}{4000^2} + \frac{h_{z'}^2}{1^2}}\right)
\end{aligned}
\tag{5}
$$

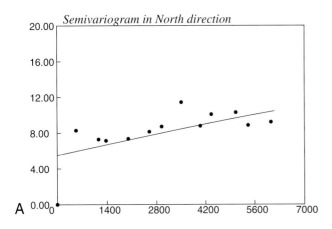

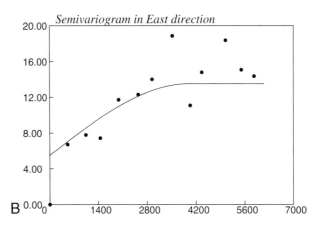

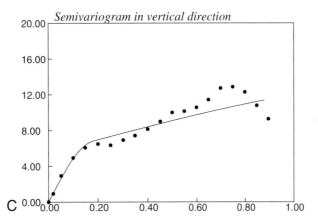

Figure 7. Experimental semivariograms (dots) of porosity along the three major directions and the corresponding 3-D model (solid lines).

with $h = (h_{x'}, h_{y'}, h_{z'})$ being the separation vector in the stratigraphic coordinates system, with x' being east and y' being north.

Simulated realizations of the 3-D porosity spatial distribution were generated using the sequential Gaussian algorithm and GSLIB program (Deutsch and Journel, 1992). The kriging grid and search strategy were the same as previously retained for the direct kriging of porosity.

Two realizations of the 3-D porosity field are displayed in Figures 11 and 12. In practice, many more simulated realizations should be obtained and visually inspected. Because all of them are constrained with the same conditioning data and variogram model, they are statistically equiprobable and can provide us quantitative measurements of the overall and local uncertainties with regard to the reservoir potential and performance. Note that zones where simulated realizations differ the most correspond to maximum uncertainty. These zones are usually not strongly influenced by well data and the local uncertainty should approach the global standard deviation shown in Figure 2B. On the other hand, the uncertainty is 0 at

data locations. These realizations should be compared to the smoother image provided by direct kriging (Figure 8). Stratigraphic continuity is apparent on both the kriging map and the simulated maps.

Reproduction of the input spatial statistics by realization 1 is shown in Figure 13. Ergodic fluctuations are expected from one realization to another when evaluating reproduction of model statistics (see discussion in Deutsch and Journel, 1992). The reproduction of the vertical variogram is very sensitive to the vertical search radius (0.18 was the value retained for this study), and the variogram reproduction (Figure 13C) is seen to be reasonably good. Figure 13D gives a probability-probability (P-P) plot of the cumulative frequencies of sample porosity data vs. the cumulative frequencies of the simulated values of realization 1; the distribution of sample porosity data is seen to be correctly reproduced.

INTEGRATING SEISMIC DATA

Seismic information is used to improve the mapping of porosity within ST. Well data (well-log-

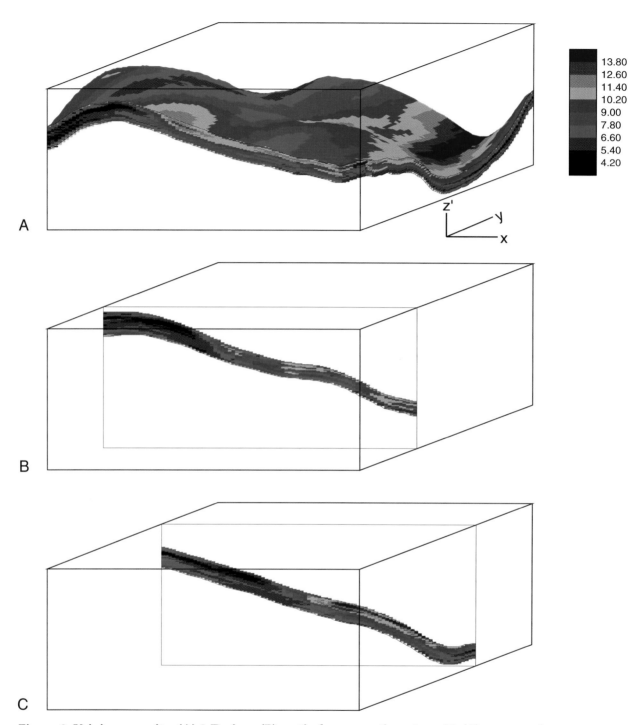

Figure 8. Kriging porosity. (A) 3-D view; (B) vertical cross section at $y = 30$; (C) cross section at $y = 60$.

derived porosity) and seismic properties reflect reservoir petrophysical properties at different scales. Seismic data are poor in resolution, especially in the vertical direction. These data represent information integrated along a thickness usually much larger than the strata under study. For the Amoco data, the ST thickness averages 50 ft and is seen in the seismic profile as a single reflection event. The seismic data are therefore 2-D and are related to some vertically averaged porosity within the ST unit rather than to individual 3-D porosity values. This difference in dimensionality calls for specific approximations.

Seismic-Porosity Calibration

Although seismic data are dense, their local porosity information content is not accurate as log-derived porosity data. The average of several seismic data, e.g., low-frequency component $\overline{Lf}$, around each well location may better reflect the vertically averaged

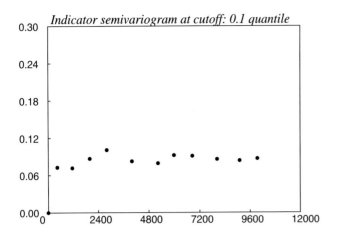

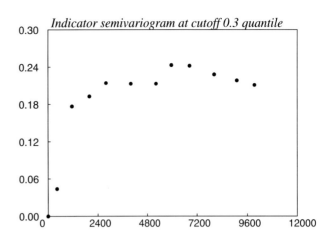

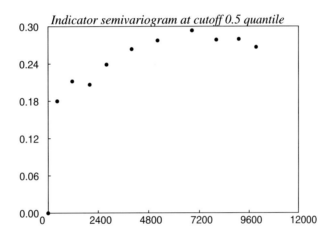

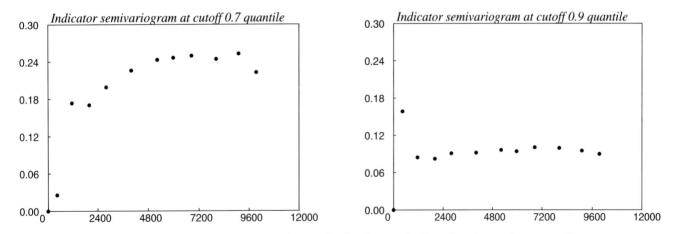

Figure 9. Experimental indicator variograms of porosity in the north direction for various cutoffs.

porosity value ($\bar{\phi}$) at that location. Two questions then arise:

- How significant is the effect of the seismic averaging window size on the correlation between the resulting $\bar{L}f$ and $\bar{\phi}$ and on the final simulation?

- If the window size does affect the correlation significantly, which averaging window size should be retained?

Figure 14A shows that the correlation $\rho_{\bar{L}f,\bar{\phi}}$ is high (0.5–0.7) and increases with window size. The larger

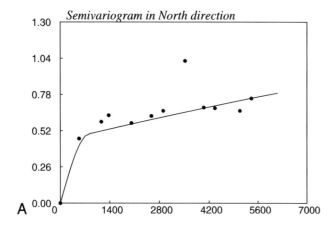

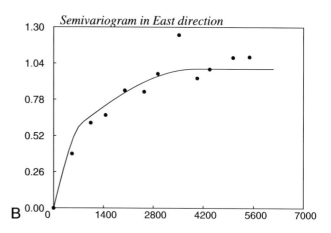

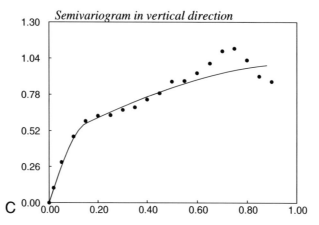

Figure 10. Experimental normal score variograms (dots) of porosity in the three major directions and the corresponding 3-D model (solid lines).

the window size, however, the more (horizontal) resolution is lost from the original seismic data as shown in Figure 14B. A compromise has to be made between high correlation and loss of resolution. A window size of 1 was retained, corresponding to the maximum correlation value of $\rho_{\overline{Lf},\overline{\phi}} = 0.614$, and a 5% loss of overall variance. That window size corresponds to averaging the nine seismic CDP locations closest to each well location.

The other three seismic parameters (after averaging over the same window size), amplitude, energy, and high-frequency component ($\overline{Hf}$) appear highly redundant with $\overline{Lf}$. They show poorer correlation with $\overline{\phi}$ (Figure 15), and thus were not considered.

Simulations of Porosity Conditional to $\overline{Lf}$

Within the context of a Gaussian model, the conditioning of porosity simulation by seismic data is theoretically straightforward. One would take the normal score transforms of both the porosity data (ϕ defined in 3-D) and the seismic data ($\overline{Lf}$ defined in 2-D), a joint Gaussian model would be assumed, and the corresponding covariance/cross-covariance functions would be inferred and modeled. Cokriging, using the normal score porosity data and the

2-D normal score seismic data, would yield the parameters of the Gaussian conditional distribution of the nodal normal score porosity. A simulated normal score porosity value would be drawn from that conditional distribution, and one would continue the analysis in the manner previously described. Such an approach, besides being tedious, would require a modification of GSLIB program SGSIM to allow for cokriging.

One may question the necessity of using a full cokriging, retaining for each simulated node a whole set of neighboring and highly redundant secondary data (the neighboring $\overline{Lf}$ normal score data). A first approximation consists of retaining only the $\overline{Lf}$ normal score datum closest or collocated with the node being simulated. A second approximation consists of retaining for the cross-covariance between normal scores of porosity and seismic, a Markov-type model (Deutsch and Journel, 1992; Xu et al., 1992).

$$C_{12}(h) = \rho_{12}(0)C_1(h) \qquad (6)$$

where $C_1(h)$ is the covariance (correlogram) of the porosity normal score transform (see equation 5), $C_{12}(h)$ is the cross-covariance (cross-correlogram) of porosity and seismic normal transforms, $\rho_{12}(0)$ is the

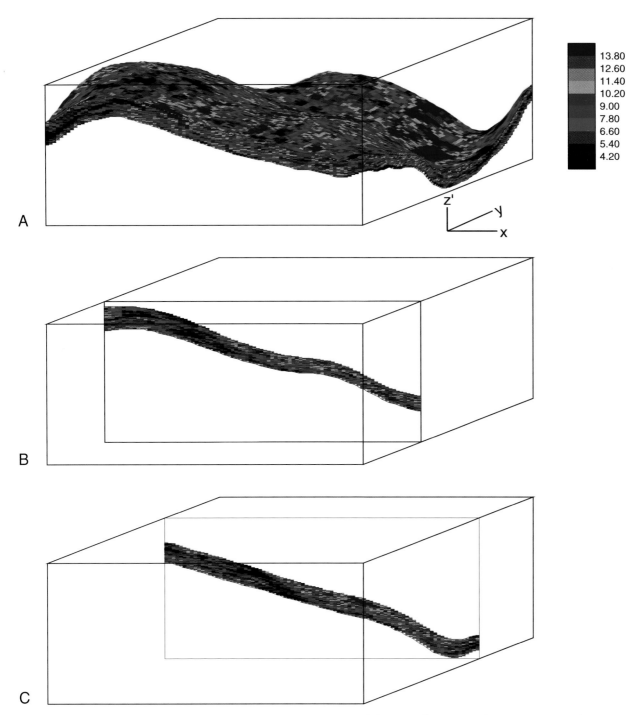

Figure 11. Realization 1 of the porosity field. (A) 3-D view; (B) vertical cross section at $y = 30$; and (C) cross section at $y = 60$.

correlation coefficient of collocated porosity and seismic normal score transforms, here $\rho_{12}(0) = 0.59$.

To justify this approximation, cross-semivariograms of $\bar{\phi}$ and $\overline{Lf}$ were calculated along the two major directions (north and east) and are shown in Figure 16C and F. Also shown are the corresponding direct semivariograms (Figure 16A, B, D, E). Note that these cross-semivariograms do exhibit structures and, more importantly, these structures appear proportional to the direct semivariogram of the primary variable porosity,

as required by the Markov model (equation 6).

One last approximation consists of spreading the original 2-D values $\overline{Lf}\,(x', y')$ in 3-D, more precisely along all nodes (x', y', z') of the vertical line at (x, y). The simulation grid was restricted horizontally to the area informed by seismic data (see shaded area in Figure 1) and the inner box in Figure 17A. Otherwise, the search strategy considered is the same as that retained for the direct porosity kriging and for the simulation of porosity using only well data.

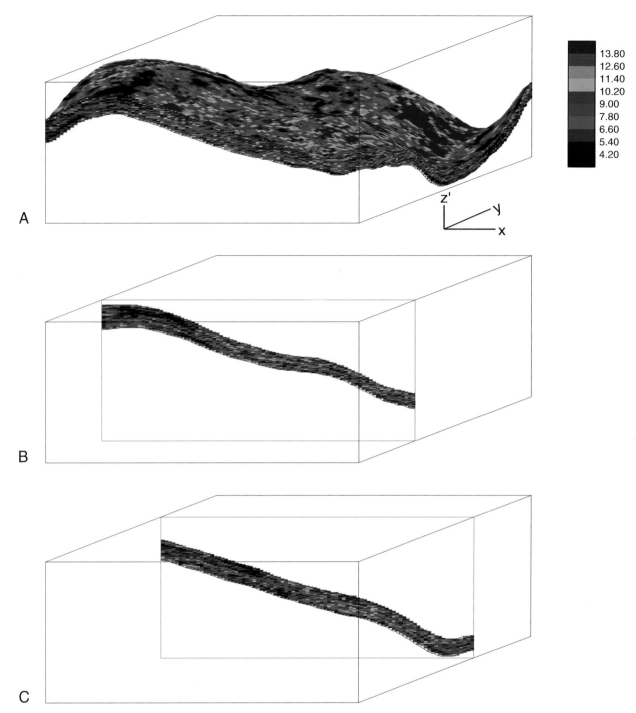

Figure 12. Realization 2 of the porosity field. (A) 3-D view; (B) vertical cross section at $y = 30$; and (C) cross section at $y = 60$.

One simulated realization is shown in Figure 17 along with several vertical 2-D cross sections. On planar views, the porosity realizations appear highly constrained by the distribution of seismic data (see the patch of high $\overline{Hf}$ values on Figure 3). On cross sections, the porosity realizations appear less laminated than the well-only conditioned realization of Figure 11 or the direct kriging image of Figure 8. This is a consequence of having spread the values along vertical lines. Consequently, the realization of Figure 17 may give a false impression of greater vertical continuity.

CONCLUSIONS

The problems encountered in this case study are typical of problems found in 3-D applications of geo-

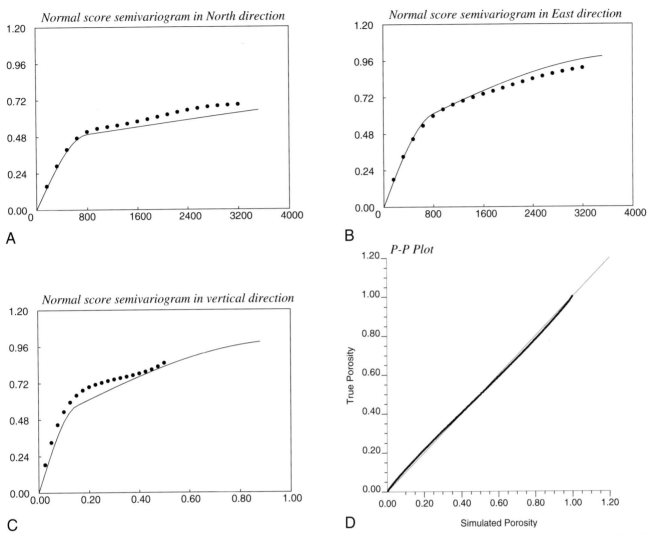

Figure 13. Reproduction of normal score variograms in the three major directions. (A, B, C) Variograms of realization 1 (dots) are compared to the input model; (D) P-P (probability-probability) plot of sample porosity data vs. simulated porosity values.

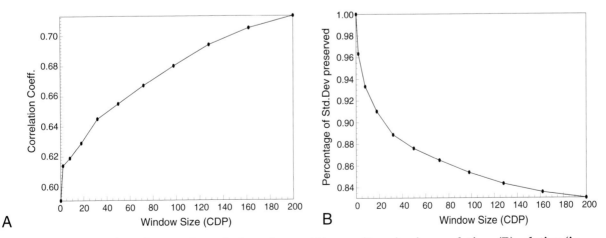

Figure 14. Impact of local averaging window size on (A) porosity-seismic correlation; (B) relative (in percent) loss of spatial variance from the original spatial variance of seismic data. CDP = common depth point.

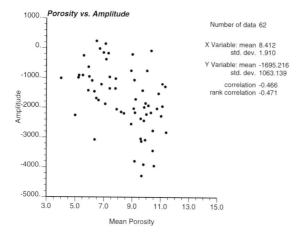
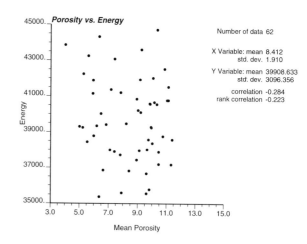
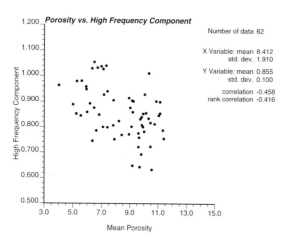
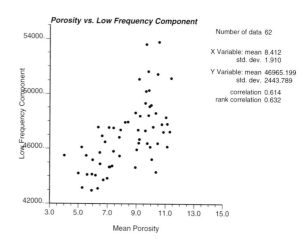

Figure 15. Scatterplots of porosity vs. horizontally averaged seismic parameters.

statistics:

- The upper and lower layer boundaries and, more generally, boundaries of major flow or lithological units are usually not known and must be estimated.
- Folding, faulting, and structural deformation may cause apparent nonstationary anisotropy. Correction can be done by either transforming the coordinates system, resetting locally the anisotropy parameters, areas subsetting, or increasingly complex unfolding of structures.
- Prior geological information is essential in identifying major anisotropy directions. Short of such information, the 3-D variogram volumes calculated from scattered data can be helpful.
- Because data are often clustered along vertical wells, a careful search strategy is needed to eliminate or reduce possible artifacts. This often requires kriging or simulation programs to be modified so that application-dependent search strategies are used.

- Seismic information does contribute to mapping of petrophysical properties (porosity). Calibration from actual field data is an absolute must. The poor vertical resolution of seismic data may limit its usefulness to 2-D applications, especially when the thickness is small.
- An interactive 3-D graphic software for volumetric visualization is essential for examining scattered data, experimental variogram volumes, and kriging and simulation results.

ACKNOWLEDGMENTS

We thank Amoco Production Company for providing the data for this case study and especially Dr. Rich Chambers for his help and comments.

REFERENCES

Chambers, R. L., and M. A. Zinger, 1990, Geostatistical approach to reservoir characterization: 3-D seismic

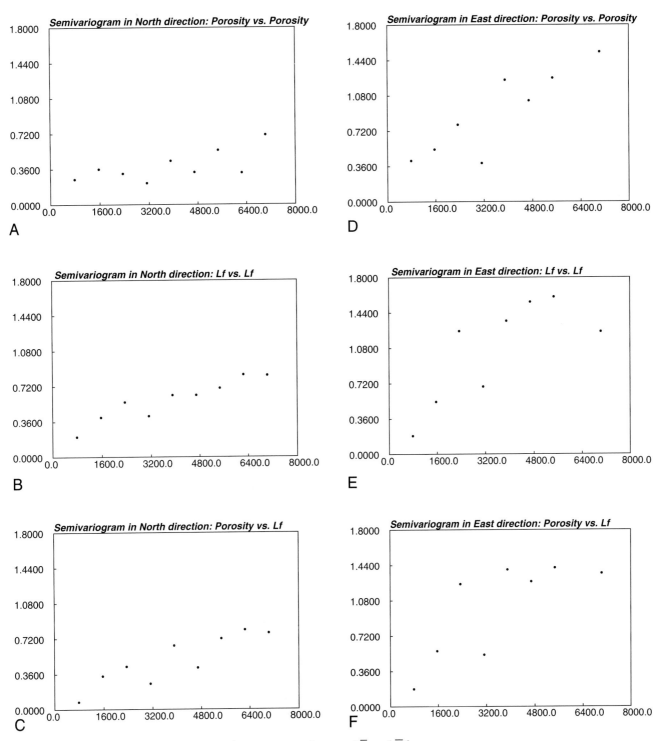

Figure 16. Cross variograms of normal scores transforms of $\bar{\phi}$ and $\bar{L}f$.

acquisition program—phase I: Proprietary material of Amoco Production Company.

Chu, J., 1993, Conditional fractal simulation, fast sequential indicator simulation, and interactive variogram modeling: Ph.D. thesis, Stanford University, Stanford, California.

Chu, J., W. Xu, H. Zhu, and A. G. Journel, 1991, The Amoco case study: Special Report of the Stanford Center for Reservoir Forecasting, Stanford University.

Deutsch, C. V., and A. G. Journel, 1992, GSLIB: geostatistical software library and user's guide: New York, Oxford University Press, 340 p.

Journel, A. G., and J. J. Gomez-Hernandez, 1993, Stochastic imaging of the Wilmington clastic sequence: SPE Formation Evaluation, March 1993, p. 33–40.

Lindseth, R. O., 1979, Synthetic sonic logs—a process for stratigraphic interpretation: Geophysics, v. 44, no. 1, p. 3–27.

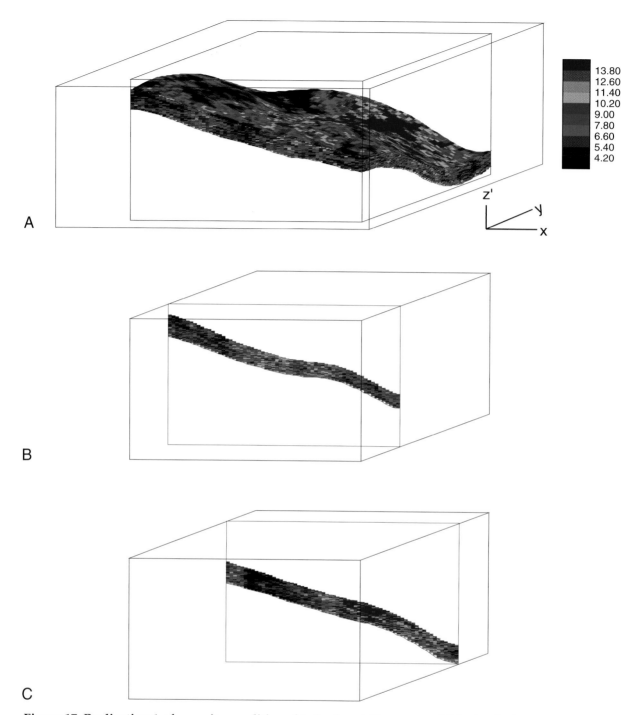

Figure 17. Realization 1 of porosity conditioned to horizontally averaged *Lf* seismic data. (A) 3-D view; (B) vertical cross section at *y* = 30; and (C) cross section at *y* = 60. Simulation was performed with seismic data only from within the inner box.

Mallet, J. L., et al., 1991, GoCAD report (confidential): Ecole Nationale Supérieure de Géologie de Nancy, France.

Soares, A., 1990, Geostatistical estimation of orebody geometry: Morphological Kriging, Mathematical Geology, v. 22, no. 7, p. 787–802.

Xu, W., T. Tran, R. M. Srivastava, and A. G. Journel, 1992, Integrating seismic data in reservoir modeling: the collocated cokriging alternative: SPE paper 24742.

Chapter 17

Stochastic Structural Modeling of Troll West, with Special Emphasis on the Thin Oil Zone

Trond Høye
Eivind Damsleth
Knut Hollund
Norsk Hydro a.s.
Oslo, Norway

INTRODUCTION

The Troll field is a huge oil and gas field situated in the North Sea approximately 100 km northwest of the city Bergen of Norway (Figure 1).

The oil leg in the Troll West oil province (TWOP, 22–28 m thick) and the Troll West gas province (TWGP, 12–14 m thick) is predicted to contain about 600×10^6 Sm^3 of oil in place; however, the combination of a layered structure with 2–6° dip and a relatively thin oil leg makes the oil prospects very uncertain with respect to optimal target locations. Only minor vertical adjustments of the highly permeable sands may lead to significant changes in the lateral position of these prospects.

In this paper, we present the stochastic modeling study (SMS) on Troll West in 1992, comprising mainly structural modeling, and describe the principles and assumptions related to the modeling. During the model building, we tried, as far as possible, to follow the same concepts as in deterministic modeling.

The main structural control of Troll West yields the seismic Top Sognefjord reflector (top reservoir) and a more blurred Top Fensfjord reflector (base reservoir). Recently an intrareservoir reflector, 4.1TR, corresponding to a near-top 3C sand, was mapped in TWOP and parts of TWGP. This framework, which is fundamental for the location of individual reservoir layers in undrilled areas, involves uncertainties related both to the seismic interpretation and processing, as well as to the depth conversion method.

The subseismic reservoir layering was sampled only at well locations. Due to this spatial undersampling, the uncertainties may be severe. Away from well control, the subseismic layering is controlled by a combination of predicted layer thicknesses and the fitting of these into the seismic framework. Note that the top reservoir reflector provides less information on the subseismic structure because this surface is erosive.

The main objective of the work described here was to highlight the target location uncertainty related to the deterministic reservoir descriptions. A procedure has been constructed to manage the structurally related uncertainties in the Troll field, including models for (1) seismically related uncertainty, (2) layer thickness uncertainty, (3) reservoir model construction aspects, and (4) fluid contact uncertainty.

There is no such thing as an objective assessment of uncertainty; there are only models based on a series of subjective decisions (A. Journel, 1994, personal communication). Also in this model, several subjective or deterministic choices have been made, e.g., how the correlations between wells were defined.

GEOLOGIC FRAMEWORK

Depositional Environments and Systems Tracts

A depositional model was established for Troll West. The reservoir correlations have been based on sequence stratigraphic principles (Van Wagoner et al., 1990).

The Sognefjord Formation in Troll West is assumed to represent a coastal deltaic environment comprising a series of stacked prograding delta lobes (Figure 2). In this model, the clean sands are explained as channelized and lobate distributary mouth-bar deposits, as well as more sheetlike delta-front deposits. This explanation is based on the observed moderate to poor sorting of the sands, the near-absence of tidal indicators, the main sedimentary dips, and the apparent positive topography seen on the three-dimensional (3-D) seismic data. The degree of tidal influence is uncertain. Most of the clean sands in Troll West are coarse grained (medium grained to pebbly). The main sedimentary dips cluster toward the sector southwest to northwest. Clinoforms seen on the 3-D seismic data may be associated with prograding sand lobes.

The clean sands (C sands) are assumed to represent lowstand deposits. The apparent lack of delta-plain sediments is due to reworking during flooding of the

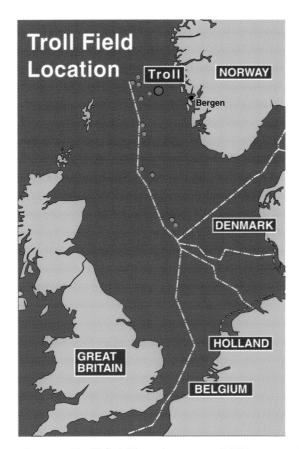
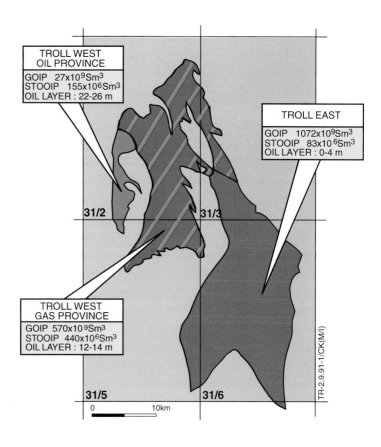

Figure 1. Troll field location map. GOIP = gas originally in place, STOOIP = stock tank oil originally in place.

delta. The overlying, laterally more extensive, fine-grained sediments (mica sands or M sands) are assumed to represent the following highstand. These units (the C and the M sands) are fundamental building blocks in the reservoir construction modeling. The surface separating the C from the above M sand, in some cases, is seismically mappable and represents an important lithological discontinuity also referred to as a major flooding surface.

Sequence Stratigraphic Reservoir Zonation

Seven sequences have been distinguished within the Sognefjord Formation in Troll West. The sequences and their numbering are based on the major flooding surfaces (FS). Each such sequence consists of a micaceous, low-permeability sand (M sand) overlain by a clean high-permeability sand (C sand). The numbering of these sequences is based on the reservoir zonation applied in the equivalent deterministic reservoir description of the field (Figure 2). Zone 1 comprises the Fensfjord Formation, therefore the numbering of the sequences in the Heather B/Sognefjord Formations starts with number 2 (oldest).

The C sand of each sequence except 4AC and 4BC may be further divided into parasequences, each consisting of a lower micaceous part (m sand) and an upper clean part (c sand), representing two reservoir zones on a sub–C/M scale (e.g. 2m1, 2c1,...,2m4, 2c4).

STOCHASTIC MODEL

The 1992 SMS activity mainly comprised the structural aspects of reservoir description modeling. Structural uncertainty was assigned to three major categories: (1) seismic uncertainty, (2) fluid-contact uncertainty, and (3) subseismic layer-thickness uncertainty.

To study the impact of these uncertainties, we decided to resort to stochastic modeling. In a stochastic approach, a stochastic model is built that explicitly caters to the different uncertainties mentioned. This stochastic model is used to generate a large number of realizations of the field, using the techniques of stochastic simulation. For each of these realizations, various key parameters are computed. Thus, a probability distribution for these parameters quantitatively describes their uncertainty, which is caused by the uncertainty in the geologic model.

The main stochastic technique used in this study is the conditional multi-Gaussian random fields technique, properly described in Omre (1987) and Deutsch and Journel (1992). The technique in this context may be looked upon as a substitute for conventional gridding techniques used in two-dimensional surface mapping. The randomness in this Gaussian field is spatially structured through a variogram model, and is used to reproduce a model for the spatial variability (frequency content) in the data. The notation "multi" refers to sev-

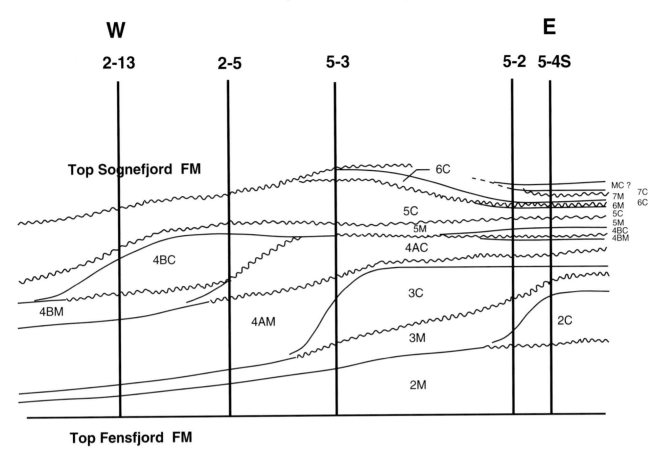

Figure 2. Cross section of Troll West field. See Figure 3 for well locations.

eral correlated parameters being cosimulated in such a way that the correlation coefficients are maintained. Such linear dependencies are easily verified and quantified in any statistical package. The conditioning guarantees that all well observations are reproduced. Haldorsen (1990) gives a good review of the various techniques and methods for stochastic modeling.

Database

The seismic interpretations used in this project are based on the 3-D seismic surveys TWOP: NH8901 and TWGP-N: NH9101. These data sets are of very good quality. The well database consists of 21 wells from blocks 31/2, 31/3, and 31/5 (Figure 1).

The Seismic Model

Several of the seismic parameters are assumed discontinuous across faults, such as interval velocities and depth to the gas/oil contact and the oil/water contact. The TWOP and TWGP area is divided into 13 major fault-block polygons (Figure 3). The major aspects of the seismic uncertainty may be divided into the following three categories: (1) uncertainty in the reflector mapping, (2) well tie uncertainty, and (3) uncertainty in the depth conversion. The first two uncertainties are modeled in the time domain, whereas the depth conversion transforms the time uncertainties to the depth domain.

Model for Uncertainty in Reflector Picking

The seismic interpreters are trying to map a visualised seismic response to a subsurface reflector, and have in this context to cope with several uncertainties. Should they follow the black peak or the white trough? What is the resolution in the seismic signal? Does it actually reflect the boundary they think they are mapping? This type of uncertainties encompasses a huge set of factors. To quantify this type of uncertainties is difficult and thus seldom done in an areal sense. Our approach to the problem is also a simplification.

Interpretation uncertainty polygons may be constructed by the geophysicist. The horizon map is divided into 3–5 classes, each assumed to represent a 95% confidence interval of a prediction estimate in milliseconds, i.e., saying that within the polygon the uncertainty is less than a fixed number of milliseconds. The classes identify regions with uncertainties of less than 4, 6, 8, 10, and 12 ms respectively. Based on these assumptions, the geophysicist constructs a standard deviation map. A simple smoothing procedure has been applied to avoid steps across class boundaries (Figure 4).

Model for Well Tie Uncertainty

Another uncertainty yields the tie between the mapped reflector and the zonation boundary defined in a well. Is this the boundary being mapped by the seismic interpreter? This is commonly verified

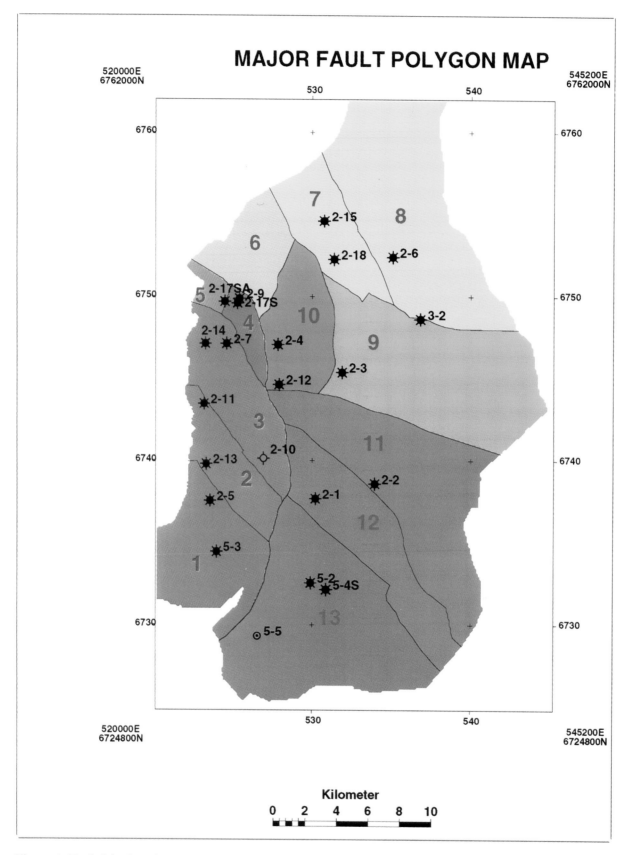

Figure 3. Fault-block polygon map showing the TWOP (Troll West oil province) and TWGP (Troll West gas province) divided into 13 major fault-block polygons.

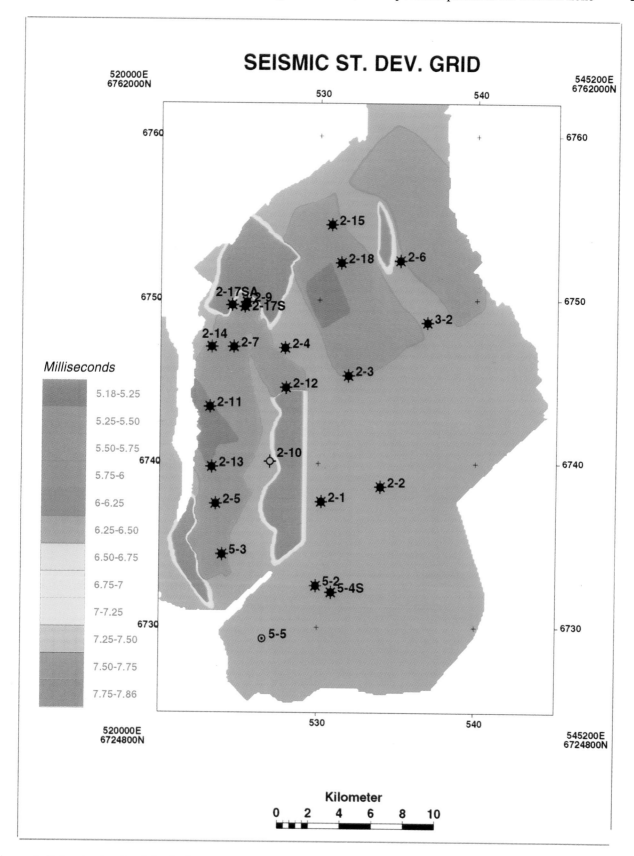

Figure 4. Seismic standard deviation map of the top of the Fensfjord representing the combined interpretation and well-tie uncertainty.

through constructing a synthetic seismogram. The well tie uncertainty is then estimated as the time shift between the mapped reflector and the equivalent event on the synthetic seismogram. This time shift encompasses several uncertainties, such as seismic processing effects, seismic acquisition uncertainties, well-calibration problems, and the problem of resolving the seismic response model.

To get the total seismic uncertainty the interpretation uncertainty (σ_1) is combined with the estimated well-tie uncertainty (σ_2) according to the formula

$$\sigma_{seismic} = \sqrt{\sigma_1^2 + \sigma_2^2}$$

In this way, a seismic standard deviation grid is generated, which, multiplied by a normalized stochastic uncertainty grid (Figure 5), gives one possible realization of the seismic error dT. Note that the normalized stochastic parameter is not spatially correlated across the major faults. These seismic error maps are added to the original seismic interpretation. Repeated applications of this procedure give a spectrum of equiprobable seismic surfaces, all within the specified error boundaries and consistent with the available data.

Model for Internal Reflectors

Partly mapped internal seismic reflectors may be hard to include in conventional reservoir description modeling. For example, the internal reflector 4.1MFS (top of 3C) was mapped in October 1992 in the TWOP area only. This seismic control for the top of 3C was accounted for in the stochastic model by resampling the seismic map to a 200 × 200 m grid and treats these gridnodes as well observations. The well observations and the resampled seismic values were transformed to a normal distribution using the normal score transform [the normal score transform is a nonlinear transform that, by construction, creates a univariate, normally distributed data set (Deutsch and Journel, 1992)]. These transformed values were then used as conditioning points in the stochastic modeling of the whole 4.1MFS seismic surface, which, in turn, was backtransformed using the inverse normal score transform. The stochastic surface thus generated was subsequently treated exactly like the top and base reservoir surfaces.

The base reservoir to top of 3C isochore is assumed to be representative of the general paleotopography in Troll West, thus carrying information about the spatial dependence (see the section on the layer model). More information on these aspects will become available as new internal reflectors are mapped seismically.

Depth Conversion

The same approach is used for the depth conversion as in the conventional reservoir description (TROLL, 1991); however, several parameters are treated as stochastic.

The model for depth to the top of the reservoir is base on the Linvel method, which is described in

TROLL (1991):

$$Z = \frac{V_0}{K}\left(e^{KT} - 1\right)$$

where V_0 represents the surface velocity, K represents the velocity-depth gradient, Z is the depth, and T is the one-way traveltime for seismic waves. The stochastic parameters are V_0, which is modeled as a Gaussian random field conditioned on the well locations (Figure 6), and T, which is modeled according to the procedure described in the previous sections. The velocity-depth gradient K is regarded as deterministic (a value of $K = 0.426$ is used). The model assumes a linear velocity-depth relationship on the form

$$V(Z) = V_0 + KZ$$

Note that the traveltime T is laterally discontinuous across faults. By choosing a K such that V_0 is least correlated with T, however, a map of V_0 may be assumed laterally continuous. V_0 is modeled independently of any other parameters. Only the well information has been used. Stacking or migration velocities are not included. An example of a V_0 realization is shown in Figure 6.

To model the time-thickness conversion in the gas and oil/water zones, the interval velocity in the gas zone (V_{ig}) and in the water zone (V_{iw}) is estimated according to the formulas

$$V_{ig} = dZ_g/dT_g$$
$$V_{iw} = dZ_w/dT_w$$

where dZ_g or dZ_w are the well-derived gas or water zone thicknesses, respectively, and dT_g or dT_w are the well-derived traveltimes through the gas or water zone, respectively. V_{ig} is assumed to be correlated with V_{iw}, the traveltime to the gas-oil contact and various well tie uncertainties (previously listed). These apparent correlations are assumed partly due to human factors like the effect of mispicking the contact and the assumed tendency of going too high or low at several reflectors simultaneously (Table 1). All of these parameters are assumed discontinuous across major faults. An example of a V_{iw} map is shown in Figure 7.

Fluid Contact Models

The fluid contacts are modeled in 13 independent fault-block areas (Figure 3). In each of these polygons, a constant horizontal contact level is assumed. The depth is assumed normally distributed with a mean predicted from the available well interpretations, and a standard deviation estimated to be ap-proximately 0.7 m. Note that this model is updated continuously in parallel with the seismic mapping of internal reflectors.

Layer Model

The layer model used in this study is based on sequence-stratigraphic concepts (see the section on geologic framework). The building blocks represent a

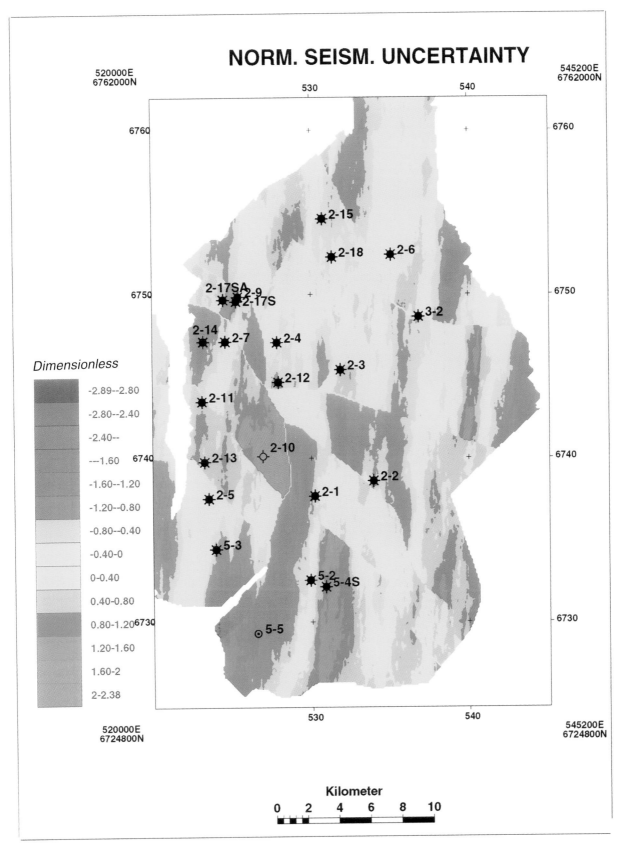

Figure 5. Stochastic realization map of the normalized well-tie and interpretation uncertainty for the top of the Fensfjord.

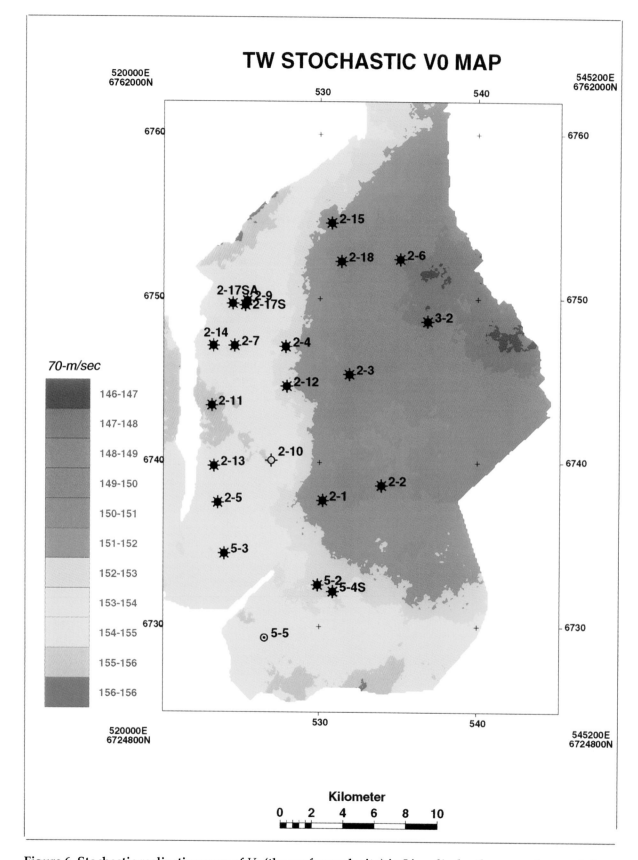

Figure 6. Stochastic realization map of V_0 (the surface velocity) in Linvel's depth conversion model.

Table 1. Correlation matrix for the depth-conversion parameters*

	twt_{goc}	V_{ig}	V_{iw}	dT_s	dT_3
V_{ig}	−0.40				
V_{iw}	0.48	−0.40			
dT_s	−0.24	0.16	0.24		
dT_3	−0.08	−0.48	0.64	0.56	
dT_f	−0.16	0.16	0.16	0.56	0.56

*twt_{goc} = two-way traveltime to the gas/oil contact, V_{ig} = interval velocity in the gas zone, V_{iw} = interval velocity in the water zone, dT_s = the top reservoir seismic error, dT_3 = internal reflector seismic error, dT_f = base reservoir seismic error.

hierarchy of four sequence-stratigraphic scales: (1) the sequences, comprising (2) a highstand micaceous (M) sand and a lowstand clean (C) sand; the C sand consists of (3) a set of parasequences or subsequences that are further divided into (4) lower order clean (c) and micaceous (m) sand units. The latter subdivision can be looked upon as the flow units where the boundary between the c and m sands is more litho-stratigraphic or petrophysically defined. Note that the zonation is deterministic in the model, meaning that the zone picks in the wells are used as hard data. Only the thickness distributions of these zones are treated stochastically. These thicknesses are modeled stochastically in a hierarchical way. At stage 1, the seven major sequences, with their associated C and M sand boundaries, are generated and adjusted to fit into the depth-converted top, middle, and bottom seismic surfaces. At stage 2, the parasequences are generated and adjusted to fit into their respective C sands. Finally, the fourth-order sequence boundaries for the c and m sands within each parasequence are generated. The different steps in the procedure are illustrated in Figure 8 and described in some detail in the following sections.

Individual Isochore Models

Every isochore, whether it represents a sequence, a parasequence, or a single sand, is modeled as a Gaussian field with a prescribed mean and variance, and a prescribed variogram structure. The underlying assumption of normality in the Gaussian simulations has been accounted for as follows. In our first approach, we used a truncated normal probability distribution function (pdf), truncated at 0, to represent the thickness distributions, implying that whenever the stochastic simulation generates a negative thickness for a layer at a particular location, this is interpreted as a 0 thickness (not present). The pdf and truncation are shown schematically in Figure 9.

The mean and variance parameters for each isochore were estimated from the well data using an estimation technique that used the observed thicknesses, as well as the 0 values to find the parameters in the underlying nontruncated Gaussian distribution [NAG routine G07BBF in NAG (1991)]. As an example, the resulting parameters for all the sequences and their associated C sands are presented in Table 2. In Table 2, the "coverage" column gives the probability of a layer being present at a given location, which equals the relative area where the layer thickness is nonzero. The "cond. mean" column gives the conditional mean, i.e., the expected thickness of a layer, given that it is present.

In each well it was necessary to assign negative thicknesses to the unobserved layers to be used as conditioning data in the stochastic modeling. A simple, ad hoc method was chosen, where all the unobserved thicknesses were given the same negative value, such that the mean of all the (positive and negative) well data equaled the unconditional mean given by the previously discussed estimation technique.

The model has been improved by implementing the normal score transform (Deutsch and Journel, 1992), but still using the estimates in defining the unsampled tails. Thickness distributions prior to the normal score transform are illustrated in Figure 10, indicating that we have data sets with fairly simple distributions, however far from normal. The shapes may be explained as a response to a systematic infilling of an accommodation space where the water depths have varied, overprinted by later erosions.

The frequency content in the thickness maps is assumed reflected in the frequency content of the seismic top Fensfjord to top 3C isochore map (Figure 13), revealed by variogram analyses of these seismic thicknesses. The results of these analyses are shown in Figure 11. We decided to use a 60/40 mixture of a Gaussian and a spherical variogram model. For the major sequences, the maximum correlation length was set at 30 km at 5° east of north, and the anisotropy ratio was set to 5:1, giving a minimum correlation length of 6 km at 95° east of north. For the parasequences, the maximum correlation length was set to 9 km for the 2c, 5c, and 7c sands, and 13 km for the 3c and 6c sands. The direction of maximum range was chosen perpendicular to the associated major sequence, and the anisotropy ratio was set to 1.5:1. The selection of these parameters was based on evaluations of possible Troll analogs, but will be subjects for future adjustments as more data become available. The shapes of the different variograms are shown in Figure 12.

Correlation Between the Different Isochores

From the observed well data, we can estimate a correlation matrix between the various thicknesses. Due to the limited amount of data, the correlation estimates are rather uncertain, and it was necessary to adjust the matrices slightly to make them conform with the geologic understanding (a thick sequence should coincide with a thick C sand, and adjacent sequences should be negatively correlated). As an example, the correlation matrix for the sequences and their associated C sands are shown in Table 3.

Adjustment of the Isochores

As was discussed, a hierarchical structure exists in the model-building procedure. When a realization of the sequences or parasequences has been generated, it

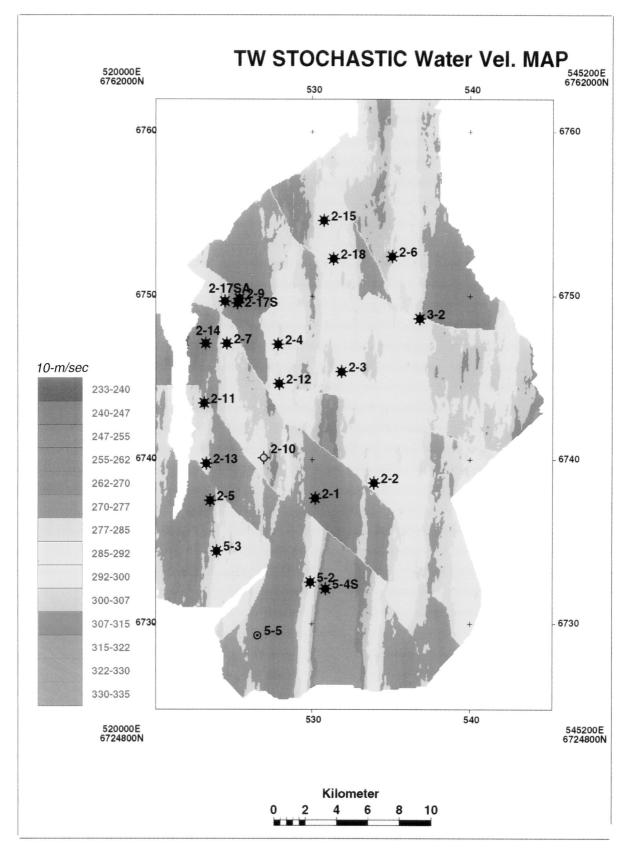

Figure 7. Stochastic realization map of V_{iw} (the interval velocity) in the water zone.

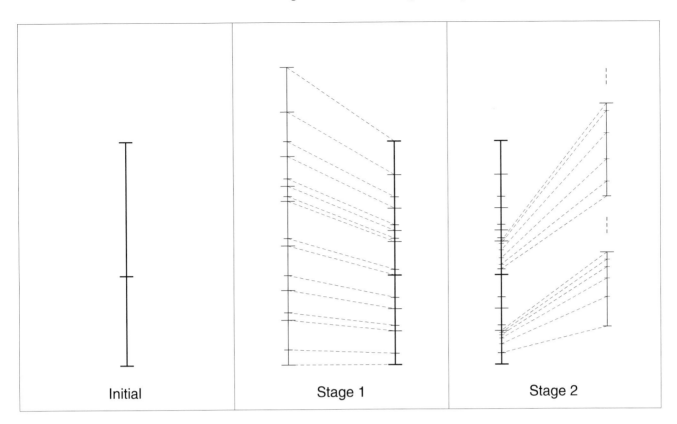

Figure 8. The different steps in the layer modeling procedure.

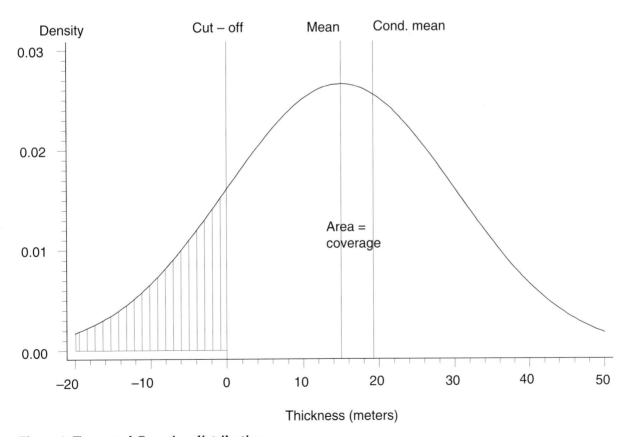

Figure 9. Truncated Gaussian distribution.

228   Høye et al.

Table 2. Parameters for all sequences and their associated C sands*

	Mean	Stand. Dev.	Coverage	Cond. Mean
Seq. 2	42.89	24.50	0.96	45.08
Seq. 3	34.08	23.52	0.93	38.16
Seq. 4A	26.98	9.29	0.99	27.03
Seq. 4B	16.98	14.72	0.81	21.89
Seq. 5	28.52	19.00	0.94	31.16
Seq. 6	4.07	10.05	0.66	9.67
Seq. 7	−2.91	15.28	0.42	11.20
2C	19.11	14.33	0.77	24.69
3C	23.01	31.26	0.77	35.32
4AC	7.08	12.41	0.72	12.96
4BC	3.47	11.15	0.62	10.29
5C	19.33	11.83	0.94	20.65
6C	2.37	9.43	0.60	8.45
7C	−5.20	14.00	0.36	9.46
8C	−21.93	13.72	0.06	5.84

*Means and standard deviations in the underlying Gaussian distribution, coverage, and conditional mean, given that the layer is present. The mean and the standard deviation are in the unobserved, underlying Gaussian distribution.

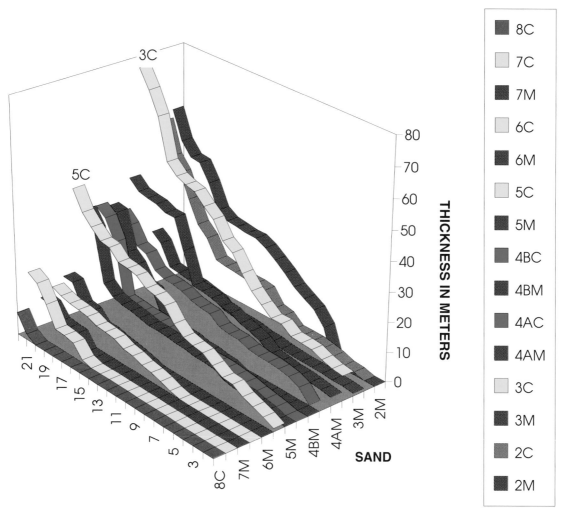

Figure 10. Sorted sand isochore thickness.

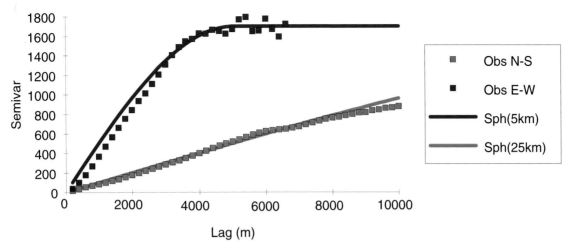

Figure 11. Variogram modeling based on the seismic isochore. Obs N-S = observed north-south, Obs E-W = observed east-west, Sph (5 km) = spherical variogram model with correlation length 5 km, Sph (25 km) = spherical variogram model with correlation length 25 km.

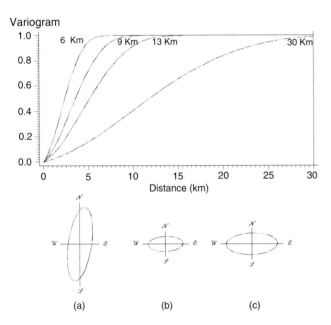

Figure 12. The different variograms and aniso-tropies used during the modeling; (a) shows sequences and C sands, (b) shows parasequences 2c, 5c, and 7c, and (c) shows parasequences 3c and 6c.

generally will not fit into the seismic envelope that has been generated higher up in the hierarchy; this realization thus needs to be adjusted to make the model consistent.

In the wells, each layer has the right thickness by construction. If all the wells were vertical, the total thicknesses would also be consistent, and there would be no need for adjustments at the well positions; however, several deviated wells were used in this study, which constitutes an additional problem. An algorithm has been developed that adjusts the thickness of each layer proportional to the kriging variance of the thickness at each location. This algorithm guarantees

that the observed thicknesses in the wells are never adjusted. The same algorithm is used both on the sequence level to fit the sequences in between the top, middle, and bottom seismically mapped surfaces, and on the parasequence level to fit the parasequences between their associated sequence boundaries. A description of the algorithm is given in Appendix 1.

Finally, if the thickness of a sequence or parasequence has been adjusted, the thickness of the associated C or c sand is adjusted accordingly. The C or c sand is assumed to cut into the sequence or parasequence from the top. If, by chance, the C or c sand is thicker than the sequence or parasequence to which it belongs, the C or c sand is taken to fill the sequence or parasequence completely, and the remainder of the C or c sand is truncated.

The combined effect of the applied thickness pdf, the spatial correlation structure, the covariance model, the well control, and the adjustments to the seismic envelope, constrain the outcome of the layer thickness simulations. Figure 14 gives a schematic cross section of the different steps in the modeling and adjustment procedure. As an example, Figure 15 shows a map of one stochastic realization of the 3C isochore. The north–northeast-south–southwest structure is clearly visible.

APPLICATION AND RESULTS

Predrilling Applications

A stochastic model was generated parallel with the deterministic model for the 31/5-5 well area. The objective was to highlight uncertainties related to the best appraisal well location and the expectations for this well.

Probability maps for the occurrence of high-permeability sand in the oil leg were generated for all of the different layers. Two examples are shown in Figures 16 and 17. A probability map is simply a map show-

Table 3. Correlation matrix for the sequences and their associated C Sands

	Seq. 2	Seq. 3	Seq. 4A	Seq. 4B	Seq. 5	Seq. 6	Seq. 7	2C	3C	4AC	4BC	5C	6C	7C
Seq. 3	0.502													
Seq. 4A	−0.604	−0.257												
Seq. 4B	−0.726	−0.783	0.434											
Seq. 5	−0.749	−0.521	0.421	0.496										
Seq. 6	0.064	0.022	−0.057	−0.128	0.069									
Seq. 7	0.165	−0.051	−0.335	−0.303	0.047	−0.301								
2C	0.740	0.105	−0.554	−0.451	−0.461	0.241	0.060							
3C	0.631	0.946	−0.320	−0.779	−0.647	−0.029	−0.051	0.213						
4AC	0.612	0.800	−0.111	−0.719	−0.708	−0.181	−0.061	0.220	0.833					
4BC	0.108	−0.109	0.350	0.222	−0.416	−0.363	−0.190	−0.049	0.095	0.261				
5C	−0.672	−0.344	0.400	0.450	0.858	0.098	−0.126	−0.474	−0.450	−0.569	−0.309			
6C	0.045	0.028	−0.026	−0.067	0.078	0.952	−0.342	0.163	−0.016	−0.140	−0.350	0.088		
7C	0.056	−0.074	−0.236	−0.227	0.165	−0.327	0.915	0.025	−0.100	−0.133	−0.230	−0.038	−0.353	
8C	0.030	0.098	−0.090	−0.163	−0.267	0.349	−0.179	−0.039	0.114	0.039	0.065	−0.351	0.219	−0.158

ing, for each location, the probability that a certain objective is fulfilled. The objective in this case was defined as areas with more than 9 m of oil in clean sands, a threshold selected with attention to the TWGP. The objective function is defined as 1 when the objective is fulfilled, and as 0 elsewhere. The maps are generated by summing the objective function from several realizations. Thus, high values represent areas where many realizations have met the objective criteria. Note that in this context, high-permeability sands are defined as C sands, i.e., not subdivided into smaller scale c and m sands, nor are there any petrophysics included, which could have given a better discriminator. The coarse layering (C and M sands only) was regarded as sufficient in a predrilling exercise, whereas finer scale models are intended to be used in a postdrilling model.

Note that these maps represent the probability of the specified objective, and must be interpreted as such. The maps do not show cases where two C sands combined, with no M sands in between, meet the criteria. Extensive areas with low probability may imply that the actual location is uncertain, while there is less uncertainty related to the existence of a thick sand in the area.

The same model is used as support for planning the drilling program, such as deciding where to start coring to secure that the top of the reservoir is sampled. Figure 18 shows histograms of the depth to top for the various layers of the planned 31/5-5 location. Figure 19 shows the associated distributions of the layer thicknesses for the C and M sands at the same location.

The well 31/5-5 was drilled early in 1993 with the objectives of appraising the Sognefjord clean sand intervals in TWGP south, and providing geological, geophysical, and petrophysical data necessary for evaluating potential cluster areas, confirm oil column thickness and contacts, and gain additional data to assist in aquifer modeling.

The stochastic model showed that the layer model uncertainties were significant in this area. Thus, pre-diction of oil/no oil in a specific sand unit was rather speculative; however, the well has fulfilled its objective in tying down this uncertainty.

The deterministic prognosis, depth to top reservoir, was about 1524 m below mean sea level. Actual depth was encountered at around 1542 m below mean sea level. The stochastic modeling predicted the top of the reservoir at about 1532 m below mean sea level (mean) with a standard deviation of about 12.5 m, explaining both the relatively unsuccessful deterministic prognosis and the actual depth (both within one standard deviation from the stochastic mean). The main source of uncertainty may be addressed to the V_0 maps, a key parameter in the depth conversion.

The stochastic model predicted 5C as the most probable target sand at well 31/5-5 (Figures 17, 20), but with less than 50% certainty of having more than 9 m of the sand in the oil leg. The deterministic model predicted a prospect in 4BC sand.

The new zonation confirmed that the upper part of the 5C sand and the lower part of the 6C sand was in the oil leg. An updated stochastic map made it possible to select a fairly robust pilot well location for horizontal drilling in these prospects (see the right side of Figure 20). The left side of Figure 20 is the lower left portion of Figure 17.

The 31/5-5 well also has reduced the uncertainties related to the 3C sand prospects to the north. This is illustrated in the updated uncertainty map (Figure 21) showing that the 3C prospects have been better constrained west of the 31/5-2 well. The left side of Figure 21 is the lower left portion of Figure 16. Remember that the search criteria is only 9 m of sand in the oil leg, and that the oil column is expected to be only 13 m in this area.

Postdrilling Applications

The drilling of a pilot well will tie down the major uncertainty in the area near the well. When the well data have been collected, a more detailed model may

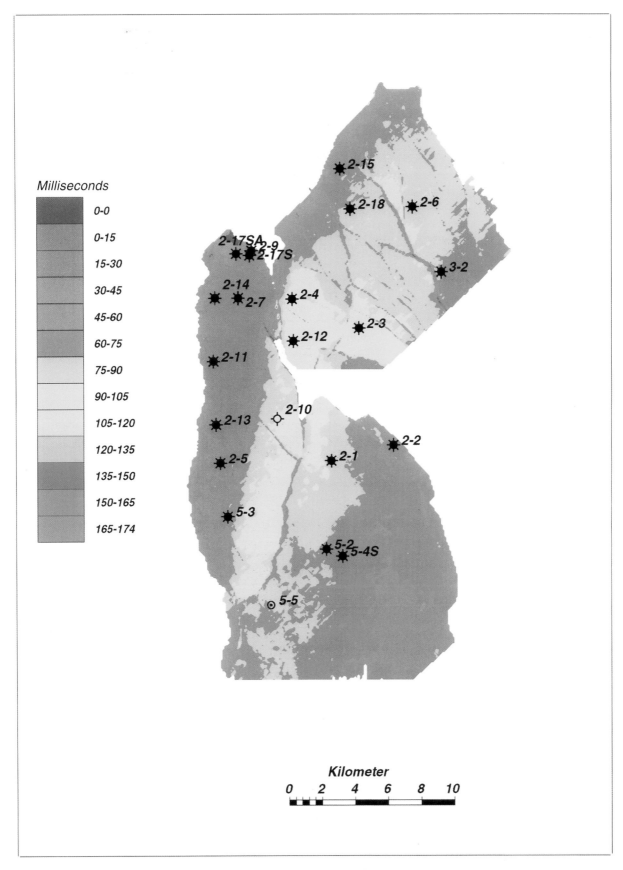

Figure 13. Seismic time isochore between base reservoir and the top of the 3C sand.

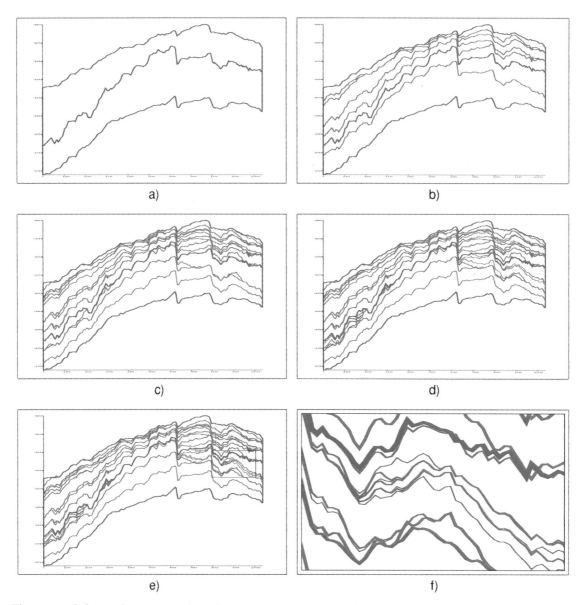

Figure 14. Schematic cross section showing the steps in the layer modeling: (a) Seismic surfaces (green), (b) includes the sequence tops (red), (c) includes the M sand and C sand boundaries (blue), (d) includes the parasequence tops in sequences 2 and 5 (thin red), (e) includes the c sand and m sand boundaries (thin blue) [box shows area enlarged in (f)], (f) detail of (e).

be applied. The objective function can then be related to problems such as best orientations for horizontal drilling, structural dip uncertainty, etc.

Further Plans for the Model

The structural modeling described in this chapter may be looked upon as a stochastic layer-cake model, or a stochastic flexible-grid model designed to follow geologic boundaries. The flexible grid cells are regular in the X and Y directions, but vary in the Z direction.

Given such a stochastic structural model, the next step is to bring petrophysical data into the model. The enormous size of the Troll field makes it practically impossible to build a full 3-D grid model on a fine

scale, even with today's most advanced computers. For the full-field model, the layer model would be used as a framework. If necessary, some of the stochastic layers would be further refined for higher resolution. For each layer, stochastic maps of porosity, permeability, etc. would be generated, all conditional on properly scaled-up well information.

Petrophysical properties, in particular permeability, are measured on the core-plug scale and related to a particular representative volume. A procedure is required to scale-up these core plug measurements to obtain effective properties for the actual volumes used in the different cells in the different layers. We propose to assign a fine local grid to each well for scaling-up to the description grid scale. Special facies

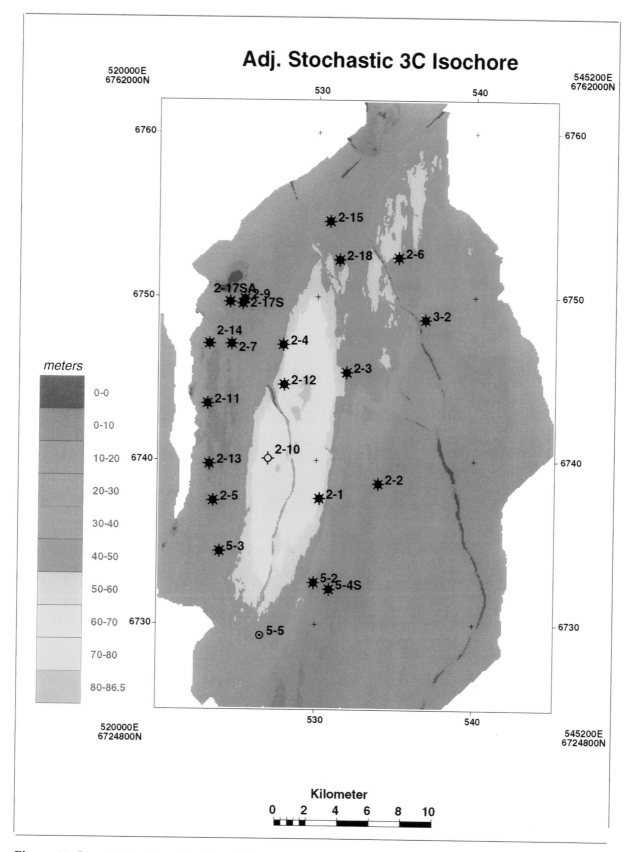

Figure 15. One stochastic realization of the 3C isochore.

234 Høye et al.

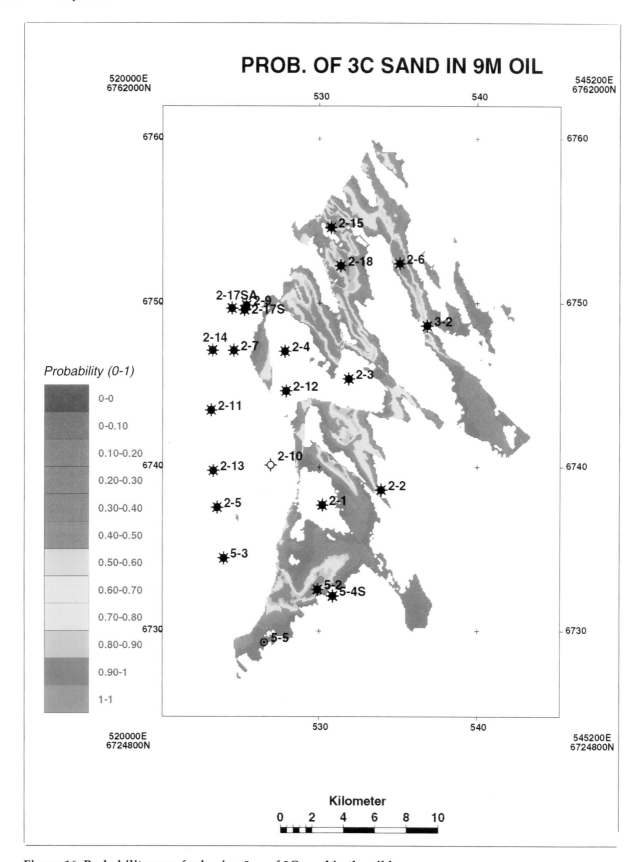

Figure 16. Probability map for having 9 m of 3C sand in the oil leg.

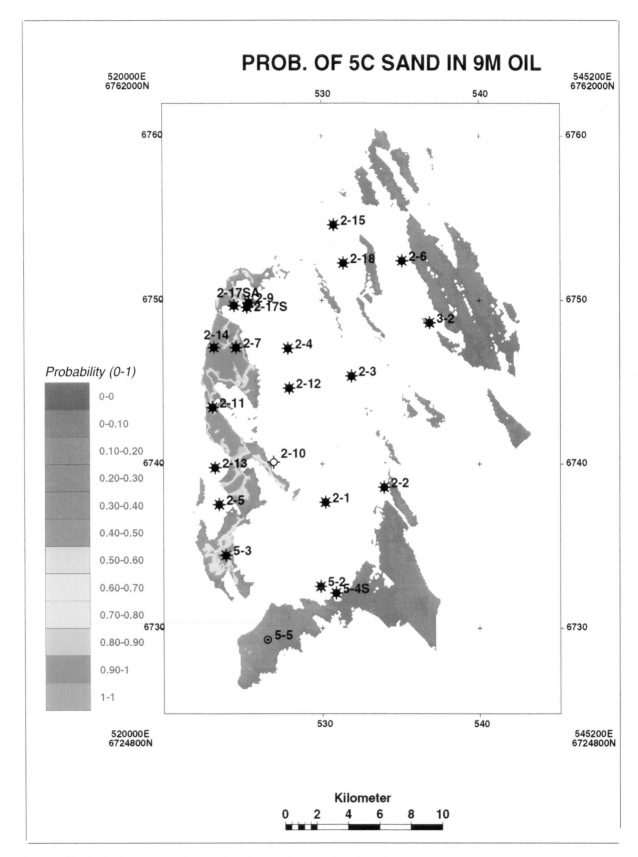

Figure 17. Probability map for having 9 m of 5C sand in the oil leg.

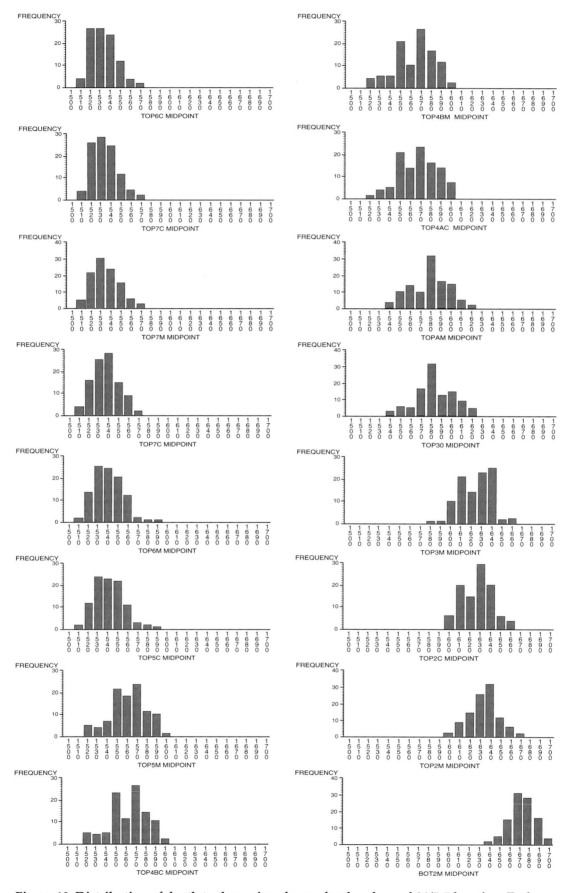

Figure 18. Distribution of depth to the various layers for the planned 31/5-5 location. Each bar represents a group of observations within 10-m intervals plotted by the group midpoints.

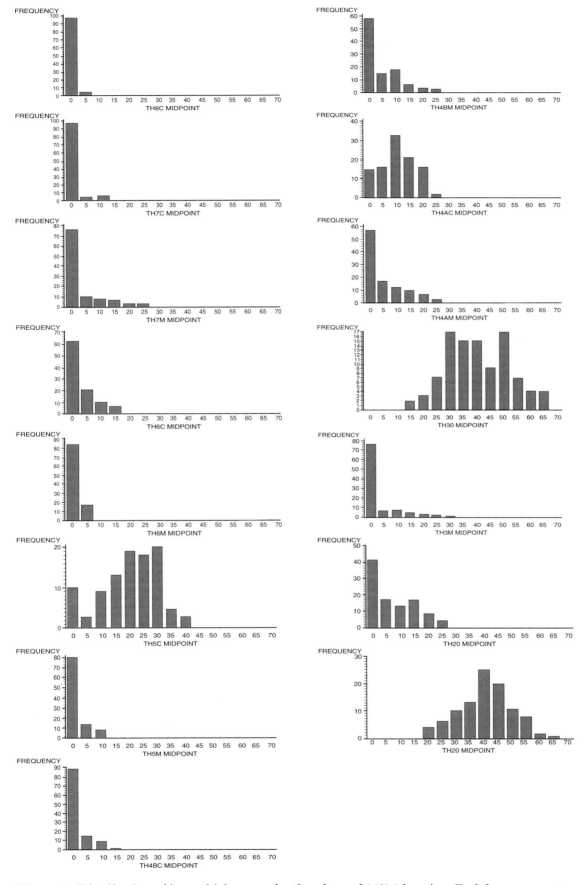

Figure 19. Distribution of layer thicknesses for the planned 31/5-5 location. Each bar represents a group of observations within 5-m intervals plotted by the group midpoints.

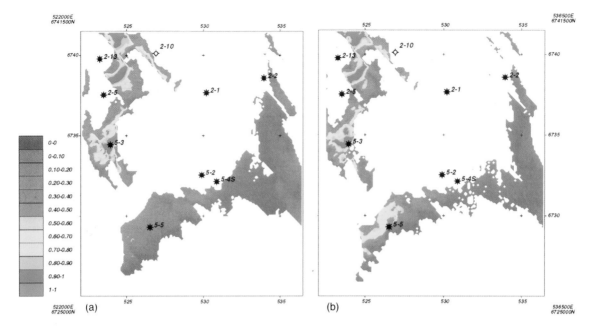

Figure 20. Probability map for having 9 m of 5C sand in the oil leg in the 31/5-5 area (a) before and (b) after the information from well 31/5-5 has been taken into account.

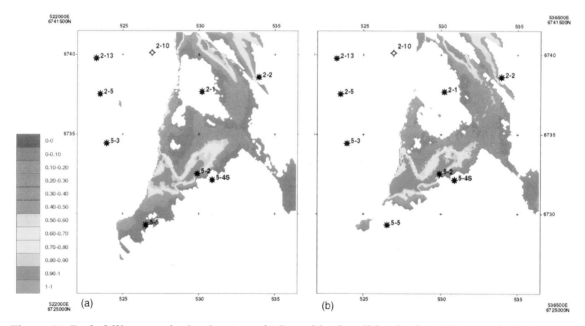

Figure 21. Probability map for having 9 m of 3C sand in the oil leg in the 31/5-5 area (a) before and (b) after the information from well 31/5-5 has been taken into account.

complexes may need special scaling-up techniques. These scaled-up well values would then constitute the input to the stochastic model.

Before flow simulations could be performed, the description grid would have to be transformed into a suitable grid, and the petrophysical properties scaled-up accordingly. This transformation would constitute mostly a merging of several layers to reduce the number of grid cells; however, in the near-well areas, local grid refinement also would be highly relevant.

REFERENCES

Deutsch, C. V., and A. G. Journel, 1992, GSLIB: geo-statistical software library: Oxford, Oxford University Press, 340 p.

Haldorsen H. H., and E. Damsleth, 1990, Stochastic modelling: Journal of Petroleum Technology, v. 42, n. 4, p. 404–412.

NAG, 1991, The NAG Fortran library manual, mark 15: Oxford, NAG Ltd.

Omre, K. H., 1987, Introduction to geostatistical theory and examples of practical applications: NCC-Note SAND/14/87, Norwegian Computing Centre.

TROLL, 1991, Plan for development and operations; v. 1, Supp. A, Norsk Hydro a.s. Oslo, Internal, 165 p.

Van Wagoner, J. C., R. M. Mitchum, and V. D. Rahmanian, 1990, Siliciclastic sequence stratigraphy in well logs, cores, and outcrops: concepts for high-resolution correlations of time and facies: AAPG Methods in Exploration Series, n. 7, 55 p.

APPENDIX 1

Layer Thickness Adjustment

Let $t_1(x)$, $t_2(x)...t_n(x)$ be a realization of the stochastic thicknesses of the n layers at a specific lateral location x in the reservoir area. Some of the $t_1(x)$ may be negative, representing a 0 thickness. Furthermore, let $T(x)$, the total thickness of the reservoir at location x, be given as well, either as a deterministic value given from the difference between two seismic surfaces or as a realization from an independent stochastic model. For simplicity, the reference to the location x is not referred to again.

Because t_1, $t_2...t_n$ are generated independently of T, in general

$$t_1^+ + t_2^+ + ... + t_n^+ \neq T$$

where $t_i^+ = \max(t_1, 0)$. The problem is then, how to adjust t_i to squeeze the layers into the envelope defined by T. Or, more formally, how to find $e_1, e_2..., e_n$ so that

$$t_1'^+ + t_2'^+ + ... + t_n'^+ = T$$

where $t_i' = t_i + e_i$, $i = 1,...,n$.

Assuming that the t_i are realizations from a multivariate Gaussian distribution, with variances σ_i^2, respectively, it is easy to show that the optimal (in the least-square sense) adjustment is to have each e_i proportional to its variance, i.e., $e_i = k\sigma_i^2$, where the proportionality constant k is chosen so that

$$t_1'^+ + t_2'^+ + ... + t_n'^+ = T \qquad (1)$$

when $t_i' = t_i + k\sigma_i^2$.

In the simple case where the t_i all are positive and large compared to their standard deviations, k is explicitly given by

$$k = \left(T - \sum_{i=1}^{n} t_i \right) / \sum_{i=1}^{n} \sigma_i^2$$

Generally, when some of the t_i are negative, or close to 0 compared to their standard deviation, k must be found from equation 1 by an iterative technique.

The conditioning of the thicknesses at the well locations is assumed to be consistent, so that $t_1(x_j)^+ + t_2(x_j)^+ + ... + t_n(x_j)^+ = T(x_j)$ for all well locations x_j, by construction. Thus, no adjustment will be necessary at the well locations. To avoid abrupt adjustments close to the well locations, and to take well data configuration into account, we used the kriging variances from the stochastic models for the different layers for the variances in the preceding formulas.

Chapter 18

Simulating Geological Uncertainty with Imprecise Data for Groundwater Flow and Advective Transport Modeling

Sean A. McKenna
Eileen P. Poeter
Dept. of Geology and Geological Engineering
Colorado School of Mines
Golden, Colorado, U.S.A.

ABSTRACT

A known cross section composed of a heterogeneous mixture of three hydrofacies is sampled along vertical lines representing ten wells. These wells provide conditioning data for stochastic geostatistical simulations. Steady-state flow and advective transport are modeled across each resulting realization to evaluate effective hydraulic conductivity, first arrival time, and duration of particle arrivals. Further ensembles of realizations are produced with both the original ten wells of conditioning data and additional soft data in the form of imprecise estimates of the hydrofacies at every location within the simulated domain. Use of soft data decreases the ensemble variances with respect to the first arrival times, bulk hydraulic conductivity of the domain, and duration of the particle arrivals relative to the ensemble variance resulting from only ten wells of hard conditioning data. The accuracy of the resulting distributions is influenced by biases in the conditioning data.

INTRODUCTION

For water supply problems, the use of relatively simple models of subsurface geology and bulk values of hydrologic parameters is sufficient. As the focus of the groundwater industry has shifted to contaminant transport problems, more complex models of the subsurface are necessary. Determination of the continuity of high hydraulic conductivity facies across a site is critical to predicting contaminant transport and designing effective remediation systems. A small but continuous geologic feature having only one order of magnitude higher hydrau-

lic conductivity than the surrounding medium can have a dramatic effect on the behavior of a contaminant plume (Figure 1). If these continuous high hydraulic conductivity pathways go undetected, it is possible that contaminants will migrate offsite at unexpected locations and remediation efforts will be thwarted.

The need to define these features presents a new challenge in site characterization. It is not possible to drill enough holes to confidently state that such features do not exist at a site. Consequently, soft data and stochastic simulation can be used to improve the subsurface characterization.

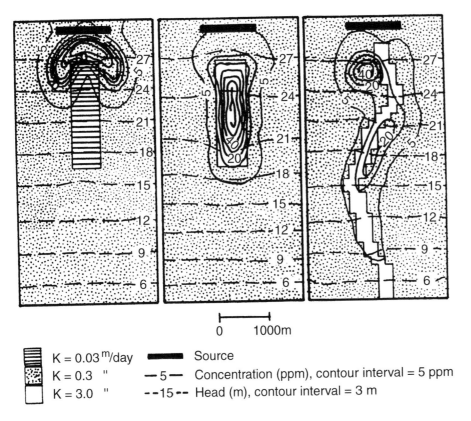

Figure 1. Computer simulated examples of the influence of aquifer heterogeneity with respect to hydraulic conductivity on contaminant transport. The contaminant is introduced for 5 yr at 1360 L/min. The contaminant is conservative, with a concentration of 1000 ppm. The hydraulic gradient is 4×10^{-3}, effective porosity is 0.20, longitudinal dispersivity is 30 m, transverse dispersivity is 3 m, and the plumes are pictured 35 yr after the injection began (after Poeter and Gaylord, 1990).

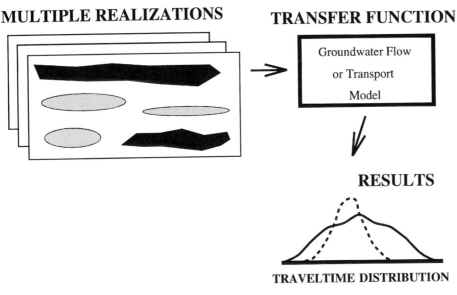

Figure 2. Conceptualization of stochastic subsurface imaging coupled with hydrologic modeling. The dashed curve demonstrates the reduction in uncertainty of traveltimes across a site due to a reduction in subsurface uncertainty through the use of soft information.

Stochastic modeling of heterogeneities within aquifers has emerged as a means of producing distributions on contaminant traveltimes or other parameters of interest (Rautman and Treadway, 1991; Nichols and Freshley, 1993). Generally, when more subsurface data are available, the response of the transfer function exhibits a narrower distribution (Figure 2). Supplementary data are usually obtained by further drilling into the subsurface at the site.

Another source of subsurface information at many sites is soft data in the form of geophysical information, uncertain drillers' logs, and knowledge concerning the depositional environment of the subsurface.

Imprecise data are defined as a type of soft data that give an uncertain estimate of the parameter under study at a location. The amount of imprecision must be quantified through a calibration procedure. This paper examines the utility of one type of soft data, imprecise information, in reducing the uncertainty associated with generating stochastic realizations of a subsurface domain. This reduction in subsurface uncertainty should result in reduced uncertainty associated with the information used to address flow and transport questions (Figure 2).

The deterministic image of the section used in this paper is a discrete interpretation of the geology of a

Yorkshire cliff face described by Matheron et al. (1987). The cross sectional domain is 600 m long × 30 m high. Each cell is 2 m long and 1 m high, yielding a total of 9000 cells. Each cell is assigned a single lithology, shale, shaly-sand, or sandstone (Figure 3A), corresponding to the lithologies described in this fluvio-deltaic deposit (Ravenne et al., 1987). Each lithology represents an indicator class in the geostatistical simulations. Lithologies are converted to hydrofacies by assigning each lithology a hydraulic conductivity and porosity value based on appropriate values reported in the literature (Freeze and Cherry, 1979) for those lithologies (Table 1).

Steady-state flow and advective transport are modeled across the two-dimensional cross sectional domain using the U.S. Geological Survey flow model MOD-FLOW (MacDonald and Harbaugh, 1988) and the particle tracking postprocessor MODPATH (Pollock, 1989). The top and bottom of the domain are modeled as impermeable boundaries and the ends of the domain are fixed head boundaries imposing a gradient of 0.01 across the section. The particle tracking is accomplished by examining the paths and velocities of 30 particles across the domain. The flow modeling and particle tracking provide information to analyze the bulk hydraulic conductivity of the domain, the fastest flowpath across the domain (first arrival time), and the length of time during which the majority of the particles arrive. The goal of this study is to examine the utility of imprecise data in reducing the uncertainty associated with these measures of flow and transport parameters relative to the uncertainty associated with those same measures when realizations are produced with limited hard data. Use of a synthetic, known section allows us to assess accuracy as well as precision.

THEORY OF IMPRECISE DATA

The type of soft data employed herein is imprecise data as discussed by Alabert (1987). Because the geology of the actual cross section is known, it can be reproduced with a given amount of uncertainty. In this study, imprecise reproduction of the geologic facies is accomplished by adding noise to the actual section. The uncertain reproduction of the cross section (Figure 3B) is used as imprecise conditioning data in indicator geostatistical simulations. Indicator geostatistical simulation is well suited to simulating facies types because it does not require the variable being simulated to have a numerical value. In addition, indicator simulation preserves the occurrence of thin but continuous units that can be critical to groundwater transport of contaminants.

The generated imprecise data in this study are considered an analog for the field situation in which a geophysical technique yields an uncertain image of facies in the subsurface. The amount of uncertainty in the imprecise data is quantified through a calibration process that compares the actual facies type to the estimated facies type at locations where both types of

A)

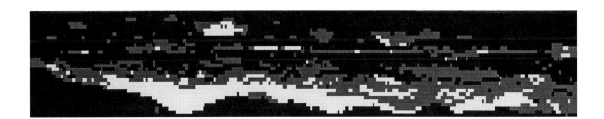

B)

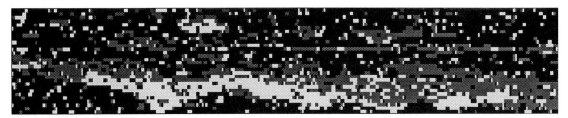

Figure 3. (A) The deterministic (known) cross section. (B) An imprecise reproduction of the deterministic image used as imprecise conditioning data. In both images, shale is black, shaly-sand is gray, and sandstone is white.

Table 1. Hydraulic properties assigned to facies

Facies	Hydraulic Conductivity (m/yr)	Effective Porosity (fraction)
Shale	0.32	0.05
Shaly-Sandstone	3.2	0.10
Sandstone	32.0	0.15

Table 2. Imprecision of soft data

Threshold	p_1	p_2	$(p_1 - p_2)$
1	0.8500	0.1513	0.6987
2	0.9206	0.1702	0.7504

data exist. Results of the calibration process are a pair of misclassification probabilities (p_1 and p_2) for each indicator threshold. For nonnumeric data, such as facies type, the indicator thresholds are abstract boundaries that divide the data into indicator classes (e.g., when mapping this section, the geologists used a mental threshold to divide sand from shaly-sand). By definition, p_1 is the probability that the imprecise estimate of the facies is less than or equal to the indicator threshold given that the actual facies is less than or equal to the indicator threshold at that location (equation 1). Similarly, p_2 is the probability that the imprecise estimate of facies type is less than or equal to the indicator threshold given that the actual facies is greater than the threshold (equation 1).

$$p_1 = P\left[\hat{Z}(x) \leq z_c \mid Z(x) \leq z_c\right]$$
$$p_2 = P\left[\hat{Z}(x) \leq z_c \mid Z(x) > z_c\right] \tag{1}$$

The variable $\hat{Z}(x)$ is the soft data estimate of the facies. $Z(x)$ is the actual facies (hard data) at that location. The indicator threshold is denoted by z_c. Note that the probabilities do not necessarily add up to 1.0. A simple measure of the quality of the imprecise data for a given threshold is the quantity $(p_1 - p_2)$. If this quantity equals 1.0, the data are considered to be hard data (i.e., the imprecision in the soft data is negligible). If $(p_1 - p_2)$ equals 0, the imprecise data provide no information on the actual facies. This quantity is equivalent to the "B" parameter in the Markov-Bayes formulation of soft data (Deutsch and Journel, 1992).

Two types of information are essential in creating geostatistical simulations: (1) an estimate of the covariance function for each threshold in the indicator study and (2) conditioning data. Imprecise data can be used to improve estimates (over use of hard data alone) of the covariance functions modeled as variograms and to provide conditioning data for the simulations. In both uses, the uncertainty of the imprecise data, as quantified by p_1 and p_2, must be taken into account. Software that uses imprecise data in both covariance modeling and in simulation of the cross sectional domain has been developed for this project.

VARIOGRAPHY AND SIMULATION

Two data sets are used in this study. The first data set consists of ten wells of hard data with a well spacing of 60 m. The other data set includes the original ten wells of hard data and soft data in the form of imprecise measurements of the facies at all other points in the domain as shown in Figure 3B. The probabilities, p_1 and p_2, have been determined for the imprecise data and are shown in Table 2.

Horizontal variograms built with the ten wells of hard data show a strong nugget effect (Figure 4A). Nugget effect variograms are common in site investigations where the well spacing is larger than the length of correlation for the parameter of interest. For each threshold, the nugget value determined for the vertical variogram was used in fitting a model to the horizontal variogram. All variograms were fit with two nested spherical models.

To build variograms with a combination of hard and imprecise data, three estimates of the true variogram gamma value are combined: an estimate calculated from the hard data at lags where hard data are present, an estimate from a hard-imprecise data cross-covariance, and an estimate calculated from the imprecise data covariance at intermediate lags (equation 2). The final estimate of spatial covariance derived from equation 2 is then converted to a variogram using the nonergodic relationship developed by Isaaks and Srivastava (1988).

$$C_I^* = \omega_1 [C_I]_{N_h} + \omega_2 \left[\frac{C_{I\hat{I}}}{p_1 - p_2}\right]_{N_h N_s} + \omega_3 \left[\frac{C_{\hat{I}}}{(p_1 - p_2)^2}\right]_{N_s} \tag{2}$$

The variables N_h and N_s refer to the number of hard and soft conditioning data in the search neighborhood. The three components of the covariance equation are weighted by the ω's, which sum to 1. The ω's are determined to account for the relative numbers of hard and soft data. One simple method to calculate ω is

$$\omega_1 = N_h / N_{total}$$
$$\omega_3 = N_s / N_{total}$$
$$\omega_2 = 1.0 - (\omega_1 + \omega_3)$$

The hard-soft cross-covariances and the soft data estimates are scaled by the quantity $(p_1 - p_2)$ to provide an estimate of the true hard data gamma value at intermediate lag spacings (Alabert, 1987).

Improvement of the variograms with spatially dense imprecise data can be substantial as demonstrated in Figure 4B. A field example of spatially dense imprecise data is an estimate of facies from a surface or cross-borehole geophysical survey.

The geostatistical simulation software developed for this project is a soft data extension of the program

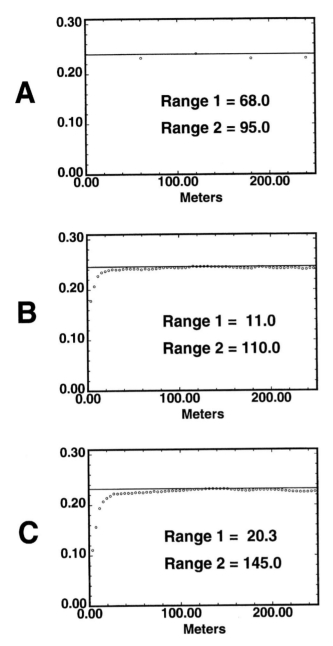

Figure 4. Variograms calculated with (A) ten wells of hard data, (B) ten wells of hard data and exhaustive imprecise data, and (C) the exhaustive hard data set. All variograms were fit with two nested spherical models.

ISIM3-D (Gomez-Hernandez and Srivastava, 1990). Both use the sequential indicator simulation (SIS) algorithm. The heart of the SIS algorithm is the estimation of a posterior cumulative distribution function (pcdf). An estimate of the pcdf is calculated for every indicator threshold at each location being simulated. The pcdf estimate is based on the proximity and values of conditioning data located within a search neighborhood. In the modified algorithm, the pcdf value at each threshold level is a linear combination of the hard data points within the search neighbor-

hood weighted by their respective kriging weights and the imprecise data points weighted by a spatial declustering weight and scaled by the p_1 and p_2 values (Alabert, 1987):

$$pcdf(z_\kappa) = \omega_1 \left[\sum_{\alpha=1}^{N_h} \lambda_\alpha i(x_\alpha, z_\kappa) \right]$$
$$+ \omega_2 \left[\sum_{\beta=1}^{N_s} v_\beta \frac{i(x_\beta, z_\kappa) - p_2(z_\kappa)}{p_1(z_\kappa) - p_2(z_\kappa)} \right] \qquad (3)$$

where ω_1 and ω_2 are weights that account for the relative quantities of the hard and imprecise data within the search neighborhood and are calculated as in equation 2. The i's are the indicator value (0 or 1) of each conditioning point at the current threshold. The kriging and declustering weights are represented by λ and v, respectively. The kriging weights are derived from solving the kriging matrix, and the declustering weights are determined by the cell declustering technique (Journel, 1983). Once the pcdf is constructed, a random number between 0 and 1 is drawn, and by examining the values of the pcdf at the indicator thresholds, the indicator class to which the random number belongs is determined. The point being simulated is assigned to that indicator class and is now considered a hard conditioning datum for the simulation of subsequent locations within the domain.

The simulator built for this project has several options that can be used to control the amount of hard and imprecise data found in the search neighborhoods and the type of kriging process (simple or ordinary) to be used. Imprecise data of the level illustrated in Table 1 are used to build 100 realizations with ordinary kriging (OK) and a maximum of 12 (4 hard and 8 imprecise) points in the search neighborhood. An example realization is shown in Figure 5. Realizations produced using simple kriging (SK) are not useful for this situation as discussed in the following section. One hundred realizations were also generated using only the hard data.

GROUNDWATER FLOW MODELING AND PARTICLE TRACKING

Steady-state flow and advective transport are simulated across each of the two ensembles of 100 realizations. Thirty particles are tracked across the domain and the results of this particle tracking give a distribution of first arrival time and duration of arrivals, which indicates the amount of spreading developed within the particle cloud. Duration time is defined as the traveltime difference between the 10th percentile arrival and the 90th percentile arrival (i.e., the 3rd and 27th particles). An effective bulk hydraulic conductivity of each realization is calculated from the volumetric discharge estimated by the flow model and knowledge of the boundary conditions. Arrival times and duration times are measured in years and bulk hydraulic conductivity is measured in meters/year (Figure 6).

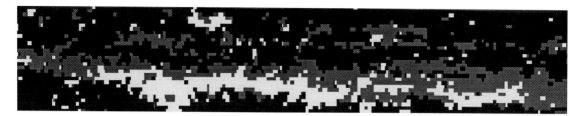

Figure 5. An example realization of the cross section produced with both hard and imprecise conditioning data. Compare this figure with Figure 3.

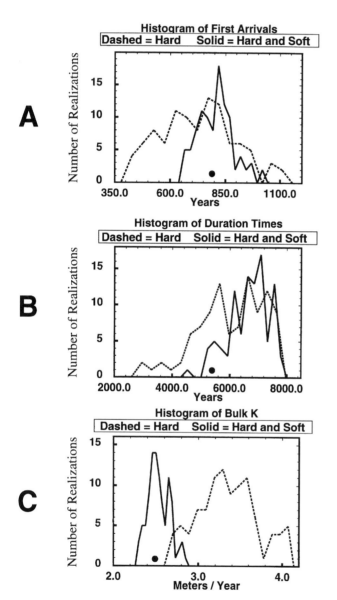

Figure 6. Histograms showing results of flow and transport modeling. For comparison, the actual first arrival time is 805 yr, the actual duration time is 5440 yr, and the actual bulk hydraulic conductivity is 2.5 m/yr. Black dots denote actual values.

The resulting distributions of bulk hydraulic conductivity, first arrival time, and duration time are shown as histograms in Figure 6. The values deter-mined by conducting flow and transport modeling in the deterministic section are provided for compari-son. The reduction in variance of each resulting distri-bution due to the incorporation of imprecise data is provided in Table 3.

DISCUSSION

The overall effect of supplementing the hard data with imprecise data is to tighten the distributions of the selected measures of flow and advective transport (Figure 6). This reduction in variance is an increase in precision (Deutsch and Journel, 1993). All of the dis-tributions are accurate; that is, all distributions in-clude the true values of flow and transport measures in the actual (deterministic) section, except for the dis-tribution of bulk hydraulic conductivity as calculated from the hard data ensemble (Figure 6C). Ideally, the resulting distributions from a field site will be both precise and accurate.

The actual bulk hydraulic conductivity does not occur within the distribution of bulk hydraulic con-ductivity produced using only the hard data. The hard data include a slightly higher percentage of sandstone than the actual section (17 vs. 14%) and a lower percentage of shale (58 vs. 61%). The resulting increased frequency of occurrence of high hydraulic conductivity facies in the realizations causes the over-estimation of bulk hydraulic conductivity. It is possi-ble that if more realizations are generated, the tail of the distribution may extend beyond the actual value.

Several attempts were made to employ simple kriging (SK) in the simulation process. SK forces a sta-tionary mean value into the simulation process. Ordinary kriging (OK) does not force an a priori sta-tionary value into the simulation, but rather corrects for local departures from the overall mean. For a number of experiments, SK was used for all points in the search neighborhood, both throughout the simu-lations and up to specified levels of simulation com-pletion at which time the process was switched to OK. The simulation program is written to allow the origi-nal hard data to be searched separately from the imprecise data and previously simulated data. This option is available to facilitate experimenting with different combinations of obtaining an SK estimate of the pcdf from the original hard data and an OK esti-mate from the soft and previously simulated data. All attempts at incorporating SK failed and the resulting

Table 3. Reduction in variance due to incorporation of imprecise data*

Parameters	Coefficient of Variation	
	Hard	Soft
First Arrivals	0.23	0.10
Duration Time	0.19	0.10
Bulk Hydraulic Conductivity	0.11	0.05

*Coefficient of variation = standard deviation/mean.

realizations were geologically unrealistic such that they were not used for flow and transport modeling.

The data set used in this study is somewhat artificial in that every point in the domain is a conditioning point. It may be realistic to have a two-dimensional section entirely composed of conditioning data from a geophysical survey; however, this would most likely be used in simulating a three-dimensional subsurface. SK could be more effectively employed in the early stages of a simulation with sparse conditioning data to impose an ergodic mean on the realizations. As the simulation process progresses, a switch to OK allows the mean to vary smoothly through the domain as the remaining locations are simulated. The abundance of conditioning data in this study negates the effective use of SK.

CONCLUSIONS

In this study, imprecise data are used to reduce uncertainty associated with groundwater flow and advective transport. Imprecise data of moderate uncertainty but extensive coverage improved the calculation of experimental variograms leading to more accurate covariance models. The improved covariance models coupled with the increase in conditioning data provided by the imprecise information reduces subsurface uncertainty in geostatistical simulations. This reduction in subsurface uncertainty leads to a reduction in uncertainty associated with flow and transport results. In a simulation study, the resulting flow and transport distributions should be both accurate and precise. The incorporation of soft data into the simulation process helps achieve this goal. The least desirable result would be an inaccurate but precise distribution. The incorporation of soft conditioning data safeguards against this result.

A thorough testing of the effects of the simulation process on the flow/transport results using a known data set is necessary before applying a simulator to a field problem. Even simulators that function properly on a test data set may not capture all the desired features of a site if the conditioning data are significantly biased. As this study has shown, it is possible to include the true first arrival time in the distribution of arrival times from an ensemble of realizations, but completely miss the true value of bulk effective

hydraulic conductivity in that same ensemble. This is an apparent conflict because the traveltime across the fastest path is not directly correlated to the bulk hydraulic conductivity of the section; however, it is disappointing that the simulation ensemble did not include the true bulk hydraulic conductivity.

When limited or biased data are available, the user of the simulator must consider a broader range of possible outcomes through the conceptual development of the simulation model. This range of outcomes can be constrained by including soft information in the form of geophysical data, hydrologic measurements, conceptual models of depositional environment, and other information correlated to the parameters being simulated.

ACKNOWLEDGMENTS

This work is supported by U.S. Bureau of Reclamation cooperative agreement 1-FC-8117500. This manuscript has benefited from thoughtful reviews by Kadri Dagdelen and Mary C. Hill.

REFERENCES

Alabert, F. G., 1987, Stochastic imaging of spatial distributions using hard and soft data: M.S. Thesis, Stanford University, Stanford, California, 197 p.

Deutsch, C. V., and A. G. Journel, 1992, GSLIB: geostatistical software library and user's guide: New York, Oxford University Press, 340 p.

Deutsch, C. V., and A. G. Journel, 1993, Entropy and spatial disorder: Mathematical Geology, v. 25, n. 3, p. 329–355.

Freeze, R. A., and J. A. Cherry, 1979, Groundwater: Englewood Cliffs, New Jersey, Prentice-Hall, 604 p.

Gomez-Hernandez, J. J., and R. M. Srivastava, 1990, ISIM3-D: an ANSI-C three-dimensional multiple indicator conditional simulation program: Computers in Geosciences, v. 16, n. 4, p. 395–414.

Isaaks, E., and R. M. Srivastava, 1988, Spatial continuity measures for probabilistic and deterministic geostatistics: Mathematical Geology, v. 20, n. 4, p. 313–341.

Journel, A. G., 1983, Nonparametric estimation of spatial distributions: Mathematical Geology, v. 14, n. 3, p. 445–468.

MacDonald, M. G., and A. W. Harbaugh, 1988, A modular three-dimensional finite difference ground-water flow model: U.S. Geological Survey Techniques of Water Resources Investigations, Book 6.

Matheron, G., H. Beucher, C. de Fouquet, A. Galli, D. Guerillot, and C. Ravenne, 1987, Conditional simulation of the geometry of fluvio-deltaic reservoirs: SPE 16753, SPE Annual Technical Conference and Exhibition, Proceedings, p. 591–599.

Nichols, W. E., and M. D. Freshley, 1993, Uncertainty analysis of unsaturated zone travel time at Yucca Mountain: Ground Water, v. 31, n. 2, p. 293–301.

Poeter, E. P., and D. R. Gaylord, 1990, Influence of aquifer heterogeneity on contaminant transport at the Hanford Site: Ground Water, v. 28, n. 6, p. 900–909.

Pollock, D. W., 1989, Documentation of computer programs to compute and display pathlines using results from the U.S. Geological Survey modular three-dimensional finite-difference ground-water flow model: U.S. Geological Survey OFR 89-381, 188 p.

Rautman, C. A., and A. H. Treadway, 1991, Geologic uncertainty in a regulatory environment: an example from the potential Yucca Mountain nuclear waste repository site: Environmental Geology and Water Sciences, p. 171–184.

Ravenne, C., R. Eschard, A. Galli, Y. Mathieu, L. Montadert, and J-L. Rudkiewicz, 1987, Heterogeneities and geometry of sedimentary bodies in a fluvio-deltaic reservoir: SPE 16752, SPE Annual Technical Conference and Exhibition, Proceedings, p. 115–124.

Chapter 19

Fractal Methods for Fracture Characterization

Thomas A. Hewett
Department of Petroleum Engineering
Stanford University
Stanford, California, U.S.A.

ABSTRACT

The geometric characteristics of self-similar nested fractal structures are reviewed and the scaling laws are presented in terms of the fractal dimension, D. These scaling laws provide the basis for the determination of the "box dimension," equivalent to D, of a set of fractures from fracture maps or photographs. The results of this procedure are applied to a number of natural fracture sets, showing that fracture networks exhibit fractal characteristics over a wide range of scales. The widespread occurrence of fractal sets in fragmented materials is explained by a hierarchical model of fragmentation processes that leads to a fractal distribution of the resulting fragments. A method of generating synthetic fractal fracture networks based on a probabilistic form of iterated function systems is described and illustrated in realistic synthetic examples. The implications of a fractal character in fracture networks on flow processes in such networks is shown to result in "anomalous diffusion" of the pressure response in well tests, which results in pressure and pressure derivative plots with straight parallel lines on double logarithmic axes. The results of several well tests performed at The Geysers geothermal field are presented and shown to exhibit the characteristics expected for fractal fracture networks.

INTRODUCTION

Mandelbrot's (1983) fractal geometry provides the proper mathematical framework for describing many of the complex and irregular shapes found in nature. These distributions range from the microscopic structure of porous solids (Avnir et al., 1985; Bale and Schmidt, 1984; Katz and Thompson, 1985), percolation clusters (Lenormand and Zarcone, 1985), and viscous fingers (Daccord et al., 1986), to the distribution of ore in mineral deposits (Turcotte, 1986a) and the shapes of coastlines and mountain ranges (Pentland, 1983). Fractal geometry has also proven very useful in characterizing the geometry of faults and fractures over a wide range of scales, from microfractures (Long et al., 1991) to the San Andreas fault (Aviles and Scholz, 1987).

Fractals are characterized by the fact that they exhibit variations at all scales of observation, and have partial correlations over all scales. Every attempt to divide such a geometry into smaller, more uniform regions results in the resolution of even more structure. The first fractals studied by mathematicians were exactly self-similar mathematical curiosities (such as the Koch curve and Sierpinski gasket) generated by repetitively rescaling and mapping a fixed geometric shape onto itself over a cascade of scales. These are useful for elucidating the scaling relations found in fractal distributions, but are too regular for modeling natural property distributions. For describing natural property distributions, the most useful fractal sets are those displaying a statistical self-similarity.

PROPERTIES OF FRACTAL DISTRIBUTIONS

The geometries of fractal distributions are characterized by their intermittent nature; they do not fill space. This characteristic is quantified by a parameter called the fractal dimension. The significance of this parameter can be seen by first considering the properties of distributions that fill space. In a d-dimensional space (d = Euclidean dimension) it takes a number

$$N = r^{-d}$$

of objects of scale rL to fill a space of scale L. As an example, for $r = 1/2$, it takes two line segments of length $1/2L$ to fill a line segment ($d = 1$) of length L. Similarly, it takes four squares with sides $1/2L$ to fill a square ($d = 2$) of side L, and eight cubes ($d = 3$) with sides $1/2L$ to fill a cube of side L.

Fractal distributions do not fill space and are characterized by a relation between number density and scale of the form (Mandelbrot, 1983)

$$N = r^{-D} \qquad (1)$$

where D is the fractal dimension and is less than d. An example of a regular fractal distribution is shown in Figure 1 where the Sierpinski gasket shown is formed by dividing the largest triangle into smaller triangles with sides equal to $r = 1/2$ of the original and retaining only three of the four required to fill the original. This process is repeated at successively finer scales. The resulting fractal dimension is

$$D = -\frac{\ln N}{\ln r} = \frac{\ln 3}{\ln 2} = 1.585$$

This relationship between scale and number density can be seen by considering the properties of self-simi-

lar nested structures. For a hierarchy of structures with a ratio $r < 1$ between adjacent scales

$$l_1 = rl_0$$
$$l_2 = rl_1 = r^2 l_0$$
$$l_3 = rl_2 = r^3 l_0$$
$$\vdots$$
$$l_n = rl_{n-1} = r^n l_0$$

with

$$n = \frac{\ln(l_n / l_0)}{\ln r}$$

If there are N structures of scale l_n embedded within each structure of scale l_{n-1}, then

$$Nr(l_n) = N^n = N^{\ln(l_n/l_0)/\ln r}$$

or

$$\ln Nr(l_n) = n \ln N = \frac{\ln(l_n / l_0)}{\ln r} \ln N$$

where $Nr(l_n)$ is the number of structures of scale l_n. Defining a similarity dimension, D, as

$$D = -\frac{\ln N}{\ln r} \qquad (2)$$

we see that

$$\ln Nr(l_n) = -D\ln(l_n/l_0)$$

and

$$Nr(l_n) = \left(\frac{l_0}{l_n}\right)^D \qquad (3)$$

FRACTAL ANALYSIS OF FRACTURE SETS

The relationship derived in equation 3 forms the basis for determining the "box dimension" of a set (Barton and Larsen, 1985), which is the result of counting the number of boxes of size l required to cover the distribution and noting that

$$Nr(l) = \left(\frac{l_0}{l}\right)^D \propto \frac{1}{l^D} \qquad (4)$$

In practice, grids of different sizes are laid down over a map or photograph of a fracture pattern and the number of boxes containing fractures are counted. When the number of boxes intersected by the fracture network is plotted as a function of the box size on double logarithmic coordinates, for distributions obeying equation 4 the points will plot as a straight line with a slope equal to $-D$. Similarly, when the number of intersected boxes is plotted against the inverse of the box size, the points plot as a straight line with slope D. Both approaches have been used in

Figure 1. Sierpinski gasket. Fractal dimension D = 1.585.

studies reported in the literature. This idea is illustrated schematically for the determination of the box dimension of a set of fractures in Figure 2 (Sammis et al., 1991). As shown in Figure 3 (Sammis et al., 1991), the validity of this procedure is easily verified for deterministic, recursively generated fractals. Because of the spatial clustering in this pattern, the number of boxes required to cover it increases only by a factor of three each time the box size is reduced by a factor of one-half. This clustering at all scales is one of the characteristic features of fractal sets.

This approach to demonstrating the fractal character of sets of natural fractures was first explored in 1985 in a study of fractures at the site of a proposed nuclear waste repository (Barton and Larsen, 1985). That study revealed that the fracture patterns at Yucca Mountain, Nevada, were self-repetitive fractals over a range of scales, spanning 0.2–25 m, within which several generations of fractures were observed. More recently, fracture patterns in graywacke outcrops at The Geysers geothermal field have been subjected to a similar analysis (Sammis et al., 1991). That study looked at fracture maps prepared from photomosaics of outcrop fracture patterns ranging from 0.5–2 m² and found that they obeyed equation 4 over

nearly two orders of magnitude. Examples of the fracture patterns analyzed and the corresponding "box plots" are shown in Figures 4 and 5.

Numerous other investigators have demonstrated the fractal nature of natural fracture sets (Aviles and Scholz, 1987; Velde et al., 1990; Barton and Hsieh, 1989; Chilès, 1988; Hirata, 1989; LaPointe and Hudson, 1985; LaPointe, 1988; Sammis and Biegel, 1989; Turcotte, 1989). The most extensive review of the fractal character of natural fractures is given by Barton (in press). He analyzed maps of fracture traces spanning nearly 10 orders of magnitude ranging from microfractures in Archean Albites (~10^{-3} m) to fractures in the South Atlantic sea floor (~10^5 m). In all cases, the data could be fit with straight lines on double logarithmic plots with correlation coefficients of 0.99 or better. The resulting fractal dimensions ranged from 1.32 to 1.70. Several of the fracture maps are shown in Figures 6–8. The results of the box-counting procedure applied to the 17 different data sets are shown in Figure 9 and summarized in Table 1.

The statistics of fractal sets are a function of the mismatch between the fractal dimension of the set and the Euclidean dimension within which it is embedded. This can be seen by noting that in a self-similar nested

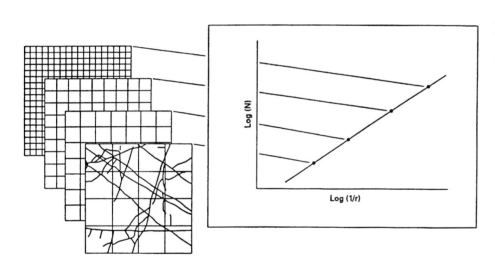

Figure 2. Schematic of the "box counting" procedure for determining the fractal dimension of a fracture set.

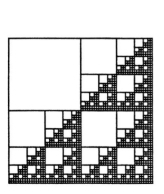

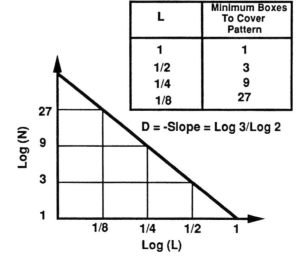

L	Minimum Boxes To Cover Pattern
1	1
1/2	3
1/4	9
1/8	27

D = -Slope = Log 3/Log 2

Figure 3. Example of the "box counting" method applied to a deterministic self-similar fractal.

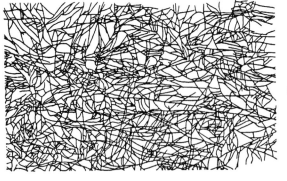

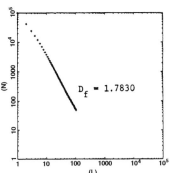

Figure 4. Fracture patterns in graywacke outcrops at The Geysers geothermal field and fractal "box plot" (Sammis et al., 1991).

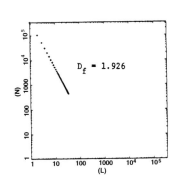

Figure 5. Fracture patterns in graywacke outcrops at The Geysers geothermal field and the corresponding fractal "box plot" (Sammis et al., 1991).

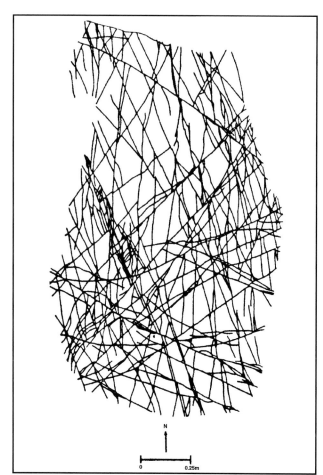

Figure 6. Fracture patterns in Permian Lyons Sandstone.

Figure 7. Fracture patterns in Oligocene quartz monzonite.

The probability that a point lies within a structure of scale l is the fraction of the total volume, l_0^d, occupied by structures of scale l, i.e.,

$$\Pr(l) = \left(\frac{l_0}{l}\right)^D l^d \left(\frac{1}{l_0^d}\right) = \left(\frac{l}{l_0}\right)^{d-D} = \left(\frac{l}{l_0}\right)^H \quad (5)$$

where $H = d - D$ is the codimension of the distribution. The codimension gives the mismatch between the dimension of the available space and the fractal dimension of the distribution.

The codimension of a sample of a distribution is independent of the dimensionality of the sample. This means that although we have only one-dimensional or two-dimensional samples of distributions that are inherently in a three-dimensional space, we can still draw conclusions about their three-dimensional character. For example, for a three-dimensional fracture network with a fractal dimension of $D_3 = 2.7$,

$$H = 3 - D_3 = 0.3$$
$$D_2 = 2 - H = 1.7 \quad \text{for a two-dimensional sample}$$
$$D_1 = 1 - H = 0.7 \quad \text{for a one-dimensional sample}$$

As an example of this, the spatial distribution of quartz/gold veins in exploratory cores from the Perseverance Mine in Juneau, Alaska, was analyzed (Barton, in press). Figure 10 shows a vertical cross section of the mine, the drilling pattern, and the location of quartz-filled fractures above a specified gold assay value along the boreholes. A one-dimensional box-counting analysis was performed on 20 cores, each approximately 90 m long and intersecting approximately 40–60 veins. A typical plot of the results of this type of analysis is shown in Figure 11. The results for all of the cores are summarized in Table 2. The values of the fractal dimension for these one-dimensional samples ranged from 0.41 to 0.62 with correlation coefficients greater than 0.98. This implies that the codimension ranged from 0.38 to 0.59, giving a range of fractal dimensions of $D = 2.41$ to $D = 2.62$ for the three-dimensional distribution.

The fractal character observed in sets of natural fractures is just one example of a larger class of phenomena

Figure 8. Fracture patterns in Miocene Paintbrush Tuff.

fractal structure in a Euclidean space of dimension d, the volume occupied by structures of scale l is given by

$$Nr(l)l^d = \left(\frac{l_0}{l}\right)^D l^d$$

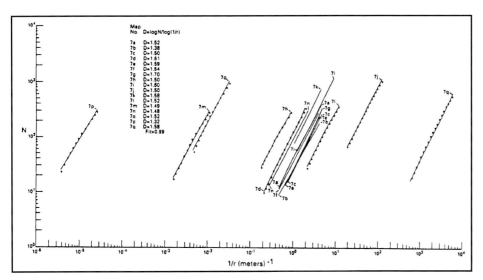

Figure 9. Box plots of the 17 fracture data sets analyzed by Barton (in press).

Table 1. Fractal Dimensions and Parameters of Fracture-Trace Maps

Fractal Dimension Box Flex Method	Location	Rock Unit/Type*	Age	Map Scale	Shortest Fracture Length (m)	Longest Fracture Length (m)
1.52	Yucca Mt., NV	TCMPB	Miocene	1:54	0.25	12
1.38	Yucca Mt., NV	TCMPB	Miocene	1:104	0.50	17
1.50	Yucca Mt., NV	TCMPB	Miocene	1:65	0.39	15
1.61	Yucca Mt. NV	TCMPB	Miocene	1:205	0.59	42
1.59	Yucca Mt., NV	TCMPB	Miocene	1:63	0.23	8.9
1.54	Yucca Mt., NV	TCMPB	Miocene	1:74	0.24	13
1.70	Yucca Mt., NV	ToSMPB	Miocene	1:55	0.20	12
1.50	Cedar City, UT	Quartz Monzonite	Oligocene	1:500	1.7	110
1.60	Lannon, WI	Niagran Dolomite	Silurian	1:33	0.091	6.4
1.50	Morrison, CO	Lyons Sandstone (orient. unknown)	Permian	1:3	0.12	9.7
1.58	Valley of Fire, NV	Aztec Sandstone	Triassic(?)/Jurassic	1.96	0.2	17
1.52	Alhambra Rock, Mexican Hat, UT	Rico Limestone	Pennsylvanian/Permian	1:5	0.08	4.8
1.49	Yucca Mt., NV	Paintbrush Tuff	Miocene	1:12,000	53	6300
1.48	Juneau, AK	Perseverance Slate (subvertical exp.)	Triassic	1:78	0.4	8
1.52	Goldhill, CO	Boulder Creek Granodiorite	Precambrian	1:12,000	26	970
1.32	South Atlantic sea floor	Basalt	Cretaceous/Holocene	$1:4.4 \times 10^7$	9.9×10^4	6.3×10^6
1.58	Timmins, Ont.	Albite, quartz, scheelite	Archean	1.65	5×10^{-4}	1.3×10^{-2}

*TCMPB = Densely welded upper lithophysal unit, Tiva Canyon Member, Paintbrush Tuff; ToSMPB = Densely welded orange brick unit, Topopah Spring Member, Paintbrush Tuff; orient. unknown = orientation unknown; subvertical exp. = subvertical exposure.

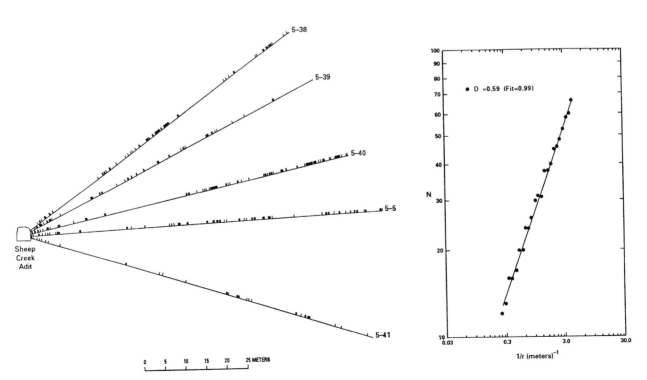

Figure 10. Vertical cross section of the Perseverance Mine.

in which brittle materials are observed to fragment into fractal distributions when subjected to stress. Numerous examples have been cataloged by Turcotte (1986b) and shown to be a very general phenomenon. Several examples are shown in Figure 11.

Although the evidence for a fractal character in all of these sets is widespread, a geometric description alone does not provide insight into the physical mechanisms that might produce such a distribution. The observation that brittle materials frequently fragment into a fractal distribution of particle sizes has led to the postulation of physical models to account for this observation. These models are all based on fragment distributions produced by a series, or hierarchy, of brittle failures of the material. A simple model proposed by Sammis et al. (1986) is illustrated in Figure 12. It is based on the hypothesis that a particle is most likely to fracture when it is loaded by particles of the same size because this geometry maximizes the stress concentrations on its surface. The result of a process that eliminates same-size neighbors at all scales is a distribution in which no two particles the same size are nearest neighbors at any scale. This is a property of the fractal pattern shown in Figure 12; at each scale, two diagonally opposed blocks are retained and no two blocks of equal size are in direct contact with each other. For the cube of dimension h shown in Figure 12, there are two blocks of dimension $h/2$ and 12 blocks of dimension $h/4$; thus, if the process is scale invariant

$$D = \frac{\ln 6}{\ln 2} = 2.585$$

This value is similar to that measured for many of the fracture networks and fragment distributions previously reported.

SYNTHETIC FRACTURE NETWORKS

The importance of the fractal character of fracture networks on fluid flow processes in naturally fractured

porous media can be evaluated by flow simulation in synthetic fracture networks. To do this, we must be able to synthesize fracture networks having the same fractal characteristics observed in natural fracture sets. Barnsely (1988) proposed a method for constructing fractal images using what is referred to as an iterated function system (IFS). The fractal image is constructed by applying a set of iterative numerical transformations to an initial set of points. At each iteration, each func-

Table 2. Summary of Fracture Data Analyzed by Barton (in press)

Drill Hole Number	Fractal Dimension
1-26	0.52
1-27	0.45
1-28	0.52
1-29	0.41
2-2	0.43
2-23	0.58
3-3	0.55
3-20	0.54
3-21	0.49
4-4	0.51
4-12	0.46
4-13	0.48
4-14	0.48
4-202	0.51
5-5	0.45
5-38	0.58
5-39	0.51
5-40	0.62
5-41	0.42
7-7	0.47
7-18	0.59
7-19	0.55
9-31	0.47

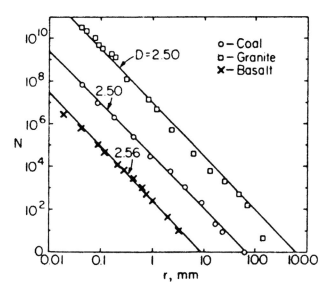

Figure 11. Examples of fractal distribution of fragments in several natural brittle fragmented materials.

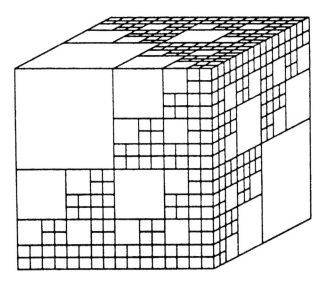

Figure 12. Fractal model for the comminution of a cube (Steacy and Sammis, 1991).

tion in the system operates on the set of points and, according to the parameters in the function, translates, reflects, rotates, contracts, or distorts the set of points. Over many iterations, the points in the picture coalesce toward an "attractor" (independent of the starting set), which is a fractal object. One of the advantages of this approach is that only a small number of parameters are required to reproduce very complex patterns. It also has the appeal that the process for constructing a pattern of fractures using an IFS mimics the hierarchical character of fracture generation suggested by rock mechanics models.

This approach has been employed by several investigators to generate realistic fracture networks with fractal characteristics (Long et al., 1991; Acuna and Yortsos, 1991). One example is shown in Figure 13, which was produced by the iterative splitting of a quadrilateral (Acuna and Yortsos, 1991). This figure was produced by starting with a quadrilateral and dividing it by a single fracture into two pieces. The IFS equations then take the whole pattern, rotate it, and map it back onto the two halves. This is repeated for many generations. In this instance,

$$D = \frac{\ln 2}{\ln \sqrt{2}} = 2$$

and the fracture network would fill the two-dimensional plane after enough generations were produced. Fracture networks with fractal dimensions of less than two can be produced by letting blocks be fractured with a probability p_f at each level of iteration. On average, then

$$N = 2p_f$$
$$r = 1/\sqrt{2}$$
$$D = \frac{\ln 2p_f}{\ln \sqrt{2}}$$

Examples of the types of fracture networks which result from this process are shown in Figures 14–17. The values of p_f range from 0.95 to 0.75, producing fracture patterns with fractal dimensions ranging from $D = 1.85$ to $D = 1.17$. The clustering and intermittence exhibited by those with values of D in the range of 1.5 to 1.7 look similar to those observed in maps of actual fracture networks.

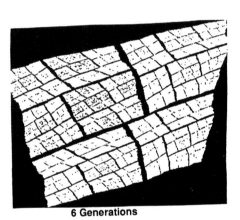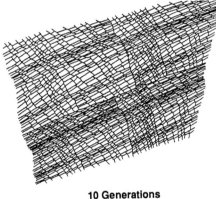

Figure 13. Synthetic fracture network produced by iterative splitting of a quadrilateral (Acuna and Yortsos, 1991).

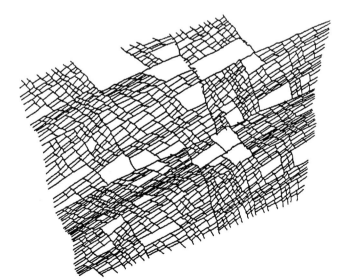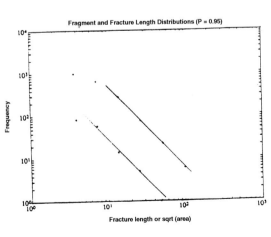

Figure 14. Synthetic fracture network corresponding to $p_f = 0.95$, $D = 1.85$ (Acuna and Yortsos, 1991).

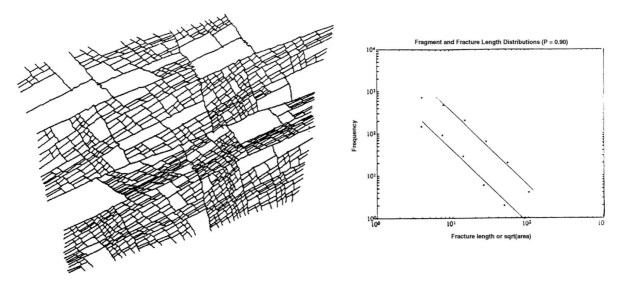

Figure 15. Synthetic fracture network corresponding to p_f = 0.90, D = 1.70 (Acuna and Yortsos, 1991).

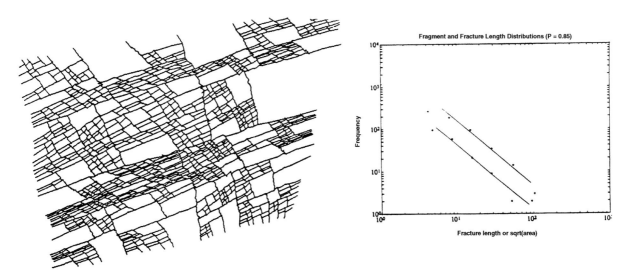

Figure 16. Synthetic fracture network corresponding to p_f = 0.85, D = 1.53 (Acuna and Yortsos, 1991).

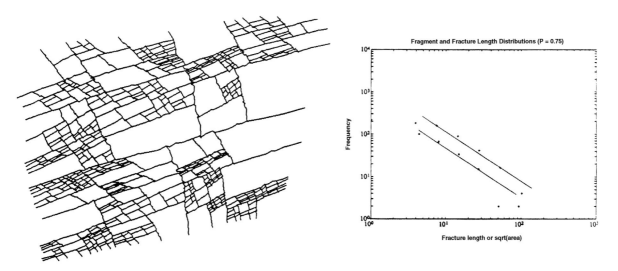

Figure 17. Synthetic fracture network corresponding to p_f = 0.75, D = 1.17 (Acuna and Yortsos, 1991).

FLUID FLOW IN FRACTAL NETWORKS

The consequences for fluid flow of the existence of a fractal fracture network have been investigated both theoretically and by simulation (Acuna and Yortsos, 1991; Chang and Yortsos, 1990a; Beier, 1990; Acuna et al., 1992). Because fractal distributions exhibit variations and clustering over all scales, the concept of representative elementary volumes (REV), or an equivalent continuum, loses meaning when describing fractal networks. In a fractal fracture system, the effective porosity and permeability exhibit a power-law scaling behavior of the form (Acuna et al., 1992)

$$\phi(r) = \phi_0 \left(\frac{r}{r_0}\right)^{D-d}$$

$$k(r) = k_0 \left(\frac{r}{r_0}\right)^{D-d-\theta} \quad (6)$$

where D and θ (spectral dimension) are fractal parameters and r_0 is the minimum size fracture considered. With the effective properties dependent on the sampling volume, the effective pressure diffusivity is also dependent on sampling volume, or radius. This variation of diffusivity with radius gives rise to a phenomenon referred to as anomalous diffusion. The fractal dimension, D, is a purely geometric parameter. The spectral dimension, θ, is a property depending on both the geometry and the connectivity of the medium. θ can be measured by taking a large number of walks of t steps each and calculating the average square distance from the origin. Contrary to the normal behavior of networks with Euclidean geometries, where $<r^2> \approx t$, fractal systems exhibit a slowdown due to the sparseness of the network, with $<r^2> \approx t^{2/(2 + \theta)}$. This is a key feature of processes involving anomalous diffusion (Orbach, 1986).

It can be shown (Chang and Yortsos, 1990a, b; Acuna et al., 1992) that at late times the pressure obeys a diffusivity equation with variable properties, namely,

$$\frac{\partial^2 p}{\partial r^2} + \frac{D-\theta-1}{r}\frac{\partial p}{\partial r} - r^\theta \frac{\partial p}{\partial t} = 0 \quad (7)$$

Based on a similarity solution, the well-bore pressure is given by

$$p_w(t) = C + \frac{(2+\theta)^{2(1-\delta)}}{\Gamma(\delta)(2+\theta-D)} t^{1-\delta} \quad (8)$$

where

$$C = \frac{\Gamma(\delta-1)}{\Gamma(\delta)(2+\theta)}$$

and

$$\delta = \frac{D}{2+\theta}$$

For a two-dimensional embedding medium, the previous relation results in linear plots with the same slope on log-log paper for both the pressure and pressure derivative. This common slope is given by $m = 1 - \delta$.

When dealing with transient tests from only a single well, the information at the producing well can determine only the ratio $\delta = D/(2 + \theta)$. Additional information is required to specify D and θ independently. From numerous experiments on synthetic fractal fracture networks, workers have found that θ ranges between 0 and 0.5 (Acuna et al., 1992).

These concepts were used in analyzing several well tests at The Geysers geothermal field, which is known to be highly fractured (Acuna et al., 1992). These concepts, applied to a pressure buildup test in well A, give the results shown in Figure 18. Both the pressure and pressure derivative plots are linear and parallel for a period of time before boundary effects become significant. The slope measured from the derivative curve is $1 - \delta = 0.40$. This corresponds to

$$\delta = \frac{D}{2+\theta} = 0.60$$

Assuming the value of θ for this network is similar to that observed in synthetic networks means that $1.20 < D < 1.50$. A synthetic fracture network with a similar simulated buildup response is shown in Figure 19. This is one of the many possible realizations that could match the observed well response.

A second example of a well test showing fractal behavior is provided in Figure 20. Pressure and pressure derivative curves are shown with the associated straight-line best-fit value of δ being equal to 0.84 for the early data and 0.87 for the later data. This change may be due to variations in the fractal structure or to boundary effects. Again, assuming the same range for θ as before, the corresponding range of fractal dimensions is $1.68 < D < 2.0$. A synthetic fracture network with a similar simulated buildup response is shown in Figure 21. This is one of many possible realizations that could match the observed well response.

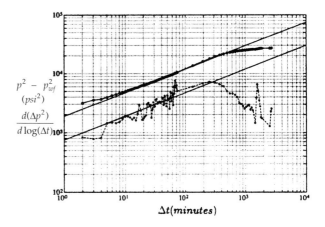

Figure 18. Pressure (solid dots) and pressure derivative (dashed line) and best-fit power law ($\delta = 0.6$) at well A (Acuna et al., 1992).

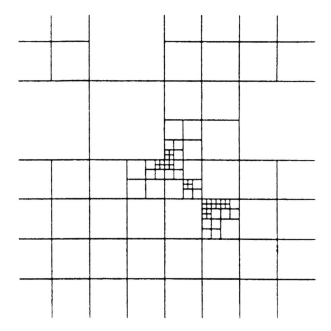

Figure 19. Synthetic fracture network with a pressure-transient response similar to that of well A (D = 1.26).

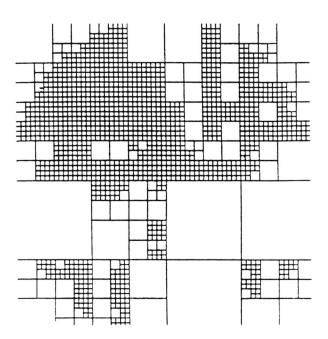

Figure 21. Synthetic fracture network with a pressure-transient response similar to that of well B (D = 1.84).

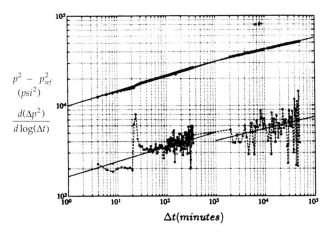

Figure 20. Pressure (solid dots) and pressure derivative (dashed line) and best-fit power law (δ = 0.84 and 0.87) at well B (Acuna et al., 1992).

- Realistic-appearing fractal fracture networks may be synthesized by probabilistic iterated function systems.
- Flow tests in fractured reservoirs have been shown to exhibit the characteristics expected for fractal fracture networks.

CONCLUSIONS

From this review of the properties of self-similar nested fractal structures and the characteristics observed in real fracture networks, one can draw the following conclusions.

- Fractal geometry provides a good framework for characterizing the intermittency and clustering over the wide range of scales observed in real fracture networks.
- Hierarchical models of brittle rock fragmentation produce fractal distributions of blocks and fracture lengths.

REFERENCES

Acuna, J. A., and Y. C. Yortsos, 1991, Numerical construction and flow simulation in networks of fractures using fractal geometry: SPE Annual Technical Conference Proceedings.

Acuna, J. A., I. Ershaghi, and Y. C. Yortsos, 1992, Fractal analysis of pressure transients in The Geysers geothermal field: 17th Annual Workshop on Geothermal Reservoir Engineering, Stanford University, Stanford, California.

Aviles, C. A., and C. H. Scholz, 1987, Fractal analysis applied to characteristic segments of the San Andreas fault: Journal of Geophysical Research, v. 92, no. B1, p. 331–344.

Avnir, D., D. Farin, and P. Pfeifer, 1985, Surface geometric irregularity of particulate materials: the fractal approach: Journal of Colloid and Interface Science, v. 103, no. 1, p. 112–123.

Bale, H. D., and P. W. Schmidt, 1984, Small-angle x-ray scattering investigation of submicroscopic porosity with fractal properties: Physical Review Letters, v. 53, no. 6, p. 596–599.

Barnesly, M., 1988, Fractals everywhere: Boston, Academic Press, 391 p.

Barton, C. C., and P. A. Hsieh, 1989, Physical and hydrologic-flow properties of fractures: American

Geophysical Union, 36.

Barton, C. C., and E. Larsen, 1985, Fractal geometry of two-dimensional fracture networks at Yucca Mountain, southwestern Nevada: International Symposium on Fundamentals of Rock Joints, Bjorkliden, Sweden.

Barton, C. C., in press, Fractal analysis of the scaling and spatial clustering of fractures, *in* C. C. Barton and P. R. La Pointe, eds., Fractals in earth sciences: Plenum Press, New York.

Beier, R. A., 1990, Pressure-transient model of a vertically fractured well in a fractal reservoir: Paper SPE 20582, 65th Annual SPE Technical Conference and Exhibition.

Chang, J., and Y. C. Yortsos, 1990a, Pressure-transient analysis of fractal reservoirs: SPE Formation Evaluation, v. 5, n. 1, p. 31–38.

Chang, J., and Y. C. Yortsos, 1990b, A note on pressure-transient analysis of fractal reservoirs: SPE Advanced Technology Series, v. 1, no. 2, p. 170–171.

Chilès, J. P., 1988, Fractal and geostatistical methods for modeling of a fracture network: Mathematical Geology, v. 20, no. 6, p. 631–654.

Daccord, G., J. Nittmann, and H. E. Stanley, 1986, Radial viscous fingers and diffusion-limited aggregation: fractal dimension and growth sites: Physical Review Letters, v. 56, no. 4, p. 336–339.

Hirata, T., 1989, Fractal dimension of fault systems in Japan: fractal structure in rock fracture geometry at various scales: Pure and Applied Geophysics, v. 131, no. 1, p. 290–305.

Katz, A. J., and A. H. Thompson, 1985, Fractal sandstone pores: implications for conductivity and pore formation: Physical Review Letters, v. 54, no. 20, p. 1325–1328.

La Pointe, P. R., 1988, A method to characterize fracture density and connectivity through fractal geometry: International Journal of Rock Mechanics, Mining Science and Geomechanical Abstracts, v. 25, no. 6, p. 421–429.

LaPointe, P. R., and J. A. Hudson, 1985, Characterization and interpretation of rock mass joint patterns: Geological Society of America.

Lenormand, R., and C. Zarcone, 1985, Invasion percolation in an etched network: measurement of a fractal dimension: Physical Review Letters, v. 54, no. 20, p. 2226–2229.

Long, J. C. S., C. Doughty, K. Hestir, and S. Martel, 1991, Modeling heterogeneous and fractured reservoirs with inverse methods based on iterated function systems: Lawrence Berkeley Laboratory.

Mandelbrot, B. B., 1983, The fractal geometry of nature: New York, Freeman.

Orbach, R., 1986, Dynamics of fractal networks: Science, v. 231, p. 814–816.

Pentland, A., 1983, Fractal-based description of natural scenes: Proceedings of the IEEE Computer Society Conference on Computer Vision and Pattern Recognition, p. 201–209.

Sammis, C. G., and R. L. Biegel, 1989, Fractals, fault gouge, and friction: Pure and Applied Geophysics, v. 131, no. 1/2, p. 255–272.

Sammis, C. G., L. J. An, and I. Ershaghi, 1991, Fractal patterns in graywacke outcrops at The Geysers geothermal field: 16th Workshop on Geothermal Reservoir Engineering, Stanford University, Stanford, California.

Sammis, C. G., R. H. Osborne, J. L. Anderson, M. Banerdt, and P. White, 1986, Self-similar cataclasis in the formation of fault gouge: Pure and Applied Geophysics, v. 123, p. 53–78.

Steacy, S. J., and C. G. Sammis, 1991, An automaton for fractal patterns of fragmentation: Nature, v. 353, p. 250–252.

Turcotte, D. L., 1986a, A fractal approach to the relationship between ore grade and tonnage: Economic Geology, v. 81, p. 1528–1532.

Turcotte, D. L., 1986b, Fractals and fragmentation: Journal of Geophysical Research, v. 91, no. B2, p. 1921–1926.

Turcotte, D. L., 1989, Fractals in geology and geophysics: PAGEOPH, v. 131, no. 1/2, p. 171–196.

Velde, B., J. Dubois, G. Touchard, and A. Badri, 1990, Fractal analysis of fractures in rocks: the Cantor's dust method: Tectonophysics, v. 179, p. 345–352.

Chapter 20

Description of Reservoir Properties Using Fractals

M. Kelkar
S. Shibli
The University of Tulsa
Tulsa, Oklahoma, U.S.A.

ABSTRACT

This work investigates the utility of applying fractal geometry for describing the spatial correlation structure of rock properties. It also provides a proper framework within which to apply a stochastic conditional simulation technique to generate equiprobable distributions of rock property.

A normalization technique is developed to remove the seasonal effects found in the well-bore data. Where the seasonal effects are unusually strong, as in the case of one carbonate reservoir, removal of the local means will result in a narrower distribution of the Hurst (H) exponent values. Theoretical and analytical variogram plots of the results of these estimations indicate that the box-counting method gives slightly improved estimates over rescaled range (R/S) analysis.

Conditional simulation runs using simulated annealing are performed for three fields. The results indicate that a combination of the greedy algorithm and a bilinear interpolation technique to generate the initial pattern decreases computer usage by as much as 85%. More important, the simulations also indicate that an fGn (fractional Gaussian noise) horizontal correlation structure gives better predictions of interwell distributions in a carbonate environment. In contrast, in a sandstone field, the use of an fBm (fractional Brownian motion) horizontal correlation structure gives more reliable results.

INTRODUCTION

One of the major difficulties in describing reservoir properties is lack of information at the interwell scale. This study is an attempt to determine whether a representative spatial correlation structure can be obtained by analyzing well-bore data with the help of fractal models. The motivation for using fractal statistics is that several studies have been conducted in the past that show it to be a promising and simple tool for determining the spatial correlation structure of geo-logical data (Hewett, 1986; Hewett and Behrens, 1990; Emanuel et al., 1989). Although different geostatistical techniques could have been used, only fractals offer the convenience of describing the spatial correlation structure in terms of only one major variable: the H or intermittency exponent of the data (see Appendix 1 for a nomenclature list).

One of the major drawbacks associated with determining the H exponent is the limited data typically available at well-bore locations. For large sample sizes, many methods exist for determining the intermittency exponent; however, as the sample size decreases, these methods become unreliable (Wallis and Matalas, 1970). We want to use a method that provides a reliable estimate of H with a limited amount of data.

Another drawback with respect to determining H using well-bore data is the commonly used practice of analyzing data without taking into account alternating periodic sequences that might be present in the zone of interest. Mixing zones together is tantamount to mixing two different populations, and would lead to biased estimates of the intermittency exponent. As shown by Feder (1988), a normalization technique is needed to analyze the sample data in the presence of local trends.

The first part of this study involves analysis of well-bore data to estimate the intermittency exponent, H. Over 90 well logs from both carbonate and sandstone environments are analyzed, and a systematic method is proposed to properly characterize the well-bore data.

The second part of this study deals with the conditional simulation of reservoir properties based on the simulated annealing technique (Farmer, 1987, 1989; Perez, 1991). The characteristic spatial correlation obtained from the fractal models is used as an input into the conditional simulator. The simulated annealing method includes the constraints of the cumulative distribution function of the conditioning data, the conditioning data itself, and the spatial correlation structure of the conditioning data (Perez, 1991). Different seed values for the random number generator creates different realizations of the reservoir properties.

Once the alternate descriptions of reservoir properties are created, validation of the reliability of the predictions involves comparing the actual well log data with the simulated data. A close agreement between actual and simulated logs indicates a reliable prediction for a particular set of constraints. In this study, several comparisons are done to determine to what extent fractal correlation structures can be used in predicting rock properties at unsampled locations.

DETERMINATION OF INTERMITTENCY EXPONENT

Hewett (1986) laid the groundwork for the application of fractal geometry to reservoir description. Using different methods of analysis, e.g., rescaled range (R/S), variogram, spectral density, etc., he showed that well-bore data can be described by fractional Gaussian noise (fGn) type of behavior. He further postulated that the horizontal data can be described by fractional Brownian motion (fBm) type of behavior.

Although several methods exist for analyzing well-bore data for determining the intermittency exponent, most of these methods become unreliable as the sample size decreases. Our own study (Shibli, 1992) indicated that the two methods that are most robust are the R/S analysis (Hewett, 1986) and the box-counting method (Feder, 1988). Out of the two methods, the box-counting method provides more reliable prediction of the intermittency exponent. We used only R/S analysis and the box-counting method for this study.

Little has been done in the literature that shows the effects of seasonal variations of rock properties on the value of the intermittency exponent. For example, alternating sand and shale sequences creates a nonstationarity in the resulting sequence of values, thus making an analysis based on fractional Gaussian noise invalid. One solution is to remove the local means from the trace to be studied to eliminate most of the nonstationarity.

To achieve this, zones of different properties based on geology can be visually identified. Each of these different zones have a local mean, which can be easily calculated, associated with it. For each point in a zone, the local mean is subtracted from the value. The total number of zones is arbitrary and dependent on the judgment of the person doing the adjusting. However, if seasonal or "hole effects" are to be removed from a trace, it is important that the width of a chosen zone does not exceed the actual width of the anomaly on the well log.

An example of a porosity log where each zone is determined and adjusted for seasonal effects is given in Figure 1. In this well log, taken from a sandstone field, one can discern three zones of changing local means; each zone has an average thickness of 40 feet, with a zone of moderate porosity at the top, a zone of high porosity in the middle, and a zone of low porosity at the bottom (regions A, B, and C, respectively, on Figure 1). A preliminary study of the geology of the region also indicated that three different flow units are present in the reservoir. Figure 1 also shows the log after seasonal adjustments were done based on the three flow units.

The analysis of Figure 1 data, which includes 100 sample points, by the box-counting method indicates that the value of the intermittency exponent H without seasonal adjustment was 0.88, whereas with the adjustment, the value was reduced to 0.78 (Shibli, 1992). This behavior is consistent with what is observed by Feder (1988) for seasonally corrected wave height statistics.

Perhaps a better indicator of the ability of the different methods to reliably estimate the H exponent would be to compare the fGn (or fBm) variogram fit to the actual data being studied. This also serves another purpose: to observe the effects of removing the seasonal trends on the shape of the variogram itself. A zone with severe "hole effects" indicates the presence of a series of different porosity or permeability zones in the reservoir. If a zone has been seasonally adjusted properly, most, if not all, of the hole effects would be eliminated.

The fGn variogram model can be written as

$$\gamma(h) = \frac{1}{2} V_H \delta^{2H-2} \left[2 - \left(\frac{|h|}{\delta} + 1 \right)^{2H} + 2 \left| \frac{h}{\delta} \right|^{2H} - \left(\frac{|h|}{\delta} - 1 \right)^{2H} \right] \quad (1)$$

where V_H is a constant, δ is a smoothing factor (resolution size of a measurement) and h is the lag distance.

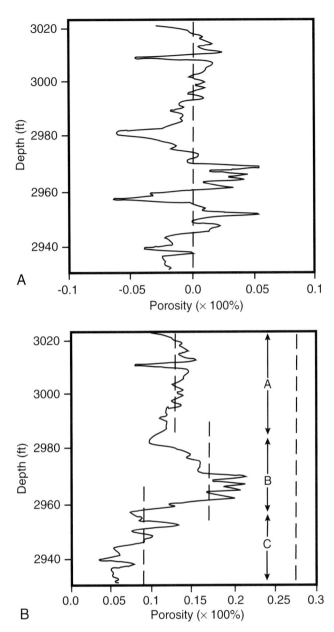

Figure 1. Porosity trace for Burbank field (a) with and (b) without trend removal.

For *f*Bm, the variogram can be written as

$$\gamma(h) = V_H h^{2H} \qquad (2)$$

where V_H is a constant and h is the lag distance.

To apply the *f*Gn model, we need to estimate three parameters, V_H, δ, and H. The value of δ can be assumed to be the resolution of the log data. We have assumed it to be equal to 3 ft. The value of H is determined from either R/S analysis or the box-counting method. As a first approximation, V_H can be assumed to be equal to the variance of the data. This value can be changed based on a good match.

Figure 2 shows the comparison between the estimated and model *f*Gn variograms for data from a Burbank

(sandstone) field. This figure shows the estimated variogram based on the raw data. Either *f*Gn or *f*Bm models fail to predict the estimated variogram values. The *f*Gn model (predicted by the box-counting method) does a slightly better job. Figure 3 shows the comparison between the estimated and model variograms after seasonal adjustments. The value of H predicted by the box-counting method does a much better job of describing the variogram. Although not shown here, similar results are observed for other well logs as well.

ANALYSIS OF FIELD DATA

Using this method, well log data from three fields were analyzed for two carbonates and one sandstone.

Hugoton Field

The Hugoton field is composed of repeated sequences of interlayered carbonates, siliciclastics, and evaporites deposited on a gently dipping, low-relief, shallow-marine carbonate ramp. The producing formation consists of three reservoir flow units or layers mainly composed of dolostone, each capped with siltstones, mudstones, and shale intervals that act as barriers to vertical flow; thus, no cross-flow occurs between the separate flow units (Ben-Rached, 1991).

Wells generally produce from depths between 2700 and 3000 ft and are located within a 640-ac spacing pattern. In the study area, layer 1 ranges from 60 to 65 ft thick, layer 2 ranges from 50 to 60 ft thick, and layer 3 ranges from 40 to 50 ft thick. Figure 4 shows a field map of the study area.

The data studied were obtained from the 31 porosity logs. Typical log data were collected at 0.5-ft intervals, resulting in at least 200 vertical data points for each well log. H exponent estimations were done for all the logs before and after adjustments for seasonal effects. For this field, the zones of varying local means chosen for adjustments are largely affected by the presence of three flow units in the reservoir. For brevity, results based on only the box-counting method are shown. Results based on R/S analysis are shown elsewhere (Shibli, 1992).

Table 1 presents the results of the field-wide analyses of Hugoton field using box-counting. A total of 31 porosity logs are studied and their H exponents estimated. The porosity logs are then adjusted for seasonal effects (using the procedure delineated before) and analyzed again. The mean value of the H exponent is significantly affected by the normalization procedure. In particular, a decrease in the mean value is observed. Because of the strong seasonal trends observed in Hugoton field, removal of the means results in a narrower range of H values. The variance without adjustment is found to be 0.0014. After applying seasonal adjustments, the variance decreased significantly to 0.0004. A similar behavior is noted for R/S analysis as well (Shibli, 1992).

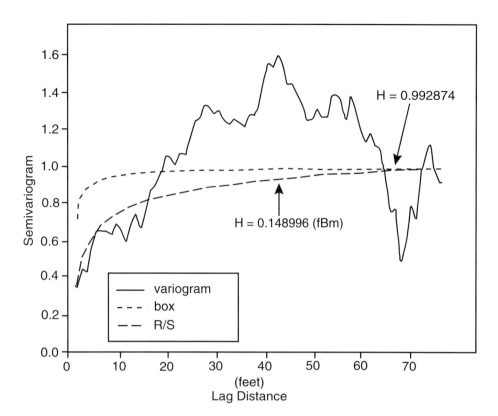

Figure 2. Theoretical and analytical matches for original porosity trace in Burbank field, well 41.

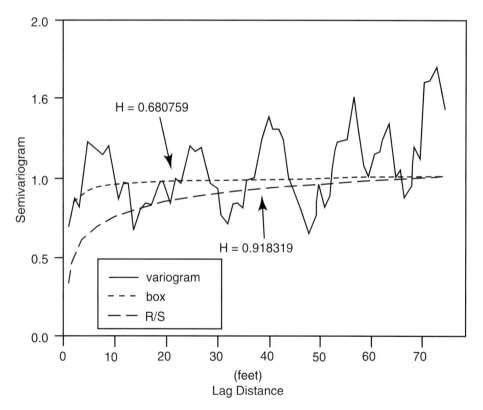

Figure 3. Theoretical and analytical matches for adjusted porosity trace in Burbank field, well 41.

Burbank Field

The reservoir produces from the Burbank Sandstone member of the Cherokee Shale at a depth of about 3000 ft. The field lies in a north–northwest trend and is 12 mi long, 4–5 mi wide, and covers an area of

23,000 productive acres (Hagen, 1972). In 1937, a study concluded that the Burbank sand was deposited as a series of overlapping beach deposits during the Teeter-Quincy and Sallyards-Lamont stages of the Cherokee Sea. However, recent geologic study has shown the reservoir to be of fluvial origin with some

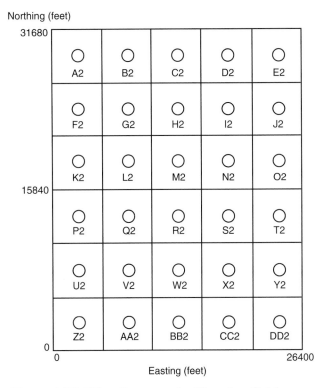

Figure 4. Well location map for Hugoton field.

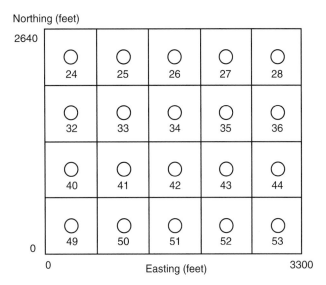

Figure 5. Well location map for Burbank field.

Table 1. Summary statistics of intermittency exponent values for Hugoton field, box-counting method

Parameter	Without Normalization	With Normalization
Number	31	31
Mean	0.885157	0.771484
Variance	0.001374	0.000388
Standard Deviation	0.037074	0.019708

Table 2. Summary statistics of intermittency exponent values for Burbank field, box-counting method

Parameter	Without Normalization	With Normalization
Number	45	45
Mean	0.931220	0.773067
Variance	0.007247	0.006684
Standard Deviation	0.085128	0.081758

contribution from marine incursions having occurred during deposition. The massive sand body was created by the superimposed effect of a series of channels cut laterally into each other. The resulting structure is an undulating monocline dipping at a rate of 35 ft/mi in a westerly direction (Szpakiewicz et al., 1986). The limits of the field are determined by an abrupt sand-to-shale lithology change on the east and a tilted oil-water contact in combination with a permeability barrier on the west.

The producing zone consists of fine-grained quartz sand, loosely cemented by magnesium, iron, and calcium carbonate, and locally by silica, dolomite, or calcite. In addition to quartz, the sand contains 1% mica; traces of feldspar, zircon, chlorite, glauconite, hornblende, rutile, magnetite, pyrite, and epidote; 10–20% detrital rock fragments (chert, shale, and schist); and a trace to 10% of carbonaceous material.

The sand averages 42 ft in thickness, with an average porosity of 16.8% and a permeability of 50 md. The formation is vertically and areally heterogeneous. Also, east-west–trending natural fractures are thought to exist throughout the field. Figure 5 shows a field map of the zone to be simulated.

Core data for 45 wells were available for analysis. The average number of data points ranged between 50 to 100 points. For some wells, the missing data points were substituted with interpolated values. The final analysis of well data is shown in Table 2.

The mean value of the H exponents before seasonal adjustments are performed is 0.93 using the box-counting method. After seasonal adjustments are made, the mean value of the H exponent decreases to 0.77 for the box-counting method. The average value after seasonal adjustment is remarkably similar to the mean value of the H exponent after seasonal adjustments for the Hugoton field, which is a carbonate environment.

Sable Field

The Sable unit (San Andreas) reservoir lies in a structural anticline trap with the oil column reaching a maximum thickness of 108 ft in Sable no. 12 well (MacAllister, 1987). Production data, supported by resistivity logs, indicate that the water-oil contact dips slightly from west to east across the reservoir. The reservoir itself is capped by a dense dolomite. The carbonate rock is heterogeneous at every scale, micro to macro, and that randomness of the unit also occurs areally, aside from vertically. Figure 6 presents a field location of the wells chosen for the study.

All 16 core porosities were analyzed and adjusted for seasonal effects. For each well, at least 300 vertical data points were available. Table 3 shows the summary statistics of the analysis. As before, the value of H reduces after seasonal adjustments. This value is close to the other values in the previous two fields.

To summarize, we observed that the box-counting method does a good job of estimating the intermittency exponent for a limited number of data points. The seasonal trends have a significant impact on the value of intermittency exponent. With removal of the trends, the values of intermittency exponents show a narrower range for a given field. Also, the average H value tends to fall between 0.75 and 0.8. This value is independent of the type of depositional environment.

STOCHASTIC CONDITIONAL SIMULATION

This section provides a set of procedures to be followed in the stochastic conditional simulation of reservoir properties. Factors to be taken into account are the removal of the nonstationarity in the vertical well data (described in the previous section) and the selection of a correct horizontal spatial correlation structure based on the structure determined from vertical well data.

Frequently, determining a reliable horizontal correlation model can be a trial-and-error procedure, primarily due to the fact that not much is known about the distribution of properties in the horizontal direction; therefore, different models have been assumed by numerous researchers. In this section we propose a framework that will help pinpoint the most reliable structure to be used in the final three-dimensional simulation of the field.

This framework can be described as having the following attributes:

1. Uses reliable intermittency exponent estimation technique that is both robust and fairly insensitive to small sample distributions.
2. Uses a combination of the greedy algorithm and a bilinear interpolation (described in a following section) technique to improve computer usage in simulated annealing runs.
3. Determines the most reliable horizontal correlation model based on different, small, two-dimensional simulations of the field.
4. Uses the same correlation model determined from point 3 to perform a regular three-dimensional stochastic conditional simulation of the entire field.

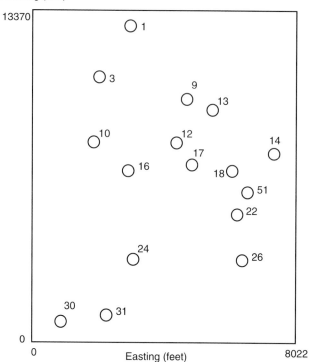

Figure 6. Well location map for Sable field.

Table 3. Summary statistics of intermittency exponent values for Sable field, box-counting method

Parameter	Without Normalization	With Normalization
Number	16	16
Mean	0.835006	0.793954
Variance	0.003973	0.005922
Standard Deviation	0.063029	0.076951

The importance of point 1 has been discussed in the previous section. Point 2 of the procedure provides an efficient way of generating a description of the reservoir properties. As discussed, we used the simulated annealing procedure for generating alternate descriptions of reservoir properties (Perez, 1991). Briefly, the method of simulated annealing requires defining an objective function that needs to be minimized. For a geostatistical procedure, this objective function is the difference between the model variogram and the observed variogram. By randomly swapping two points within the distribution and selectively accepting the swaps that will either reduce or temporarily increase the objective function, the method of annealing provides an acceptable solution. The solution is assured to reach a global minimum by using a Metropolis condition, which allows an acceptance of a swap that may result in a temporary increase in the objective function. (For details see Perez, 1991.) If we eliminate the Metropolis condition, the algorithm accepts only a swap that will reduce the objective function. This specialized type of annealing is called a greedy algorithm (Shibli, 1992; Doyen et al., 1989; Deutsch and Journel, 1992).

One advantage of the greedy algorithm is that it converges to a solution at a much faster rate. The disadvantage is that it may not reach a global minimum of a solution. To achieve a speedy convergence and obtain an acceptable solution, we adopted a modified approach.

In our approach, we generated an initial distribution of reservoir properties using a simple bilinear interpolation routine. In this routine, we first adjust the tops of individual pay zones and generate interwell properties layer by layer. For example, if we were generating a two-dimensional cross section between two wells, we would start with the first (top) point from each of the two wells and generate values between the two wells by a linear interpolation. We would repeat the procedure at the next level until we reached the bottom point of each of the two wells. By stacking line by line, we would create a two-dimensional surface. One of the requirements of this method is that each well have an equal number of data points. This requirement can be satisfied by adjusting the number of values at each well by varying the distance between two successive points. In the alternative, depending upon the geologic interpretation, for the well having the smaller number of points, the remaining values are assumed to be zero. This may indicate nonpermeable rock below a certain level at a particular well (e.g., due to erosion of part of the rock). We did not use this second alternative in our simulations.

The method of linear interpolation can also be extended to a three-dimensional description. For generating a three-dimensional description, we would first pick the top points for all the sample wells and generate a surface at interwell locations using a bilinear interpolation. The procedure would be repeated at subsequent depths until we reached the bottom point of each of the wells. Stacking the surfaces over each other would result in a three-dimensional structure. As in the case of two-dimensional cross sections, we are limited by the requirement that each well contains the same number of sample data points.

Once the interwell distributions are created using the bilinear interpolation method, to create more variability within the data we add a random component to each of the estimated values at interwell locations. This random component is taken from a Gaussian distribution with a mean of 0 and a variance equal to the variance of the difference between the two wells located farthest apart. This type of a random addition allows us to reasonably match the histogram of the sample data set with the histogram of the estimated values.

This modified interwell distribution of properties is used as an initial distribution for the simulated annealing program. Realizing that a better correlation exists in the horizontal direction than the vertical

direction, we believe that our initial guess for simulated annealing is reasonably close to the final solution.

At this point, instead of using the conventional annealing procedure, we use the greedy algorithm to match the suggested variogram models in both the horizontal and vertical directions with the estimated variograms based on the generated distribution. Because a conventional simulated annealing method also behaves as a greedy algorithm as the final solution is approached, we believe that application of the greedy algorithm is a reasonable approximation to reach a global minimum. Such a combination of bilinear interpolation and greedy algorithm can result in substantial savings in terms of computation time. In some cases, we observed as much as 85% less computer time using this method rather than the conventional simulated annealing method.

Point 4 in our procedure requires that a correct variogram model for the horizontal variogram needs to be identified. We used either fGn or fBm models for describing the variograms in the horizontal directions. Although this choice is arbitrary, previous researchers have indicated success in using fractal models in describing horizontal data (Hewett, 1986; Beier and Hardy, 1993). In most instances, we did not observe any perceptible spatial relationship in the areal direction based on the well data. This means that beyond the minimum distance between two well pairs, the variogram exhibits a pure nugget. As a result, we calculated the value of V_H in the fGn or fBm variograms by assuming that the variogram at a distance equal to the distance between the well pair is equal to the variance of the difference between two wells. For example, if we are simulating a cross section between two wells, A and B, then we define a new variable

$$\delta_{Vi} = V_{Ai} - V_{Bi} \tag{3}$$

where V_{Ai} is the value of a given variable at location A at a level i, and V_{Bi} is the value of a given variable at location B at the same level. The procedure is repeated at every depth level until we reach the bottom pair. If $\overline{\delta V}$ is the arithmetic mean of all the values δV_i, we can define the variance of difference, $\delta\sigma^2$, as

$$\delta\sigma^2 = \frac{1}{N}\sum_{i=1}^{N}\left(\delta V_i - \overline{\delta V}\right)^2 \tag{4}$$

By equating $\delta\sigma^2$ to $\gamma(h)$ at the distance h equal to the distance between the well pair, we can calculate the value of V_H. For the fGn model, we assumed the value of δ to be equal to the scale of resolution of the measurement. Referring to equation 1, the value of δ is known, the value of H is known based on the analysis of vertical well-bore data, and the value of V_H can be computed by knowing the value of the variogram, $\gamma(h)$, at the distance h equal to the distance for a well pair. Similar calculations can be done for the fBm model.

To confirm the type of model we wish to use for three-dimensional conditional simulations, we first conducted two-dimensional simulation between different well pairs. We selected the pair such that we have at least one or two wells between that well pair. After the simulations are complete, we compare the "simulated" well logs with the actual well logs at the in-between wells to investigate which model provides a better solution in terms of generating the interwell distribution of the properties. To quantitatively evaluate the comparison, we use a certainty coefficient defined as (Perez, 1991)

$$C = \frac{\sum_{i=1}^{N}\left[V(xi)_{sim} - \overline{V}_s\right]\left[V(xi)_{true} - \overline{V}_t\right]}{\sqrt{\sum_{i=1}^{N}\left[V(xi)_{sim} - \overline{V}_s\right]^2 \sum_{i=1}^{N}\left[V(xi)_{true} - \overline{V}_t\right]^2}} \quad (5)$$

where $V(xi)_{sim}$ is the simulated value at location xi; $V(xi)_{true}$ is the observed value at location xi; $\overline{V}_s$ and $\overline{V}_t$ are the means of the simulated and observed values, respectively; and N is the total number of points at a well location. Ideally, if we predict the exact value at each location, the certainty coefficient would be equal to 1. We would expect that a correct model in the horizontal direction would consistently predict a higher value of certainty coefficient for different realizations.

We use both fGn and fBm models in horizontal directions and calculate the certainty coefficients by comparing the simulated well logs with the actual well logs. We repeat the procedure for several well pairs within the field. The model that consistently predicts a better certainty coefficient is selected for a three-dimensional conditional simulation.

There are several advantages associated with this procedure. We realize that one of the biggest uncertainties in conditional simulation methods is the spatial relationship in the horizontal direction. Some type of "history match" is needed to verify the choice of the model in the horizontal direction. Rather than conducting a full field-scale conditional simulation that may be computationally expensive, it is much easier to conduct two-dimensional, cross sectional simulations. Also, by conducting several of these two-dimensional simulations and comparing the simulated data with the actual data, the horizontal variogram model can be properly validated. Once the horizontal variogram is chosen, it can be used for three-dimensional simulation.

For three-dimensional simulations, we already know the type of variogram model to be used based on our two-dimensional results. We make one adjustment in the calculation of the V_H value for the three-dimensional simulation. We select a well pair with the maximum distance between the two wells, and by following the procedure outlined in equations 3 and 4, calculate the variance of the difference between the data at these two wells. By equating the variance of the difference equal to the variogram at the maximum distance, we recompute the value of V_H. Although this adjustment does change the model for three-dimensional simulations from the two-dimensional simulation in practice, the change is very small when the variogram is plotted as a function of the distance.

ANALYSIS OF FIELD DATA

Applying the method described, we evaluated three fields: Hugoton, Burbank, and Loudon.

Hugoton Field

We chose two wells from the northern section of the field (wells A2 and E2, Figure 4). The spacing between the well pair is 21,120 ft. Well B2 was chosen as the test well for the simulation. Based on the analysis of the well logs, three regions of porosity can be discerned. The same logs were seasonally adjusted to remove the local trends, and an H exponent analysis was performed. Table 4 summarizes the simulation parameters used in the simulations.

In addition to simulations performed using a combination of interpolation followed by the greedy algorithm, two other algorithms were also studied: a conventional annealing algorithm, and interpolation followed by the conventional annealing algorithm. In using the conventional algorithm, the program was instructed to generate its own initial pattern based on the distribution of the conditioning data using the conventional algorithm. In using interpolation followed by the conventional algorithm, an initial pattern was generated using linear interpolation with added random elements; however, the same conventional algorithm was also used. The two horizontal correlation structures studied are fGn and fBm.

Figure 7 shows a comparison between the simulated and the observed values at well B2. Certainty coefficients were also calculated for two horizontal correlation structures, fBm and fGn. From the figure, predictions obtained using an fGn horizontal structure seem to be more reliable than those obtained using an fBm structure. The plots for the fBm horizontal structure display erratic behavior. As shown in Table 5, the certainty coefficients for fGn are also markedly higher as the figures indicate. The calculated computer time on the Apollo 10000 workstation for the greedy algorithm is 611.12 s, compared to 4248.50 s for the simulation with interpolation and 5573.60 s for normal simulation without any interpolation.

Simulations in other principal directions preferably should be done to determine the best horizontal cor-

Table 4. Summary of simulation parameters for two-dimensional stochastic conditional simulation: Wells A2 and E2, Hugoton field

Distance Between Well Pair	21,120 ft
Variance of Difference	34.2956
Average H Value	0.7874
Blocks in x Direction	51
Blocks in z Direction	399
Blocks in y Direction	1
Smoothing Factor, δ	3 ft
Scaling Factor (fGn), V_H	56.204
Scaling Factor (fBm), V_H	5.305×10^{-6}

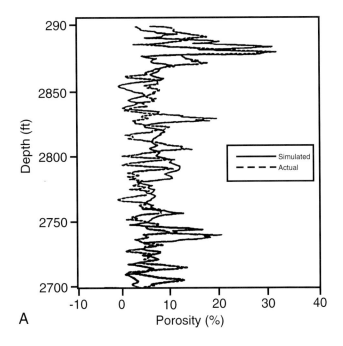

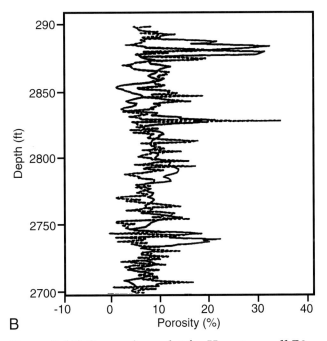

Figure 7. (A) Comparison plot for Hugoton well B2 using an *f*Gn structure and greedy algorithm. (B) Comparison plot for Hugoton well B2 using a combination of *f*Bm structure and greedy algorithm.

relation structure for the entire region. The previous example represents only the direction along the east-west plane of the field. Two other directions, north-south and diagonally, are also simulated in two dimensions. Referring to Figure 4 again, two well pairs were taken from each of these directions. In the north-south direction, the region between well pair A2 and U2 was simulated. In the diagonal direction, the region between well pair J2 and V2 also was simulated. Wells F2 and N2 represent the test wells for

these two-dimensional simulations. Table 6 presents the results of the certainty coefficients for the simulations using one of the realizations. Interestingly, the correlation coefficients for both directions show that *f*Gn is the best horizontal correlation structure to use for this field, where the correlation coefficients are larger than that for the *f*Bm model.

We can observe from these simulation runs that the best horizontal model to use for a carbonate environment seems to be the *f*Gn structure. This result is consistent with the results reported by Perez and Kelkar (1992) for the Sable field. The greedy algorithm also results in considerable savings in computer time, with the savings estimated from the previous runs amounting to as high as 85%. The conventional annealing algorithm takes the greatest amount of computer time. The greedy algorithm with interpolation retains the best overall certainty coefficients for the *f*Gn model, resulting in the best predictions.

Burbank Field

Preliminary two-dimensional simulations for the Burbank field were performed to have a better feel for the horizontal spatial correlation structure. The correlation structure in the vertical direction is assumed to be *f*Gn, with a smoothing interval of 3 ft. Two well pairs, 25, 28 and 50, 28 (Figure 5) were chosen for the simulations.

Table 7 shows the typical simulation parameters used for the two-dimensional simulation using a combination of linear interpolation and the greedy algorithm. The comparisons between the simulated and the test well logs are shown in Table 8 for one of the realizations.

Although not shown here (Shibli, 1992), the visual comparison between the simulated and the observed porosity values indicates that the *f*Bm model captures the major geologic trends more accurately than the *f*Gn model. This is consistent with the higher values of correlation coefficients observed for the *f*Bm model.

In addition to two-dimensional, cross sectional simulations, we also conducted three-dimensional field-scale simulations for the Burbank field. To compare the usefulness of additional data, two sets of simulations were conducted. One set had four conditioning wells; the other set had eight conditioning wells. The simulation using four conditional wells used wells 24, 49, 28, and 53 (Figure 5) as conditional wells and wells 27, 43, and 40 as test wells. We used both *f*Gn and *f*Bm structures for the horizontal variogram models. We also compared the results with a simple linear interpolation. The results of certainty coefficients are shown in Table 9.

As can be seen in Table 9, the correlation coefficients are very small, indicating a very poor match in the presence of sparse data. The linear interpolation method performs as good as the *f*Bm model method.

To assess the importance of additional data, we selected eight conditioning wells, 24, 28, 33, 35, 41, 43, 49, and 53 (Figure 5), and five test wells, 25, 26, 27, 34, and 42. We used the same procedure as was

Table 5. Certainty coefficients and computer use for two-dimensional stochastic conditional simulations, well B2, Hugoton field

Technique	Correlation fGn Structure	Correlation fBm Structure	Computer Use(s)
Interpolation with Normal Algorithm	0.136443	0.040206	4248.50
Interpolation with Greedy Algorithm	0.165463	0.014462	611.12
Normal Simulation	0.114589	0.002425	5573.60

Table 6. Correlation coefficients for two-dimensional stochastic conditional simulation runs for Hugoton field

Test Well	fBm Structure	fGn Structure
F2	0.102891	0.281018
N2	0.091213	0.132636

Table 7. Summary of simulation parameters for two-dimensional stochastic conditional simulation, well pair 28 and 50, Burbank field

Distance Between Well Pair	933 ft
Variance of Difference	0.0015
Average H Value	0.8051
Blocks in x Direction	40
Blocks in z Direction	24
Blocks in y Direction	1
Smoothing Factor	3 ft
Scaling Factor (fGn), V_H	2.3739E-03
Scaling Factor (fBm), V_H	1.8614E-08

Table 8. Correlation coefficients for two-dimensional stochastic conditional simulation runs, Burbank field

Test Well	fBm Structure	fGn Structure
35	0.529878	0.353443
26	0.386122	0.291828

Table 9. Correlation coefficients for three-dimensional stochastic conditional simulations for four conditioning wells, Burbank field

Well	Correlation fGn Structure	Correlation fBm Structure	Correlation Interpolation
40	−0.45	−0.04	−0.32
27	−0.22	0.09	0.08
43	−0.01	0.02	0.02

Table 10. Correlation coefficients for three-dimensional stochastic conditional simulations for eight conditioning wells, Burbank field

Well	Correlation fBm Structure	Correlation fGn Structure	Correlation Interpolation
25	0.23	−0.08	−0.02
26	0.05	−0.06	0.07
27	0.13	−0.21	0.06
34	0.49	0.21	0.43
42	0.12	−0.05	0.09

used for the four conditioning wells. The results of correlation coefficients are shown in Table 10. The fBm model in the horizontal direction outperforms the other two methods. Surprisingly, the linear interpolation does a better job than the fGn model in the horizontal direction.

Based on these results, observe that the fBm model is a good choice for modeling the variogram in the horizontal direction. Also, an increase in the amount of data increases the certainty coefficient prediction significantly.

Loudon Field

In addition to the Burbank field, another sandstone field was selected for further validation. Loudon field, located in Illinois, is in a fluvially dominated deltaic region.

The two-dimensional results are reported in Shibli (1992). Based on the two-dimensional results, we decided to use the fBm model to describe the horizontal variogram structure. Figure 8 shows the well loca-

tions. The solid well symbols indicate the conditioning wells (433, 437, 453, and 504), and the cross-hatched well symbols indicate the test wells. Table 11 presents the results of conditional simulation for this field in terms of correlation coefficients.

Although the results of correlation coefficients are not impressive, the fBm model does a better job than the fGn model in describing the horizontal variogram. Also, although the correlation coefficients

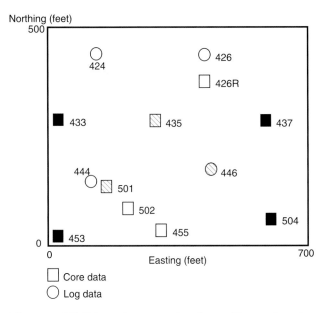

Figure 8. Well location map for three-dimensional Loudon field simulation.

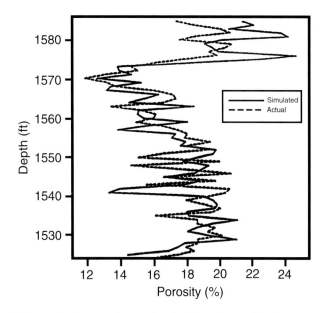

Figure 9. Comparison plot for Loudon well 501 using an fBm horizontal correlation structure and greedy algorithm.

Table 11. Correlation coefficients for three-dimensional stochastic conditional simulation runs, Loudon field

Well	Correlation fBm Structure	Correlation fGn Structure
426r	0.083683	0.060136
446	−0.105691	−0.116193
501	0.225909	−0.144635

indicate smaller values, the visual comparison indicates a good match between the simulated and the observed porosity values (Figure 9).

Based on the simulations of two sandstone and two carbonate fields, it seems more likely that the use of fGn structure for carbonate reservoirs to describe the horizontal spatial relationship will result in a better prediction of interwell distribution of the reservoir properties; however, the use of fBm structure for sandstone reservoirs to describe the horizontal spatial relationship will result in a better prediction of the interwell distribution of the reservoir properties.

CONCLUSIONS

Estimating the intermittency exponent for the vertical well-bore data is an important step for using fractal models to describe reservoir properties. Where seasonal effects are observed in the data trace, the seasonal bias is removed using a normalizing technique that involves eliminating the local means. The distribution of exponent values H before and after removing the sample means is compared.

The results indicate that if the seasonal effects are unusually strong, removing the local means will result in a narrower distribution of H exponent values. Actual plots of variogram values favor the box-counting method to give the most reliable fit. The results indicate that after seasonal adjustments, the range of H exponent values calculated using the box-counting method narrows to between 0.75 and 0.80 regardless of the type of sedimentary environment; therefore, in the absence of data, this value can be used for the stochastic conditional simulation of a reservoir.

The section on conditional simulations presents results for the stochastic conditional simulations of three fields, a carbonate reservoir and two sandstone reservoirs. This section also delineates a set of procedures to be followed in the stochastic conditional simulation of the rock properties. Factors that should be taken into account include removing the nonstationarity in the vertical well data and selecting a suitable horizontal correlation model based on the correlation structure determined from vertical well data. The results seem to indicate that using an fGn model in the horizontal direction for a carbonate field results in better predictions than using an fBm model; however, in a sandstone environment, the reverse is apparent. Techniques for decreasing the extensive computer time associated with most simulated annealing algorithms are also presented. These techniques include generating an initial pattern using bilinear interpolation and adding a random Gaussian element, and using only the greedy version of the annealing algorithm. The results obtained using these enhancements seem to provide accurate and reliable predictions of the actual rock properties, as well as saving as much as 85% computing time.

REFERENCES

Beier, R. A., and H. H. Hardy, 1993, Comparison of different horizontal fractal distributions in reservoir simulations, reservoir characterization III: Tulsa, Oklahoma, PennWell, p. 271–296.

Ben-Rached, L., 1991, An application of geostatistical techniques for reservoir performance prediction: M.S. Thesis, The University of Tulsa, Tulsa, Oklahoma.

Deutsch, C. V., and A. G. Journel, 1992, GSLIB: geostatistical software library and users' guide: New York, Oxford Press, p. 154–160.

Doyen, P. M., et al., 1989, Monte Carlo simulation of lithology from seismic data in a channel sand reservoir, SPE Paper 19588, presented at the SPE Annual Technical Conference and Exhibition.

Emanuel, A. S., G. K. Alameda, R. A. Behrens, and T. A. Hewett, 1989, Reservoir performance prediction methods based on fractal geostatistics: SPE Reservoir Engineering Journal, August, p. 311–318.

Farmer, C. L., 1988, The generation of stochastic fields of reservoir parameters with specified geostatistical distributions, Mathematics in Oil Production (S. Edwards and P. King, editors), p. 235–252.

Farmer, C. L., 1991, Numerical rocks, the mathematical generation of reservoir geology (J. Fayers and P. King, editors), Oxford Press.

Feder, J., 1988, Fractals: New York, Plenum Press.

Hagen, K. B., 1972, Mapping of surface joints on air photos can help understand waterflood performance problems at North Burbank unit, Osage and Kay counties, Oklahoma: M.S. Thesis, The University of Tulsa, Tulsa, Oklahoma.

Hewett, T. A., 1986, Fractal distributions of reservoir heterogeneity and their influence on fluid transport: SPE Paper 15386 presented at the 1986 SPE Annual Technical Conference and Exhibition, New Orleans.

Hewett, T. A., and R. A. Behrens, 1990, Conditional simulation of reservoir heterogeneity with fractals: SPE Formation Evaluation Journal, September, p. 217–225.

MacAllister, D. J., 1987, Reservoir and fluid properties for the Sable unit: ARCO Internal Report.

Perez, G., 1991, Stochastic conditional simulation for description of reservoir properties: Ph.D. Dissertation, The University of Tulsa, Tulsa, Oklahoma.

Perez, G., and B. G. Kelkar, 1992, Effectiveness of conditional simulation to describe carbonate reservoir properties: SPE Paper 23971, presented at the SPE Permian Basin Oil and Gas Recovery Conference, Midland.

Shibli, S., 1991, An approach to generating reservoir properties distribution using stochastic conditional simulation method: M.S. Thesis, The University of Tulsa, Tulsa, Oklahoma.

Szpakiewicz, M. J., K. McGee, and B. Sharma, 1986, Geological problems related to characterization of clastic reservoirs for enhanced oil recovery: SPE/DOE Paper 14888 presented at the 1986 SPE/DOE Fifth Symposium on Enhanced Oil Recovery, Tulsa.

Wallis, J. R., and N. C. Matalas, 1970, Small sample properties of H and K estimators of the Hurst coefficient H: Water Resource Research, v. 6, p. 1583–1594.

APPENDIX 1

Nomenclature

C = certainty coefficient
h = lag distance
H = intermittency exponent
N = number of sample points in a vertical well
V = simulation variable
V_H = scaling factor for fGn and fBm models
x = location of a point
γ = semivariogram
σ^2 = variance

Geostatistical Modeling of Chalk Reservoir Properties in the Dan Field, Danish North Sea

Alberto S. Almeida
Stanford Center for Reservoir Forecasting
Stanford University
Stanford, California, U.S.A.[1]

Peter Frykman
Geological Survey of Denmark
Copenhagen, Denmark

ABSTRACT

A geostatistical study of the uppermost Maastrichtian chalk reservoir unit in the Dan field was performed. The ready availability of directional and horizontal wells, uncommon in reservoir characterization studies, allows better inference of the horizontal correlation structure in this special type of reservoir.

The objective was to obtain stochastic images of porosity and permeability properties of the reservoir. Because these parameters are spatially interrelated, a cosimulation algorithm in which these properties are jointly simulated was applied. This algorithm, called the Gaussian collocated cosimulation algorithm, was built on a Markov-type hypothesis, whereby collocated secondary information is assumed to screen out farther away secondary data. This method allows us to perform the direct cosimulation of several interdependent variables integrating several sources of soft information.

The abundantly available wireline log–calculated porosity data from the wells were used as soft data, being completed into the regular grid by using a separate simulation step. The resulting stochastic models reflect both the short- and long-range structures, and the high correlation between porosity and permeability. The fine-scale stochastic images are useful for studies on upscaling techniques.

INTRODUCTION

Danish oil and gas production comes exclusively from the North Sea Central Trough, a system of Jurassic half-grabens located 250 km west of the coast of Denmark.

The Dan field is a broad domal anticline comprising Upper Cretaceous to lower Danian chalk limestone (Danian and Maastrichtian), largely unfractured. The discovery well was drilled in 1971 and production started in 1972. Daily production averages

[1]Present address: Petrobras S.A./Cenpes, I. do Fundao, Rio de Janeiro, RJ 21949-900, Brazil.

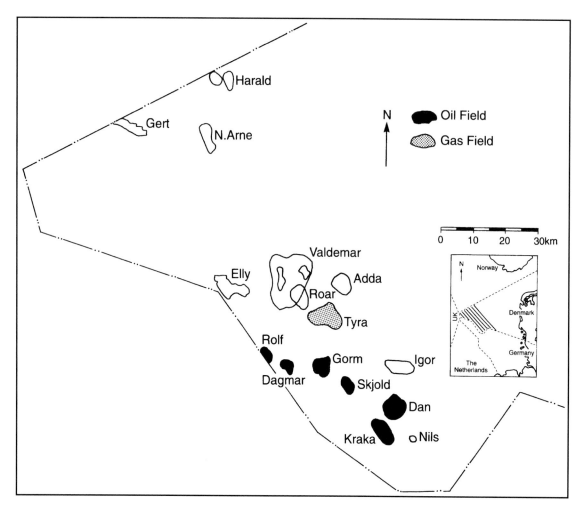

Figure 1. Location map for producing fields (oil or gas indicated) and discoveries in the Danish Central Trough. Dan field is located in the southeastern part of the Danish Central Trough (after Andersen and Doyle, 1990).

30,000 BOPD. Ultimate recovery is expected to be low due to limited natural fracturing and very low permeability (average 1 md). Induced hydraulic fracturing and various stimulation techniques have been tried to enhance recovery. The first North Sea horizontal well (500 m) was drilled there in 1986, followed by many others (Andersen et al., 1990; Fine et al., 1993; Damgaard and Wæver, 1993). Not only do these horizontal wells improve production rates, but they also provide crucial information about horizontal continuity of fracturing and heterogeneities.

GEOLOGY

The Dan field (Figure 1) is a broad domal anticline resulting from uplift by a salt pillow of Zechstein age. The anticline is dissected by a major northeast-southwest normal fault downthrown to the northwest (Figure 2). The fault has approximately 100 m of throw near the crest of the structure and dies out in the synclines on the northeastern and southwestern edges of the field. Smaller associated normal faults are present in the downthrown block.

The reservoir is found at a depth of about 2000 m in Upper Cretaceous to lower Danian chalk limestone. The reservoir contains a saturated oil accumulation overlain by an initial gas cap. Ultimate recovery is expected to be low (present estimate about 18%) due to the low permeability and a low degree of fracturing of this field. Other chalk fields in the North Sea have much higher levels of fracturing, and accordingly higher recovery estimates, as well as production rates.

The hydrocarbon-bearing rocks of the Dan field are chalk limestones similar to the Ekofisk and Tor formations of the Norwegian North Sea. The porosity average of 25–30% shows a significant decrease from the crest of the field toward the flanks of the dome.

Figure 3 shows a typical vertical porosity profile across the producing zone of the Dan field. This profile shows the clear discontinuity in porosity between the Danian (depth above 6450 ft or 1965 m in the M-10x well) and the Maastrichtian chalk.

Chalk sediments are interpreted to be deep shelf deposits of pelagic carbonate materials derived from microscopic calcareous planktonic algae, Coccolithophorides. The calcareous skeletons of these algae are

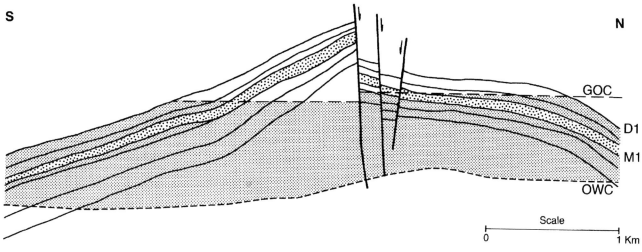

Figure 2. South-north cross section showing main structure, the units in the uppermost chalk, and the tilt of matrix oil/water contact (OWC) in the Dan field (after Megson, 1992). The unit designated M1 on the figure corresponds closely to the R1 unit described in this paper. GOC = gas/oil contact.

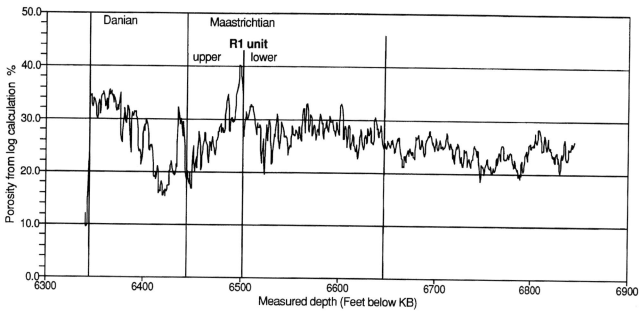

Figure 3. Vertical porosity profile of well M-10x in the Dan field across the producing zone showing the two-part (upper and lower) R1 unit in the uppermost Maastrichtian.

called coccospheres, and are composed of wheel-shaped coccoliths built of calcite platelets that are 0.5–2 μm in size (Figure 4). These algae were pelagic, and constituted an important nutrient source for the other pelagic organisms like small crustaceans. Eaten and digested by these, the coccolithophorid remains probably arrived at the sea bottom as fecal pellets wrapped in thin organic films.

The sea bottom had extremely high porosity in its upper layer due to the loose packing of these pellets; the coccolith ooze was very unstable, and easily activated into mud flows. The unstable sediment combined with the structural development of salt structures at

early stages of the sedimentation, creating a pattern where a very low angle of slope could cause the chalk sediment to move causing nondeposition or erosion (Nygaard et al., 1983).

The chalk sediment has a very homogeneous lithology, and the difference between sedimentary facies is very subtle. This homogeneity makes it difficult, if not impossible, to divide a sequence into easily recognizable facies that can be used to outline lithologic geometry; furthermore, the diagenetic overprint is considered to be very important, and probably controls much of the present petrophysical properties of the chalk rock. The sediment no doubt is highly layered

Figure 4. Scanning electron microscope photomicrograph of chalk from Dan field well showing the very fine grain size, and the very narrow pores in this type of rock. Remnants of coccospheres are still visible. White scale bar is 10 μm long.

with a high width:thickness ratio, which is expected to be reflected in the distribution of porosity and permeability values.

The present work on modeling porosity and permeability is limited to the uppermost Maastrichtian

reservoir unit. This reservoir unit, designated R1, corresponds closely to the most productive interval, including the M1a and M1 units as shown by Megson (1992) and Jørgensen (1992). Picks for the top and bottom of R1 are taken on the porosity wireline logs along the wells, and are used as the main data for reservoir division and correlation in this chapter.

The uppermost part of R1 [comparable to M1a of Jørgensen (1992)] is present only on the western part of the structure. Only a minor lithological difference seems to exist between the two parts. Generally, a porosity variation exists within the upper part of R1, being less porous in the upper section and more porous in the lower section, when compared with the porosity in the lower part of R1.

The data for this study come from 36 selected wells on the structure (Figure 5), one of which is horizontal. Of the 35 normal wells, 8 have additional core data within the R1 unit. The horizontal wells have the potential to supply information on the horizontal correlation lengths of the parameters investigated. Although they were not drilled consistently and precisely along the layers present in the chalk, they are close enough to being parallel to these layers that they can be used to analyze horizontal continuity.

The area chosen for this study is the northwestern part of the Dan field. The Dan field is divided by a main fault that has uplifted the southeastern block,

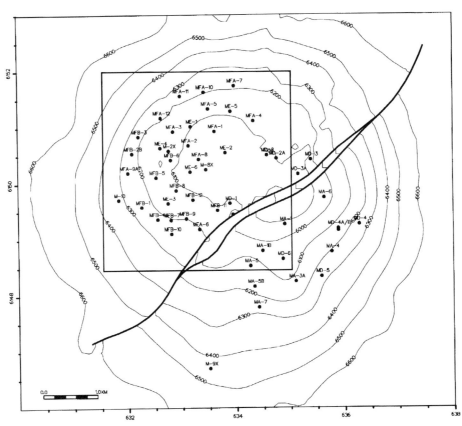

Figure 5. Simplified structural map showing the main fault, well locations, and the area selected for the detailed modeling. Contour = 100 ft.

thus restricting communication between the reservoirs on the two blocks. The study area has been limited to cover 3500×3500 m^2 of the northwestern part of the field (Figure 5).

STRUCTURAL TRAP

The Dan field dome is a simple structural trap with oil and gas in the Danian/Maastrichtian chalk. An extensive gas cap exists in the Danian reservoir, whereas the Maastrichtian contains only a limited gas cap. A variation in the level of the gas-oil contact across the major northeast-southwest fault (about 10 m) indicates the fault to be partly sealing (Megson, 1992).

The reservoirs have a long zone (more than 120 m) with high oil saturation (S_w <20%). The underlying transition zone commonly is more than 65 m long, and probably is a result of the low permeability and the capillary effect of the chalk. The oil-water contact is dipping gently (less than 1° dip) toward the south–southwest (Jørgensen, 1992).

EXPLORATORY DATA ANALYSIS

The objective was to perform a basic statistical spatial analysis to detect any particular aspect of the data, such as trends, in the study area that could offer valuable information for the subsequent geostatistical study.

The available data for this study include:

1. Thirty-five normal wells (some of these "vertical" wells are deviated up to 52° in the chalk)
2. One horizontal-well
3. Core analysis (porosity and permeability) from eight vertical wells
4. Porosities for the 36 wells obtained from interpretation of density well logs
5. Geologic information (structural maps, geologic sections, and interpretation)

Figure 6 shows the histograms of core permeability, core porosity, and porosity values derived from log interpretation. These data are related only to the R1 unit. The permeability appears lognormally distributed with mean 2.45 md and coefficient of variation 0.68. Core porosity and log porosity have Gaussian-like distributions with mean 29.62% and coefficient of variation 0.15 for core porosity, and mean 29.60% and coefficient of variation 0.16 for log-derived porosity.

In analyzing core data, we noted some abnormal values for permeability. These values are above 10 md, falling outside the main correlation trend with the core measured porosity. These five permeability measurements were reevaluated, and are considered to be caused by fractures in the core material. These values have been excluded from the data set throughout the study.

Figure 7 shows a relatively good relationship among the three petrophysical parameters, core porosity, core permeability, and log porosity. The correlation coefficients are 0.70 or better.

The data have been analyzed to detect any possible trend related to depth or structural position. The trends seen are not substantial enough to justify special treatment during simulation. The soft (log-calculated porosity) and the hard data conditioning the simulations will ensure that any general trends are reflected in the final realizations.

STRATIGRAPHIC COORDINATES

The simulation of the petrophysical variables (core permeability, core porosity, log-calculated porosity) are done with stratigraphic coordinates to ensure that the horizontal spatial continuity measure refers to the layer geometry rather than to some constant vertical elevation.

For this purpose, we have interpreted the thickness variations to be caused by erosion of the sequence in parts of the reservoir. Missing values are inserted at the stratigraphic positions where the erosion took place. In this case, the stratigraphic coordinate is determined by

$$z' = z - bot(x,y)$$

Where $bot(x,y)$ is the elevation of the bottom of the sequence at a point with horizontal coordinates (x,y). The simple case where the sequence has been eroded only from the top is handled by building up the data from the base level, and inserting missing values if data do not reach the maximum elevation (maximum thickness). Once the simulation is completed the simulated values are relocated back to the Cartesian coordinates.

To find the elevation relative to the base of R1 for all the data, the structural maps of the top of the Maastrichtian and the base of R1 were used. These maps (comparable to Figure 5) were supplied as gridded data in 125×125-m grid blocks. The area mapped is larger than the area used for the simulation.

For the transformation of coordinates into stratigraphic grid, we need top and bottom R1 maps on a fine grid of 10×10 m. To this purpose, the maps at 125-m resolution (Figure 8) were further interpolated by kriging into a resolution of 10×10 m. When compared with the original maps, the smoothing effect of kriging is evident particularly along the fault.

VARIOGRAPHY

Figure 9 shows the variograms in the vertical direction for the normal scores of the core porosity, core permeability, and log-calculated porosity data from the whole data set. In all these variograms a first range of 12 m is evident. The model adopted is a spherical model with 20% nugget and a range of 12 m.

The hole effect observed at distances around 24 m is due to vertical layering. This effect was not modeled, however, because the vertical effective search

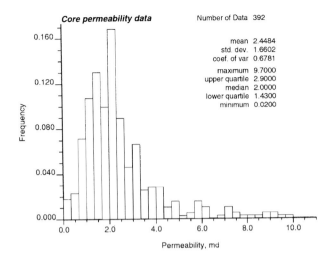

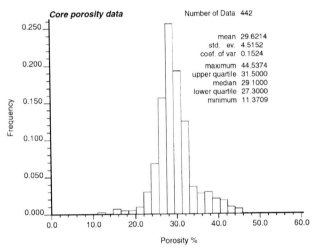

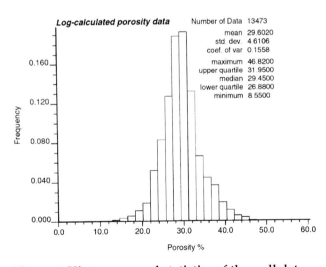

Figure 6. Histograms and statistics of the well data used in the simulations.

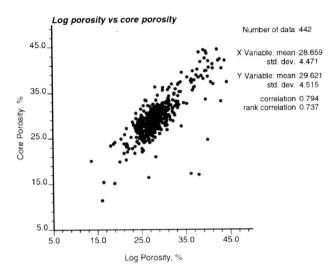

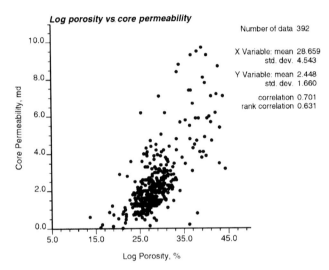

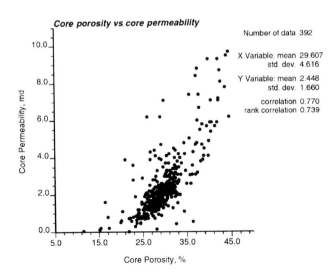

Figure 7. Crossplots of the petrophysical data in the wells. The correlation between log-calculated porosity and core analysis porosity is fairly good, despite the inherent averaging in the wireline log measurement of porosity.

Top of unit R1

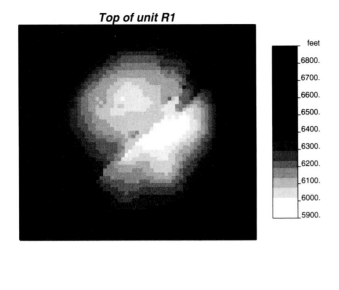

Base of unit R1

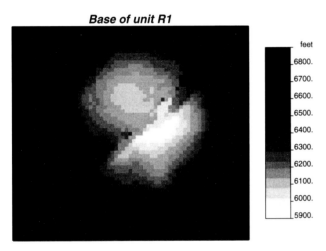

Figure 8. Gray scale maps of top and base of R1 unit in the 125-m block mapping grid. Slight smoothing of the main fault relief will be induced by the kriging used to obtain the fine-gridded map data from this coarse-grid original map.

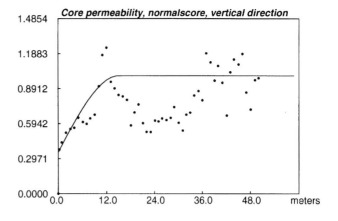

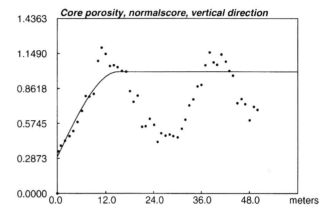

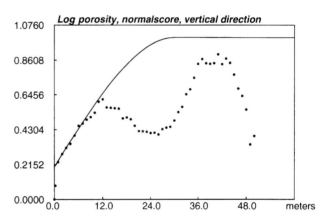

Figure 9. Vertical variograms for the normal scores of core porosity, permeability, and log porosity all showing a range of around 12 m.

distance for future kriging will be limited to 10 m. Note that the sill (variance) of normal scores variograms has to be 1.

The well spacing (approximately 500 m) and the lack of core porosity and core permeability data make a conventional three-dimensional (3-D) variogram analysis impossible; however, there is enough log-calculated porosity data to provide an idea about the long-range horizontal correlation structure. Figure 10 shows a horizontal variogram of the log porosity data set from the 35 wells. The behavior of the variograms in different directions is similar, indicating horizontal isotropy. The long-range structure can be fitted with a range of approximately 1100 m.

To supply additional information on this horizontal structure, a horizontal well was analyzed for spatial correlation structure in the log porosity data. A pseudo-"horizontal" variogram has been calculated

using drilled depth as the variable for lag distance to account for the slightly varying angle of the near-horizontal well bore relative to the layers.

The well shows a long correlation range that was modeled by a spherical structure with 320-m range (Figure 11). This gives an anisotropy ratio of approximately 1:27 when compared to the vertical range of 12 m.

Considering that this well does not provide a fair representation of the correlation range in the direction exactly parallel to the layers, this ratio was regarded as too small. To model the horizontal correlation, that

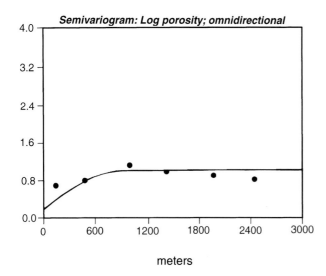

Figure 10. Horizontal omnidirectional variogram of log porosity derived from full data set of vertical wells.

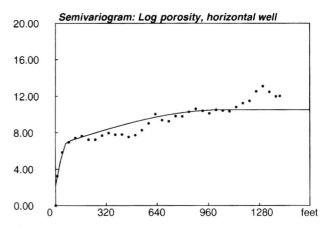

Figure 11. Log porosity pseudo-"horizontal" variogram from horizontal well data, fitted with a model having a short-range structure of 24 m and a long-range structure of 320 m.

anisotropy ratio was increased to 1:120 for the longest range structure.

Based on this analysis, and because the diagenetic process is deemed to be the same for porosity and permeability, a unique variogram model was adopted for all petrophysical properties within the R1 unit of the Dan field:

$$\gamma\left(h_x, h_y, h_z\right) = 0.20$$

$$+ 0.20 sph \sqrt{\left(\frac{h_x}{24}\right)^2 + \left(\frac{h_y}{24}\right)^2 + \left(\frac{h_z}{12}\right)^2}$$

$$+ 0.60 sph \sqrt{\left(\frac{h_x}{1440}\right)^2 + \left(\frac{h_y}{1440}\right)^2 + \left(\frac{h_z}{12}\right)^2}$$

This 3-D anisotropic model has 20% nugget, a single short-range 12 m in the vertical direction and two ranges of 24 m and 1440 m in the horizontal directions. This model is not to be used for vertical distances greater than 12 m.

SELECTING THE SIMULATION TECHNIQUE

One important aspect of a geostatistical study is selecting the appropriate approach to perform the study. Depending upon the characteristics of the data, an indicator (nonparametric) or a Gaussian (parametric) approach or other techniques may be more suitable for stochastic imaging (simulation) of a field. For instance, if the variable under study presents different correlation ranges depending on whether it is low, median, or high valued, most likely an indicator approach will be more suitable. Otherwise, Gaussian techniques, which are quite simple and have a history

of successful applications, are in order (Deutsch and Journel, 1992).

When applying a Gaussian technique, the first step is to perform the normal score transform of the data. This transformation defines a new variable that is, by construction, (univariate) normally distributed. The Gaussian algorithm actually requires that this new variable is also multivariate normally distributed; therefore, before going ahead with the simulation, we checked this last hypothesis. This check consisted of comparing the experimental indicator variograms of the normal scores of the data to the theoretical ones, assuming a binormal distribution, for different thresholds. The check did not allow us to reject the hypothesis of multi-Gaussianity for the normal scores of the variables.

JOINT CONDITIONAL SIMULATION

Using the previously adopted Gaussian model, stochastic imaging of the petrophysical variability within the unit R1 of the Dan field can proceed. The petrophysical parameters here considered are porosity and isotropic permeability as measured from cores.

Because these parameters are spatially interrelated, simulating them independently would be inappropriate. A cosimulation algorithm, in which the spatial variability of the variables are conditioned by a prior model of covariances and cross-covariance, must be applied.

One has various options in performing such cosimulation. An algorithm based on a generalization of the sequential Gaussian technique for the simulation of a vectorial random function $z(u) = \{z_1(u), z_2(u), ..., z_k(u)\}$ could be derived. This algorithm would call for vector cokriging and simulation (Verly, 1992) and would have as main drawbacks (1) a usually tedious inference and modeling of the covariance/cross-covariance matrix, (2) the limitation of the linear model of coregionalization required to ensure positive definiteness of the cokriging matrix, and (3) computer-intensive numerical solution of a large system of equations.

Another procedure consists of drawing from successive conditional distributions (Almeida et al., 1992). This approach consists of simulating first the most important or the best auto-correlated variable, $z_1(u)$. Then, the other covariates $z_k(u)$ are simulated by drawing from the specific conditional distribution of $z_k(u)$ given the simulated collocated primary value $z_1(u)$. This approach is much simpler, but has the following severe drawbacks: (1) the spatial auto-correlation of the secondary variables, $z_2(u)$, $z_3(u)$,..., $z_k(u)$, are only indirectly reproduced through that of the primary variable $z_1(u)$, (2) the cross-correlation between the first secondary variable, $z_2(u)$, and the primary variable is reproduced by conditioning on only the collocated value $z_1(u)$, and (3) the cross-correlation between any two secondary variables, say $z_2(u)$ and $z_3(u)$, is indirectly approximated through their cross-correlation with the primary variable $z_1(u)$.

A new approach for joint simulation of several variables was used in this study (Xu et. al, 1992; Almeida, 1993). This approach, called sequential Gaussian cosimulation algorithm (SGCOSIM), allows direct cosimulation of several interdependent random variables without the tedious inference and modeling of a full cross-covariance matrix as required by a full cokriging-based algorithm. It also allows the integration of different sources of soft information.

The algorithm, which can be seen as a generalization of the traditional sequential Gaussian simulation, incorporates two key ideas. (1) The collocated cokriging idea reduces the computational effort involved in the simulation process, and makes the cokriging matrix more stable. This idea amounts to retaining only the secondary datum closest to the location where the primary variable is to be estimated, e.g., the collocated secondary datum. (2) The Markov-type model of coregionalization simplifies the inference and modeling of the cross covariances. This model relies on that the closest secondary-type datum screens the influence of all further away data of the same type.

This approach requires the soft (secondary) information to be available at all locations being simulated. When these data are not dense enough, a presimulation of this soft information is done before applying the collocated cosimulation algorithm.

APPLICATION

For the present study, the core-based isotropic permeability is considered as the first primary variable to be simulated because it is the most crucial parameter for the flow simulations to be performed later. The other primary variable to be jointly simulated is the porosity. The wireline log–calculated porosity available from all wells used in this study is considered as the soft or secondary data. The study area of 3500 × 3500 m, with a maximum thickness of 63 m, was covered by a grid with 100 × 100 × 42 nodes for the stochastic simulation, corresponding to a cell size of 35 × 35 × 1.50 m.

For the application of the present cosimulation algorithm, the soft data must be available at all points to be simulated. The well log–derived porosity was used here as soft information because it reflects averaged valued information. Because the log-calculated porosity is available only along the 35 wells, it was necessary to complete that information with a prior simulation. A sequential Gaussian simulation algorithm (Deutsch and Journel, 1992) was applied to simulate the normal scores of log porosity into the same fine grid size as was used for the later cosimulation.

Because the size of the simulated area (3500 m) is not much larger than the maximum horizontal correlation range (1440 m) we decided to perform the simulation in two steps to guarantee that the larger scale features of the spatial continuity model (variogram) were correctly reproduced (Deutsch and Journel, 1992). These two steps are as follows:

1. Create a conditional simulation of the log porosity over a coarse grid (175 ¥ 175 ¥ 3 m) conditioned on the well data using a large data search ellipsoid (1600 ¥ 1600 ¥ 12 m). This large search neighborhood allows reproduction of the large-scale structure of the horizontal variogram, yet, because of the coarse grid, it does not contain too many nodes to be searched (Figure 12).
2. Create a conditional simulation of porosity values over the final dense grid (35 ¥ 35 ¥ 1.50 m) using a smaller search neighborhood (800 ¥ 800 ¥ 12 m) and only the values simulated in the first step as conditioning data. This second step ensures the reproduction of the small scale structures of continuity (Figure 13).

The variogram model adopted for the normal scores is spherical with 20% nugget effect, an isotropic short horizontal range of 24 m, an isotropic long horizontal range of 1440 m, and a vertical range of 12 m.

The variogram of the simulated log porosity values is given in Figure 14. The variogram correctly reproduces the short- and long-scale ranges of the input model. Once the log-calculated porosity is available at all grid points the cosimulation algorithm can be applied. The joint simulation of permeability and porosity was performed in two steps for the reason previously stated. In the coarse grid, the search ellipsoid radii are 1600 × 1600 × 12 m, whereas for the fine grid the search ellipsoid radii are 800 × 800 × 12 m.

Core porosity data and core permeability data were not great enough to provide a reasonable inference and modeling of the horizontal variogram; therefore, the variogram model used for all the variables involved in the process (core permeability, core porosity, log-calculated porosity) was that used for log porosity.

The cross-covariance between these variables was obtained through the Markov-model relation and the correlation coefficients used for this Markov model are given in Table 1.

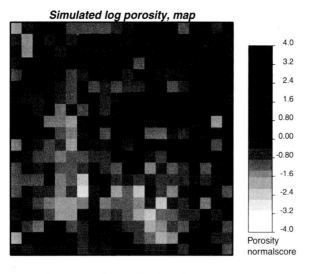

Simulated log porosity, map

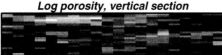

Log porosity, vertical section

Figure 12. Map and a vertical section (10× exaggeration) of log-porosity soft-data simulation in coarse grid with an anisotropy ratio of 1:120. These data are simulated in a first step with conditioning from the log-calculated porosities from all 35 wells using the same correlation structure as in the later cosimulations.

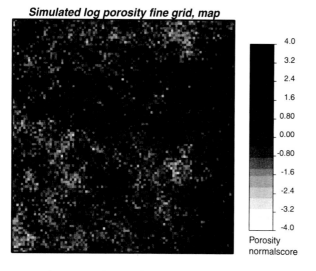

Simulated log porosity fine grid, map

Log porosity, vertical section

Figure 13. Map and a vertical section (10× exaggeration) of log-porosity soft-data simulation in fine grid with an anisotropy ratio of 1:120. These data are simulated in a second step with conditioning from the porosities simulated in the coarse grid using the same correlation structure as in the later cosimulations. The long-range correlation structure is still reflected and the short-range variations have been added.

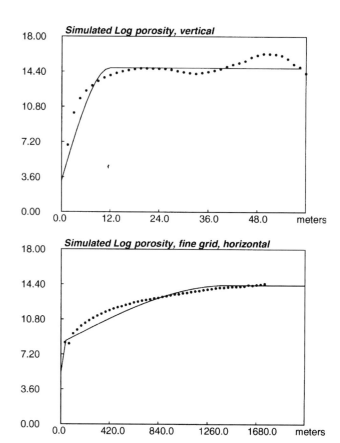

Figure 14. Variograms of simulated log-porosity soft data in fine grid in the horizontal and vertical direction (dots = experimental, line = model). The variograms show good reproduction of the structure.

Table 1. Comparison of sample correlation values and correlation values of the simulated values in the original space

Variables	Sample Correlation	Simulation
Permeability vs. porosity	0.77	0.71
Permeability vs. log porosity	0.70	0.61
Porosity vs. log porosity	0.79	0.73

DISCUSSION OF RESULTS

Figure 15 shows the histogram of the sample permeability data used for conditioning the simulation and the histograms of the simulated permeability values in both the coarse- and the fine-grid models. As can be seen, the distribution is correctly reproduced accounting for expected statistical fluctuations.

Figure 16 shows a comparison of the scatter plots for the sample data and the simulated values of porosity/permeability. As can be seen also from Table 1, the reproduction of the original data correlation is reasonable.

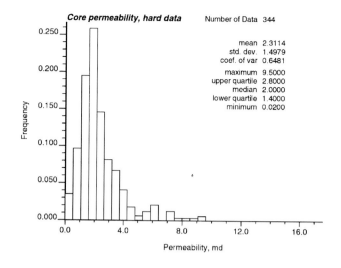

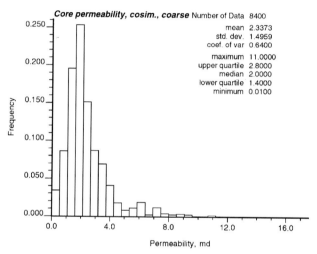

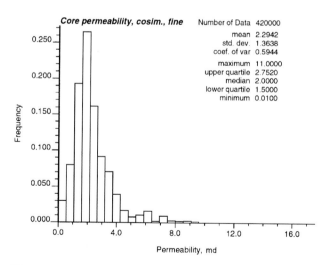

Figure 15. Histogram statistics of the original permeability data set and of the simulated values in both the coarse and the fine grid, and both showing good reproduction.

The scatter plot of the simulated values in the original space after backtransform (Figure 16) shows a pronounced banding in the high values. This banding

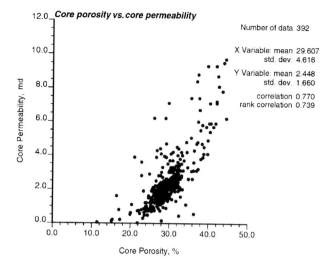

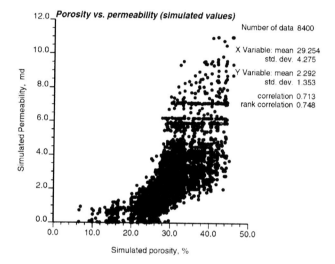

Figure 16. Comparison of crossplots of porosity/permeability from the original conditioning core data and the simulated values showing very good reproduction of the trend and correlation statistics. The banding in the simulated data set is an artifact created by the backtransformation procedure.

is an artifact due to the algorithm used for the backtransformation. This algorithm uses a transformation table, created in the initial step of transforming the original data to normal scores, for interpolating between original data values. If there are not enough classes or values in this table, then simulated normal scores that differ little will be backtransformed to the same original value.

Figure 17 shows an example of the simulated porosity distribution for a given vertical section. In this figure, the color scale corresponds to a twentieth percentile. Also in Figure 17, the geologic grid has been reestablished, and the steeply dipping part in the middle, in fact, is the major fault plane that has been smoothed by kriging of the structural maps used for grid relocation. The relocation also includes erosion of some of the upper part of the unit with

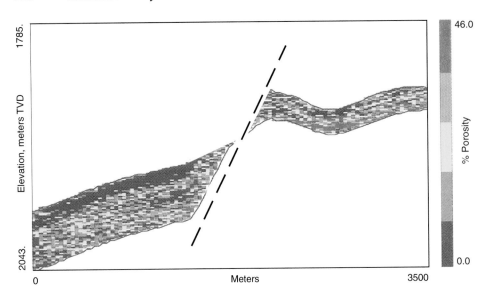

Figure 17. Simulated porosity distribution in vertical
section deformed back to
structural position via the
structural maps for the top
and the bottom of the unit.
The steep middle section
represents the fault zone,
which was smoothed dur-
ing the kriging of the original structural maps.

poorer reservoir properties. Reproduction of the
high correlation between the porosity and permeability is observed when comparing corresponding
realizations.

For this study just one realization was performed;
however, we do not expect much variability from one
realization to another due to the homogeneity of the
R1 unit.

The values for the log-calculated porosity were
taken from the well trace and plotted against the
nearest simulated value in the model (Figure 18).
This plot shows the ability of the simulation in
reproducing the major trends in the porosity pattern
and, at the same time, the amount of very short
scale variability. Observe that in some wells the fit
is not as good because the well is not exactly coincident with the grid simulation. Also, the well
log–derived porosity used here as comparisons
were considered as soft information in the simulation procedure.

CONCLUSIONS

This study presents a typical geostatistical
approach to reservoir modeling. The study was
restricted to the geologically homogeneous R1 unit of
the Maastrichtian formation of the Dan field. The data
available for this study came from 36 wells. The data
consisted of log-calculated porosity from wireline
logs in wells, core porosity and core permeability data
available from 8 wells, structural maps, geologic sections, and geologic interpretation.

The exploratory analysis identified that the permeability sample data appears lognormally distributed,
whereas the core porosity and log-calculated porosity
have Gaussian-like distributions. No major trends
were identified from the data available. Based on geologic information that shows evidence of erosion in
the upper sequence of the unit, the stratigraphic coordinate transformation assumed an erosion model.
According to this model, each sample information has

its "z" coordinate computed as the vertical distance
from the base of the unit.

The data used in this study satisfied checks of
multi-Gaussianity, making it possible to apply the collocated cosimulation algorithm.

The availability of directional and horizontal wells,
uncommon in reservoir characterization studies,
allows better inference of the horizontal variograms.
The vertical variogram showed a hole effect at distances around 24 m due to vertical layering; however,
this effect was not modeled because the vertical effective search distance was limited to 12 m.

A better reproduction of this feature would be
obtained by modeling the hole effect, but we considered the conditioning information enough to reproduce the vertical layering.

To apply the collocated cosimulation algorithm, the
soft information, in this case the log-calculated porosity, must be available from all over the field; therefore,
this parameter was completed by simulation prior to
the cokriging required by the collocated cosimulation
algorithm. Because the size of the simulated area is not
much larger than the maximum horizontal correlation
range, the simulation was performed in two steps.

The first step considered a conditional simulation
over a coarse grid conditioned on the well data,
using a large data search ellipsoid. The second step
considered a conditional simulation over the final
dense grid, conditioned to the values simulated in
the first step, using a smaller search neighborhood.
This procedure guarantees that the larger scale features of the spatial continuity model are correctly
reproduced. Yet, the procedure is quite fast because
the step that considers the large search neighborhood does not have too many points to calculate
(coarse grid).

The main reason for using a conditional simulation
technique was to avoid the smoothing effect inherent
in the estimation techniques. The result of this study
confirmed the homogeneous character of the reservoir
with a high correlation between porosity and permeability. We do not expect different realizations (differ-

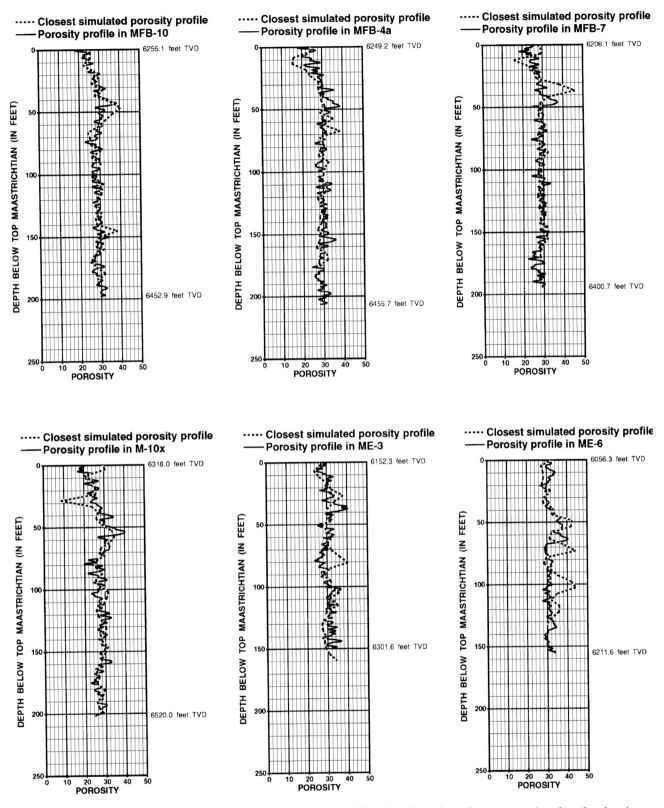

Figure 18. Vertical profiles of log porosity from selected wells plotted against the nearest simulated value in the gridded model. Although the detailed variations are not reflected, the general trends are reproduced in the simulation.

ent images) of the petrophysical parameters to have much of an impact on the flow simulation study because of the extreme homogeneity of the R1 unit; however, the availability of such fine-scale stochastic images would be useful for applying any upscaling techniques, a necessary step in any flow simulation.

ACKNOWLEDGMENTS

We thank Andre Journel and Mohan Srivastava at Stanford University and Lars Kristensen and Thomas Dons at the Geological Survey of Denmark for support with discussions, programming, and data supply.

REFERENCES

Almeida, A. S., 1993, Joint simulation of multiple petrophysical properties using a Markov-type coregionalization model: Stanford Center for Reservoir Forecasting, Report 6, Stanford, 81 p.

Almeida, A. S., L. Cosentino, and T. Tran, 1992, Generation of Stanford-1 reservoirs: Stanford Center for Reservoir Forecasting, Report 5, Stanford, 31 p.

Andersen, C., and C. Doyle, 1990, Review of hydrocarbon exploration and production in Denmark: First Break, v. 8, no. 5, May 1990, p. 155–165.

Andersen, S. A., J. M. Conlin, K. Fjeldgaard, and S. A. Hansen, 1990, Exploiting reservoirs with horizontal wells: the Mærsk experience: Schlumberger Oilfield Review, v. 2, no. 3, p. 11–21.

Damgaard, A. P., and M. M. Wæver, 1993, Reservoir exploitation using multipurpose horizontal wells—case studies from Danish chalk fields: 5th International Conference on Horizontal Well Technology, Proceedings, 20 p.

Deutsch, C. V., and A. G. Journel, 1992, GSLIB: geostatistical software library User's Guide: Oxford, Oxford University Press, 340 p.

Fine, S., M. R. Yusas, and L. N. Jørgensen, 1993, Geological aspects of horizontal drilling in chalks from the Danish sector of the North Sea, in J. R. Parker, ed., Petroleum geology of north-west Europe: Geological Society of London, p. 1483–1490.

Jørgensen, L. N., 1992, Dan field—Danish North Sea, in E. A. Beaumont and N. H. Foster, compilers, Structural Traps VI, Atlas of oil and gas fields: AAPG Treatise of Petroleum Geology, p. 199–215.

Jørgensen, L. N., and P. M. Andersen, 1991, Integrated study of the Kraka field: SPE paper 23082, Offshore Europe Conference, Aberdeen, p. 461–474.

Journel, A., W. Xu, and T. Tran, 1992, Integrating seismic data in reservoir modeling: the collocated cokriging alternative: Stanford Center for Reservoir Forecasting, Report 5, Stanford, 34 p.

Megson, J. B., 1992, The North Sea chalk play: examples from the Danish Central Graben, in R. F. P. Hardman, ed., Exploration Britain: geological insights for the next decade: Geological Society Special Publication No. 67, p. 247–282.

Nygaard, E., K. Lieberkind, and P. Frykman, 1983, Sedimentology and reservoir parameters of the Chalk Group in the Danish Central Graben: Geologie en Mijnbouw, v. 62, p. 177–190.

Nygaard, E., C. Andersen, C. Møller, C. K. Clausen, and S. Stouge, 1990, Integrated multidisciplinary stratigraphy of the Chalk Group: an example from the Danish Central Trough: Proceedings of the International Chalk Symposium, p. 195–201.

Verly, G., 1992, Sequential Gaussian cosimulation: a simulation method integrating several types of information: Proceedings of 4th International Geostatistical Congress, 11 p.

Xu, W., T. Tran, R. M. Srivastava, and A. G. Journel, 1992, Integrating seismic data in reservoir modeling: the collocated cokriging alternative: SPE paper 24742, p. 833–842.

Chapter 22

Integrated Modeling for Optimum Management of a Giant Gas Condensate Reservoir, Jurassic Eolian Nugget Sandstone, Anschutz Ranch East Field, Utah Overthrust (U.S.A.)

Dennis L. Cox
Sandra J. Lindquist
Connie L. Bargas
Amoco Production Company
Denver, Colorado, U.S.A.

Karen G. Havholm
Department of Geology
University of Wisconsin
Eau Claire, Wisconsin, U.S.A.

R. Mohan Srivastava
FSS International
Vancouver, British Columbia, Canada

ABSTRACT

Innovative modeling techniques were combined to generate a quantitative, realistic characterization of the eolian Jurassic Nugget Sandstone for an ongoing reservoir study. Dimensions of four different eolian stratification types, infilling cross-set frameworks derived from synthetic bed-form modeling, provided statistical parameters for conditional simulations. Permeability fields generated in the conditional simulation were scaled up for flow simulation in cross section and three-dimensional models. Sensitivities to geologic uncertainties and to conditional simulation parameters were studied. A representative geologic model, with appropriate delineation of reservoir heterogeneity for this field, was attained by matching historical performance of a field segment.

INTRODUCTION AND GOALS OF STUDY

The Anschutz Ranch East (West Lobe) field produces a retrograde gas condensate by use of a full pressure maintenance program. As prescribed by the initial plan of depletion for this field, reinjection of hydrocarbon gas and nitrogen maintain pressure above the original dew point to maximize liquid recoveries. Declining liquid rates, high operating costs, and increased natural gas prices prompted the current review of depletion plans. Optimum operation of this field through the remaining full pressure and ensuing partial pressure maintenance periods requires understanding reservoir fluid flow and effects of potential operational changes. Compositional flow simulations for a segment of the reservoir are necessary to address complex relationships between fluid properties, geology, and operating conditions. A critical part of this process is adequate representation of interwell heterogeneities. A multicompany, multidisciplinary team worked to integrate a large variety of data into a workable solution.

GEOLOGIC SETTING

Anschutz Ranch East field is an overturned anticline on the eastern leading edge of the Absaroka thrust in northeasternmost Utah and southwesternmost Wyoming (Figures 1, 2). Field characteristics and development history have been discussed in Lelek (1982), Kleinsteiber et al. (1983), White et al. (1990), and Lindquist and Ross (1993).

The producing reservoir is the eolian Jurassic Nugget Sandstone, approximately 305 m (1000 ft) thick. Although the reservoir is texturally all sandstone, different stratification types make the formation quite heterogeneous (Lindquist, 1988), with porosity ranging from several percent to approximately 22%, and permeability ranging from hundredths to hundreds of millidarcys. Fractures and faults also overprint the reservoir to varying degrees. For background information on eolian sedimentology, refer to Kocurek (1991) and his references.

AVAILABLE DATA

The West Lobe structure currently has 48 wells (29 producers, 18 injectors, and 1 shut-in producer) on approximately 32-ha (80-ac) spacing. Each well has conventional log suites and dipmeters of varying quality. Eleven wells contain cores with whole-core porosity, horizontal orthogonal permeability (kh_{max} and kh_{90}), and vertical permeability (kv) data. A 1500-sample kh_{max} plug data set supplements the whole-core data.

Prior to this study, cores had been described for sedimentologic, diagenetic, and tectonic features. Special core tests and petrographic analyses delineated mineralogic and textural variability, as well as relative permeability characteristics.

Equity determination resulted in defining field structure (Figure 2); a gross tripartite stratigraphic zonation (Figure 3); and log-calculated porosity, permeability, and water saturation distributions. Injection and production profile logs, pressure transient tests, and as much as ten yr of performance history add to the database.

A ten-component, Amoco Redlich-Kwong equation of state represented fluid properties for flow simulations (Chaback, 1991). The description was tuned to match experimental data from recombined fluid samples, nitrogen-enriched samples, and simulated one-dimensional displacement behavior using nitrogen. The rich retrograde condensate has a dew point 1586 kPa (230 psia) below the initial reservoir pressure of 36,610 kPa (5310 psia) at a subsea datum of –1623 m (–5324 ft). Average initial gas/oil ratio (GOR) was 800 m^3/m^3 (4500 scf/bbl), but it ranged from 1060 m^3/m^3 (6000 scf/bbl) at the structural crest to 620 m^3/m^3 (3500 scf/bbl) at the gas-water contact because of a compositional gradient over the 640-m (2100-ft) hydrocarbon column. For background information on the experimental data and significance to reservoir performance, refer to Metcalfe et al. (1985) and to Kleinsteiber et al. (1983).

Figure 1. Location of Anschutz Ranch East field and adjacent production on Absaroka thrust in western U.S. overthrust belt (after Lelek, 1982).

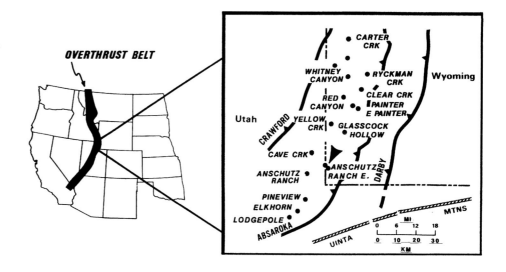

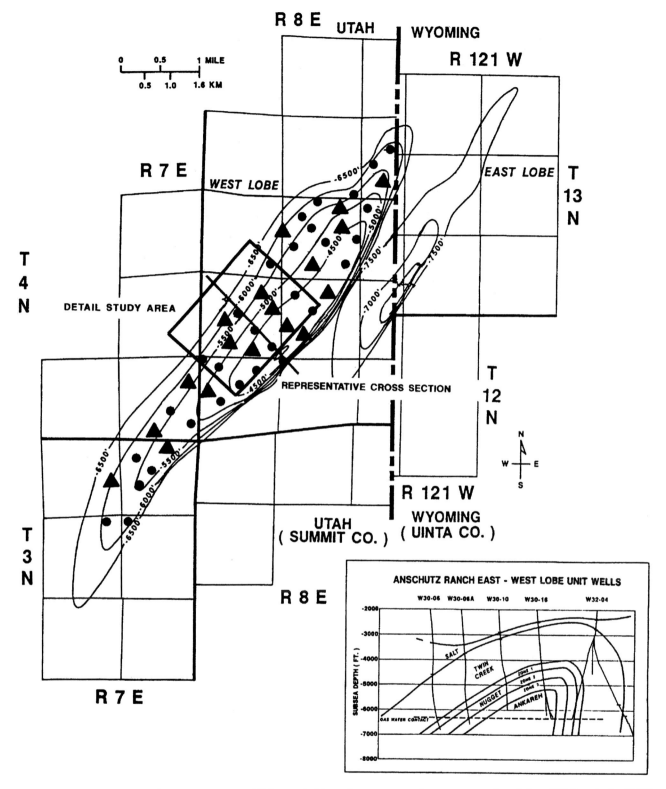

Figure 2. Anschutz Ranch East structure (C.I. = 500 ft) and representative cross section (after White et al., 1990). Present injection wells are the open triangles. Study area is further detailed on Figure 5.

PHILOSOPHY AND STUDY METHODS

Reservoir fluid-flow patterns are complex, as evidenced by nonsymmetrical well interaction and movement of injection fluids over large areal and vertical distances. Earlier flow simulations were unsuccessful at matching performance without large, arbitrary adjustments to reservoir properties. A more realistic geologic description was needed to quantify depositional and postdepositional heterogeneities. It

290   Cox et al.

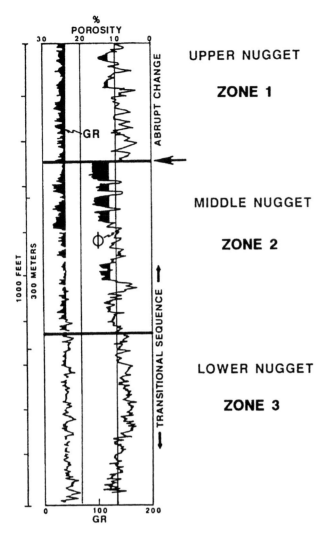

Figure 3. Nugget gamma-ray/porosity type log for Anschutz Ranch East (after Lindquist, 1988).

ability contrast between different stratification types is likely to impact performance more than the smaller-scale heterogeneities within any one stratification type [cross-sectional flow simulation of the eolian Page Sandstone of northern Arizona (Kasap and Lake, 1989)].

There are no suitably large, three-dimensional Nugget outcrops near this field to serve as direct analogs, and correlation of subsurface log character is problematic. Simple interpolation between wells is not likely to be representative of true stratification architecture. Thus, we decided to attempt a genetic reconstruction approach, as per Kocurek et al. (1991), to establish reasonable dimensional ranges for these architectural elements.

A bed-form modeling program (Rubin, 1987a, b) was used to build a plausible, synthetic Nugget cross-set framework that could be infilled with a reasonable distribution of stratification types (Havholm, 1992). Information needed to construct and calibrate these models came from core, dipmeter, and outcrop data; empirical relationships of dune dimensions; and paleoclimatic studies indicating Jurassic wind regime. Nugget vertical mineralogic and diagenetic variability was incorporated in the depositional model by honoring stratigraphic position and vertical trends of permeability and stratification proportion. The genetic reconstruction and final infilled Rubin models are discussed in the section on the quantitative geologic model.

Conditional indicator simulation (Gomez-Hernandez and Srivastava, 1989; Journel, 1990; Journel and Gomez-Hernandez, 1989) was used to construct interwell reservoir descriptions honoring well data and statistical parameters of the geologic model. A geostatistical approach was used because the connectivity pattern of high-permeability rock is critical and because deterministic correlation between wells is not possible at current well spacing (average 488 m or 1600 ft). Indicator simulation was chosen over alternative geostatistical methods because it can adapt flexible coding to satisfy a specific problem; handle discrete data (stratification types) and discretized continuous data sets (permeability); specifically honor multiple structures of extreme valued data; and easily honor available conditioning data. Procedures and results of the conditional simulations are discussed in the geostatistical reservoir characterization section.

The suitability of a statistically generated reservoir description was evaluated by comparing fluid-flow simulation results with field performance. Flow simulations were controlled by specifying field-actual gas production and injection rates. A flow-simulator match of injection-fluid breakthrough timing, produced-oil rate and recovery, GOR, reservoir pressures, and well-bore pressures was desired to validate the geologic description.

The ten-yr Anschutz production history provided significant additional information useful in narrowing the range of geologic uncertainty. Our objective was to modify geologic parameters (and dependent

was important to account for the appropriate scale at which observable heterogeneities most impacted field performance. Thus, we decided to develop a quantitative geologic description in which basic depositional building blocks of stratification type would reflect critical geologic scale.

Stratification type is a significant control on porosity and permeability in this (Figure 4) and other eolian reservoirs (Chandler et al., 1989; Goggin et al., 1989). Reservoir compartments for eolian rocks are defined not only by the dimensions of genetic compartments (preserved cross-sets delineated by bounding surfaces), but also by dimensions of their infilled stratification types (grain flow, grain fall, wind ripple, and other). Cross-set dimensions alone do not necessarily indicate reservoir compartment size, nor do modern dune dimensions in themselves. Dimensions and spatial arrangement of these reservoir compartments define the primary interwell heterogeneity critical to reservoir fluid flow. Fine-scale flow simulation suggested that spatial distribution of stratification types and the perme-

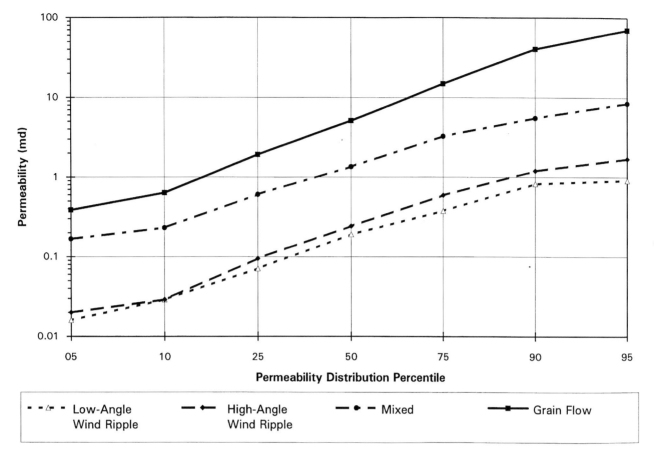

Figure 4. Permeability distribution percentiles by stratification type for Nugget zone 2 (mixed stratification is grain flow interlaminated with high-angle wind ripples). Five-ft-block average, log-calculated kh$_{max}$ and HORIZON stratification estimates for 14 wells.

conditional simulation parameters) to produce a close representation of actual field performance. A large range of possible values for characteristics at various scales reflected uncertainties associated with the initial geologic model. Simulated flow performance was found to be more sensitive to these basic geologic uncertainties than to the random component of multiple realizations generated with a fixed set of parameters; therefore, rather than focusing on the range of response from multiple realizations, at each step in the modification of the geologic model we selected (based on subjective geologic expertise) a single realization for flow simulation. The multidisciplinary study team was critical to this process of modifying parameters and selecting an agreeable, most realistic, single realization.

Several different models were incorporated in this study, including a two-dimensional (2-D), three-well cross section model; a three-dimensional (3-D), six-well small field segment model; and a 3-D, sixteen-well field segment model (Figure 5). Flow simulations included only zones 1 and 2 because these zones dominate reservoir performance. Objectives and major conclusions drawn from these models are discussed in the flow simulation and history matching section.

QUANTITATIVE GEOLOGIC MODEL

Genetic Reconstruction

Several cores were reexamined for depositional process characteristics. Information on grain-flow thicknesses (here 0.5–4.5 cm) could be related to original dune height (Kocurek, unpublished data) and thus widths and spacing (Kocurek et al, 1992; Breed and Grow, 1979). New core descriptions also added data for cross-set thicknesses, vertical stratal progressions, stratal proportions, and stratal thicknesses. Major stratification classes described in detail included grain flow (GFW), high-angle wind ripple (HWR), low-angle wind ripple (LWR) and "other." Nugget grain fall was rarely preserved. A mixed type (MXD), alternating GFW and HWR, was also used in defining a scaled-up genetic composite description used for log calibration with 30-cm (1-ft) data. A sample core description is shown in Figure 6.

Stratigraphic dipmeter data analyzed by Phil Winner defined an overall unimodal cross-set azimuth of 240°, thus a southwestward (west–southwest) dune lee-slope orientation. Zone 3 exhibits more azimuth variability relative to zones 1 and 2. Paleoclimatic studies suggest Jurassic summer winds in this area were directed

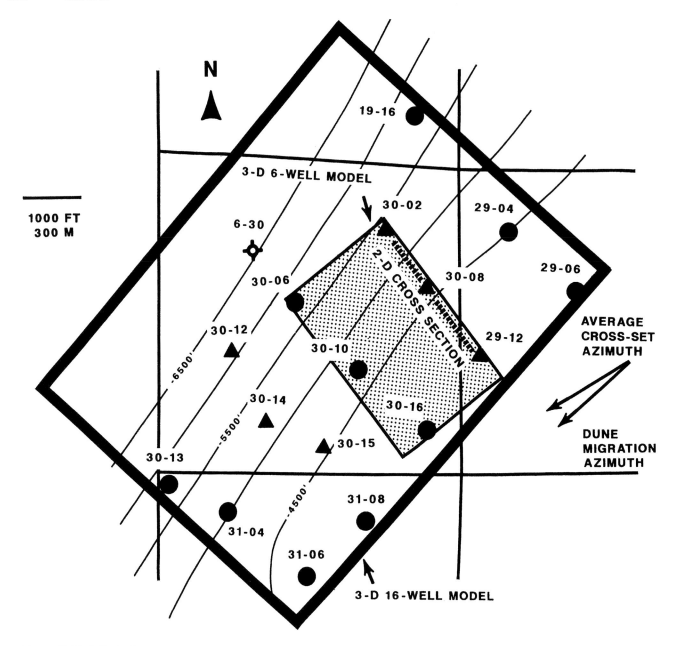

Figure 5. Detail study area map.

approximately 245° (west–southwest) and winter winds 215° (south–southwest) (Parrish and Peterson, 1988). Based on a method for predicting bed-form orientation from wind direction and strength (Rubin and Hunter, 1987), Nugget summer winds were inferred to be able to transport six times as much sand as winter winds. These seasonal changes add variability and asymmetry to the preserved distribution of stratification types.

Outcrop observations [Hanna, Utah, 80 km (50 mi) south and Devil's Slide, Utah, 40 km (25 mi) west of the field] confirmed out-of-phase Nugget dune migration behavior (trough cross-sets in depositional-strike section; planar cross-sets in depositional-dip section) and an oblique sand transport direction relative to bed-form azimuth (locations of trough axes shift laterally up and through cross-sets; troughs are asymmetrically preserved).

Based on core descriptions, we concluded that zone 1 and zone 2 Nugget dunes were similar, described as simple crescentics with maximum original heights of nearly 61 m (200 ft). The size and morphology of dunes in zones 1 and 2 were fairly consistent throughout the section; however, the empirical relationships on which these dimensions were based introduced considerable uncertainty in the actual width dimensions. Initially, an average dune width of 730 m (2400 ft) was assumed from a possible range of 305 to 2134 m (1000 to 7000 ft). Relative to zones 1 and 2, zone 3 Nugget dunes were more sinuous and smaller [maximum heights of 40 m (130 ft), with widths of less than 610 m (2000 ft), and averaging 490 m (1600 ft)]. Although small spurs migrated along the base of zone 3 dune lee slopes, they were not included in the model because

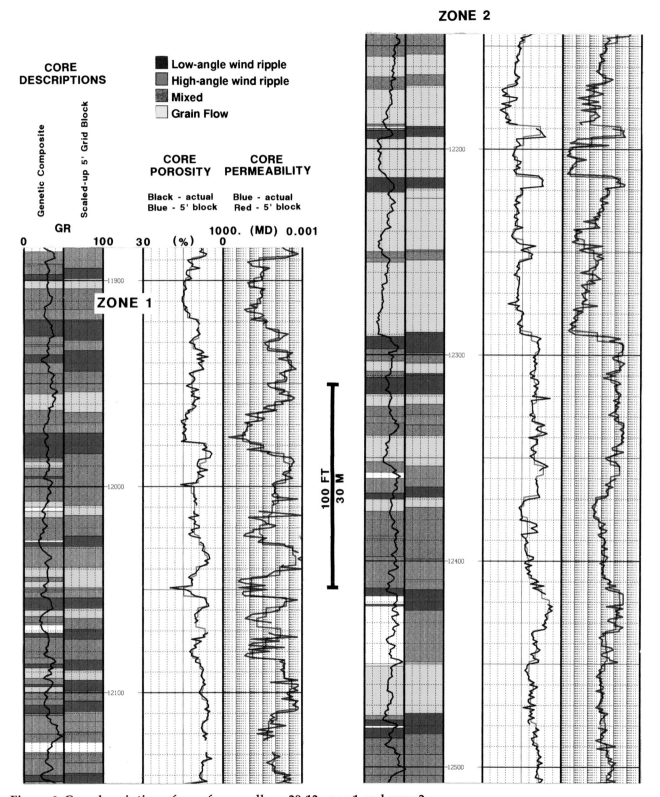

Figure 6. Core description of core from well no. 29-12 zone 1 and zone 2.

of their small size relative to the model cross-sets. Because of the seasonal wind variability, Nugget dunes were determined to have migrated at an azimuth approximately 10° counterclockwise to that of their lee slopes (migration azimuth 230°).

The zone 3/zone 2 transition perhaps records a fore-erg to central-erg progression, with the zone 2/zone 1 porosity break possibly being a super surface of erosion (Figure 3). Overlying zone 1 would then have been deposited with the reestablishment of a succeeding erg.

Final "Infilled Rubin" Models

Based on these considerations, one Rubin model was constructed for zones 1 and 2 and another Rubin model for zone 3. Rubin input parameters were adjusted until a reasonable match between Rubin rose-diagram outputs and dipmeter rose diagrams was attained (Figure 7). A varying climb angle of 0.04–0.44° was modeled for zones 1 and 2; and 0.015–0.17° for zone 3. Average values for dune width were used, and dune height was allowed to vary by ±10%.

The depositional-strike cross section of the selected Rubin framework was then infilled with stratification types. For each wind season, a one-wavelength section of dune lee-slope sinuosity was divided into segments; the resulting amounts and style of erosion, sand transport, and sand deposition were calculated for each segment. These calculations took into account the effects of dune surface slopes and wind incidence angle on various segments of the dune crest [combined sand transport equations of Hardisty and Whitehouse (1988) and Sweet and Kocurek (1990)].

Resulting seasonally dependent, areal lee-slope stratification distributions were then used to make templates of predicted vertical stratification distributions within preserved cross-set troughs. The templates were used to infill the final Rubin frameworks, from which stratal proportions, vertical stratal progressions, and stratal thicknesses could be compared to "real" subsurface values from cores and logs (Table 1). Stratification contacts on the templates could be adjusted vertically and laterally to some degree; thus, the stratification infill was fine-tuned iteratively to obtain a reasonable match of infilled Rubin parameters with subsurface data.

The final base-case infilled Rubin models for zones 1 and 2 (Figure 8) and for zone 3 (Figure 9) provide dimensional statistics for the genetic reservoir compartments in depositional-strike view. These models represent average zone characteristics with quantifiable uncertainty. This genetic reconstruction provides relative dimensions for Nugget eolian architectural elements that are geologically plausible and not available from any other source. Note how the zones 1 and 2 Rubin model accommodates vertical profiles with significantly variable stratification stacking, a situation confirmed in field wells.

Sensitivities to the possible range of original dune width and length were evaluated in flow simulation and history matching. "Average" dune width for the zones 1 and 2 model was expected to be between a minimum of 305 m (1000 ft) and a maximum of 2134 m (7000 ft). Depositional-dip dimensions could not be generated directly from the Rubin models or from subsurface data. Based on limited Nugget outcrop observations of cross-set framework geometries, depositional-dip continuity of stratification types was estimated to range from two to eight times that of the strike dimension.

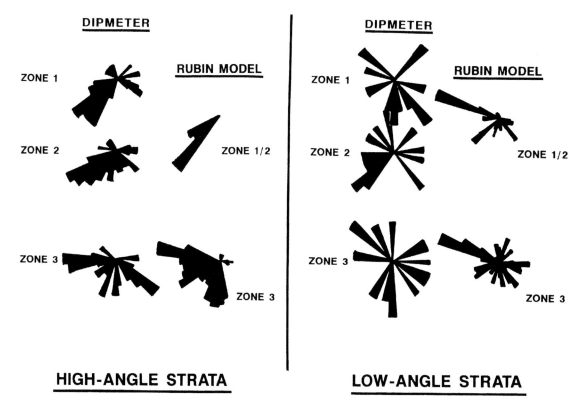

Figure 7. Rose diagrams of high-angle (> 12°) stratal azimuths and low-angle (< 12°) stratal and bounding-surface azimuths for final Rubin models, compared by zone to stratal dipmeter data from well no. 29-12.

Table 1. Comparison of stratification-type statistics from core, infilled Rubin model, well logs, and realization

| | Stratigraphic Thickness (Within-Set Average) (ft) | | | |
	Zone 1 (Core/Rubin)	Zone 2 (Core/Rubin)	Zone 2 Realization	Zone 3 (Core/Rubin)
Cross-set average	19.2 / 24.4	22.7 / 24.4	—	19.4 / 20.2
LWR average	6.1 / 5.2	6.4 / 5.2	6	8.1 / 7.2
HWR average	6.5 / 8.4	8.4 / 8.4	11	9.8 / 12.1
MXD average	9.7 / 12.6	10.0 / 12.6	10	10.9 / 10.5
GFW average	15.0 / 20.4	18.0 / 20.4	16	12.5 / 11.8

| | Well-Log Stratigraphic Proportion (HORIZON Estimates) (%) | | | |
	Zone 1	Zone 2	Zone 2 Realization	Zone 3
LWR average (range)	18 (13–50)	14 (14–14)	13	14 (11–28)
HWR average (range)	21 (7–35)	29 (7–42)	30	51 (37–56)
MXD average (range)	24 (15–34)	19 (19–19)	19	17 (4–26)
GFW average (range)	37 (10–60)	38 (25–60)	38	19 (10–33)

| | Rubin Stratigraphic Proportion (%) | | |
	Zone 1	Zone 2	Zone 3
LWR average	10	10	16
HWR average	25	25	47
MXD average	25	25	21
GFW average	40	40	17

Vertical Transition Probabilities* (Uphole), Combined Zones 1 and 2 Data Comparison

| | Core Data | | | | Log HORIZON | | | | Infilled Rubin | | | | Realization | | | |
	L	H	M	G**	L	H	M	G	L	H	M	G	L	H	M	G
LWR	31	32	15	23	37	30	16	17	5	72	21	2	23	41	15	22
HWR	7	47	28	18	16	52	18	14	7	54	23	16	17	55	13	15
MXD	12	12	55	21	11	21	38	30	7	10	63	20	7	22	52	20
GFW	9	8	8	74	8	9	16	66	14	5	4	77	9	12	9	70

*Probability that stratification type on each line is overlain by stratification type in column, from four different data sets. For example, from core data there is a 28% probability that HWR is overlain by MXD.

**L = LWR = low-angle wind ripple; H = HWR = high-angle wind ripple; M = MXD = mixed, alternating GFW and HWR; G = GFW = grain flow.

GEOSTATISTICAL RESERVOIR CHARACTERIZATION

The application of indicator simulation in this study was a three-step process. First, realizations of stratification types were generated. These realizations were constrained to well data and statistical parameters obtained from the infilled Rubin model. Next, independent permeability distributions (one each for GFW, MXD, HWR, and LWR stratification types) were generated, based on respective spatial correlation parameters and conditioning data. Direct simulation of permeability (rather than deriving permeability from a porosity simulation) improved control over correlation of extreme values (critical to fluid flow) and allowed use of permeability conditioning data calculated with a porosity and gamma-ray algorithm. Finally, the four independent permeability realizations were merged with a stratification-type realization for assignment of block permeabilities. During the merge step an average porosity was also assigned, based on the selected stratification type and permeability class. These steps were repeated for each reservoir zone.

Conditioning Data

A grid-block size of 1.5 m (5 ft) vertical and 15 m (50 ft) on a side was chosen to fit the rectangular grid necessary for conditional simulation. Based on zone 1/2 infilled Rubin model dimensions (Figure 8) and observed vertical variability of stratification type and permeability (Figure 6), these block dimensions represented a maximum that preserved critical geologic scale. Well-log and core data were depth shifted and normalized to a field-average, true stratigraphic thickness for each of the three major correlation zones. The top of zone 2, the only possible "correlative" event on logs, was used as a datum. Stratification type, permeability,

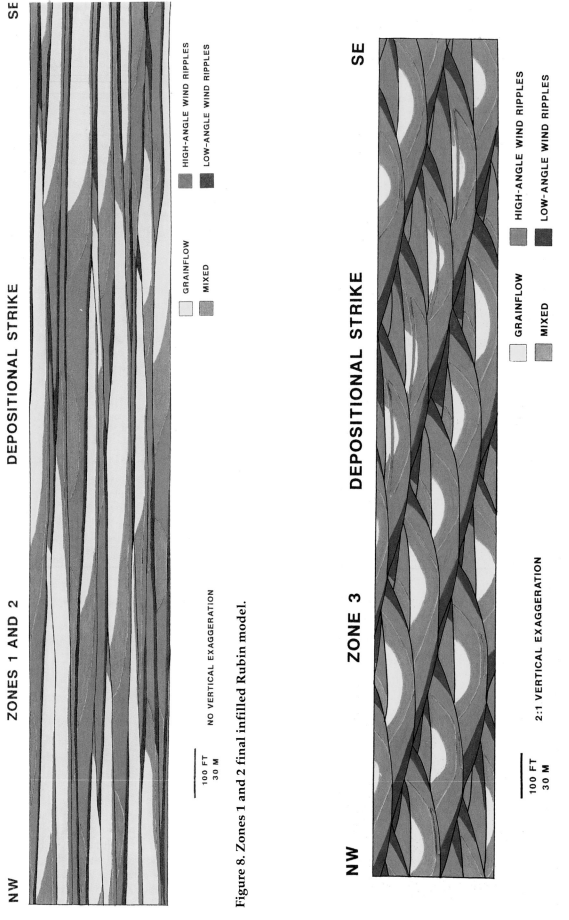

Figure 8. Zones 1 and 2 final infilled Rubin model.

Figure 9. Zone 3 final infilled Rubin model.

and porosity data were integrated to reflect effective block properties.

Stratification type was estimated from well logs using the Atlas Wireline WDS HORIZON software package (Tetzloff et al., 1989) with core descriptions as a learning set. Overall, 70% of GFW, 50% of MXD, and 60% of all WR (wind ripple) were correctly estimated. Critical estimations (highest permeability GFW and lowest permeability WR) were reproduced well enough that the infilled Rubin model was concluded to offer a reasonable estimation of spatial distribution of HORIZON-estimated stratification types. Approximately 90% of all well-bore matrix permeability is in the GFW stratification type.

Log-calculated permeability [a geometric average of 30 cm (1 ft) data was used to represent larger 1.5 m (5 ft) block permeability] was used for conditioning data and to define permeability histograms for simulation. Log-calculated data were based on multiple linear regression of the $\log_{10}$ of core permeability on porosity and gamma ray. These data were used rather than core data for the following reasons. Log-calculated data were more abundant, thus the primary source of conditioning data. A 76% correlation is found between log-calculated and core permeabilities. Many differences between core and log-calculated permeability could be rationalized due to fault/fracture presence or scale of measurement. Log-calculated data reflected a greater separation between GFW and WR permeability.

During the step of merging permeability with stratification-type realizations, log-calculated air permeabilities were adjusted for net overburden, initial water saturation, and directional permeability anisotropy. Correction for overburden is insignificant

(–10%) at high permeability and severe (–90%) at low permeability, which adds to contrasting permeability between GFW and WR rocks. A 30% reduction was applied for initial irreducible water saturation. Grid-block permeability anisotropy was estimated for each stratification type. Whole-core vertical and horizontal permeability anisotropies (kh_{90}/kh_{max}, kv/kh_{max}) range from less than 0.1 to 1, with no clear clustering by stratification type. Questionable sample integrity (fracturing), unknown sample orientation with respect to bedding at the time of measurement, and assumptions made to transfer these measurements to the grid-block scale were sources of uncertainty in assigning grid-block properties.

Conditional Simulation Parameters

Stratification-type correlation parameters in the depositional-strike cross section were obtained by analysis of the infilled Rubin model. Gridded and digitized representations of these models were analyzed to develop variogram models for each stratification type and for the intermediate threshold of combined HWR and LWR. Values were obtained for maximum strike-direction correlation, anisotropy (maximum strike range/vertical range), and rotation from horizontal (Table 2).

The zones 1 and 2 variogram model compares well with vertical variograms derived from subsurface data, adding validation to the geologic model (Table 2). The 1.5 m (5 ft) vertical range of the LWR variogram model is somewhat less than the range from well data. This discrepancy is due to the assumption of a 1.5 m (5 ft) thickness for LWR in constructing the

Table 2. Stratification-type variogram parameters from infilled Rubin models*

	Zones 1 and 2 Model				
	Maximum Strike Range (ft)	Vertical Range (ft/Anis.)	Dip Range Factor × Strike (Factor/Anis.)	Rotation from Horiz. (°)	Well Data Vert. Range (ft)
GFW	460	15 / 31	2 / 0.5	–1.4	15
MXD	140	15 / 9	2 / 0.5	–7.7	12
CWR**	500	14 / 36	2 / 0.5	–1.7	15
HWR	180	12 / 15	2 / 0.5	–1.8	15
LWR	640	5 / 128	2 / 0.5	–1.2	12
	Zone 3 Model				
GFW	130	13 / 10	5 / 0.2	0.0	15
MXD	90	15 / 6	5 / 0.2	±6.7	12
CWR**	138	15 / 9	5 / 0.2	±6.7	15
HWR	90	15 / 6	5 / 0.2	±6.7	12
LWR	170	9 / 19	5 / 0.2	±6.7	12

* The vertical range from well data is shown for comparison. Dip correlation direction is oriented at an azimuth of 230°. The zones 1 and 2 base case included a 0.25° climb angle, and the rotation matrix accounted for this rotation angle as well as the indicated rotation from horizontal in the strike direction (see Isaaks and Srivastava, 1989, p. 388–390). With increasing correlation lengths associated with wider dunes and longer dip correlation, these angles were reduced. Zone 3 models had two variogram structures with equivalent ranges and opposing rotation matrices. No dip angle was used.

** CWR = combined LWR and HWR (all wind-ripple stratification types).

infilled Rubin model and limited occurrences of stacking this stratification type in the Rubin model. Realizations commonly contained multiple instances of 3–6 m (10–20 ft) LWR stacks, so the discrepancy was not a problem.

Proportion curves described the vertical trend of stratification-type proportions (Figure 10). Initially, constant proportions (no depth trends) were assumed, based on data from four described cores. As the database of log-derived parameters was developed, the existence of significant vertical trends in stratification proportion became apparent. The indicator simulation used simple kriging for all estimates, making proportion curves the source of a vertically variable, local mean.

Permeability indicator-class thresholds for each stratification type were developed at distributions of 5, 10, 25, 50, 75, 90, and 95%, to provide eight indicator classes with good definition of extreme values. Vertical correlation parameters for these thresholds were obtained from well data (Table 3). Vertical variogram ranges of permeability median thresholds are similar to those of respective stratification types, and decrease toward the extreme value thresholds.

Without a model for lateral correlation of permeability or suitable outcrop permeability-correlation data, we had to assume strike and dip permeability-correlation lengths. In a direction parallel with depositional strike (where deposition is essentially coeval), it was assumed that similar grain-size distributions of available sand (and most subsequent diagenetic changes) would result in permeability-correlation lengths approximately equivalent to those of stratifi-

cation type. In a direction parallel with depositional dip (where deposition is time transgressive), there was more uncertainty about constancy of grain size and the consequences of fluid flow normal to bedding laminae. The initial assumption was that dip permeability-correlation lengths would be significantly less than those used for the stratification-type simulations. Thus, the dip permeability-correlation lengths initially were estimated at 1/10 the strike permeability-correlation range.

Permeability (especially GFW permeability) also exhibited a vertical trend, decreasing with depth in each zone (Figure 11) because of subtle overall decreases in grain size. Vertical proportion curves for permeability indicator-class thresholds of each stratification type were calculated and used in permeability simulations.

Results of Conditional Simulation

A depositional-strike–oriented, full-Nugget stratification simulation and matching final merged-permeability distribution are shown in Figures 12 and 13, respectively. Conditioning data points are outlined in black. Statistical validity of the simulated stratification-type realizations was judged by comparing variograms, vertical transition probabilities, and frequency distributions of the length and thickness of each stratification type with the same measures made on the infilled Rubin model. (Zone 2 stratification thickness and transition probabilities are compared with core, log, and Rubin-model data in Table 1.) The random component contributed by simulated realizations improved upon the "too perfect" stratification fabric

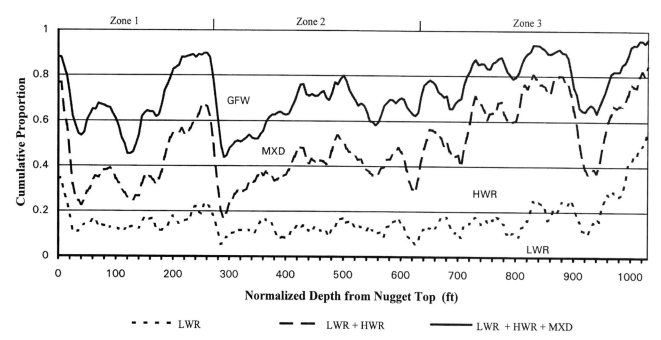

Figure 10. Vertical variability of stratification-type proportion curves (25-ft moving average of HORIZON-estimated stratification types for 42 wells). GFW = grain flow; MXD = mixed, alternatingGFW and HWR; HWR = high-angle wind ripple; LWR = low-angle wind ripple.

Table 3. Variogram parameters for permeability distribution thresholds (seven thresholds for each stratification type)*

	Zone 1	Zone 2	Zone 3
LWR			
Rotation (°)	−1.2	−1.2	±6.75
Strike Range (ft)	640	640	170
1 Vert. Range (ft) / Anis.	5 / 128	5 / 128	5 / 34
2 Vert. Range (ft) / Anis.	5 / 128	15 / 43	10 / 17
3 Vert. Range (ft) / Anis.	10 / 64	15 / 43	15 / 11
4 Vert. Range (ft) / Anis.	15 / 43	20 / 32	20 / 9
5 Vert. Range (ft) / Anis.	10 / 64	15 / 43	10 / 17
6 Vert. Range (ft) / Anis.	5 / 128	10 / 64	5 / 34
7 Vert. Range (ft) / Anis.	5 / 128	5 / 128	5 / 34
HWR			
Rotation (°)	−1.8	−1.8	±6.75
Strike Range (ft)	180	180	90
1 Vert. Range (ft) / Anis.	10 / 18	15 / 12	10 / 9
2 Vert. Range (ft) / Anis.	10 / 18	10 / 18	10 / 9
3 Vert. Range (ft) / Anis.	15 / 12	20 / 9	10 / 9
4 Vert. Range (ft) / Anis.	20 / 9	15 / 12	15 / 6
5 Vert. Range (ft) / Anis.	10 / 18	10 / 18	10 / 9
6 Vert. Range (ft) / Anis.	10 / 18	5 / 36	5 / 18
7 Vert. Range (ft) / Anis.	10 / 18	5 / 36	5 / 18
MXD			
Rotation (°)	−7.7	−7.7	±6.75
Strike Range (ft)	140	140	90
1 Vert. Range (ft) / Anis.	10 / 14	5 / 28	5 / 18
2 Vert. Range (ft) / Anis.	10 / 14	10 / 14	5 / 18
3 Vert. Range (ft) / Anis.	20 / 7	15 / 9	10 / 9
4 Vert. Range (ft) / Anis.	15 / 9	30 / 5	15 / 6
5 Vert. Range (ft) / Anis.	20 / 7	25 / 6	10 / 9
6 Vert. Range (ft) / Anis.	10 / 14	20 / 7	5 / 18
7 Vert. Range (ft) / Anis.	10 / 14	20 / 7	5 / 18
GFW			
Rotation (°)	−1.4	−1.4	0
Strike Range (ft)	460	460	130
1 Vert. Range (ft) / Anis.	5 / 92	5 / 92	5 / 26
2 Vert. Range (ft) / Anis.	10 / 46	10 / 46	10 / 13
3 Vert. Range (ft) / Anis.	20 / 23	15 / 31	10 / 13
4 Vert. Range (ft) / Anis.	35 / 13	30 / 15	20 / 7
5 Vert. Range (ft) / Anis.	25 / 18	15 / 31	15 / 9
6 Vert. Range (ft) / Anis.	20 / 23	5 / 92	10 / 13
7 Vert. Range (ft) / Anis.	15 / 31	5 / 92	5 / 26

*Permeability distribution threshold percentiles: 1 = 5%, 2 = 10%, 3 = 25%, 4 = 50%, 5 = 75%, 6 = 90%, 7 = 95%. Correlation range in dip direction was initially assumed to be 0.1 times the strike range and later was tested at 0.5 times. GFW permeability ranges in dip direction were further modified to be 4 times strike at the median threshold, 0.4 times at the 25th and 75th percentile, 0.2 times at the 10th and 90th percentile, and 0.1 times at the extreme 5th and 95th percentile thresholds. In the final description, dip-direction permeability range was 6 times strike for GFW and 2 times strike for MXD. See Table 4 and text discussion of 3-D models.

derived directly from the infilled Rubin, resulting in a more "realistic" appearance (as perceived by the geologists).

Several features in these realizations are of interest. There is a striking difference in stratification-type distributions in zones 1 and 2 as compared to zone 3 (Figure 12) because of different depositional models. The slight rotation of major correlation axes from hor-izontal can be seen in zones 1 and 2. Zone 3 LWR and HWR show the effect of a double-structure variogram model and oblique orientation of major correlation axes. The influence of conditioning data is stronger in zones 1 and 2 than in zone 3. Multiple realizations of zones 1 and 2 showed that the area within about 91 m (300 ft) of the wells is fairly consistent, but that the interwell description is highly probabilistic.

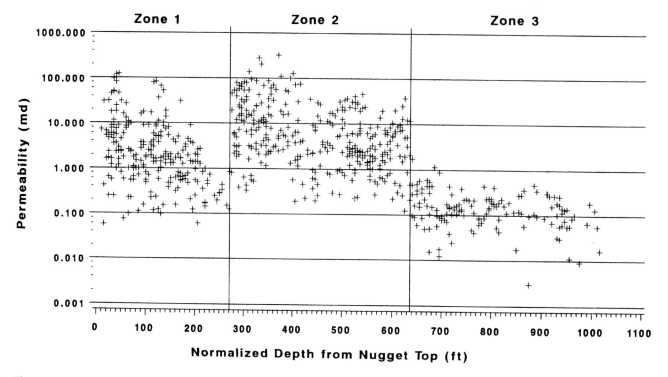

Figure 11. Vertical trends of grain-flow permeability versus depth (log-calculated, 5-ft-block, geometric average kh_{max} for 14 wells).

The final permeability field is displayed using five levels of permeability (Figure 13). Of interest here is the high-permeability streak near the top of zone 2. This interval dominates flow performance in parts of the field. The predominance of low-matrix permeability at the base of zone 1 and throughout zone 3 is also consistent with field production data.

Fracture-Enhanced Vertical Permeability

Both high-permeability open (or intermittently open) fractures and low-permeability, gouge-filled faults and microfaults are common in Nugget cores (Nelson, 1985). Initial reservoir descriptions did not include these fractures and faults because they are difficult to model in terms of extent and quantitative impact on performance.

Open-fracture–dominated flow intervals occur in a few wells, but are obvious mostly in zone 3 and the steep-to-overturned forelimb (regions excluded from flow simulations addressed in this chapter). Field pressure-transient tests do not indicate flow from a dual-porosity system. Thus, most open fractures are thought to influence performance as "matrix assists" (with effective horizontal permeability comparable to that of better matrix rock) rather than as major conduits. Dominant open-fracture orientation (as determined from borehole-washout elongation and limited oriented core data) is field-axial (northeast-southwest), although oblique and transversely-oriented open fractures also exist.

Gouge faults and microfaults are ubiquitous in field cores, but are difficult to document as impacting

field performance. Their expected negative effect on whole-core matrix permeability was not apparent unless massive gouge zones were present. No attempt was made to represent these major gouge zones in the reservoir description realizations because too few were recognized to permit quantified spatial extrapolation into a model.

Production history, selective well testing, and pressure-transient analysis all provide strong indications of vertical communication in the reservoir, greater than that which can be explained by vertical matrix permeability alone. Early flow simulations in both 2-D and 3-D demonstrated that, without vertical permeability enhancement, low permeability at the base of zone 1 would be a barrier to zone 1 and zone 2 pressure communication. The apparent field presence of enhanced vertical permeability is attributed here to natural open fractures, although we recognize that other phenomena, such as poor cement jobs or operations-induced opening of fractures, could contribute to this communication.

To add the improved vertical communication necessary in the 3-D reservoir descriptions, an enhanced vertical permeability was superimposed on a limited number of grid blocks. This permeability enhancement was accomplished by examining the porosities in a vertical column and assigning a vertical transmissibility multiplier to the 50% of blocks having the lowest porosity in the column (core and borehole data confirm the presence of more open fractures in lower-porosity rock). Fractured intervals were extended vertically by 10% out of those low-porosity strata to feather into higher-porosity rock, a situation observed

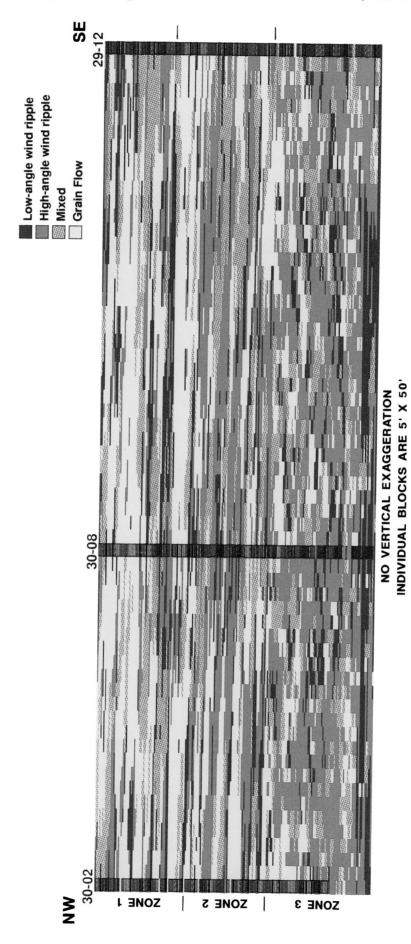

Figure 12. Full Nugget stratification realization, depositional strike cross section.

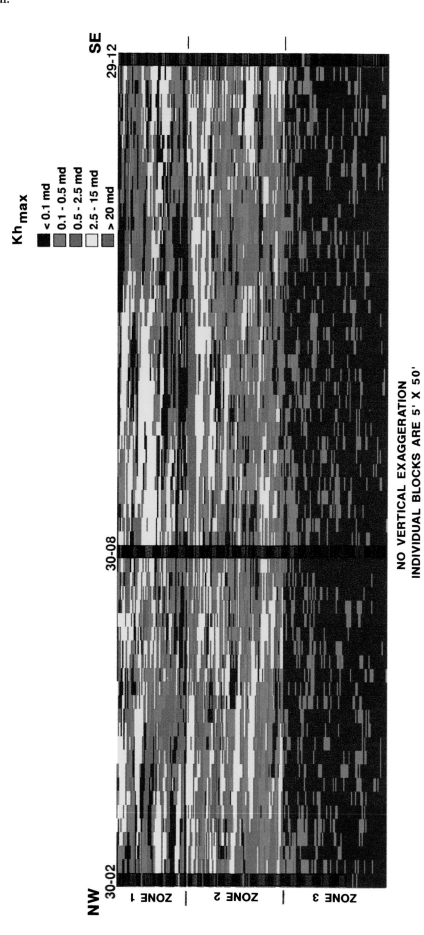

Figure 13. Full Nugget merged permeability realization, depositional strike cross section.

in outcrop and core. Fracture permeability enhancement was applied on a zone-by-zone basis at each well location and at a limited number of locations offsetting some wells. The vertical permeability enhancement affected less than 5% of the reservoir volume, which was consistent with the overall volume of open fractures described in core. Such enhancement around each well bore is possible because drilling and production operations could have artificially fractured the formation or opened existing fractures.

FLOW SIMULATION AND HISTORY MATCHING

2-D Cross Section Models

Cross section models contained three wells in a depositional-strike orientation (Figures 5, 12, 13). These models were useful in establishing procedures and methodology for generating realizations and conducting flow simulations. Main objectives were to (1) refine and validate geostatistical procedures discussed previously, (2) evaluate fluid-flow characteristics at the fine-grid scale of the geologic description, and (3) address sensitivities to certain geologic and petrophysical parameters. Significant observations regarding fluid flow and parameter sensitivity are as follows:

- A low-permeability cutoff could not be determined. Ability to contribute to production was more a function of relative position with respect to higher-permeability rock than to absolute block permeability.
- 2-D cross section models without vertical permeability enhancement had insufficient vertical pressure communication.
- Incorporating the trend of decreasing GFW permeability with depth significantly improved recovery comparisons and the model-versus-actual correlation of vertical distribution of injection and production fluids.
- Use of vertical stratification proportion curves in the conditional simulation improved the comparison of simulated flow profiles with actual flow profiles.
- Oil recovery [at 1.15 hydrocarbon pore volumes (HCPV) injected] from multiple realizations ranged from ± 15% of the midpoint.
- Increasing stratification and permeability strike-correlation lengths by 1.5 times produced a rate increase of 23%, whereas increasing the length by 2.5 times resulted in just a 25% rate increase. Recovery (at 1.15 HCPV injected) was reduced by 5 and 10%, respectively, from the 2-D base case.
- Petrophysical property values drawn at random from within a permeability class distribution could be used to reproduce the observed scatter of porosity and permeability, but flow simulation results were no different from results acquired using the class midpoint.

- The practical degree of permeability scale-up was limited because of the small-scale heterogeneity present in these eolian rocks. Scale-up of 2 × 2 (horizontal × vertical) gave results close to base-case rate and recovery. Scale-up of 4 × 2 reduced rates approximately 10%, but had no effect on recovery. Scale-up of 4 × 4 reduced rates approximately 10% and increased ultimate recovery by about 8% at 1.15 HCPV injection.

3-D Six-Well Models

The six-well model contained one row of producers and one row of injectors, each in a depositional-strike orientation (Figure 5). This small field segment was modeled to investigate 3-D flow characteristics. Of primary concern were volumetric sweep, vertical communication, and breakthrough timing of injected fluids, all as related to geologic parameter sensitivity. Review of the geologic model and results of the cross section flow simulations identified critical geologic uncertainties as follows:

- Stratification correlation lengths in depositional dip and strike directions.
- Permeability correlation lengths in the dip direction.
- Permeability magnitude and anisotropy within a block.
- Fracture impact on fluid flow and pressure communication.

Three cases of six-well model description and flow-simulation results are discussed in the following paragraphs (base case, case B, and case E).

The base case geologic description was constructed with parameters described in previous sections. Parameters reflecting the critical geologic uncertainties are summarized in Table 4. Cross section and map view distributions of stratification are shown in Figure 14 and corresponding permeability distributions in Figure 15. Enhanced vertical permeability (described previously in the section on fracture-enhanced vertical permeability) is included in these descriptions because initial 3-D flow simulations proved it necessary to obtain adequate vertical pressure communication. Prior to scale-up and flow simulation, the rectangular fine-grid description was converted to a true vertical thickness and structural depth.

Scale-up procedures in this study used a steady-state, single-phase simulation at the fine-grid scale. Flow simulation in each of three principal directions determined diagonal terms of the scaled-up transmissibility tensor. Calculation of effective transmissibility for the coarse grid used Darcy's equation and the steady-state flow rate and pressure gradient in each direction. Additional calculations determined the effective permeabilities at grid blocks containing wells. Porosity was scaled as a volumetric average (Mansoori, this volume). Scale-up reduced the total number of grid blocks from 272,801 to 11,180. Areal scale-up was by a factor of 4 × 4. A vertical scale-up factor of 2 was used for layers containing the largest permeability contrast and a factor of

Table 4. Conditional simulation parameters investigated in 6-well and 16-well models

		6-Well Models			16-Well Models	
		Base	B	E	Initial	Final
Stratification						
Strike Rng × Rubin Base		1.0	1.5	2.0–2.5	2.0–2.5	2.0–2.5
Dip/Strike Range		2.0	2.0	6.0	6.0	6.0
Permeability						
Dip/Strike Range	GFW*	0.1	0.5	0.1–4.0	0.1–4.0	6.0
	MXD	0.1	0.5	0.5	0.5	2.0
	HWR	0.1	0.5	0.5	0.5	0.5
	LWR	0.1	0.5	0.5	0.5	0.5
kv/kh_{max}	GFW	0.1	0.5	0.5	0.5	0.5
	MXD	0.1	[0.01]	[0.001]	0.1	0.025
	HWR	0.1	[0.01]	[0.001]	0.1	[0.01]
	LWR	0.1	[0.01]	[0.001]	[0.001]	[0.001]
kh_{90}/kh_{max}	GFW	0.8	0.8	1, 0.75, 0.5	1, 0.75, 0.5	1, 0.75, 0.5
	MXD	0.3	[0.01]	[0.001]	0.25	0.10
	HWR	0.2	[0.01]	[0.001]	0.10	0.01
	LWR	1.0	[0.01]	[0.001]	1.00	1.00
Histogram: multiplier times original std dev.		1.0	1.2	1.2	1.2	0.75–1.35 0.75
Other (first occurrence denoted by x)						
Well	30-06 perm × 4		x	=	=	=
	30-12 perm × 5				x	=
	30-15 perm × 7.5				x	=
	19-16 perm × 4				x	=
	30-13 perm × 2					x
Reduce overburden adjustment.			x	=	=	=
Edit GFW proportion curve.				x	=	=
Add variable Sw.				x	=	=
Conditioning data edits.				x	x	x
Adjust rotation matrix for wider trough and no climb.				x	=	=
Add gas hysteresis curve.						x
Increase kg (swi) from 0.7 to 0.9.						x

*Dip-direction GFW permeability correlation is described in text and Table 3.

4 was used for remaining layers, reducing the number of layers from 127 to 43.

Base-case performance is shown compared to allocated field actual in Figure 16. Pessimistic gas production and injection rates, and optimistic oil rate (high) and GOR (low) implied a need for increased system permeability and reduced volumetric sweep efficiency.

Case B model performance improved (Figure 16). Gas production and injection rates were reasonable, but oil rate and GOR remained optimistic. Although not shown, one producer matched historical performance, whereas the other two recovered too much oil. The geologic model of case B (Figures 17, 18) had increased strike-correlation lengths (stratification type increased 1.5 times and permeability 1.5 times), increased dip-correlation lengths (stratification type increased 1.5 times and permeability 7.5 times), and increased contrast between stratification-type kh_{90} and kv (Table 4). A wider range of permeability values was assigned to each stratification type, resulting

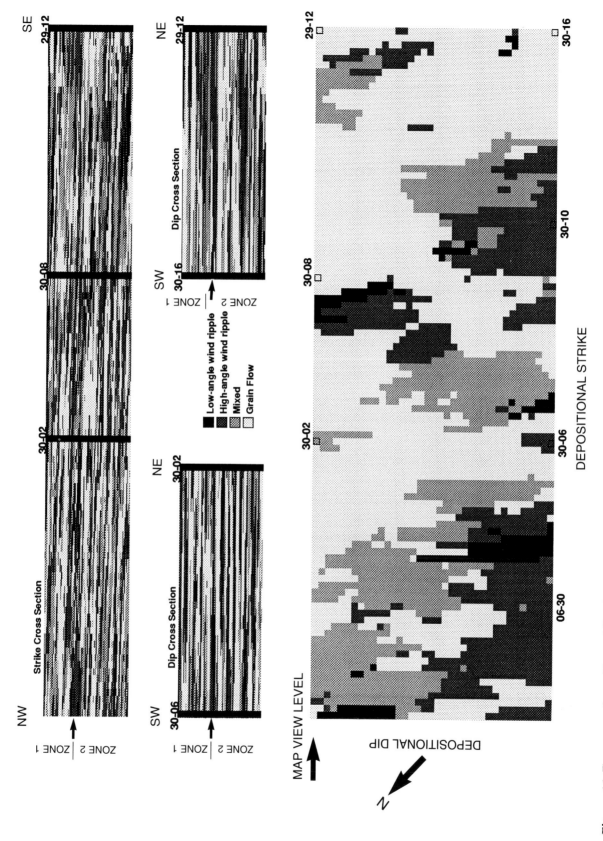

Figure 14. Base case six-well stratification realization.

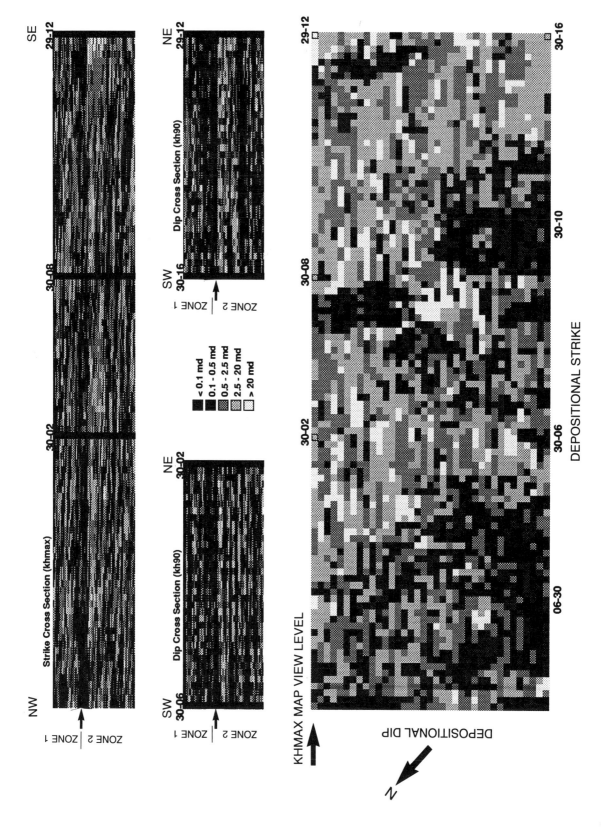

Figure 15. Base case six-well merged permeability realization.

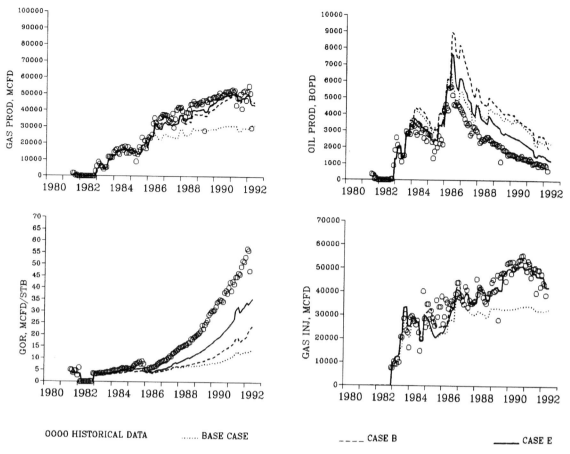

Figure 16. Six-well performance plots. Comparison of three cases with historic actuals (gas, oil, GOR, and gas injection plots).

in larger contrast between high- and low-permeability values. Other changes included increasing permeability in well no. 30-06 and correcting the permeability adjustment due to overburden. (Note that permeability adjustments made to well no. 30-06, and others to be made in the sixteen-well model, were supported by performance and pressure-transient data. Permeability was not increased beyond the range of the porosity-permeability relationship observed in field core.)

The case E geologic description (Figures 19, 20) had additional increases in lateral correlation lengths and greater contrast between stratification type kh_{90} and kv. Compared to case B, strike correlation was increased 1.67 times for GFW stratification and permeability, and 1.33 times for WR and MXD stratification and permeability. Dip-correlation length was increased 5 times for GFW stratification and 4 times for WR and MXD stratification and permeability. GFW permeability dip-correlation lengths were reduced by 0.75 times at the 5th and 95th percentile thresholds, but increased by 6.64 times at the median threshold. These changes resulted in increased depositional-dip continuity of moderate and lower GFW permeability, but restricted dip continuity of extremely high or low GFW permeability. WR and MXD kh_{90} and kv were reduced to 0.001 md. GFW kh_{90} was

adjusted to equal $1 \times kh_{max}$ at low permeability, $0.5 \times kh_{max}$ at high permeability, and $0.75 \times kh_{max}$ at moderate permeability (Table 4).

Flow simulation of case E (Figure 16) was a closer match to field performance. Total gas rates were comparable; however, recovery was too optimistic (higher oil rates and lower GOR). No individual producing well matched historic performance. Two wells overproduced oil and another underproduced, raising a significant question of whether the description was poor or whether allocated production from the six-well area was incorrect. Two producers were known to be affected by injectors outside this model area, so the larger sixteen-well model was implemented to evaluate further changes to geologic parameters.

An interim realization (case B) was used to investigate two alternatives of matching performance history. One alternative superimposed long vertical fractures, and a second alternative applied layer transmissibility multipliers. Simulating extensive fractures improved breakthrough timing; however, the effect on GOR was limited to a quick initial increase, and recovery remained optimistic. Such open fractures do not appear to dominate lateral flow in this reservoir. Layer transmissibility multipliers could be used to match the allocated historical

308   Cox et al.

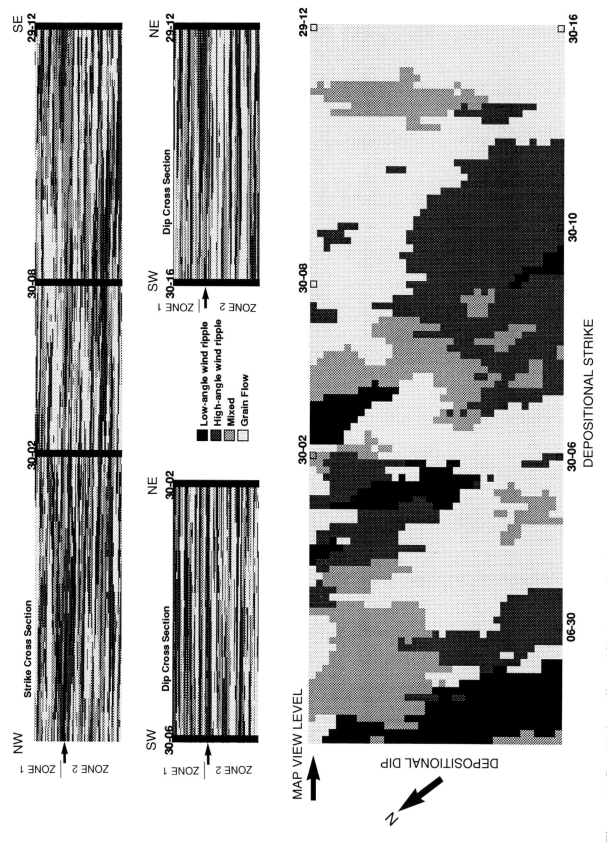

Figure 17. Case B six-well stratification realization.

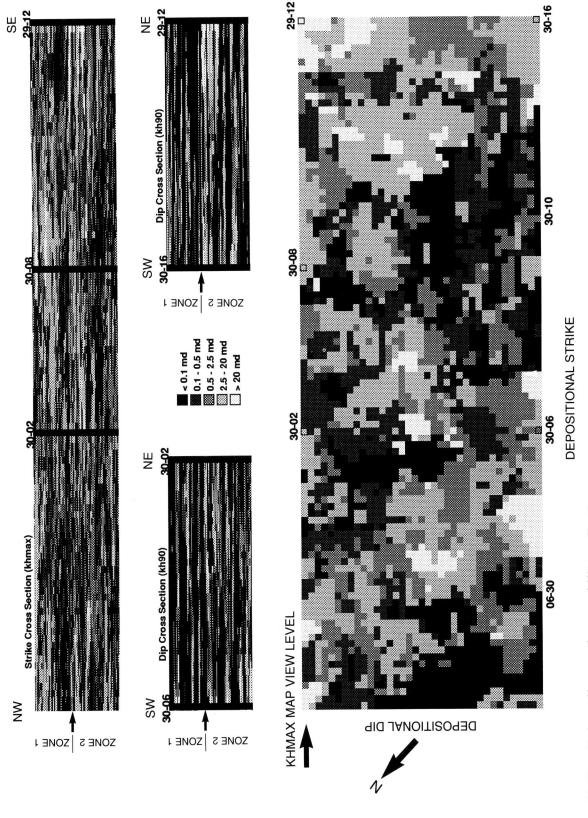

Figure 18. Case B six-well merged permeability realization.

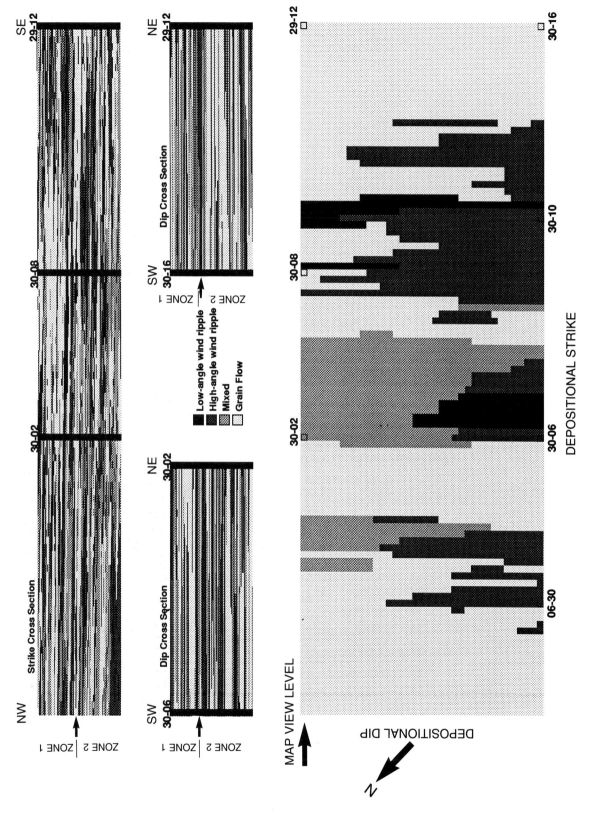

Figure 19. Case E six-well stratification realization.

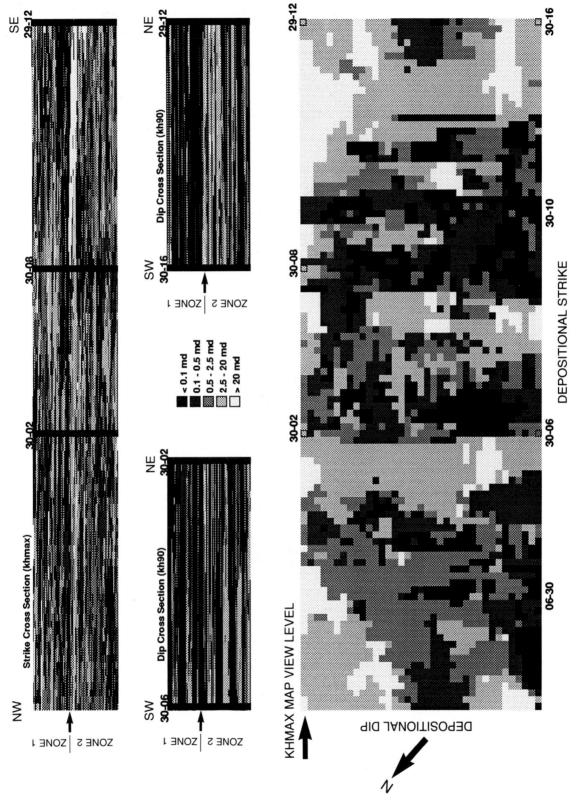

Figure 20. Case E six-well merged permeability realization.

rates and recovery. Typical multipliers increased transmissibility in the high-average permeability layers by 2 to 3 times, and reduced the remaining layer transmissibilities by 0.1 times. Such transmissibility changes highlighted the need for increased connectivity of higher-permeability rock between injectors and producers.

One other significant conclusion from the six-well models was that rapid depositional-strike communication (injection gas breakthrough) could occur in unfractured, high-permeability matrix. Several pseudo–conditioning points were required to cause each conditional simulation to connect enough high-permeability matrix between the two wells involved. Subsequent realizations included these pseudo–conditioning points.

3-D Sixteen-Well Models

The six-well model was expanded to add one row of producing wells to the northeast and one row of injectors and one row of producers to the southwest (Figure 5). Primary emphasis with this model was to match historical performance of the center row of producers and the two adjacent rows of injectors through additional modifications to the geologic model. Scale-up prior to flow simulation was the same as in the six-well model, reducing the number of grid blocks from over 2.2 million to 51,987.

The sixteen-well initial description (Figures 21–24) used parameters similar to case E of the six-well model (Table 4). WR and MXD kh_{90} and kv were modified by multipliers (relative to maximum block permeability) rather than fixed at 0.001 md. Also, variogram model rotation matrices were adjusted to reflect the spatial orientation of the sixteen-well model grid at $10°$ counterclockwise from the six-well model.

Comparison of this initial sixteen-well model versus actual field performance for the inner three rows (six injection and four producing wells) is shown in Figure 25. The model GOR trend was similar to that observed in six-well case E, and oil rates were close to actual; however, gas production and injection rates were low. These comparisons implied that an additional increase in system permeability was required and that volumetric sweep efficiency had to be further reduced.

Changes to geologic-model and conditional-simulation parameters up to that point had seemed significant, but they had not adequately improved flow simulation. Stratification correlation lengths approached 85% (strike) and 65% (dip) of the maximums relative to geologic-model uncertainty. Further increases were possible, but of questionable justification or benefit. The most significant contributors to improved performance comparisons in the six-well models were increased permeability and increased dip correlation of permeability. These parameters were less well defined and seemed to offer the greatest potential for the significant changes yet required.

The final sixteen-well description (Figures 26–29) reflects significant changes in assumptions governing permeability distribution. The most striking difference is increased depositional-dip continuity of GFW permeability. Depositional-dip correlation lengths for all GFW permeability classes were increased to 6 times the strike correlation, making them equivalent to stratification correlation length.

The geologic basis for reversing the initial assumption of short depositional-dip permeability correlation relates to the grain-flow process, which commonly concentrates the coarsest available grains continuously at the base of the dune slipface as it advances downwind. Sinuous crescentic crestlines would create downwind-elongate GFW avenues through time, with the best depositional-dip flow properties in the basal, coarser-grained portions. Such vertical permeability trends within individual GFW compartments can be significant (see Figure 6, upper zone 2 for examples). The increased depositional-dip correlation lengths for GFW permeability assume the persistence of these trends observed in the conditioning data. Greater grain-size heterogeneity inherent in the laminated structure of WR and MXD stratification results in considerable permeability variation in their dip directions. Permeability dip correlation was increased to 2 times the strike correlation for MXD and kept at 0.5 times the strike correlation for HWR and LWR.

Other changes affected permeability magnitude in the final description. Permeability histogram standard deviations were reduced, representing an expectation of effective property averaging. GFW permeability histograms were also adjusted as a variable function of stratigraphic depth to better reflect the magnitude of the permeability trend (Figure 11). In upper zones 1 and 2, high-permeability GFW had been understated by treating GFW permeability as a single distribution with vertically variable proportions. In retrospect, GFW permeability could be simulated more appropriately by removing the vertical trend, simulating residuals, then adding back the trend. Lastly, permeability modifications related to initial water saturation were changed to increase system permeability.

The simulated flow performance of this final case showed significant improvement and a good match of the total of the inner well rows compared to historical (Figure 25). The inner row wells all had acceptable gas rates; however, two producing wells were optimistic and two were pessimistic in oil recovery and GOR performance. Flow simulation of two different realizations (generated with the same parameters) produced similar results, suggesting that individual well problems were due to imprecise model representation of specific local characteristics, rather than to a random difference within the set of all possible realizations. The difficulty in attaining consistent, individual well performance matches illustrates the limitations of the methods used, as well as this reservoir's complex hierarchical scale of heterogeneities. Nevertheless, the final geologic description presented here is believed to be reasonably realistic and representative of the Anschutz Ranch East Nugget reservoir.

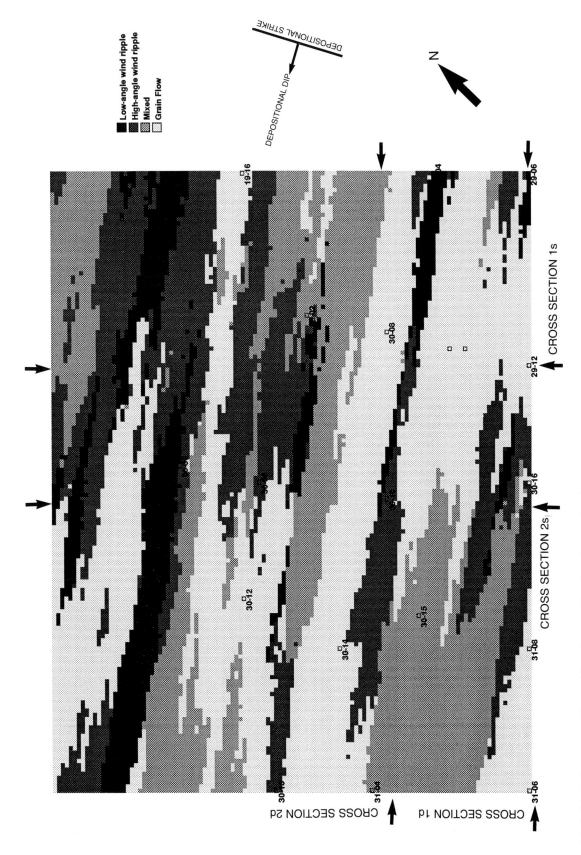

Figure 21. Initial sixteen-well stratification realization (map view).

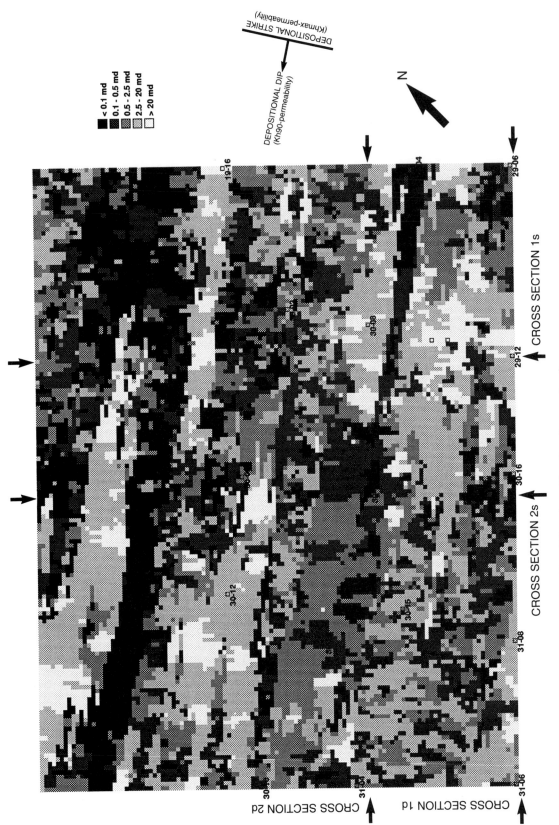

Figure 22. Initial sixteen-well merged permeability (kh_{max}) realization (map view).

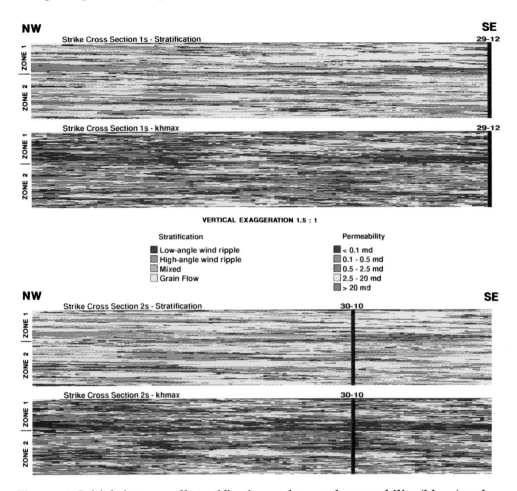

Figure 23. Initial sixteen-well stratification and merged permeability (kh_{max}) realizations (depositional-strike sections).

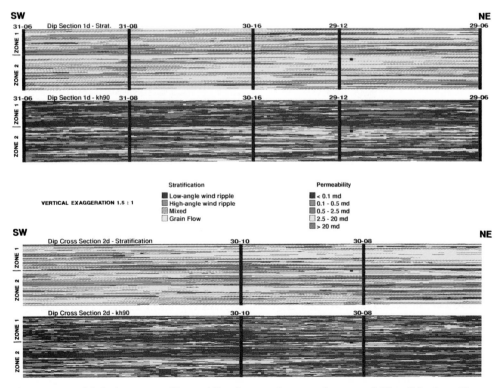

Figure 24. Initial sixteen-well stratification and merged permeability (kh_{90}) realizations (depositional-dip sections).

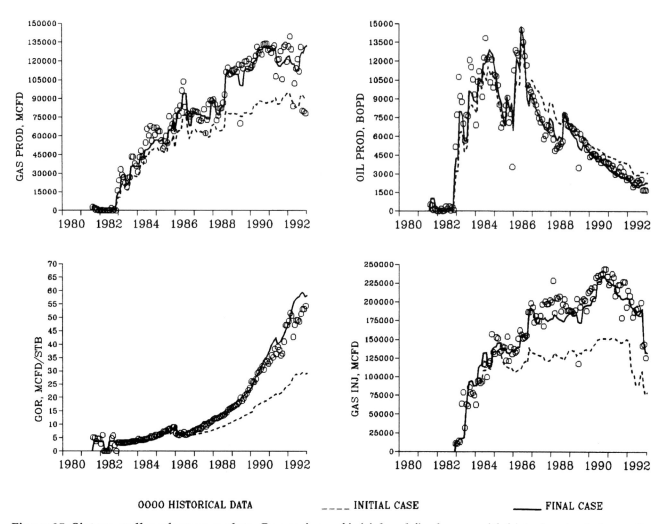

Figure 25. Sixteen-well performance plots. Comparison of initial and final cases with historic actuals (gas, oil, GOR, and gas injection plots) for inner three rows of wells (see Figure 5).

CONCLUSIONS

Results of this meticulously integrated study have confirmed some previous qualitative Anschutz Nugget performance expectations, modified others, and completely changed still others. Local compounded stratigraphic, structural, and mechanical factors have impacts on individual well performance that proved difficult to resolve.

The geologic approach to modeling this reservoir results in a more realistic stratigraphic reservoir description than traditional correlation methods, although uncertainty still exists with regard to the large ranges of values possible for some model parameters. Interwell connectivity is primarily controlled by geometry of stratification types. The spatial distribution of permeability within a stratification type is equally important, particularly in highest-permeability (coarsest) grain-flow packages.

Preferred matrix permeability parallel with depositional strike (parallel with bedding) commonly has been proposed for eolian reservoirs with relatively straight-crested crescentic ("transverse") dunes. Such

a directional (northwest-southeast) preference is confirmed in some Anschutz Nugget wells, but its importance is somewhat diluted in the field because Nugget dune-crestline sinuosity resulted in limited depositional-strike extent of constant stratification types. Thus, the depositional-dip (northeast-southwest) continuity of stratification types (especially grain flow), resulting from the migration of dunes and their preservation through time, becomes of nearly equal importance here even though flow through such units must move across rather than along strata. We make this conclusion with caution because depositional-dip direction is approximately the same trend as potential fracture strike of most field open fractures.

Fractures are recognized as a possible complication in matching individual well performance. Significant open-fracture overprinting is more rare in the field backlimb than in the steep and overturned forelimb. At most, open fractures have a secondary performance impact for most of the backlimb in zones 1 and 2. Open fractures likely improve vertical reservoir communication and add to well capacity in

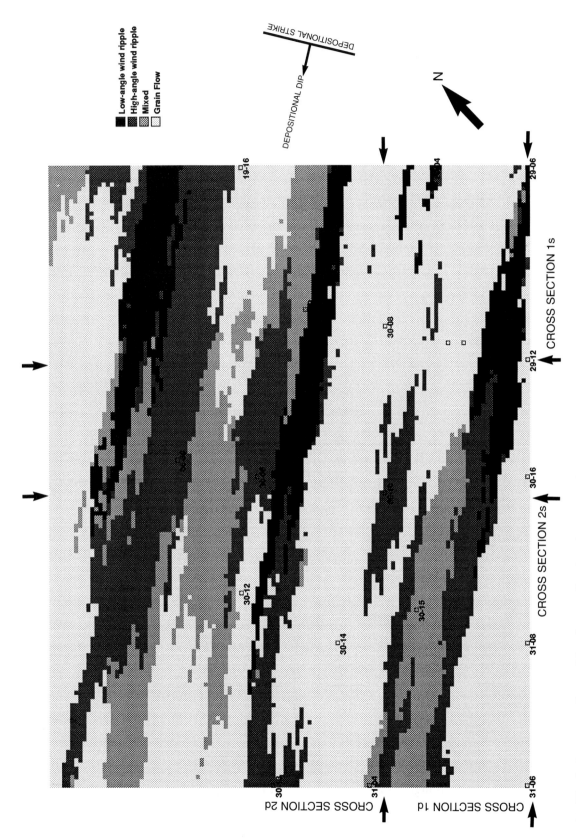

Figure 26. Final sixteen-well stratification realization (map view).

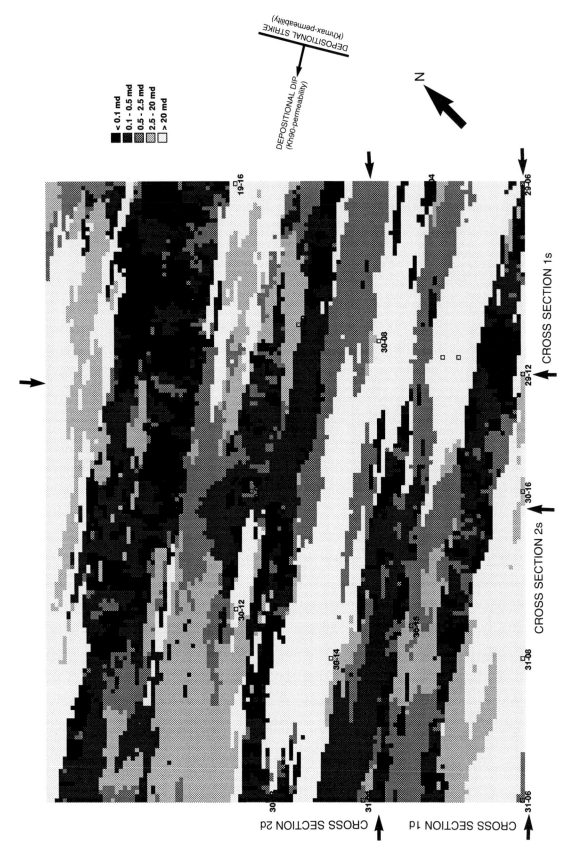

Figure 27. Final sixteen-well merged permeability (kh$_{max}$) realization (map view).

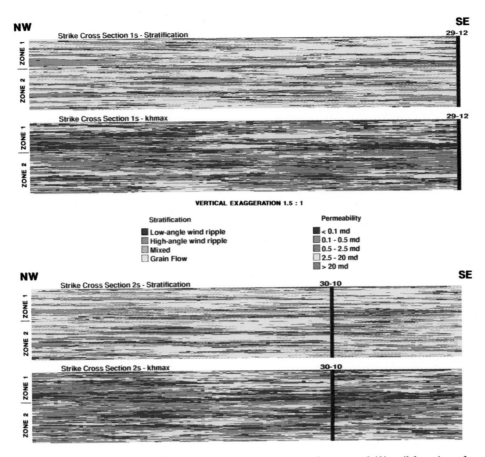

Figure 28. Final sixteen-well stratification and merged permeability (kh$_{max}$) realizations (depositional-strike sections).

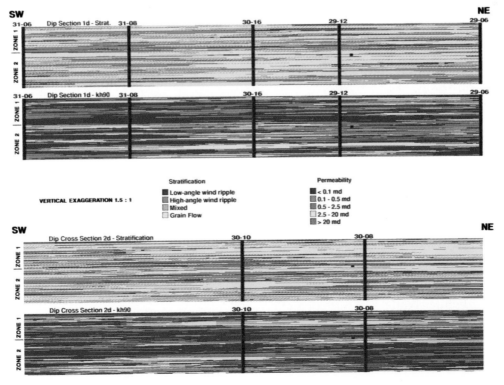

Figure 29. Final sixteen-well stratification and merged permeability (kh$_{90}$) realizations (depositional-dip sections).

some instances, but they do not appear to have considerable negative effect on reservoir sweep.

The uncertainty in numerous geologic parameters highlights the importance of obtaining the best-possible direct measurements of subsurface interwell variability and the importance of using all available performance data. Without historical performance, one must consider the full range of geologic possibilities. In this study, the iterative comparison of flow simulations with field performance reduced such uncertainty and led to significant changes in the initial geologic model.

Reservoir characterization should always include as much determinism as possible before applying geostatistical methods. Development of the Anschutz Nugget geologic description was complicated by a lack of deterministic features. Only the tripartite reservoir zonation and some well-bore conditioning data were modeled deterministically. Pressure-transient and production-profile information was helpful in improving conditioning data quality. The experience with highly probabilistic interwell descriptions illustrates the potential value of including more than just statistical geologic constraints in conditional simulations [e.g., reservoir connectivity factor (Hird and Kelkar, 1992)]. Similar efforts could also be aided by the use of methods to rank multiple realizations (Alabert and Modot, 1992; Guerilot and Morelon, 1992).

Realistic representation of critical eolian geologic characteristics requires small grid blocks. Necessary scale-up for flow simulation creates a problem in capturing small-scale reservoir heterogeneities in "averaged" grid blocks. Improvements in the speed of flow simulators or flexible gridding techniques adaptable to stratification or permeability distributions (such as those described by Hove et al., 1992) could optimize flow simulations at the geologic scale used here.

ACKNOWLEDGMENTS

The authors thank Gary Kocurek for early geological advice on this project. Geologists Phil Winner (Mobil), Rick Davis (The Anschutz Corporation), Lee Krystinik (Union Pacific Resources), and Peter Frykman (Geological Survey of Denmark) enthusiastically assisted these efforts, as did engineers Richard Burns (Mobil), Hal Koerner (The Anschutz Corporation), Rob Dunleavy (Union Pacific Re-sources), and John Mansoori (Amoco Production Research). We thank the following Working Interest Owners of Anschutz Ranch East (West Lobe Participating Area) for permission to publish this work: Amoco Production; The Anschutz Corporation; Mobil USA; Union Pacific Resources; Chevron USA; Norcen Explorer; and Brownlie, Wallace, Armstrong and Bander.

REFERENCES

Alabert, F. G., and V. Modot, 1992, Stochastic models of reservoir heterogeneity: impact on connectivity and average permeabilities: SPE Paper 24893 presented at 67th Annual Technical Conference and Exhibition of the SPE, Washington, D.C.

Breed, C. S., and T. Grow, 1979, Morphology and distribution of dunes in sand seas observed by remote sensing, in E. D. McKee, ed., Global sand seas: U.S. Geological Survey Professional Paper 1052, p. 257–302.

Chaback, J. J., 1991, Development of compositionally compressed Amoco Redlich-Kwong equation of state descriptions for fluids from the Anschutz Ranch East field, Summit County, Utah, for application in 3-D field segment simulations: unpublished Amoco Research Report No. 91212art0002, 6 p.

Chandler, M. A., G. Kocurek, D. J. Goggin, and L. W. Lake, 1989, Effects of stratigraphic heterogeneity on permeability in eolian sandstone sequence, Page Sandstone, northern Arizona: AAPG Bulletin, v. 73, no. 5, p. 658–668.

Goggin, D. J., M. A. Chandler, G. Kocurek, and L. W. Lake, 1989, Permeability transects in eolian sands and their use in generating random permeability fields: SPE Paper 19586 presented at 64th Annual Technical Conference and Exhibition of the SPE, San Antonio, TX.

Gomez-Hernandez, J. J., and R. M. Srivastava, 1989, ISIM3D: an ANSI-C three-dimensional multiple indicator conditional simulation program: Computers in Geosciences, v. 16, no. 4, p. 395–440.

Guerilot, D. R., and I. F. Morelon, 1992, Sorting equiprobable geostatistical images by simplified flow calculations: SPE Paper 24891 presented at 67th Annual Technical Conference and Exhibition of the SPE, Washington, D.C.

Hardisty, J., and R. J. S. Whitehouse, 1988, Evidence for a new sand transport process from experiments on Saharan dunes: Nature, v. 332, p. 532–534.

Havholm, K., 1992, Geological interpretation of the Nugget Sandstone, Anschutz Ranch East field: Unpublished consultant report for Amoco Production Company, 13 p.

Hird, K. B., and M. G. Kelkar, 1992, Conditional simulation method for reservoir description using spatial and well performance constraints: SPE Paper 24750 presented at 67th Annual Technical Conference and Exhibition of the SPE, Washington, D.C.

Hove, K., G. Olsen, S. Nilsson, M. Tonnesen, and A. Hatloy, 1992, From stochastic geological description to production forecasting in heterogeneous layered reservoirs: SPE Paper 24890 presented at 67th Annual Technical Conference and Exhibition of the SPE, Washington D.C.

Isaaks, E. H., and R. M. Srivastava, 1989, An introduction to applied geostatistics: New York, Oxford University Press, 561 p.

Journel, A. G., 1990, Geostatistics for reservoir characterization: SPE Paper 20750 presented at 65th Annual Technical Conference and Exhibition of the SPE, New Orleans, LA.

Journel, A. G., and J. J. Gomez-Hernandez, 1989, Stochastic imaging of the Wilmington clastic sequence: SPE Paper 19857 presented at 64th Annual Technical Conference and Exhibition of the SPE, San Antonio, TX.

Kasap, E., and L. W. Lake, 1989, An analytical method to calculate the effective permeability tensor of a grid block and its application in an outcrop study: SPE Paper 18434 presented at the SPE Symposium on Reservoir Simulation, Houston, TX.

Kleinsteiber, S. W., D. D. Wendschlag, and J. W. Calvin, 1983, Study for development of a plan of depletion in a rich gas condensate reservoir, Anschutz Ranch East Unit, Summit County, Utah, Uinta County, Wyoming: SPE Paper 12042 presented at 58th Annual Technical Conference and Exhibition of the SPE, San Francisco, CA.

Kocurek, G., 1991, Interpretation of ancient eolian sand dunes: Annual Review of Earth and Planetary Sciences, v. 19, p. 43–75.

Kocurek, G., J. Knight, and K. Havholm, 1991, Outcrop and semi-regional three-dimensional architecture and reconstruction of a portion of the eolian Page Sandstone (Jurassic), in A. D. Miall, ed., Facies architecture of terrigenous clastic sediments and its implications for hydrocarbon discovery and recovery: Concepts in Sedimentology and Paleontology, v. 3, Tulsa, Oklahoma, SEPM, p. 25–43.

Kocurek, G., M. Townsley, E. Yeh, M. L. Sweet, and K. Havholm, 1992, Dune and dunefield development stages on Padre Island, Texas: effects of lee airflow and sand saturation levels and implications for interdune deposition: Journal of Sedimentary Petrology, v. 62, p. 622–635.

Lelek, J. J., 1982, Anschutz Ranch East field, northeast Utah and southwest Wyoming: Geological Studies of Cordilleran Thrust Belt Field Conference Guidebook, Rocky Mountain Association of Geologists, v. 2, p. 619–631.

Lindquist, S. J., 1988, Practical characterization of eolian reservoirs for development: Nugget Sandstone, Utah-Wyoming thrust belt, in G. Kocurek, ed., Late Paleozoic and Mesozoic eolian deposits of the western interior of the United States: Sedimentary Geology, v. 56, p. 315–339.

Lindquist, S. J., and R. Ross, 1993, Anschutz Ranch East field, in B. G. Hill and S. R. Bereskin, eds., Oil and gas fields of Utah: Salt Lake City, Utah, Utah Geological Association, v. 22, 5 p.

Metcalfe, R. S., J. L. Vogel, and R. W. Morris, 1985, Compositional gradient in the Anschutz Ranch East field: SPE Paper 14412 presented at the 60th Annual Technical Conference and Exhibition, Las Vegas, NV.

Nelson, R. A., 1985, Geologic analysis of naturally fractured reservoirs, in G. V. Chilingar, series ed., Contributions in petroleum geology and engineering, v. 1: Houston, Texas, Gulf Publishing Company, 320 p.

Parrish, J. T., and F. Peterson, 1988, Wind directions predicted from global circulation models and wind directions determined from eolian sandstones of the western United States—a comparison, in G. Kocurek, ed., Late Paleozoic and Mesozoic eolian deposits of the western interior of the United States: Sedimentary Geology, v. 56, p. 261–282.

Rubin, D. M., 1987a, Cross-bedding, bedforms and paleocurrents: SEPM Concepts in Sedimentology and Paleontology, v. 1: Tulsa, Oklahoma, SEPM, 187 p.

Rubin, D. M., 1987b, Bedforms 2.0: computer programs for simulation and animation of bedform and cross-bedding: Tulsa, Oklahoma, SEPM, diskette.

Rubin, D. M., and R. E. Hunter, 1987, Bedform alignment in directionally varying flows: Science, v. 237, p. 276–278.

Sweet, M. L., and G. Kocurek, 1990, Empirical model of aeolian dune lee-face airflow: Sedimentology, v. 37, p. 1023–1038.

Tetzloff, D. M., E. Rodriguez, and R. L. Anderson, 1989, Estimating facies and petrophysical parameters from integrated well data: SPWLA Symposium for Log Analysis Software and Review, 17 p.

White, R. R., T. J. Alcock, and R. A. Nelson, 1990, Anschutz Ranch East field—U.S.A. Utah-Wyoming thrust belt, in E. A. Beaumont and N. H. Foster, compilers, Structural traps III, tectonic fold and fault traps: AAPG Treatise of Petroleum Geology, Atlas of Oil and Gas Fields, p. 31–55.

Chapter 23

Identification and 3-D Modeling of Petrophysical Rock Types

Christopher J. Murray
Stanford Center for Reservoir Forecasting
Department of Applied Earth Sciences
Stanford University
Stanford, California, U.S.A.[1]

ABSTRACT

Because the reservoir properties of various rock types can differ considerably, it is best to simulate the distribution of reservoir properties in two steps: simulate the rock types, then simulate the reservoir properties conditional on the rock types; however, identifying the rock types themselves can be difficult, due to the interaction of depositional processes and diagenesis, and the variable amount and quality of data available from well to well. This chapter presents a methodology using cluster analysis for identifying petrophysical rock types in a training data set of cored wells. A case study from the Cretaceous Muddy Sandstone of the Powder River basin illustrates the methodology. After identifying the rock types, analyzing the core properties provided geologic information concerning each rock type. Discriminant function analysis extended the rock type classification to wells without core data. Sequential indicator simulation was used to model and simulate the spatial distribution of the rock types in three dimensions. An annealing method was used to postprocess the simulations so that they honor the rock type transition frequencies seen in the well data.

INTRODUCTION

Petroleum reservoirs result from the complex interaction of several geologic processes. The interaction of erosion, deposition, and diagenesis controls the distribution of the reservoir properties. Although those properties exist in a deterministic fashion beneath the ground, only a limited sample of well log and core data is available from the reservoir. Reservoir flow simulations, which can be used to examine the uncertainty attached to the future performance of the reservoir, require complete grids of the important reservoir properties, and stochastic simulation provides one way to generate these simulations (Haldorsen and Damsleth, 1990). Because the reservoir properties of various rock types can be quite different, some researchers have proposed that the geologic architecture of the reservoir be simulated first, then infilled with reservoir properties (Journel and Gomez-Hernandez, 1989; Suro-Perez and Journel, 1990).

Petrophysical rock types can be defined as classes of rock characterized by differences in reservoir properties, especially porosity and permeability. Petrophysical rock types cannot be defined solely on the basis of the original depositional environments, because diage-

[1]Current address: Pacific Northwest Laboratory, Geology and Geophysics Group, MSIN K9-48, Battelle Blvd., Richland, WA 99352.

nesis often alters the original reservoir properties. Core data are normally available for a limited number of wells, so petrophysical rock types must be distinguishable using well log data. This chapter presents a methodology for defining rock types using multivariate statistical analysis of well log and core data. The spatial distribution of those rock types is then simulated in three dimensions using sequential indicator simulation and simulated annealing (Journel and Gomez-Hernandez, 1989; Deutsch and Journel, 1992).

GEOLOGIC SETTING

The Lower Cretaceous Muddy Sandstone is the major producing formation in the study area (Figure 1), which is centered on Amos Draw field in the northern Powder River basin of Wyoming. The Muddy Sandstone in the study area averages 65 ft (19.8 m) thick and dips gently to the southwest. Marine shale encases the Muddy Sandstone, with the Skull Creek Shale below and the Mowry Shale above. The Rozet sandstone, which is the reservoir horizon at Amos Draw, is the basal member of the Muddy Sandstone in the Amos Draw area. The average thickness of the Rozet is 30 ft (9.1 m). The Rozet is a progradational sandstone deposited during a regression of the Cretaceous sea (Odland et al., 1988). After deposition of the Rozet, a further drop in sea level led to the development of an unconformity that truncates the Rozet member (Figure 2).

Relief on the post-Rozet unconformity and subaerial weathering that developed below the unconformity directly control the permeability distribution and pro-

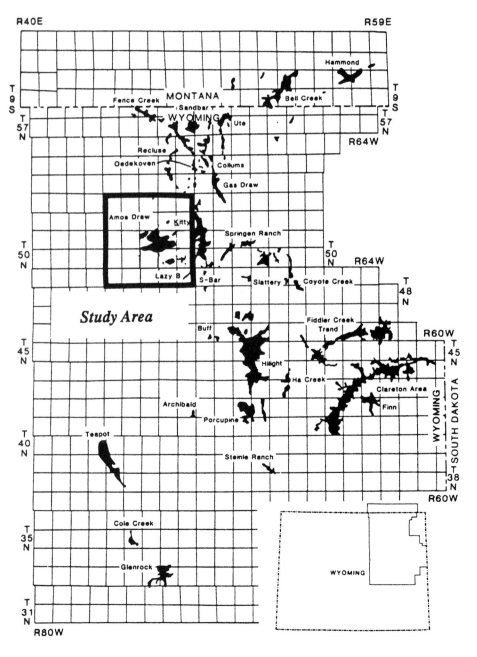

Figure 1. Index map showing the distribution of Muddy Sandstone oil and gas fields in the Powder River basin, with the study area highlighted (after Odland et al., 1988).

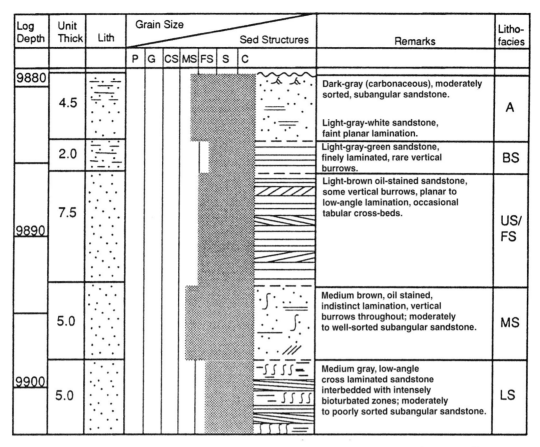

Log Depth	Unit Thick	Lith	Grain Size						Sed Structures	Remarks	Litho-facies
			P	G	CS	MS	FS	S	C		
9880	4.5									Dark-gray (carbonaceous), moderately sorted, subangular sandstone. Light-gray-white sandstone, faint planar lamination.	A
	2.0									Light-gray-green sandstone, finely laminated, rare vertical burrows.	BS
9890	7.5									Light-brown oil-stained sandstone, some vertical burrows, planar to low-angle lamination, occasional tabular cross-beds.	US/FS
	5.0									Medium brown, oil stained, indistinct lamination, vertical burrows throughout; moderately to well-sorted subangular sandstone.	MS
9900	5.0									Medium gray, low-angle cross laminated sandstone interbedded with intensely bioturbated zones; moderately to poorly sorted subangular sandstone.	LS

Figure 2. Core description of the Rozet member from the Amoco Bertolet #1 well at Amos Draw field. A = altered zone, BS = backshore, US/FS = upper shoreface/foreshore, MS = middle shoreface, LS = lower shoreface.

duction at Amos Draw (Chambers, 1986; Odland et al., 1988). Several subtle anticlinal noses exist in the Amos Draw area, but structure does not appear to control production. The Rozet member at Amos Draw has an average of 12–14 ft (3.7–4.3 m) of net pay with porosity averaging 12–18%, and low permeability, averaging 1–2 md (VonDrehle, 1985). As of 1988, 98 producing wells had been drilled within the field (Odland et al., 1988). Figure 3 is a map of the cumulative production in thousands of barrels of oil equivalent (with 6 mcf converted to 1 bbl of oil equivalent) measured at the end of the fourth year of production of each well. Amos Draw is the large productive area that occupies the center of the map.

METHODS

The study area encompasses 900 mi^2 (2500 km^2) in the northern Powder River basin of Wyoming, and is roughly centered on Amos Draw field (Figure 1). Drilling depths to the top of the Rozet range from 8500 to 11,700 ft (2591 to 3566 m). Well log data were available for 70 wells within the study area (Figure 4). Log data included gamma ray, bulk density, and resistivity measurements for all of the sample wells. Core data from the Rozet were available from 183 plug samples taken in 9 of the 70 wells (Figure 4).

The core data included measurements of the porosity and permeability. Three of the cored wells were available for description and sampling (Figure 4). Detailed core descriptions were made for those three wells, with interpretations of the depositional environments. A total of 26 thin sections were taken, including samples from all of the depositional environments and petrophysical rock types. Point counts (200 per slide) were made of the sandstone composition for each thin section.

Statistical analysis was used to define several petrophysical rock types. After identifying the rock types and studying their geological characteristics and spatial distribution in the study area, stochastic three-dimensional (3-D) simulations of the rock types were generated. For that phase of the study, a subset of the study area was used (Figure 4). The 3-D study area is a square measuring 30,000 ft (9,144 m) on a side that occupies the heart of Amos Draw field. The 3-D study area includes 22 of the 70 wells, including 7 of the 9 cored wells, and all 3 of the described cores.

ANALYSIS OF THE DEPOSITIONAL ENVIRONMENTS

The goal of the study is the conditional simulation in 3-D of rock types that would control the fluid flow

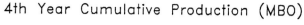

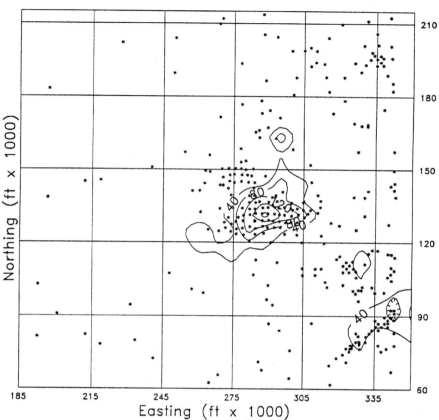

Figure 3. Cumulative production map for the Amos Draw area. Production from each well is in thousands of barrels of oil equivalent (BOE, with 6000 ft^3 of gas = 1 BOE), measured at the end of the fourth year of a well's production.

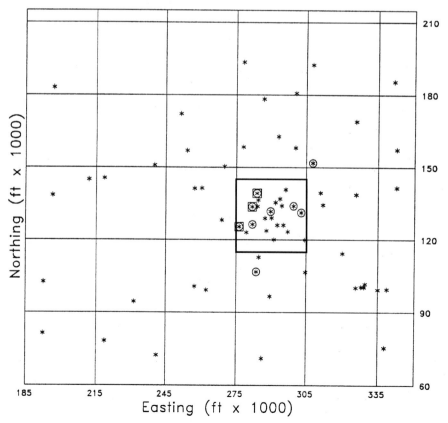

Figure 4. Sample location map for the Amos Draw area. Study area highlighted by box from y = 115–145 and x = 275–305. Open circles = cored wells, open squares = described cores, asterisks = wells having geophysical log data (gamma ray, bulk density, and resistivity).

properties in the study area. In this study, one difficulty common to many reservoir studies is the distribution of data. Core data (porosity and permeability) were available for a small proportion of the wells in the study area, and only three cores were available for description of the depositional environments and thin section analysis. For most of the wells, only well log data were available. Thus, a two-step method for identifying the rock types was developed. The rock types were first defined on the cored wells, so that the geologic attributes of the rock types could be examined in detail and understood in terms of depositional and diagenetic processes. Once the rock types were defined in the cored wells, the classification was extended to wells for which log data alone were available. The multivariate techniques of cluster analysis and discriminant function analysis provided tools for performing this second analysis.

Rozet Depositional Environments

Three Rozet sandstone cores in the study area were measured and described. The interpretations of the depositional environments within the Rozet were based on the sedimentary structures, grain size, and stratigraphic relationships identified in the cores, and on comparisons with published depositional models (Berg, 1986). The depositional environment of the Rozet appears to be a prograding beach or barrier bar adjacent to the Cretaceous Interior Seaway (Odland et al., 1988; Chambers, 1986).

Six lithofacies were identified in the described cores (Figure 2). From bottom to top, the first four lithofacies were the lower shoreface, middle shoreface, upper shoreface/foreshore, and backshore environments associated with a regressive shoreline. They were succeeded by a severely altered zone whose original depositional environment is unclear.

The middle shoreface, which is 1.2–1.5 m (4–5 ft) thick, is the main reservoir at Amos Draw field. The middle shoreface consists of well sorted, fine- to medium-grained feldspathic sandstone, which is brown in hand sample due to oil staining (Figure 2). Sedimentary structures are not well preserved in the middle shoreface, which is not unusual because the middle shoreface commonly is massive (Berg, 1986). Some indistinct planar to low-angle cross-laminations are present, with occasional higher angle laminations near the base of the unit. Vertical trace fossils are common. This unit had been interpreted as an upper shoreface deposit by Chambers (1986); however, the lack of well-defined sedimentary structures, the common trace fossils, and the stratigraphic position between intensely bioturbated muddy sandstones of the lower shoreface and well-laminated sandstones of the upper shoreface indicate that the unit has the characteristics of the middle shoreface, as defined by Berg (1986).

The top of the Rozet consists of an altered zone in which the original sedimentary structures have been destroyed. At the top is a dark-gray carbonaceous zone containing possible root structures, which overlies a whitish zone with a bleached appearance (Figure 2). The altered zone appears to represent a paleosol developed at the Rozet unconformity (Odland et al., 1988). Based on the well-sorted medium quartz grains remaining in portions of these altered zones, the original deposits may have included clean well-sorted sandstones, perhaps with good reservoir properties. Possible depositional settings for those sandstones might include eolian sand dunes or washover fans in the backshore environment.

The well log data and the production data indicate that the reservoir rock is in the middle shoreface. However, the rock quality appears to be highly variable, with some of the middle shoreface appearing to be of marginal quality. The upper shoreface ranges from marginal reservoir to nonreservoir. The lower shoreface, backshore, and the altered zone all appear to be nonreservoir, and the well log properties do not appear to provide any separation between those four lithofacies.

IDENTIFYING PETROPHYSICAL ROCK TYPES USING CLUSTER ANALYSIS

Although it may have been possible to identify the depositional lithofacies in the noncored wells using detailed log correlations, that approach was not pursued because the variation in their reservoir properties indicated that the depositional lithofacies alone did not determine the reservoir properties due to diagenetic alteration of the sandstones. As an alternative, cluster analysis (Davis, 1986; Johnson and Wichern, 1988) of the well log and core data was used to identify petrophysical rock types having different reservoir properties.

In cluster analysis, observations are grouped based on their similarity to other observations. This similarity is often measured using the Euclidean distance between observations:

$$d_{ij} = \sqrt{\frac{\sum_{k=1}^{m}\left(X_{ik} - X_{jk}\right)^2}{m}}$$

where d_{ij} is the distance between observations i and j, and X_{ik} is the measurement on the kth of m variables for observation X_i (Davis, 1986). Because the scale of the variable affects the Euclidean distance, standardized variables are normally used in cluster analysis (Davis, 1986). Hierarchical clustering methods, which are the most common in the earth sciences, start by clustering the two observations that are the closest to one another. The distance between an observation and a cluster, or between two clusters, can be calculated by several methods (Johnson and Wichern, 1988), including the shortest distance between samples in two clusters (nearest neighbor or single linkage), the average distance between all pairs of samples in two clusters (average linkage), or the greatest distance between samples in two clusters (farthest neighbor or complete linkage).

The cluster analysis done in this study used the average linkage method in the SAS CLUSTER procedure (SAS Institute Inc., 1988). The variables input to the CLUSTER procedure were the gamma ray, bulk density, resistivity, grain density, core porosity, and core permeability, all of which were standardized. Because the histogram of the core permeability had a very high positive skewness, logarithms were taken to normalize the marginal distribution of the permeability.

More than 180 samples were used in the analysis, so a display of the dendrograms from the cluster anal-ysis is impractical here. Instead, the results of the cluster analysis are displayed on a canonical discriminant plot (Figure 5). Canonical discriminant analysis summarizes the between-class variance for grouped observations in a manner similar to the way in which principal components analysis summarizes total variance. Three petrophysical rock types were identified by the cluster analysis, and the canonical coefficients indicate good separation between the clusters (Figure 5). To ensure that the clustering results were not bi-ased by the choice of a poor clustering algorithm for the data set, comparisons were made with the results from two other clustering methods available in the SAS CLUS-TER procedure, the farthest neighbor and Ward's methods (SAS Institute Inc., 1988). Each method classified over 95% of the observations into the same clusters assigned by the average linkage method, indicating that the results were stable over the three clustering algorithms.

After identifying the three petrophysical rock types in the cored wells using cluster analysis, discriminant function analysis was used to extend the classification

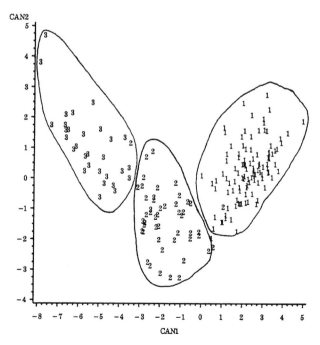

Figure 5. Canonical discriminant plot showing the results of the cluster analysis of the well log and core data. Three petrophysical rock types were identified. CAN1 and CAN2 are the first two canonical variables (SAS Institute Inc., 1988).

of rock types to wells that did not have core data. This was accomplished using the SAS procedure DISCRIM (SAS Institute Inc., 1988), using the cored wells as a training set, and wells that had only log data as the target set to be classified. Cross-validation (SAS Institute Inc., 1988) was used to test the ability of discriminant function analysis to identify the three petrophysical rock types using only the well log data (gamma ray, bulk density, and resistivity). The misclassification rate was only 8.6%.

DESCRIPTION OF THE PETROPHYSICAL ROCK TYPES

Cluster analysis assigns each sample to a cluster or group of observations. The user must determine the geologic "meaning" of the clusters, and check that the classification of the samples is geologically sound. In this study, the stratigraphic, petrographic, and petrophysical properties of the clusters were examined for the training set of three wells with described cores. Figure 6 shows the distribution of well log and core properties for the three clusters, and indicates a clear separation among the reservoir properties of the three rock types. The clusters correspond to nonreservoir rock (rock type 1), an intermediate rock type (rock type 2), and reservoir rock (rock type 3).

Rock Type 1

The core and well log properties of rock type 1 indicate that it consists of nonreservoir rock. The core porosity and core permeability are both low, with the average porosity equal to 5.9% and the median permeability equal to 0.02 md (Figure 6). The gamma-ray readings are high, averaging 68 API units, suggesting that the clay content is high. This was confirmed by thin-section analysis, with an average of 30% clay in rock type 1. This clay includes both depositional and diagenetic clay, based on textural interpretation of thin sections and cores.

Samples from rock type 1 were deposited in all of the sedimentary lithofacies present in the Rozet. Several of the lithofacies were almost completely composed of rock type 1: the lower shoreface (24 of 26 observations), the backshore (7 of 7 observations), and the paleosol at the top of the Rozet (12 of 12 observations). The source of most of the clay in the paleosol, and nearly all of the clay in the middle and upper shoreface, appears to be diagenetic, whereas much of the clay in the lower shoreface appears to be the result of bioturbation. Because clay from either source reduces the permeability and porosity, the well log and core properties cannot be used to distinguish between them; therefore, all of the nonreservoir rocks get lumped into rock type 1 during the cluster analysis.

The feldspar and clay contents of rock type 1 vary widely. This variation appears to be the result of diagenetic alteration of the original sandstone compositions. Chambers (1986) noticed a systematic variation in the composition in the Rozet. Her work showed that the lowermost sandstones in the Rozet have the

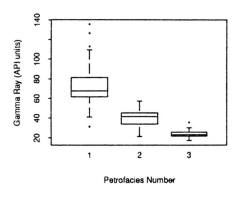

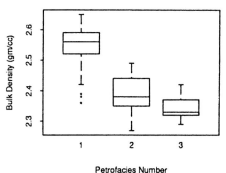

Figure 6. Box plots of five well log and core properties for the three petrophysical rock types.

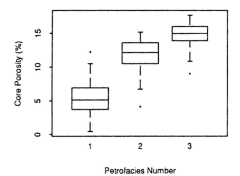

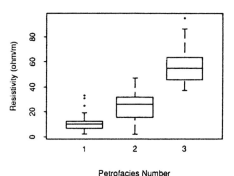

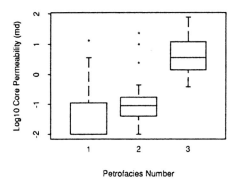

highest feldspar contents and lowest clay contents. As one moves upward through the sedimentary column, the amount of feldspar decreases and the clay content increases. Chambers (1986) suggested that the original sandstone compositions probably were similar throughout the column, and that the upward increase in clay content through the section results from the replacement of feldspar by kaolinite. The upward increase in alteration appears to be due to the proximity to the overlying Rozet unconformity (Chambers, 1986; Odland et al., 1988).

Rock Type 2

Rock type 2 occupies an intermediate class between the reservoir and nonreservoir rock types. The gamma-ray values for rock type 2 are higher than those of rock type 3 (Figure 6), with an average of 39 API units; how-

ever, all of the samples have values of less than 60 API units, a figure often used as a cutoff value for identifying clean sandstone. The core porosity for rock type 2 is also quite good, averaging 12% with 90% of the samples exceeding 10% porosity (Figure 6). Even though the gamma-ray and porosity values for rock type 2 look good enough to classify it as a reservoir rock, the permeability of rock type 2 is very poor, with a median of only 0.1 md, and 90% of the samples having permeabilities of less than 0.4 md.

About one-third of the rock type 2 samples were deposited in the middle shoreface environment, with most of the rest of rock type 2 samples deposited in the upper shoreface environment. The middle shoreface is the main reservoir at Amos Draw, and the upper shoreface would normally be expected to be good reservoir rock (Berg, 1986); however, rock type 2 averages 22% clay (from point counting), whereas rock type 3 has only 11% clay. Also, rock type 2 has an average of only 1% visible porosity, but rock type 3 has an average of 7% visible porosity.

These facts suggest that rock type 2 was originally a clean porous sandstone that has been damaged by diagenetic alteration. Most of the clay that occurs in rock type 2 is kaolinite (Chambers, 1986), which does not contain potassium. This may explain why the diagenetic alteration did not increase the gamma-ray values above the sandstone-shale cutoff (Odland et al., 1988). Much of the porosity indicated by the density log and the core porosity data may actually be microporosity in the interstices of the clay, which would not contribute to hydrocarbon production.

Rock Type 3

Of the three petrophysical rock types, rock type 3 has the best reservoir properties (Figure 6), with an average core porosity of 15%, and a median core permeability of 3.6 md (Figure 6). The gamma-ray values are low, averaging only 24 API units, indicating that the sandstones of rock type 3 are very clean. The density and resistivity values reflect the high porosity, permeability, and oil saturation in rock type 3.

All of the cored samples of rock type 3 come from the middle shoreface environment, and are composed of fine-grained, well-sorted, subangular to subrounded sandstones. Petrographic analysis of rock type 3 revealed that the sandstones are feldspathic arenites (Pettijohn et al., 1987). Visible porosity in thin sections of rock type 3 averages 6–7%. The samples contain 6.5–15% clay, but most of the clay appears to be diagenetic, not depositional, in origin (Chambers, 1986). Several forms of diagenetic alteration are present in the samples from rock type 3. Both silica and calcite cementation are present, with silica preceding the calcite (Chambers, 1986). Secondary porosity can be recognized by the presence of pores that are larger than the average grain size and the presence of partially dissolved framework grains (Hayes, 1979). Many of the pores are filled with oil-stained clay that has reduced the primary and secondary porosity and permeability. Chambers (1986) indicated that most of the clay is kaolinite, and that at least two stages of kaolinite diagenesis are present.

Geographic Distribution of the Rock Types

The cluster analysis performed for identifying petrophysical rock types used traditional clustering methods, which do not consider the spatial distribution of the samples. There was some concern that the rock types identified by the cluster analysis would not exhibit spatial continuity (A. G. Journel, 1991, personal communication). Figure 7 is a cross section through the three wells for which detailed core descriptions were available. The cross section shows that the three petrophysical rock types display a high degree of vertical continuity. Figure 8 displays isolith maps of the number of feet of rock type 3, and rock types 2 and 3 together. Again, the distribution of the rock types shows a high degree of spatial continuity. The distribution of rock types 2 and 3 closely correspond to the main productive area at Amos Draw (compare Figures 3 and 8).

The final output from the rock type classification process was a foot-by-foot determination of the petrophysical rock type in each of the wells in the study area. There were 645 samples in the 22 wells in the 3-D study area. Rock type 1, the nonreservoir rock, constituted 72% of the observations. Rock type 2, the intermediate rock, comprised 17% of the observations, and the remaining 11% of the samples were reservoir rock.

SIMULATION OF THE ROCK TYPES

Spatial models for the petrophysical rock types were developed and used to generate 3-D simulations of the

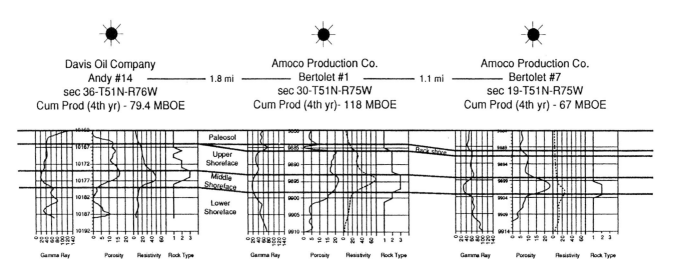

Figure 7. Stratigraphic cross section through the three cored wells that were sampled and described for this study. MBOE = thousands of barrels of oil equivalent.

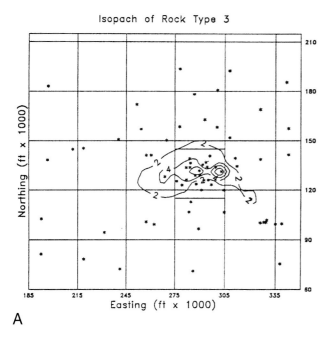

A

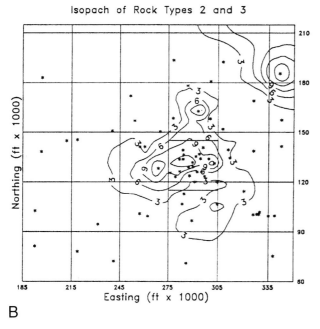

B

Figure 8. Contour maps of (A) the number of feet of rock type 3, the reservoir rock, and (B) the number of feet of rock types 2 and 3, the reservoir and marginal rock types, present at each sample well in the Amos Draw area.

rock types. In work performed for an earlier paper (Murray, 1993), indicator principal component simulation (Suro-Perez and Journel, 1990) was used for the simulations. This technique was used in the hope that it would account for cross-correlation between the rock types; however, the indicator principal component transform did not completely orthogonalize the rock-type data (V. Suro-Perez, 1992, personal communication), and the method did not successfully reproduce the cross-correlation between the rock types (Murray, 1993).

For this study, sequential indicator simulation (Journel and Gomez-Hernandez, 1989) was used to generate 3-D conditional simulations of the rock types. The cross-correlation between rock types was then imparted by postprocessing the simulations using simulated annealing (Deutsch and Journel, 1992).

Variogram Modeling

For the variogram analysis and for the simulations, the data location coordinates were transformed to stratigraphic coordinates (Journel and Gomez-Hernandez, 1989) based on an average Rozet thickness of 30 ft (9.14 m). A stratigraphic coordinate system relocates observations proportionally, based on their distance from the top and base of the stratigraphic unit. The stratigraphy of the unit is assumed to be parallel to the top and bottom contacts. The equation for calculation of stratigraphic coordinates is

$$z'(x,y) = \frac{top(x,y) - z(x,y)}{thickness(x,y)}$$

where $z'(x,y)$ is the stratigraphic coordinate, $top(x,y)$ is the elevation of the top of the unit at the location, $z(x,y)$ is the vertical coordinate of the observation being transformed, and $thickness(x,y)$ is the thickness of the unit at that location.

Figure 9 presents the experimental and model variograms for the two indicator variables in the vertical and horizontal directions (two indicator cutoffs, equal to 1.5 and 2.5, respectively, can be used to represent the three petrophysical rock types). The vertical variograms are well behaved and easily modeled using spherical variograms with ranges of 14 and 12 ft, respectively. The horizontal variograms were more difficult to model, because there were only 22 wells in the study area with an average well spacing of 2640 ft (805 m). The direction of maximum horizontal continuity of N60°E and the 3:1 horizontal anisotropy ratio were derived, in part, from geologic studies of the distribution of productivity at Amos Draw (Odland et al., 1988; VonDrehle, 1985), and from two-dimensional modeling of the distribution of hydrocarbon pore volume at Amos Draw that used a larger study area with more samples (Murray, 1992). The range of 30,000 ft for the variogram model for the first indicator cutoff was also drawn from those studies. The second indicator cutoff has a shorter range, about 12,000 ft (Figure 9). The northeasterly direction of maximum continuity can also be seen on the map of the distribution of rock types 2 and 3 (Figure 8).

Conditional Simulations Using SISIM

Several simulations of the three petrophysical rock types were generated using SISIM, a sequential indicator simulation subroutine (Deutsch and Journel, 1992). Each three-dimensional simulation contained 108,000 grid cells, 60 × 60 × 30 cells on a side. The grid cell size was 500 × 500 × 1 ft (152 × 152 × 0.3 m). The simulations capture several important geologic features present at Amos Draw (Figure 10). The marginal

Vertical Variogram - 1st Cutoff

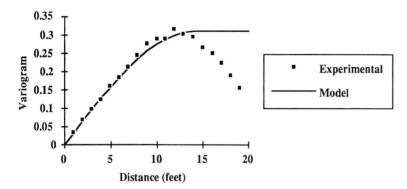

Vertical Variogram - 2nd Cutoff

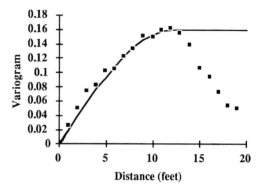

Horizontal Variogram - 1st Cutoff

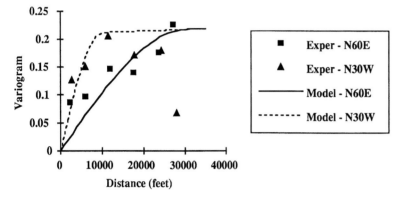

Horizontal Variogram - 2nd Cutoff

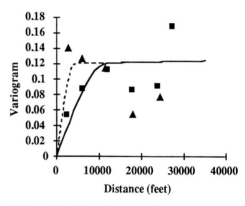

Figure 9. Vertical and horizontal variograms for the two indicator variables showing the experimental and the model variograms.

and reservoir rock types (2 and 3, respectively) are concentrated in the middle of the Rozet, with most of the reservoir rock within the intermediate rock type. The proportions of the three rock types were reproduced, with 74% of rock type 1, 16% of rock type 2, and 10% of rock type 3 for the simulation shown in Figure 10 (the input proportions were 72, 17, and 11%, respectively). The simulations do a reasonable job of reproducing the horizontal variogram model (Figure 11, SISIM #1 and #2).

Postprocessing of the Simulations

Although the simulations output by SISIM appear to be reasonable and honor several important geologic features, they are not entirely satisfactory. One undesirable feature is the presence of a large number of vertical transitions from rock type 3 to rock type 1 (Figure 10). This feature is due to the simulation of many blocks of the best reservoir rock within the worst reservoir rock, and vice-versa. Figure 7 shows that this was not common in the cored wells used for the training data set, and examination of the complete data set from all 22 wells confirmed this. This property of the simulations is

not surprising because indicator cross-correlation is not accounted for by SISIM. Another problem with the simulations produced using SISIM is the presence of a nugget in the vertical variograms from the three rock types, which does not match the vertical variograms of the rock types from the 22 well bores (Figure 11). The vertical variogram models for the two indicator cutoffs are well constrained by the log data. In order to better reproduce the vertical transitions seen in the well bores and improve the vertical variogram reproduction, simulated annealing was used to post-process the simulations produced by SISIM.

The postprocessing was performed using a modified version of the GSLIB program ANNEAL (Deutsch and Journel, 1992). ANNEAL imposes two-point histograms from a training image on the image to be post-processed. The two-point histograms, which can be calculated for several different directions and lags, measure the frequency of transitions from each integer code (e.g., a rock type) to every other integer code (Figure 12). The objective function that was minimized during the postprocessing was the sum of the squared differences between the two-point histograms from the actual and the training images.

North-South

Sisim #1 cross sec for x = 280 (NS)

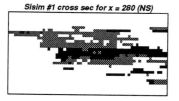

Rock Type

3
2
1

Sisim #1 cross sec for x = 285 (NS)

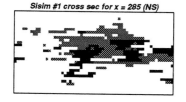

Sisim #1 cross sec for x = 290 (NS)

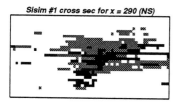

Sisim #1 cross sec for x = 295 (NS)

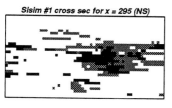

Sisim #1 cross sec for x = 300 (NS)

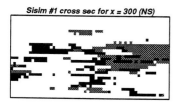

East-West

Sisim #1 cross sec for y = 120 (EW)

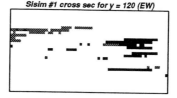

Sisim #1 cross sec for y = 125 (EW)

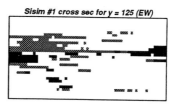

Sisim #1 cross sec for y = 130 (EW)

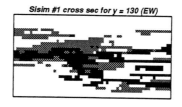

Sisim #1 cross sec for y = 135 (EW)

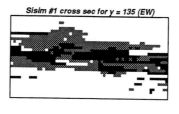

Sisim #1 cross sec for y = 140 (EW)

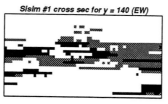

Figure 10. Vertical cross sections from a 3-D simulation of the three petrophysical rock types. Rock type 3 is the reservoir rock. Cross sections are spaced 5000 ft (1524 m) apart. The x and y coordinates are linked to Figure 4. Each cross section is 30,000 ft (9144 m) across and 30 ft (9.1 m) high. The apparent blockiness of the cross sections is due in large part to the 500:1 vertical exaggeration.

For this study, ANNEAL was modified to honor conditioning well data and to use training images from different sources. For the vertical direction, the transition frequencies for the two-point histograms were inferred from the rock types identified in the 22 well bores. The two-point histograms for the horizontal directions were taken from the SISIM simulation that was being postprocessed because the horizontal continuity is poorly constrained and the simulations did a reasonable job of honoring the input horizontal variogram model. For the horizontal directions, two-point histograms were calculated from the simulation to be postprocessed for one lag in each of four directions (north, east, northeast, and southeast).

The postprocessing improved the simulations considerably (cf. Figure 13 with Figure 10), while altering the horizontal variogram reproduction only slightly. Direct transitions from the best reservoir rock to the worst have been reduced, so that rock type 3 is usually surrounded by the transitional rock type 2. This matches the pattern seen in the well data. The post-

processed simulations also do a better job of reproducing the vertical variograms from the well data (Figure 11). One problem with the current implementation of the postprocessing method is the presence of artifacts at the top of the simulated volume (Figure 13). The artifacts are apparently due to slight differences between the rock type proportions in the two-point histograms calculated from the two training images for the horizontal and vertical directions. Current research includes a search for better methods of resolving minor variations in proportions between training data sets.

One possible question is whether the simulations of the rock types after postprocessing are too continuous, and whether the original indicator simulations might not better reflect the "true" variability of the deposit. An obvious answer is that the changes that have been made reflect the petrofacies transition frequencies calculated from the well data; these changes did not come from a subjective model. Also, the Rozet Sandstone was deposited in a coastal depositional

334   Murray

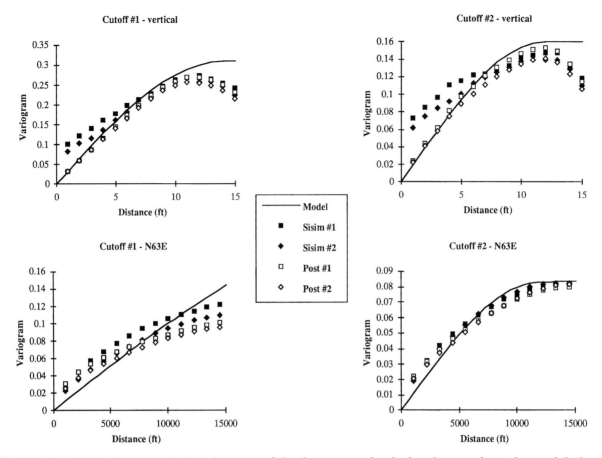

Figure 11. Horizontal and vertical variograms of the three petrophysical rock types from the model, the first two indicator simulations from SISIM (Sisim #1 and #2), and the same two simulations after post-processing (Post #1 and #2).

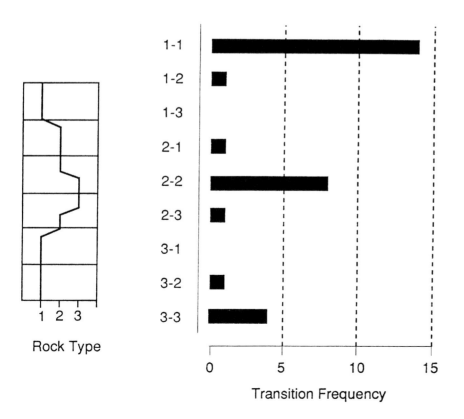

Figure 12. The curve on the left shows a plot of the rock types from one well in the study area. On the right is a plot of the transition frequencies between each of the rock types in that well. Transition frequencies of rock types are not directly captured by sequential indicator simulation. Simulated annealing can explicitly reproduce the two-point histograms calculated from the rock type transitions.

North-South

East-West

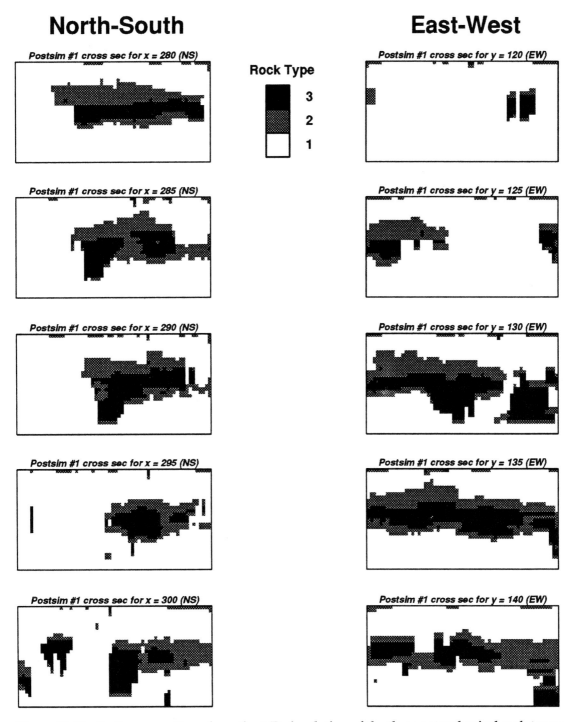

Figure 13. Vertical cross sections from the 3-D simulation of the three petrophysical rock types (shown in Figure 10) after postprocessing using simulated annealing. The scale of the cross sections is the same as in Figure 10.

environment that is often characterized by very high lateral continuity. In the Amos Draw area, that lateral continuity was modified by diagenetic alteration controlled in large part by the original depositional environments in the upper Rozet, as suggested by Chambers (1986) and confirmed by this study. This conclusion suggests that the reservoir interval should

be moderately continuous. The amount of reservoir rock varies in thickness, depending on the depth to which the diagenetic fluids penetrated below the unconformity. The postprocessed simulations appear to capture that character. Figure 14 shows that there is still a high degree of variability among simulations after postprocessing.

East-West

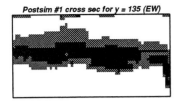

Rock Type

3

2

1

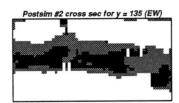

Figure 14. Comparison of results from two simulations for an east-west and a north-south cross section. Postsim #1 and #2 are two indicator simulations generated using the same parameters, but different random seeds, then each was postprocessed using simulated annealing.

North-South

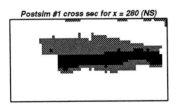

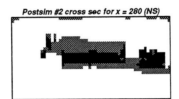

CONCLUSIONS

Within the study area, diagenesis severely altered the reservoir properties of Rozet sandstones, making identification of the original depositional facies using the well log data very difficult. Because of the overlap in their reservoir properties, simulation of the depositional lithofacies would not be sufficient to characterize the reservoir in any case. Cluster analysis provided a method of identifying three petrophysical rock types in the cored wells. The petrophysical rock types have different reservoir properties, and can be fit into a geologic model for the deposition and diagenesis of the Rozet member of the Muddy Sandstone. Discriminant function analysis was then used to extend the classification of the rock types to wells that did not have core data, using the core data as a training data set. This two-step process provided foot-by-foot identifications of the three rock types in each well in the study area.

Three-dimensional simulations of the petrophysical rock types were generated using the sequential indicator simulation algorithm. The simulations reproduced most of the important geologic features identified in the study; however, the vertical transition frequencies between the rock types found in the well data were poorly reproduced. Postprocessing by simulated annealing improved the reproduction of those transition frequencies and the character of the final simulations.

By infilling the simulated rock types with reservoir properties, including porosity, permeability, and hydrocarbon saturation, the simulations could be used as a basis for reservoir fluid-flow simulations. Running fluid-flow simulations of several simulations of the petrophysical rock types allows one to estimate the uncertainty concerning important output variables, including breakthrough times and sweep efficiencies; these variables can be used in reservoir development, risk assessment, and economic planning.

ACKNOWLEDGMENTS

I would like to acknowledge the assistance of Amoco Production Company, which provided the data for this study and partial funding of the research. I also received partial funding for the study from the Stanford Center for Reservoir Forecasting. I also would like to thank Andre Journel, Steve Graham, and Clayton Deutsch for their advice during the research.

REFERENCES

Berg, R. R., 1986, Reservoir sandstones: Engelwood Cliffs, Prentice Hall, 481 p.

Chambers, L. D., 1986, Hydrocarbon trapping mechanism for the Muddy Sandstone reservoir, Amos Draw field, Powder River basin, Wyoming: M.S. thesis, University of Colorado, Boulder, Colorado, 223 p.

Davis, J. C., 1986, Statistics and data analysis in geology, 2d ed.: New York, John Wiley and Sons, 646 p.

Deutsch, C. V., and A. G. Journel, 1992, GSLIB: geostatistical software library and user's guide: New York, Oxford University Press, 340 p.

Haldorsen, H., and E. Damsleth, 1990, Stochastic modeling: Journal of Petroleum Technology, v. 42, n. 4, p. 404–412.

Hayes, J. B., 1979, Sandstone diagenesis—the hole truth, in P. A. Scholle and P. R. Schluger, eds., Aspects of diagenesis: SEPM Special Publication n. 26, p. 127–139.

Johnson, R. A., and D. W. Wichern, 1988, Applied multivariate statistical analysis: Englewood Cliffs, Prentice-Hall, 607 p.

Journel, A. G., and J. Gomez-Hernandez, 1989, Stochastic imaging of the Wilmington clastic sequence: SPE Annual Technical Conference, Paper 19857, p. 591–606.

Murray, C. J., 1992, Geostatistical applications in petroleum geology and sedimentary geology: Ph.D. dissertation, Stanford University, Stanford, California, 301 p.

Murray, C. J., 1993, Indicator simulation of petrophysical rock types, in A. Soares, ed., Geostatistics Troia 92: Kluwer Academic Publishers, p. 399–411.

Odland, S. K., P. E. Patterson, and E. R. Gustavson, 1988, Amos Draw field: a diagenetic trap related to an intraformational unconformity in the Muddy Sandstone, Powder River basin, Wyoming: Wyoming Geological Association Guidebook, 39th Annual Field Conference, p. 147–160.

Pettijohn, F. J., P. E. Potter, and R. Siever, 1987, Sand and sandstone, 2d ed.: New York, Springer-Verlag, 553 p.

SAS Institute Inc., 1988, SAS/STAT user's guide, release 6.03: SAS Institute Inc., Cary, North Carolina, 1028 p.

Suro-Perez, V., and A. G. Journel, 1990, Simulation of lithofacies, in Guerillot and Guillon, eds., Proceedings of 2nd European conference on the mathematics of oil recovery: Paris, Editions Technip, p. 3–10.

VonDrehle, W. F., 1985, Amos Draw field, Campbell County, Wyoming: Wyoming Geological Association, 36th Field Conference Guidebook, p. 11–31.

Chapter 24

The Visualization of Spatial Uncertainty

R. Mohan Srivastava
FSS International
Vancouver, British Columbia, Canada

INTRODUCTION

For several years, we geostatisticians have fought against modern workstation software that uses slick three-dimensional (3-D) color graphics to fool the viewer into believing that some simplistic and smoothed interpolation is real. We geostatisticians would prefer that scientists and engineers think probabilistically and that they recognize the uncertainty inherent in any numerical model of a reservoir. Conditional simulation is a powerful tool in this fight to popularize the acknowledgment of uncertainty.

The growing popularity of conditional simulation in the petroleum industry is due to fact that it enhances visual realism. By presenting an image that honors a variogram, the outcome of a conditional simulation procedure successfully exploits the fact that human vision equips us with very sophisticated tools for recognizing texture and pattern. When geologists are asked to compare the result of a conditional simulation procedure to the result of an interpolation procedure, they usually have a strong preference for the conditional simulation result because it simply looks more real. Unfortunately, this visual realism is a double-edged sword; though it makes the numerical model more appealing it also provides a stronger temptation for scientists and engineers to ignore uncertainty. Unable to cope with a family of realizations, many users of conditional simulation are using a single realization as the basis for performance prediction and reservoir management. Once viewed with slick 3-D graphics tools, this single realization often is mistaken for reality and the space of uncertainty is quietly ignored.

This paper considers some of the issues surrounding the visualization of uncertainty and presents some ideas on how we can harness the sophisticated graphical tools available on modern workstations to serve the goal of educating scientists and engineers about uncertainty in spatial models. It begins with a look at the conventional displays of uncertainty information, most of which are static, and then turns to the possibility of using a dynamic and evolving display. Like good animated cartoons, visually acceptable dynamic displays of a spatial model need to evolve in a gradual fashion and cannot simply be a succession of different realizations. Probability field simulation is used to tackle the problem of producing conditional simulations that are incrementally different. The paper closes with a discussion of some ideas on how uncertainty animations could be used in actual practice.

STATIC DISPLAYS OF UNCERTAINTY

Several types of displays are conventionally used to present the uncertainty information from a family of possible realizations.

- **Probability Maps.** We can display the probability of exceeding a given threshold value at each location in our model. In such a display, the large amount of information in a family of realizations is reduced to a more manageable form by freezing one aspect, the threshold of interest, and presenting the uncertainty information in a single display linked to a specific threshold.

- **Quantile Maps.** Rather than freezing the threshold, we sometimes prefer to freeze the probability value and to present a map showing the corresponding local quantiles or percentiles. Figure 1 shows examples of this type of display for a problem involving the modeling of the structure of the top of a petroleum reservoir in a Gulf Coast reservoir (Xu et al., 1992). Figure 1A shows the values that have only a 10% chance of being too low; because the variable being modeled is the depth to the top of the structure, this map is a pessimistic map from the point of view of potential hydrocarbon volume. Figure 1B shows the values that have only a 10% chance of being too high. Figure 1C shows the median values; at any location, the depth shown on this display has a 50% chance of being too high (and a 50% chance of being too low).

- **Maps of Spread.** Another type of display used to portray uncertainty is a map of some measure of the spread of the local conditional probability distribution. The local standard deviation, variance, or

(A) 10TH PERCENTILES

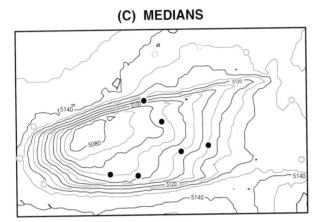

(B) 90TH PERCENTILES

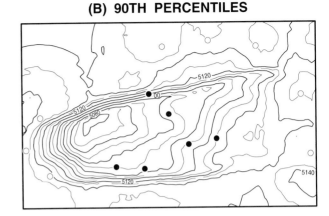

(C) MEDIANS

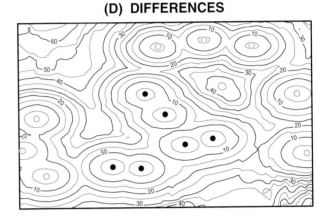

(D) DIFFERENCES

Figure 1. Quantile maps and spread maps.

interquartile range are all measures of spread that reflect some aspect of uncertainty. Where these statistics are low, the spread of possible values in the various realizations is small and there is a higher degree of certainty in the unknown true value. Where these statistics are high, the spread of possible values is large and there is less certainty in the unknown true value. Figure 1D shows the difference between the 90th and 10th percentiles; though this is not a conventional measure of spread, it clearly is similar in many regards to the interquartile range. For this problem involving the top structure of a reservoir, the uncertainty is lowest where we have actual wells that penetrate the structure and is highest where we are far away from well control. The anisotropy evident in this spread map is due to the anisotropy in the velocity field used to process the seismic information.

- **Multiple Realizations.** Another common format for displaying uncertainty is simply to show several possible realizations. Figure 2 shows four possible realizations of the top structure for the same reservoir problem shown in Figure 1. These four realizations represent a broad spectrum of possibilities, with the gross rock volume (GRV) of the reservoir ranging from a quite pessimistic value of

less than 150 million m^3 (Mm^3) to a quite optimistic value of more than 200 million m^3.

Though all of these displays have become conventional in geostatistics, their major shortcoming is that their format is not intuitive. The set of displays shown in Figure 1, for example, is not immediately obvious to most scientists and engineers—it requires some explanation of what a quantile is and why the differences between quantiles is an expression of uncertainty. The same is true of most of the other common conventions for displaying uncertainty. Though probability maps are usually a bit easier to grasp, they still require some familiarity with the format and are often misinterpreted when it is not clear if the display shows the probability above or below the threshold.

Though not yet conventional, there are some hybrid displays that could do a better job of making uncertainty more intuitively obvious by taking advantage of the color graphics capabilities of modern workstations. For example, any of the realizations from Figure 2 could be shown as a three-dimensional (3-D) object illuminated by spotlights with the intensity of illumination being linked to some measure of spread. Where there is high certainty in the unknown value, the surface would be brightly lit

(A) GRV = 145.4 Mm^3

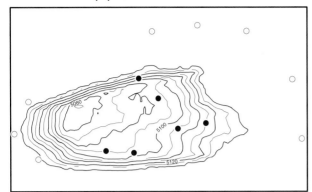

(B) GRV = 159.1 Mm^3

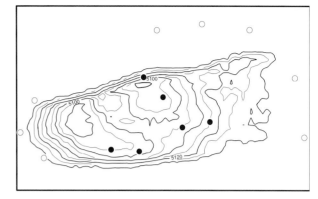

(C) GRV = 181.2 Mm^3

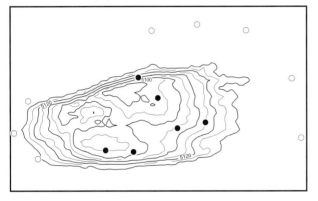

(D) GRV = 204.4 Mm^3

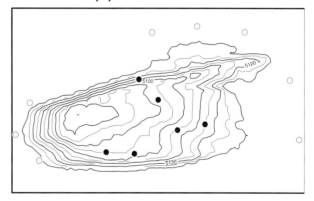

Figure 2. Multiple realizations.

and would be clearly visible; where there is less certainty, the surface would be poorly lit and difficult to discern. By using illumination as a key to uncertainty, such a display taps on the intuitive link between visibility and certainty. As humans, we have a sense of certainty about objects that are brightly lit and clearly visible and a corresponding sense of doubt about objects that recede into the shadows and are not clearly visible.

While research on such hybrid displays continues, there remains an important question: Which particular realization do we choose as the basis for rendering the underlying surface that is to be colored or illuminated? If we choose any particular realization then we do the viewers of our display a disservice by not showing them other equally likely outcomes. If we try to create some kind of average surface, then we end up smoothing away some of the spatial variability that may be a critical aspect of the problem.

DYNAMIC DISPLAYS OF UNCERTAINTY

The key to a truly successful rendering of uncertainty is to find some way of showing all possible realizations. The separate displays of multiple realizations are a start in the right direction, but they are always limited by the space available on the screen or page and by

the viewer's patience for scrolling through hundreds of different maps.

We could try to relieve the viewer of the burden of manually scrolling through hundreds of different maps by automating the process and presenting the realizations one at a time in rapid succession. Unfortunately, because successive realizations may differ considerably in their local details, the eye has difficulty settling on any stable feature and the overall effect is a flickering display that is more irritating than enlightening.

To make a rapid succession of different realizations an appealing and informative display, we need to find some way of ensuring that the successive realizations behave like successive frames in an animated cartoon. If each frame is minimally different from the ones that come immediately before and after it, then our human vision gracefully interprets the minor transition as gradual movement. If the transitions between frames are carefully minimized, a successful animation can be created from as few as 10 frames per second.

If we were able to produce a series of realizations in which consecutive outcomes were minimally different, we could then create a visual display showing a numerical model that is constantly evolving. Using the top structure example shown in Figure 2, our display would show a broadly dome-shaped structure that was perpetually undulating. At locations where we have well control, the surface would appear fixed and unmoving. As we moved away from the well control, the surface would appear to heave and swell. Small depressions would grow larger into deeper holes and valleys; small bumps would gradually grow into larger hills and ridges that would then later subside. In several seconds of such an animation, the viewer would have seen several hundred realizations and the uncertainty in the family of realizations would have been conveyed through the magnitude of the changes in the surface over time. In regions that remained relatively stable through time, the viewer would have a sense of more certainty; in regions that changed in time from valleys to ridges and back again, the viewer would recognize greater uncertainty.

Unfortunately, in geostatistics we do not yet have a way of producing the series of minimally different realizations that such an animation requires. When we want a new realization we simply select a new seed for the random number generator and rerun the program. Small changes in the seed for the random number generator produce unpredictable differences in the appearance of the final realization. The seeds for the four realizations shown in Figure 2, for example, were 12017, 12018, 12019, and 12020, respectively; despite being seeded with consecutive integers, the resulting realizations are very different. We have no way to predict which two seeds will produce realizations that are close enough to one another to serve as consecutive frames in an animation.

Fortunately, the problem of producing minimally different realizations is not intractable. The following section discusses one way of producing such realizations.

PROBABILITY FIELD SIMULATION

Froidevaux (1992) and Srivastava (1992) recently presented a conditional simulation technique known as "probability field simulation." Figure 3 shows a one-dimensional (1-D) example of how the technique works. The first step in the procedure is to create a set of local conditional probability distributions. For the permeability example shown in Figure 3, the left profile summarizes the local conditional probability distributions by showing their 10th, 50th, and 90th percentiles as three separate curves. Where these three curves are far apart, there is considerable uncertainty in the permeability value at these depths; where the three curves coincide, there is a high degree of certainty. The second step in the procedure is to create a probability field, which is a field of values between 0 and 1 that has a uniform distribution and a user-specified pattern of spatial continuity. The middle profile in Figure 3 shows the P-field for this 1-D exercise. At depths of about 6400 and 6460 ft, the P-field reaches its two lowest values of close to 0.0; at depths of about 6425 and 6500 ft, the P-field reaches its two highest values of close to 1.0. The third and final step in the procedure is the sampling of the local conditional probability distribution at each location with the corresponding probability value given

by the P-field for that same location. The conditionally simulated values near 6400 and 6460 ft come from the low end of the conditional probability distributions at these depths; the conditionally simulated values near 6425 and 6500 ft come from the high end of the local conditional probability distributions.

The probability field simulation procedure can be summarized as follows:

1. At each location, x, where a simulated value, zsim(x), is required, calculate the local conditional probability distribution, Fx(z).
2. Build a P-field, P(x), of values between 0 and 1 that are uniformly distributed and have an appropriate pattern of spatial continuity.
3. Use the P-field to sample the local conditional probability distributions:

$$z_{sim}(x) = F_x^{-1}\big[p(x)\big]$$

Froidevaux (1992) and Srivastava (1992) discussed the advantages and disadvantages of probability field simulation. One of the advantages not previously recognized in these two papers is that the technique is ideally suited to the problem of uncertainty animation.

UNCERTAINTY ANIMATION

The key to getting incrementally different realizations from a probability field simulation is to realize that the P-field is unconditional and still serves as a legitimate P-field even if it is translated. In Figure 3, for example, the P-field shown in the middle profile has the same uniform histogram and the same variogram even if it is shifted up or down.

Though the normal practice of probability field simulation is to create a P-field that is exactly the same size as the grid being simulated, we can easily create a P-field that is larger than we need for a single realization. With the example shown in Figure 3, the P-field has additional probability values at the top (for depths less than 6318 ft) and at the bottom (for depths greater than 6520 ft) that we do not need if we want to produce only a single realization. If we want to produce a second realization, however, and want to ensure that it is minimally different from the first one, then all we need to do is shift the P-field by 1 ft up or down and rerun step 3 given in the preceding section. Because the probabilities have changed from the first to the second run, the simulated permeability values will change; because the changes in the probability values are small, the changes in the simulated permeabilities will be small.

The same procedure will work in 2-D or 3-D. Figure 4 shows an example of four minimally different realizations from the 2-D example shown in Figure 1.

In Figure 4, the quantity being simulated is the thickness of the oil reservoir (the difference between the top of the structure and the oil water contact). For all four of these realizations, the local conditional probability distributions are fixed (these local conditional probability distributions are the ones summa-

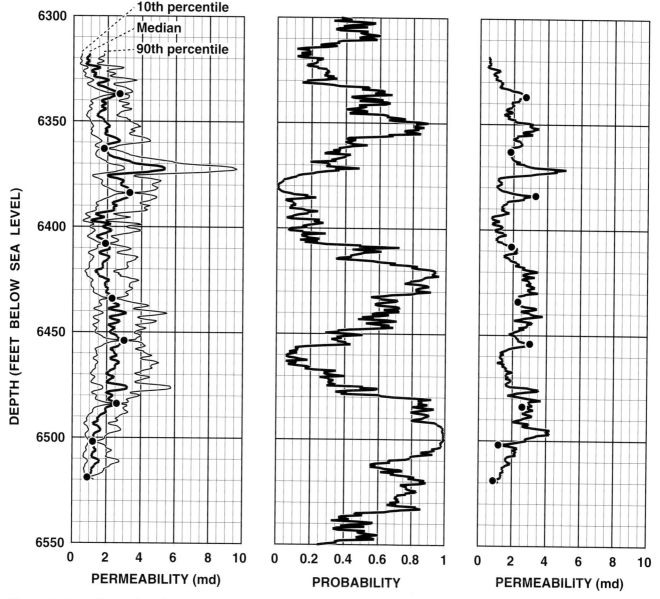

Figure 3. One-dimensional example of P-field simulation.

rized by the quantile maps in Figure 1). The grid to be simulated has 120 columns and 72 rows; for each of the 8640 grid nodes, we have a local conditional probability distribution. Rather than creating a P-field that is also 120 × 72, we intentionally create a P-field that has three additional columns. To create the first realization, we sample the 8640 conditional distributions with the 8640 probability values from columns 1–120. To create the second realization, we sample the same 8640 conditional distributions with the 8640 probability values from columns 2–121. To create the third realization, we sample the same 8640 conditional distributions with the 8640 probability values from columns 3–122. To create the fourth realization, we sample the same 8640 conditional distributions with the 8640 probability values from columns 4–123. Using a notation similar to the one presented previously, we can write the third step of the proba-

bility field simulation as follows:

$$z_{sim}(i, j, k) = F_{(i,j)}^{-1}\big[p(i + k - 1, j)\big]$$

where $z_{sim}(i,j,k)$ is the kth realization of z at node (i,j) on the grid; $F_{(i,j)}(z)$ is the local conditional probability distribution at node (i,j), and $F_{(i,j)}^{-1}(p)$ is its inverse; $p(i,j)$ is the P-field value for column i and row j on the grid.

At first glance, the four realizations in Figure 4 appear identical. A very careful examination, however, will reveal minor differences. One of the more obvious differences across the four realizations is the appearance of the ridge in the 5-m contour line in the lower right corner of the map area. Other slightly less obvious differences are the appearance of the peak of the structure between realizations in Figure 4C and D, and the appearance of the 0-m contour line in the lower right corner between realizations in Figure 4B

(A) GRV = 201.9 Mm^3

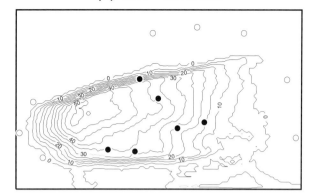

(B) GRV = 201.2 Mm^3

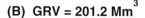
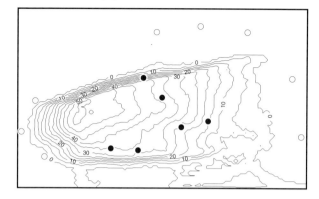

(C) GRV = 200.4 Mm^3

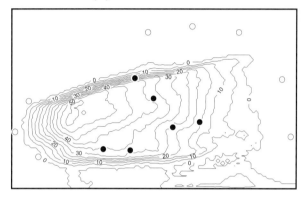

(D) GRV = 199.0 Mm^3

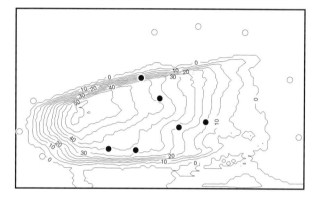

Figure 4. Incrementally different realizations.

and C. All four realizations are slightly different from one another, but in such a minor way that if they were shown in rapid succession, they would create an illusion of gradual movement.

Any conditional simulation technique that uses unconditional simulation can be used to create minimally different realizations. Though the specific example shown in the previous section was accomplished with probability field simulation, it also could have been accomplished with turning bands or with any of the other similar techniques that create the conditional simulations by adding simulated error from an unconditional simulation to an estimation model. With any such technique, we can create successive frames of our uncertainty animation simply by shifting the unconditional simulation by a single column. The only thing we have to change from the conventional practice of conditional simulation is that we need to make the unconditional simulation larger than we would if we were producing a single realization.

CONCLUSIONS AND FURTHER RESEARCH

The uncertainty animation described in this paper has been implemented on a Silicon Graphics workstation where the successive realizations can be shown as an illuminated 3-D surface. The animation can be done at about eight frames per second, so that a five-minute segment consists of 2400 incrementally different realizations.

Longer segments of uncertainty animation could be created by simply adding more columns to the unconditional P-field and producing more new frames. A more expedient solution is to make the unconditional simulation cyclic so that the last column flows naturally into the first one. With this type of P-field, the last realization will be incrementally different from the first and the entire sequence can be repeated indefinitely as an endless loop without any visual break in the animation. With this type of approach, however, the unconditional P-field must be long enough that it allows the user to see a large variety of different possible outcomes. Though the cyclic P-field could be made very short, such a P-field is not likely to produce conditional simulations that reflect the full range of uncertainty.

One of the most interesting avenues of research is the idea of allowing the user interactive control of the animation via a cursor or mouse. In the examples in this paper, the P-field had additional columns that allowed us to generate incrementally different realizations simply by offsetting the P-field by one column. The same trick could be accomplished by adding additional rows to the P-field and offsetting the P-field by one row. With a P-field that has both additional columns and additional rows, one could link the origin of the P-field used in the third step of the procedure to the coordinates of the cursor or mouse. The interesting aspect of this style of interaction is that it gives the user a sense of exploring the space of uncertainty. As the mouse is moved around, the display constantly evolves. If a particular realization looks interesting—a large hydrocarbon volume, for example—then the user could examine other similar realizations in the immediate vicinity by moving the mouse very slightly.

Another advantage of using additional rows and columns in the P-field is that it opens up the possibility of ordering the sequence of realizations according to some parameter of interest to the engineer. For example, one could find a sequence of consecutive origins for the P-field that produces a constantly increasing gross rock volume. The user could then interact with the uncertainty animation by using a sliding bar that controls whether a low-volume scenario or a high-volume scenario is displayed.

As scientists and engineers continue to do more of their design and planning directly at the workstation, the ideal use of this uncertainty animation would be to force the planner to construct a plan based on this constantly evolving model. Imagine the surprise of petroleum engineers when they go to site an additional production well in the thickest part of the reservoir, only to discover that they cannot find part of their display that settles down to a predictably good net

pay. Such surprises would help engineers make appropriate plans in the face of uncertainty. They might learn how to make their plans flexible enough to accommodate all of the possibilities or they may conclude that there is too much uncertainty to make any sensible plan and that they need more data. In either case, the uncertainty animation would have served a useful purpose in ensuring that planning was not taking place on arbitrary models that can never duplicate reality.

REFERENCES

Froidevaux, R., 1992, Probability field simulation, *in* A. Soares, ed., Geostatistics Troia '92, Proceedings of the Fourth Geostatistics Congress: Kluwer Academic Publishers, Dordrecht, Holland, p. 73-84.

Srivastava, R. M., 1992, Reservoir characterization with probability field simulation: SPE Paper 24753 presented at the 67th Annual Technical Conference of the Society of Petroleum Engineers, Washington, D.C.

Xu, W., T. Tran, R. M. Srivastava, and A. G. Journel, 1992, Integrating seismic data in reservoir modelling: the collocated cokriging alternative, SPE paper 24742 presented at the 67th Annual Technical Conference of the Society of Petroleum Engineers, Washington, D.C.

SECTION IV

Public Domain Software and Bibliography

We do not, in this volume, intend to review all of the available geostatistical software; however, for people interested in familiarizing themselves with geostatistical techniques using a computer, C. M. Clayton reviews six public domain geostatistical software programs: STATPAC, Geo-EAS, GEOPACK, Geostatistical Toolbox, and GSLIB.

Finally, J. M. Yarus supplies readers with a basic bibliography. The books and articles listed in this chapter are intended for those wishing to become more familiar with geostatistics. References that further explore the specific subject matter in this volume are found at the end of each chapter and are not necessarily repeated in the bibliography.

◆

Chapter 25

Public Domain Geostatistics Programs: STATPAC, Geo-EAS, GEOPACK, Geostatistical Toolbox, and GSLIB

C. Michael Clayton
Texaco Exploration and Production Inc.
Denver, Colorado, U.S.A.

INTRODUCTION

The means for geostatistical mapping and modeling are now available to all geoscientists and engineers with access to a personal computer. Several geostatistics programs are now readily available in the public domain and can be obtained at very little cost. Five software packages are reviewed here and information on where to obtain them is given.

The geostatistics packages STATPAC, Geo-EAS, GEOPACK, Geostatistical Toolbox, and GSLIB are discussed in their approximate chronological order of appearance in the public domain. The programs reflect, roughly, the evolution in personal computer graphics capabilities, user interface programming, and advances in geostatistical analysis techniques. Each package is reviewed for its analytical capabilities, computer requirements, user friendliness, and suitability for use and learning geostatistics. Several sources for the programs are discussed at the end of the chapter.

Although geostatistics is a discipline over 40 yr in development, geologists, geophysicists, and engineers have been timid about embracing geostatistical techniques. This reluctance is partly due to techniques that have been evolving that are commonly discussed in the language of mathematics, and due to the mistaken perception that the techniques are too difficult and esoteric to be of practical benefit.

Geostatistics can be learned outside of the university and several texts are available for self study. The petroleum, mining, environmental, and groundwater industries frequently offer short courses in the use of geostatistics. True understanding of the power of geostatistical analysis, however, comes with practical application of the theory. Personal computers are capable of doing the complex calculations and data manipulation that many people would simply avoid if it were necessary to do them manually. Thanks to university and government research and development, several geostatistics packages are available in the public domain. The following reviews should help novices select personal computer software that will help them learn geostatistical methods.

A note of caution: these programs have been put into the public domain with the understanding that the user is ultimately responsible for putting the software to proper use and is also ultimately responsible for ensuring that the computations have been done correctly. Computer programs such as these can be exceedingly complex to debug considering all of the possible hardware and operating systems on which they might be tried. Compound that with the potential for user and data errors. The list of problems a user may encounter using these programs, as well as any other sophisticated software, can be daunting. Serious errors are not likely to occur with these programs, but they may occur and the user should be aware of that.

Another caution is that different authors have adopted different nomenclature and mathematical conventions. As an example, Geo-EAS specifies variogram models with an "effective" range, whereas GSLIB uses range parameter "a" defined by the particular variogram model. Another example is that Geo-EAS uses trigonometric angles (counterclockwise from $x = 0$), but GSLIB uses map azimuth directions where $y = 0$. The users must familiarize themselves with the conventions and be aware of differences in the geostatistical software packages. Although data may be exchanged between some of the programs, pitfalls due to definition or convention differences await the unwary.

The International Association of Mathematical Geologists has attempted to standardize geostatistical jargon, and to a lesser extent conventions, with the publication of the *Geostatistical Glossary and Multilingual Dictionary*, edited by Ricardo Olea. Users of geostatistics will find this glossary an invaluable aid in understanding the precise meaning of terms used in the programs and documentation discussed in this chapter.

U.S. GEOLOGICAL SURVEY STATPAC

STATPAC (STATistical PACkage) is a collection of general purpose statistical and geostatistical programs developed over the years by the U.S. Geological Survey (USGS). The programs were compiled in their current form by David Grundy and A. T. Miesch and released as *USGS Open-File Report 87-411-A, 87-411-B, and 87-411-C*, and last updated in May 1988.

Audience

The programs in STATPAC were originally developed for use in applied geochemistry and petrology studies within the USGS. The programs can be used in analyzing any sampling program, but the geostatistical routines require, of course, sample locations to perform two-dimensional spatial data analysis.

STATPAC was developed for personal computer use early in the development of personal computing and does not take advantage of the quality graphics routines now available. Although many of the statistical and geostatistical calculations performed in STATPAC have advantages over other programs available in the public domain, the limited graphics capabilities may discourage beginning practitioners of geostatistics from using this software package.

Experienced practitioners, however, will appreciate the control and complexity the programs in STATPAC offer for complete data analysis.

Capabilities and Functions

STATPAC programs are divided into several categories: data entry, file conversion, graphics, retrieval, mathematics and statistics (data transformations, matrix operations, curve fitting, regression analysis, analysis of variance), two-dimensional spatial statistics (kriging), factor analysis, and utilities.

Standard statistical functions are complemented by one- and two-way ANOVA, Cohens estimates for censored data, discriminate analysis, R-mode factor analysis, extended Q-mode analysis of compositional data, analysis for statistically significant variable discontinuities, chi-squared tests, partial correlation coefficients, step-wise multiple linear regression, and stereographic vector display computation.

Geostatistical functions include extensive variogram analysis, ordinary and universal kriging for both punctual and block data, drift residual analysis for nonstationary data, and cross validation techniques.

Hardware Requirements

STATPAC will run on IBM™ PC/XT/AT/PS-2 and compatible computers, using DOS 2.0 or later, having at least 320 K RAM and a color or EGA monitor with graphics adapter. Hercules monochrome adapters are not supported. The programs will run from floppy drives, but a hard disk is preferred because of the large number of files used and created. The programs, as distributed by the USGS, require a math coprocessor. It would be possible (although not easy) to recompile the FORTRAN source code using a floating-point math

coprocessor emulator in the FORTRAN library or to use a coprocessor emulator TSR utility program. An IBM or Epson™ graphics dot-matrix printer is also required.

Most users will find a Microsoft FORTRAN compiler (Version 4.0 or later) and a BASIC compiler helpful for compiling and running various programs even though compiled executables are provided. Experienced users may find an advantage to modifying code, linking, and recompiling programs to incorporate user-provided customized graphics or data format routines. Microsoft's Macro Assembler (MASM) may also assist advanced users.

Language and User Interface

The programs in STATPAC were originally written in ALGOL or FORTRAN for various computers. As presented in STATPAC, most programs have been rewritten in FORTRAN 77 and supplemented with some BASIC and 8086 Assembly routines. Source code is included, but most programs are linked and compiled into executable programs.

Twenty-nine programs are accessible from a simple menu program and over 60 other programs may be called individually. An inconvenience to new users is the necessity to convert ASCII data files into formats usable by STATPAC programs. Several utilities included in the package allow the creating and editing of the various required format files.

As noted, screen and output graphics from STATPAC are crude compared with programs developed more recently; however, the graphics presented are moderately functional in allowing necessary interpretation to proceed in the analysis. STATPAC does not provide full-screen graphics in many areas, however, including when one tries to fit a model to an experimental semivariogram; this is a disadvantage to all but the most experienced geostatisticians.

User's Manual and Example Data Sets

The user manual for STATPAC is the 34-page *USGS Open-File Report 87-411-A*, "Brief Description of STATPAC and Related Statistical Programs for the IBM Personal Computer." This publication provides essential information about the program and data files, but is not an instruction manual. Two document files included with the program files provide an introduction to the use of STATPAC. Experienced users may find the disk document and open-file report sufficient to use the programs.

The STATPAC distribution disks include a demonstration data file and data files from published works of John Davis and Isobel Clark that can be analyzed in the tutorial. On-line help is nonexistent, but some of the FORTRAN files have comment lines that are helpful in understanding the programs.

Data File Structure and Size

ASCII data are generally entered from the keyboard or into binary STATPAC files using included utility programs. Output from STATPAC FORTRAN or executable programs can generally be used by other STATPAC programs. The open-file report lists the file-naming conventions used by the programs. Knowledge of FORTRAN format statements will facilitate data file preparation. Data are basically row and column format with information headers. Special characters identify where data were not present or detected, data concentrations are too low for detection or too high to be accurately measured, data are best characterized as trace, or the data are a sample blank. Another utility converts STATPAC output for use with Golden Software's SURFER™ and GRAPHER™ and Lotus 123™.

Limits on data file size for the various programs in STATPAC are not available; generally, the size will depend on available computer RAM under the execution of the individual program.

Graphics

Graphics in STATPAC are sent to either the screen or the printer. Screen graphics include histograms, scatterplots, probability plots, and stereograms. Factor variance diagrams, symbol maps, printer maps, discontinuity maps, and ternary diagrams can be printed. A major omission of the program is the ability to display semivariograms while trying to fit experimental data to a model semivariogram. The quality of the graphics in STATPAC presents an opportunity for improvement, but the programs in STATPAC instead should be judged by their individual usefulness. Example printer graphics from STATPAC are shown in Figure 1.

EPA GEO-EAS

The U.S. Environmental Protection Agency (EPA) has developed a public domain geostatistics program called Geo-EAS. Geo-EAS stands for Geostatistical Environmental Assessment Software. The program was developed by Evan Englund at the EPA Environmental Monitoring Systems Laboratory and Allen Sparks of Computer Sciences Corporation in Las Vegas, Nevada. The current release, Version 1.2.1, was compiled in July 1990. An updated manual was released in April 1991.

Audience

Geo-EAS was developed for environmental site assessment and monitoring situations where data are collected on a spatial network of sampling locations (in soil, air, or water) and where one desires contour maps of pollution concentration (or other variables). This program design is not limited to environmental pollution; any spatial sampling program (mineral sampling, bio-assays, structure mapping, petrophysical measurements, etc.) can use the geostatistical estimation tools in Geo-EAS to provide better maps of distributions and a measure of the error in the resultant mapping than is possible with ordinary computer mapping methods.

Geo-EAS was developed for practical geostatistical analysis by scientists who have a working knowledge of geostatistics and a basic understanding of geosta-

tistical concepts; however, the integrated program layout, interactive design, and the excellent user's manual combine to make this an excellent teaching or self-study tool for learning the procedures of geostatistical analysis.

Capabilities and Functions

Geo-EAS is a collection of program modules for two-dimensional (surface) analysis of spatially related data. The interactive programs make extensive use of screen graphics, such as maps, histograms, scatterplots, and semivariograms. The graphical presentation of the geostatistical functions helps to search for hidden patterns, correlations, and problems in data sets.

The basic univariate and bivariate statistical analyses make use of scatterplots, linear regression, and line graphs. Geostatistical pair comparisons, semivariogram analysis and modeling (nested semivariogram, madogram, and nonergodic or inverted covariance variogram), cross validation, error mapping, and two-dimensional kriging (ordinary and simple; point and block) are used to further analyze the data. The program also uses anisotropy analysis. Universal kriging capability, however, is not included and limits Geo-EAS use where the stationarity assumption and ordinary kriging may not apply due to gradual trends in the data.

Geo-EAS includes several data preparation and management utilities for data transformations, log and ln transform, unary and binary operations, data contouring, screen plots, plot files, DOS access, and file operations for sorting, row and column extraction, merging, and reporting data and results.

Hardware Requirements

Geo-EAS was designed for PC/XT/AT personal computers using DOS 3.1 or higher and having 640 KB RAM. CGA, EGA, VGA, or Hercules™ graphics adapters and graphics monitors are highly recommended because graphics analysis is fundamental to using the program. A hard disk with 3 MB of storage space is also recommended; the program builds many temporary and output files during execution, so more storage is desirable. Although a math coprocessor is not required, if one is installed it will increase program execution speed. The use of a virtual RAM disk will also allow the program to execute faster, especially with older or nearly full hard drives. An optional graphics printer and a plotter supporting HPGL™ will allow hardcopy of the analysis. Memory resident programs should be unloaded, if possible, to retain as much free RAM as possible.

Language and User Interface

Geo-EAS is provided as compiled executable programs but was written in Microsoft FORTRAN 77 Version 4.01. The source code is available (see following sections), but some of the graphic routines' source code is not public domain and would have to be purchased to compile the programs as they were by the EPA and Computer Sciences Corporation.

```
C:\WORK> bastat

ENTER INPUT STATPAC FILE NAME = clarkfe.uxv
Wed.  12/10/86
03:41:22pm
NO OF ROWS   =    50
NO OF COLUMNS=    7


DO YOU WANT SELECTED ROWS ? n
DO YOU WANT SELECTED COLUMNS ? n
WHAT DO YOU WANT INCLUDED IN BASIC STATISTICS ?
          1-ONLY UNQUALIFIED DATA
          2-ONLY QUALIFIED DATA
          3-ALL DATA(IGNORING QUALIFYING CODES)
TYPE 1, 2, OR 3 : 3
   50    READING DATA...


                    UNIVARIATE STATISTICS

VAR  COLUMN   MINIMUM   MAXIMUM    MEAN    DEVIATION  VALID B   L   N   G OTHER
 1 X-COORD.   .000E+00 4.000E+02 2.149E+02 1.2982E+02  50   0   0   0   0   0
 2 Y-COORD.   .000E+00 3.900E+02 1.837E+02 1.0842E+02  50   0   0   0   0   0
 3 %Fe        2.440E+01 4.410E+01 3.450E+01 4.7516E+00  50   0   0   0   0   0
 4 KRIG VAL 2.833E+01 4.292E+01 3.464E+01 3.1758E+00  50   0   0   0   0   0
 5 KRIG ERR-8.328E+00 6.394E+00-1.343E-01 3.5660E+00  50   0   0   0   0   0
 6 KRGSTDEV 2.099E+00 5.372E+00 3.833E+00 8.0561E-01  50   0   0   0   0   0
 7 ERR/STDV-1.880E+00 1.924E+00-1.092E-02 9.0375E-01  50   0   0   0   0   0

DO YOU WANT TO SEE THE CORRELATIONS ? y

CORRELATIONS -  COMPUTED USING ORIGINAL DATA

       1     2    3    4    5    6     7
 1  1.00   .11  .30  .39  .05 -.23   .05
 2   .11 1.00   .14  .22 -.01 -.08  -.01
 3   .30   .14 1.00  .66  .74 -.35   .73
 4   .39   .22  .66 1.00 -.01 -.35  -.01
 5   .05  -.01  .74 -.01 1.00 -.16   .98
 6  -.23  -.08 -.35 -.35 -.16 1.00  -.13
 7   .05  -.01  .73 -.01  .98 -.13  1.00

DO YOU WANT HISTOGRAMS FOR ANY SELECTED COLUMNS ? y
ENTER COL NO. FOR WHICH HISTOGRAM IS DESIRED  = 7
DO YOU WANT TO ENTER LWR LIMIT & CLASS INTERVAL FOR THIS COL?
    IF NOT THE MINIMUM VALUE FOR THIS COL AND STURGES' RULE WILL BE USED TO
    COMPUTE THE HISTOGRAM.
y
ENTER LWR LMT & INTERVAL (IN ORIG DATA) = -2.0,0.5
   50    READING DATA...

HISTOGRAM FOR COL NO.   7(ERR/STDV)


                     OBS CUM  PER  CUM     CLASS
     LIMIT - UPPER   FRQ FRQ  FRQ  FRQ   MIDPOINT  DISTRIBUTION

          N           0   0   .0   .0
          L           0   0   .0   .0
-2.000E+00--1.500E+00 2   2  4.0  4.0   -1.750E+00 X
-1.500E+00--1.000E+00 7   9 14.0 18.0   -1.250E+00 XXXX
-1.000E+00--5.000E-01 7  16 14.0 32.0   -7.500E-01 XXXX
-5.000E-01-  .000E+00 6  22 12.0 44.0   -2.500E-01 XXX
  .000E+00- 5.000E-01 13 35 26.0 70.0    2.500E-01 XXXXXX
 5.000E-01- 1.000E+00 9  44 18.0 88.0    7.500E-01 XXXXX
 1.000E+00- 1.500E+00 3  47  6.0 94.0    1.250E+00 XX
 1.500E+00- 2.000E+00 3  50  6.0 100.0   1.750E+00 XX
          G           0  50   .0 100.0
          B           0  50


DO YOU WANT ANOTHER HISTOGRAM ? n
DO YOU WANT PERCENTILES FOR SELECTED DATA ? n
DO YOU WANT ANOTHER SET OF COMPUTATIONS ? n
DO YOU WANT TO PROCESS DATA FROM ANOTHER FILE ? n
NORMAL END OF PROGRAM
                                                              A
```

Figure 1. (A) Example of STATPAC univariate statistics and histogram plot.

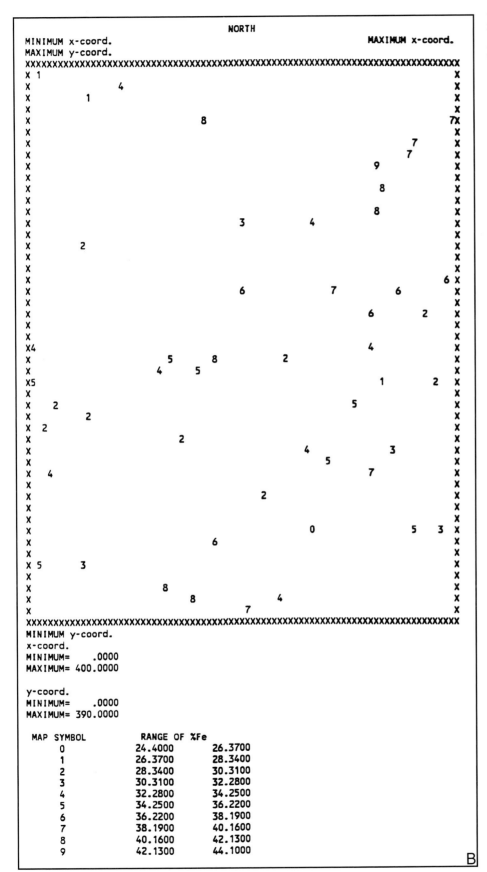

Figure 1 (continued). (B) Example of STATPAC *x,y* plot.

```
                          VARIOGRAM
                       VARIABLE ID: %Fe
       Variogram of Isobel Clark's iron data (from Practical Geostatistics)
                                          DATA USED IN CACULATIONS
   DIRECTION   =  90.   WINDOW  90.       MEAN          = .345E+02
   CLASS SIZE  =  45.0                    VARIANCE      = .226E+02
   MAX. DISTANCE =        900.            STD DEVIATION = .475E+01
   LOGARITHMS -NO      RELATIVE VARIOGRAM -NO   NO.OF SAMPLES =    50
   ROBUST -NO

      DISTANCE      PAIRS    "DRIFT"    GAMMA (H) MOMENT CENT AVER DIST
        0 -   45     38    -.324E+01   .834E+01   .837E+01   30.88
       45 -   90    111    -.527E+01   .196E+02   .203E+02   67.47
       90 -  135    159    -.496E+01   .190E+02   .188E+02  113.30
      135 -  180    186    -.565E+01   .244E+02   .243E+02  158.14
      180 -  225    178    -.521E+01   .197E+02   .197E+02  201.69
      225 -  270    170    -.585E+01   .250E+02   .252E+02  246.04
      270 -  315    133    -.631E+01   .280E+02   .280E+02  290.16
      315 -  360    127    -.546E+01   .234E+02   .234E+02  337.29
      360 -  405     88    -.590E+01   .261E+02   .261E+02  381.28
      405 -  450     27    -.606E+01   .258E+02   .259E+02  425.36
      450 -  495      6    -.617E+01   .208E+02   .209E+02  464.90
      495 -  540      2    -.440E+01   .968E+01   .968E+01  500.67

                       VARIABLE ID: %Fe
       Variogram of Isobel Clark's iron data (from Practical Geostatistics)

      .280E+02 *            X
      .266E+02 *                    X &
      .252E+02 *        X   X
      .238E+02 *              X
      .224E+02 *
   G  .210E+02 *          X          &
   A  .196E+02 *   X  X
   M  .182E+02 *
   M  .168E+02 *
   A  .154E+02 *
      .140E+02 *
      .126E+02 *
   *  .112E+02 *
   H  .981E+01 *                 &
   *  .841E+01 * X
      .701E+01 *
      .560E+01 *
      .420E+01 *
      .280E+01 *
      .140E+01 *
               +----+----+----+----+----+----+----+----+----+----+
               0.    180.    360.    540.    720.    900.

      AMPERSANDS (&) DENOTE FEWER THAN 30 PAIRS            C
```

Figure 1 (continued). (C) Example of STATPAC experimental variogram.

The 13 programs that constitute Geo-EAS may be run from a main menu program. The main menu uses a moving light-bar cursor to move among menu selections. As each selection is highlighted, a one-line description of the program appears at the bottom of the screen. After selection from the main menu, each program has a similar interactive menu screen; many of the menu choices are common to other programs in Geo-EAS, which facilitates learning to use the program. After entering the name of the data file to be used, the program reads the file and allows many choices to be made by "toggling" with the space bar. Other numeric or alphabetic data necessary to execute the program are entered at the cursor prompt. Some menu choices bring up additional nested menus or allow viewing the graphics or printing the output. Occasionally, if not enough memory is available, some of the selections may not load and execute prop-

erly from the menu program; to remedy this, exit and run the desired routine from the DOS prompt. All of the programs may be executed by bypassing the main menu program, allowing the use of the statistics or graphics routines on data when a complete geostatistical analysis is not needed.

User's Manual and Example Data Sets

The softbound user's guide documenting the Version 1.2.1 release is very well written. The manual clearly explains how to install and use Geo-EAS, and includes a four-page glossary of geostatistical terms and an example of how to use Geo-EAS to conduct a geostatistical study. The manual also clearly explains input file format and all of the program's menu selections. Liberal use of screen capture displays, explanatory diagrams, and example graphic displays punctuate the manual's explanations and discussions. The manual assumes one has a general knowledge of geostatistical concepts and techniques, but is written clearly enough to be used as a workbook in conjunction with a good geostatistical methods text, such as those by Isaaks and Srivastava (1989), Journel and Huijbregts (1978), or Hohn (1988).

The distribution disks include sample data to be used in conjunction with the example analysis in the manual. These data can be used to immediately begin using the program and try its features. The example data also serve as a reference for the data file format. Additionally, the example study has some suggested exercises designed to familiarize the user with the program's capabilities and to suggest various approaches to an investigation and analysis of the data.

The manual serves as the help resource for the program; there are no help screens in the program, nor are there tutorial or manual files on the distribution disks. Many menu choices, however, display an explanatory line when highlighted by the cursor bar. The README file contains information on bug fixes, changes, and updates to the printed documentation.

Data File Structure and Size

All programs in Geo-EAS use the same data file format. The files are simple ASCII text that can be created from spreadsheets, database programs, or with most DOS text editors. The format, clearly illustrated in the user's manual, accommodates a title, variable names, variable units, and data in a spreadsheet-like rows-and-columns matrix. FORTRAN format statements may be used if desired.

Typical maximum data file limits are 1000 samples with up to 48 variables. If more samples exist than can be read into memory, only those that fit, typically the first 1000, will be used; if more than 48 variables are attempted in the data file, an error message will be displayed and the program will terminate. One program, XYGRAPH, will work with only 500 samples. One of the most severe limitations is the number of pairs that can be computed for variogram modeling: fewer than 200 data points. This limitation can be overcome by producing pair-distance files from many randomly selected data pairs and comparing results. Another option would be to do variogram analysis using another geostatistical program, such as GEOPACK or Geostatistical Toolbox, which does not have such a small file-size restriction and has the added advantage of automatic semivariogram model fitting.

Memory allocation is fixed for samples and variables in memory. The program will use all of the available 640 KB RAM that is available to it and memory resident programs should be unloaded to maximize use of RAM. Projects with 1000 samples and 48 variables can be handled as a single data file. Larger data sets may require splitting the data into smaller sets and doing separate analyses for each set.

The program writes ASCII output files that can be used by other programs for graphics or further analysis. A significant time and effort saver is the fact that due to the program's development origins, Geo-EAS files are usable in the EPA GEOPACK and FSS Geostatistical Toolbox geostatistical programs; GSLIB also uses a simplified Geo-EAS file format.

Graphics

Visualizing the spatially related data and geostatistical analysis is the key to predicting the best fit of predicted interpretations. Geo-EAS provides good on-screen graphics of the data with histograms, probability and distribution plots, semivariograms, contour plots, line graphs, scatterplots, box plots, kriging displays, and error maps. Some example graphics from Geo-EAS are shown in Figure 2.

Most graphic displays and results can be sent directly to a printer or to a file. Metacode files are created and can be converted to HPGL plot files using the included HPPLOT utility. HP 7470, 7475, and 7450 plotters are supported directly; many graphics programs will import HPGL files and send the graphic output to a wide variety of printers and plotters. Many word processing programs will also import the HPGL graphics files. Screen dumps of semivariograms and other screens may be sent to Epson printers and HP™ laserjet and Paintjet printers, or graphics can be captured and printed later with the use of a separate screen grabber utility.

Although Geo-EAS graphics are sufficient for interactive interpretation and informal reports, the user may find that importing the metacode or HPGL files into commercial presentation graphics programs is desirable to edit, enhance, and produce publication-quality graphics hardcopy. A commercial screen capture program such as Pizazz Plus™ or any of numerous public domain utilities may also prove useful for saving and enhancing display output.

EPA GEOPACK

A public domain geostatistical program suitable for teaching, research, and project work called GEOPACK has been released by the EPA. GEOPACK was developed for the U.S. Environmental Protection Agency's Robert S. Kerr Environmental Research Laboratory in Ada, Oklahoma, by S. R. Yates at the U.S. Department of Agriculture U.S. Salinity Laboratory in Riverside, California, and M. V. Yates at the University of California–Riverside, Department of Soil and Environmental Sciences. Version 1.0 was released in January 1990.

Audience

GEOPACK is recommended for scientists in mining, petroleum, environmental, and research who do not have access to more powerful workstation or mainframe geostatistics programs and who wish to exploit the advantages of quantitative interpretation techniques for spatially related data.

GEOPACK is designed to be used by both novice and experienced geostatistics practitioners. The program

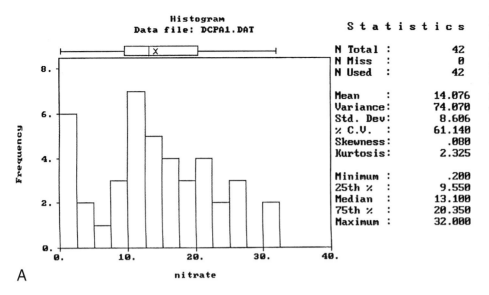

A

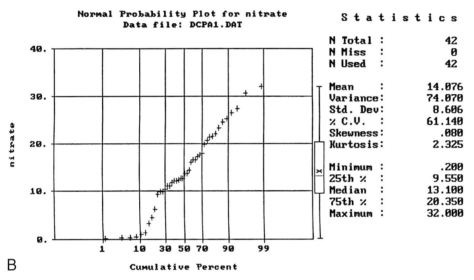

B

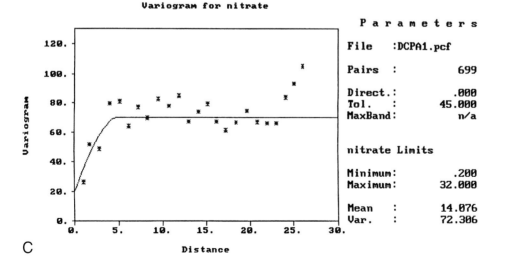

C

Figure 2. (A) Example of Geo-EAS histogram plot. (B) Example of Geo-EAS probability plot. (C) Example of Geo-EAS experimental and fitted model variogram.

may be used as a learning tool, but is sophisticated enough for serious project work. Although formal training in geostatistics is desirable, interested scientists can learn how to use many geostatistics tools with this program; however, texts such as those by Hohn (1988), Isaaks and Srivastava (1989), or Journel and Huijbregts (1978) are necessary to explain the theory and proper use of geostatistics.

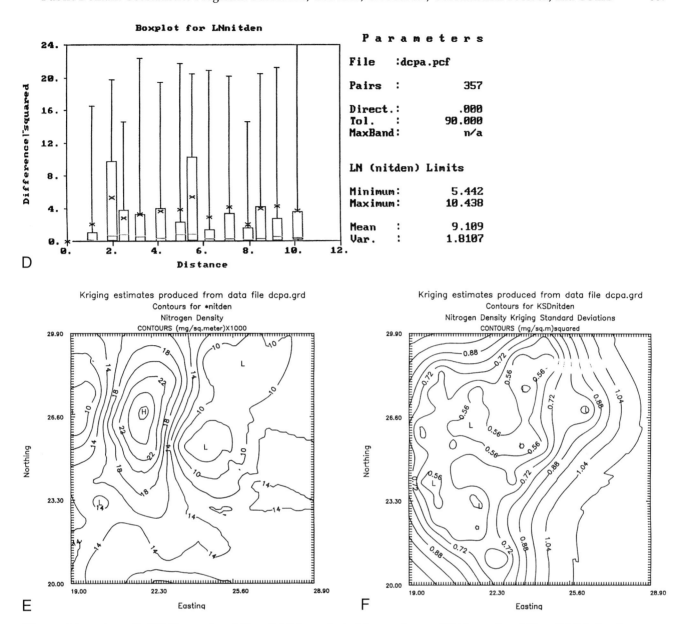

Figure 2 (continued). (D) Example of Geo-EAS box plot of lag variance. (E) Example of Geo-EAS kriged contour plot. (F) Example of Geo-EAS kriged error contour plot.

Capabilities and Functions

GEOPACK is a geostatistical software system for analysis of spatially correlated data. The programs are integrated to free the user from excessive file editing and data manipulation. GEOPACK performs basic statistical functions, including mean, median, standard deviation, skew, kurtosis, minimum and maximum, linear regression, Kolomagorov-Smirnov test for distribution and percentile calculation, and polynomial regression.

The geostatistical functions of GEOPACK include variography (semivariograms, cross-semivariograms, and semivariograms for combined random two-dimensional spatially dependent functions), least-square or iterative variogram model fitting, linear estimation (ordinary kriging and cokriging in two-dimensions with associated variance, point and block kriging, geo-

metric anisotropy, cross validation of spatially correlated structure, and data transformation for indicator kriging), and nonlinear estimation (disjunctive kriging, disjunctive cokriging with conditional probability cutoffs, point or block support, and anisotropy). GEOPACK does not support universal kriging.

GEOPACK provides several utility functions. The program allows setup of menus and incorporation of other statistical, editing, technical, and graphic programs. DOS may be accessed while the user is in the program and a file packing and unpacking utility compresses data to conserve disk space. An interface is provided to convert GEOPACK data to Geo-EAS format (Geo-EAS is another EPA geostatistical program) and convert Geo-EAS output files for use in GEOPACK. The program selectively creates files of calculated geostatistics that can be used by other user-supplied programs.

Hardware Requirements

GEOPACK requires a 80286 or 80386 processor-based personal computer with 640 KB RAM and using DOS 3.3 or later. A CGA, EGA, VGA, or Hercules graphics adapter and appropriate monitor are also required. A hard disk with a minimum 4 MB is required for the program, but data files will require additional storage space. A math coprocessor will be used if present, but is not required. Faster file manipulation can be obtained by using a virtual RAM disk.

Language and User Interface

The program was written in FORTRAN and C. Executable programs are provided, but the source code is not currently distributed. The program, however, is customizable. Provision is built in to add Geo-EAS and other programs one may wish to use, such as statistics (e.g., Microstat™), line graphics (e.g., GRAPHER™), surface graphics or mapping (e.g., SURFER™), text editing (e.g., EDT™), and database management (e.g., Foxbase™). These user programs are not provided, but the menu creation and program interface capabilities are built into GEOPACK.

The user interface for GEOPACK is a system of nested menus. A moving light bar is used for menu item selections, and function keys select menus for help, utilities, and user-defined applications. Provision is made for the user to modify and add menus.

User's Manual and Example Data Sets

The user's manual is published EPA document EPA/600/8-90/004 entitled *Geostatistics for Waste Management: A User's Manual for the GEOPACK (Version 1.0) Geostatistical Software System*. The program, however, is suitable for general applications and is not limited to waste management. The manual clearly and concisely explains installation, default settings, file-naming conventions, data file formats, menu choices, changing and editing menus and user files, the Geo-EAS interface, examples, and references. The manual is illustrated with captured screens showing menus and input data.

Three examples with data sets provide a somewhat abbreviated tutorial of the geostatistical procedures. The reference section of the manual lists more complete geostatistical studies that can be consulted for analytical procedures.

In addition to the manual, the program contains on-line, context-sensitive help at the press of the F1 function key. A stand-alone executable HELP program also provides access from outside the program.

Data File Structure and Size

Formatted ASCII data files for GEOPACK can be created or edited using a standard text editor or from formatted output from other programs, such as spreadsheets or database managers. The data format uses standard FORTRAN format instructions. Minimal contents of a data file are one or more spatially related random variables and x,y Cartesian-coordinate pairs for each variable value. The file format also provides for a title, the number of variables, the number of samples, variable names, and maximum sample value cutoffs.

Example data files are shown in the manual and are included with the program.

Data file size is constrained by computer RAM; a storage of approximately 10,000 locations is allocated to samples and variables. The program currently has a limit of 10 variables for each sample location. If a file is too large for GEOPACK, an error message occurs indicating that not enough memory is available. This can be remedied by reducing either the number of samples or variables.

Data files can be converted into Geo-EAS input data file format using a utility in GEOPACK; however, Geo-EAS data files will need to be manually edited (or a translation routine will need to be written) for them to be used by GEOPACK. Utility routines in GEOPACK will allow Geo-EAS output files to be used by GEOPACK. Spreadsheet or database managers can also be used with little effort to create the properly formatted data input files for GEOPACK.

The program creates output files for most calculations, but some displayed statistics and graphics cannot be saved to data files or graphics files for later plotting without the use of a screen capture utility.

Graphics

On-screen graphics within GEOPACK allow viewing of semivariograms, cross-semivariograms, linear and logarithmic line plots, and line or pixel-block contour maps. Screen capture or screen dump programs (not included) can be used to print the on-screen graphics.

Device drivers are included with GEOPACK to plot graphics to HP LaserJet II printers, HP plotters and Epson graphics-compatible dot-matrix printers. Other device drivers could be written and used. The quality of the graphics is intermediate, and is good enough for interpretation and use in reports. The graphics files can also be used with commercial graphic packages for enhanced appearance. Figure 3 shows examples of GEOPACK graphics.

FSS GEOSTATISTICAL TOOLBOX

FSS International, a consulting company specializing in natural resources evaluation and risk assessment, has released Geostatistical Toolbox to the public domain. The program was developed and written by Roland Froidevaux, principle partner of FSS, in part while a Visiting Scientist at Stanford University. Version 1.30 was released in December 1990. It is used in conjunction with geostatistics training offered by the company.

Audience

Geostatistical Toolbox was developed to provide personal-computer-based, interactive, user-friendly geostatistics tools to workers in the mining community. The program was designed to incorporate state-of-the-art applied geostatistical techniques. The program has been used at Stanford University as a teaching aid in the Applied Earth Sciences Department and was the model that lead to the EPA Geo-EAS program. Geostatistical Toolbox is appropriate for academic, mining, petroleum, and environmental applications.

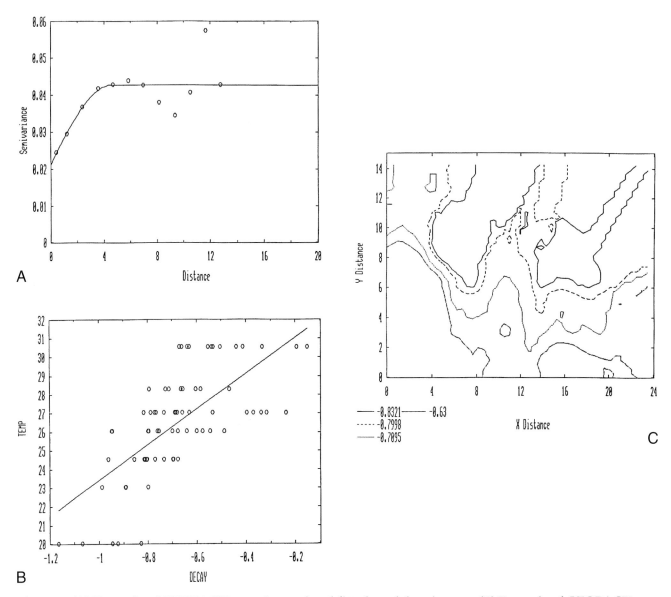

Figure 3. (A) Example of GEOPACK experimental and fitted-model variogram. (B) Example of GEOPACK covariant crossplot. (C) Example of GEOPACK kriged contour plot.

The rigorous application of tested geostatistical methods makes Geostatistical Toolbox required PC software for geostatisticians who do not have access to commercial, institutional, or governmental mainframe geostatistics programs. Students will also be able to use the program to analyze problems in conjunction with texts and formal course work in geostatistics. Users of the EPA programs Geo-EAS and GEOPACK will find the routines in Geostatistical Toolbox compatible and invaluable adjuncts.

Capabilities and Functions

Geostatistical Toolbox currently consists of 22 compiled programs that provide modeling and analysis of two-dimensional and three-dimensional spatially correlated data.

Traditional statistical functions include univariate and bivariate statistics, histograms, cumulative frequency, scattergrams, least-squares curve fitting, logarithmic transforms, mean, variance, coefficient of variation, standard deviation, skewness, kurtosis, minimum, maximum, median, quartiles, and interquartile range. Additionally, conditional statistics can be determined: class limits, class size, frequency, cumulative frequency, class mean, and averages above and below class limit. Bivariate analysis of covariance, Pierson's correlation coefficient, Spearman rank correlation coefficient, polynomial, exponential, and least-squares curve fitting are provided for complete data set analysis.

The geostatistics functions include two-dimensional and three-dimensional (downhole) variography, semivariograms with angular tolerance directional control, covariance, H-scattergrams, H-histograms, madograms, constant support, regular and irregular grids, experimental variogram modeling (spatial, exponential, Gaussian, point, and nugget effect), two-dimensional and three-dimensional simple and ordinary kriging with point or block support, two-dimensional cokriging, universal

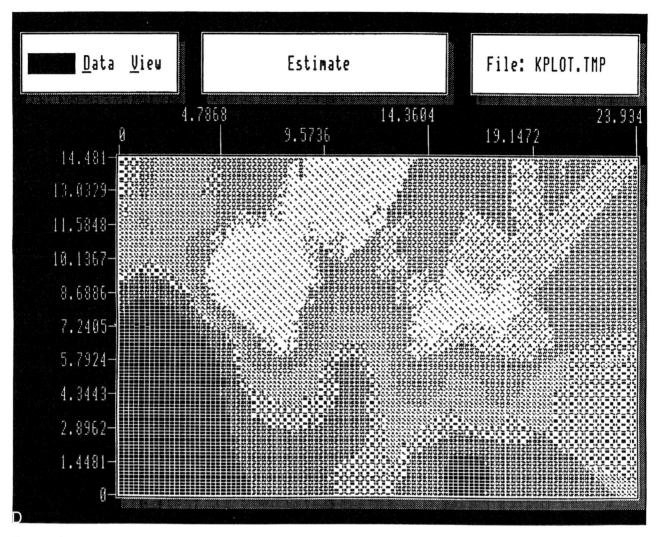

Figure 3 (continued). (D) Example of GEOPACK kriged block map.

trend kriging, kriging and cokriging cross validation, and reserves calculation (mining).

Hardware Requirements

The programs in Geostatistical Toolbox require a
PC/AT 286 or 386 CPU processor, 640 KB RAM, a
math coprocessor (80287 or 80387), an EGA graphics
adapter and monitor, and a hard disk drive with at
least a couple of megabytes of free space. An optional
graphics printer will allow hardcopy output. A screen
capture program such as Pizazz Plus or any of numerous public domain programs may also prove useful
for saving, printing, and enhancing graphics screen
displays.

Language and User Interface

The Geostatistical Toolbox programs are provided
as compiled executables and were written in Borland
Turbo Pascal. The source code is not generally distributed, but is available for noncommercial use.

Each program in Geostatistical Toolbox is executed from the DOS prompt; there is no main menu of

selections, which has the advantages of lowering
RAM memory overhead and allowing the executables to be incorporated with other programs. Each
program displays similarly organized user input
screens. Once the input data file is read, many
required parameters may be chosen by toggling
between choices with the space bar. A light-bar cursor is used to move between and select program
input options, and a one-line explanation of the
required input appears with each selection. Some
screens provide additional context-sensitive help
using the F1 function key. Optional background journal files can be created to retain program parameters
in case of program execution errors; otherwise the
programs do not retain parameter selections and
input as you go from one application to another.

The Geostatistical Toolbox programs do not allow
for customization. The program is intended to be
used in conjunction with commercially available mapping, graphics, spreadsheet, and database programs
and does not attempt to duplicate those functions.
Output files from the programs are in ASCII and
readily usable by most commercial applications.

User's Manual and Example Data Sets

The Geostatistical Toolbox user's manual is distributed with the program as disk files; both Word-Perfect™ Version 5.1 (which displays Greek and special symbols accurately) and DOS text files are included. Print the manual prior to attempting to use the program.

The manual provides definitive descriptions of each of the programs in Geostatistical Toolbox. A brief description of the program tasks and the program's constraints is followed by a detailed explanation of each program's menu items and comments on input parameters. Where appropriate, geostatistical equations of the functions to be performed are also shown. The manual assumes a familiarity with geostatistical methods and terminology. There are no illustrations or explanatory dissertations in the manual, nor are there tutorials or example exercises. Scientists using the program will appreciate the manual's completeness of instructions for supplying the proper data into the programs; however, students of geostatistics will require additional background and instructional material, such as that provided in texts by Hohn (1988), Isaaks and Srivastava (1989), or Journel and Huijbregts (1978).

The manual also clearly illustrates the two-dimensional and three-dimensional data file formats. Samples of the two file formats are included on disk, but there is no example analysis of the data sets. The files may used to test the workings of the programs, but the user will have to provide his or her own interpretation.

Data File Structure and Size

The Geostatistical Toolbox programs require that the data be arranged in one of two ASCII text formats that can easily be created by text editors or as output from spreadsheet or database programs. One format, the general purpose format, is suitable for two-dimensional surface data applications, such as soil or water samples. The second format is specific to three-dimensional borehole applications, such as mineral exploration boreholes. Both formats include provision for file description and variable names. The three-dimensional format also requires coordinate inclination and azimuths of the sample locations in the drill hole. The general purpose format of Geostatistical Toolbox is identical to that used in the EPA Geo-EAS program and facilitates using the two programs together.

The Geostatistical Toolbox programs achieve acceptable interactive response on 286-class computers by storing data in RAM. To accomplish this, the data files are generally limited to 2000 sample with 15 variables each. Certain programs impose even tighter restrictions. In common use, if at least 580 KB of 640 KB is available to the program, Geostatistical Toolbox is quite stable. Data files exceeding the limits of RAM will produce predictable memory error messages. Resampling the data set or dividing it into smaller subsets for multiple analysis are the best options for using Geostatistical Toolbox with data sets that exceed design limits.

Graphics

The graphic screen displays in Geostatistical Toolbox are accurate and adequate for analysis of geostatistical problems, but the appearance may not be considered to be presentation quality. The ASCII data output files from the analysis, explained in the manual, can be input into commercial graphics and mapping packages for enhanced appearance. The variograms, histograms, H-histograms, scattergrams, H-scattergrams, madograms, contour plots, and block models generated, however, are both necessary and usable for conducting the geostatistical analysis. The program allows some control on the annotation of the graphic displays. An interactive cursor allows identification of points on the displays so that the data set may be edited. Displays may be sent directly to a graphics printer; the resolution will be that of the EGA screen. Example graphics for Geostatistical Toolbox are shown in Figure 4.

STANFORD GSLIB

GSLIB is a library of geostatistical programs developed at Stanford University under the direction of Andre Journel. Oxford University Press has published the user's guide and FORTRAN programs authored by Clayton V. Deutsch, Exxon Production Research Company, and Andre Journel under the title *GSLIB: Geostatistical Software Library and User's Guide*.

Audience

The GSLIB programs and user's guide are particularly addressed to graduate students and advanced geostatistics practitioners, but should also be a valuable resource for those intending to learn how to apply geostatistics to their work. GSLIB presents a robust geostatistical software library and well-written documentation sufficient to serve both the student and experienced user. The programs alone are complete enough for complex investigations, yet the software modules are also designed to be easily modified and incorporated into advanced analytical programs. The programs in GSLIB represent the most advanced geostatistical methods, including three-dimensional cokriging and several forms of conditional simulation and serve as a platform upon which additional geostatistical research may be based.

The GSLIB program library does not contain executable code, but rather uncompiled ASCII FORTRAN program listings. These can be run on any computer platform that can compile FORTRAN. The user's guide presents theoretical background notes and descriptions of the programs in a well-organized textlike fashion. Students will need the guidance of additional geostatistics texts such those as by Journel and Huijbregts (1978), Hohn (1988), and Isaaks and Srivastava (1989).

Capabilities and Functions

GSLIB is a collection of geostatistical programs developed and used at Stanford University that repre-

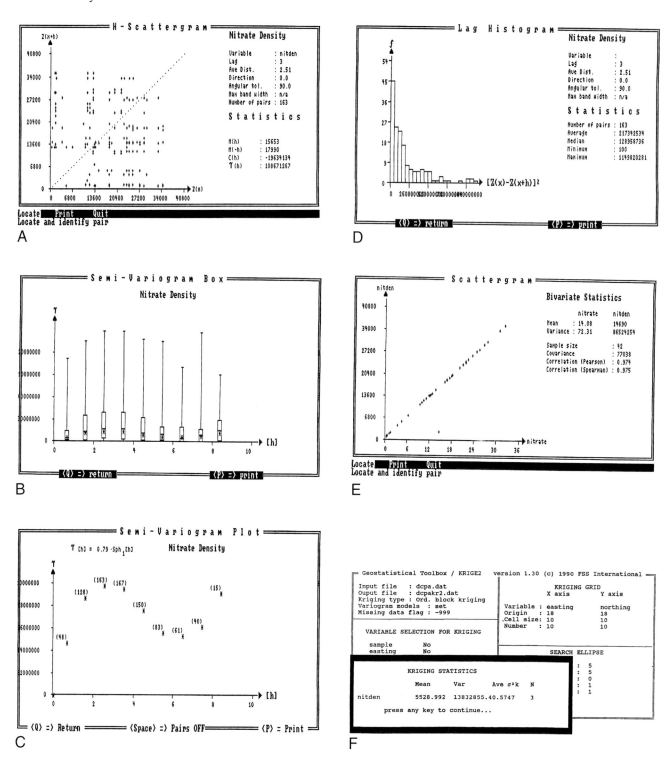

Figure 4. (A) Example of Geostatistical Toolbox H-scattergram. (B) Example of Geostatistical Toolbox variogram box plot. (C) Example of Geostatistical Toolbox experimental variogram. (D) Example of Geostatistical Toolbox lag histogram. (E) Example of Geostatistical Toolbox bivariate scattergram. (F) Example of Geostatistical Toolbox typical menu screen.

sent the most current developments in the field. GSLIB is an essential core of the programs necessary for spatial analysis: variograms, kriging, and stochastic simulation. Four variogram programs analyze irregular or gridded, two-dimensional or three-dimensional data. Ten measures of spatial variability

and continuity can be determined and analyzed. Modules for kriging include simple, ordinary, and various trend models. Cokriging facilitates the use of secondary variables, and indicator kriging allows estimation of posterior probability distribution. A variety of unique geostatistical simulation programs are available in GSLIB. In addition to the conditional tuning bands and LU-matrix techniques, four different sequential simulation programs and two annealing-based programs are provided. These programs allow practitioners the ability to generate unconditional and conditional simulation models of both continuous and categorical variables.

The two distribution diskettes contain 203 files including 37 programs, several utilities, example problem sets, parameter files, and PostScript™ postplot template files.

Hardware Requirements

According to the user's guide, GSLIB programs have been compiled to run on computers ranging from PC-XT clones to the latest multiple-processor Cray Y-MP™. This is a tribute to the ANSI standard FORTRAN 77 language in which the programs were written. Dynamic memory allocation, however, is not supported by this standard. Thus, practical dimensioning limits must be specified and coded, prior to compilation, for the specific machine on which the programs are to be run. Some FORTRAN compilers extend or supplement the ANSI 77 standard with dynamic memory management utilities; using these compilers will allow larger data sets on a given computer; however, the GSLIB programs may be used on virtually any computer for which a FORTRAN compiler is available.

Language and User Interface

No executable programs are provided with GSLIB. The source code, ANSI standard FORTRAN 77, is provided as ASCII text, which must be compiled into executable programs before use. This setup allows GSLIB to be distributed as a device-independent package. There are no restrictions on the operating systems. FORTRAN compilers are available for nearly every computing platform, allowing GSLIB to be used by researchers using personal computers, sophisticated workstations, or supercomputers. Obviously, more computing power and memory are required for more complex modeling and larger data sets, but today's personal computers are capable of serious analysis. FORTRAN compilers that allow use of extended memory are highly recommended.

The FORTRAN language was selected because of its universal use in scientific computing, ease of meaningful program commenting and distribution, and because it is supported by compilers for most computers. FORTRAN is a language easy to modify, customize, and incorporate into existing or new applications. It is flexible, versatile, modular, and device independent. The GSLIB code, as written, is not necessarily optimized for speed or efficiency, but is computationally sound. Commercial graphics libraries are widely available that could enhance the utility of the programs. No elegant user interface is included. The programs require that the user have precise knowledge of the program functions and make decisions while preparing parameter files and running the programs. The programs could be

rewritten in object-oriented languages. These features spell opportunity for both users and program developers. One expectation of this release of GSLIB is that the programs will be modified, rewritten, and incorporated into sophisticated programs as more people realize the power and utility of geostatistics. GSLIB can be considered as a toolbox of validated routines freely available for the use of programming craftsmen, custom code builders, and researchers. The Stanford students and faculty involved in the development of the routines deserve noble credit for sharing GSLIB with the scientific community.

Prior to compilation of the programs, users should read the installation notes, programming conventions and variable names, descriptions of parameter files, and application notes in the user's guide. The programs also contain nonexecutable comment lines that further explain the code logic and program purpose in plain English. As previously mentioned, the dimensioning statements must also be specified in an include file. A good working knowledge of the FORTRAN compiler program is also necessary.

The compiled programs require data files and parameter files to run. There is no error checking when the programs are run. The user must provide the appropriate information in the data and parameter files. Careful reading of the user's guide will prevent many common mistakes.

User's Manual and Example Data Sets

The user's guide is a 340-page hardbound textbook with a pocket containing the two 3½-in 1.44 MB program diskettes. Introductory chapters in the book discuss general geostatistical concepts and GSLIB programming conventions. Many figures, plots, and example file listings illustrate the concepts and terminology presented. The three main topics, variograms, kriging, and simulation, are covered and include problem sets. Partial solutions to the problems are included in the appendix and in disk files. A set of utilities is included for generating traditional statistics, plotting, postprocessing, and other data manipulation. The appendix includes notes on software installation, programming conventions and variable names, a program list, and a list of acronyms. The guide has an extensive bibliography and is well indexed.

Six problem sets cover the use of the GSLIB programs for data analysis, variograms, kriging, cokriging, indicator kriging, and simulations. The data sets for the problems, as well as partial solutions, are included on the disks. The problem sets are designed to allow users to check installation of the programs, try out the features, and learn to use the parameter files effectively. The problem answers include illustrations of maps, histograms, and other plots. Considerable thought has been given to asking problem questions that will probe the user's understanding of the programs and techniques. The user's guide, combined with geostatistics texts, will help the user to understand and appreciate the GSLIB capabilities.

Data File Structure and Size

The GSLIB programs read and write ASCII data in a format similar to Geo-EAS and Geostatistical Toolbox. This is not coincidence, because the latter two programs used developmental GSLIB routines as their foundation. The ASCII files can be created with a simple text editor or through customized user application programs. The ASCII data format provides machine independence and affords user readability, but sacrifices space; no data compression or binary format is supported by the GSLIB programs. The data files contain an information header, specification of the number of variables and their names, and numerical data. Both two-dimensional and three-dimensional data can be used in GSLIB. Both regular grids of data points or block values and irregularly spaced data may be used. The locations of data points and grid (two-dimensional) or block (three-dimensional) centers are specified in Cartesian x,y,z format. Grid ordering (specification of the number of rows, columns, and grid spacing) can also be used. The user must do any coordinate transformations or rotations outside of the programs (or else add appropriate user-supplied code prior to compilation).

Graphics

The GSLIB programs are written to be device independent and do not contain graphics routines. Utilities are included on the disks, however, to provide hardcopy plots for histograms, normal probability plots, Q-Q/P-P plots, scatterplots, semivariogram plots, and gray-scale and color-scale block maps on PostScript-compatible printers and file viewers. (No PostScript file viewer program is included, but some shareware graphics utilities have the capability to view and convert PostScript graphics files.) Users with minimal programming ability will find that output from the GSLIB programs can be directed to any number of commercial graphics packages through appropriate modification of output statements in the code prior to compilation. Geo-EAS can also be used to examine GSLIB output. Example PostScript graphics from GSLIB are shown in Figure 5.

SOURCES FOR PROGRAMS

STATPAC

STATPAC may be ordered either directly from the U.S. Geological Survey or through GeoApplications. The address is

Books and Open-File Reports
U.S. Geological Survey
Federal Center
P.O. Box 25425
Denver, CO 80225
Telephone: (303) 236-7476

The individual open-file report number and the biographical information for each report follows.

OF 87-411-A: Brief Descriptions of statpac and Related Statistical Programs for the IBM Personal Computer: W. D. Grundy and A. T. Miesch, 1987, 34 p. $5.25. Text.

OF 87-411-B: Brief Descriptions of STATPAC and Related Statistical Programs for the IBM Personal Computer—B, Source Code: W. D. Grundy and A. T. Miesch, 1987. $30.00. Five 5¼–in diskettes.

OF 87-411-C: Brief Descriptions of STATPAC and Related Statistical Programs for the IBM Personal Computer—C, Source Code and Selected Executable Modules: W. D. Grundy and A. T. Miesch, 1987. $60.00. Ten 5¼–in diskettes.

GeoApplications will send the latest release of STATPAC on 5¼–in 360 K disks, 5¼–in 1.2 MB, or 3-½–in 1.4 MB disks (please specify disk size when ordering) and the USGS report for $30.00 ($25.00 plus $5.00 shipping). Those ordering from outside the United States should contact GeoApplications for additional information on shipping charges.

GeoApplications
P.O. Box 41082
Tucson, AZ 85717-1082
Telephone: (602) 323-9170 (Please call to confirm pricing.)
Fax: (602) 327-7752

Geo-EAS

Geo-EAS is available from the Computer Oriented Geological Society (COGS). The program may be obtained on 5¼–in 360 KB or 1.2 MB disks or on 3½–in 720 KB or 1.4 MB diskettes (please specify disk size when ordering). The user manual is included. The program is offered as *Multi-Disk Set #2* and is available for $50.00, which covers the cost of duplication and mailing. COGS members receive a 20% discount.

Computer Oriented Geological Society
P.O. Box 370246
Denver, CO 80237
Telephone: (303) 751-8553 (Please call to confirm current pricing.)

Geo-EAS also may be ordered from the National Technical Information Service (NTIS) for $93, which includes shipping and handling. Specify order no. PB93-504959 for 5¼–in disks or order no. PB93-504967 for 3½–in disks.

National Technical Information Service
Springfield, VA 22161
Telephone: (703) 487-4650
Fax: (703) 321-8547

GeoApplications will send the latest release of Geo-EAS on 5¼–in 360 KB disks, 5¼–in 1.2 MB, or 3½–in 1.4 MB disks (please specify disk size when ordering) and the EPA manual for $40 ($35 plus $5 shipping). Those ordering from outside the United States should contact GeoApplications for additional information on shipping charges.

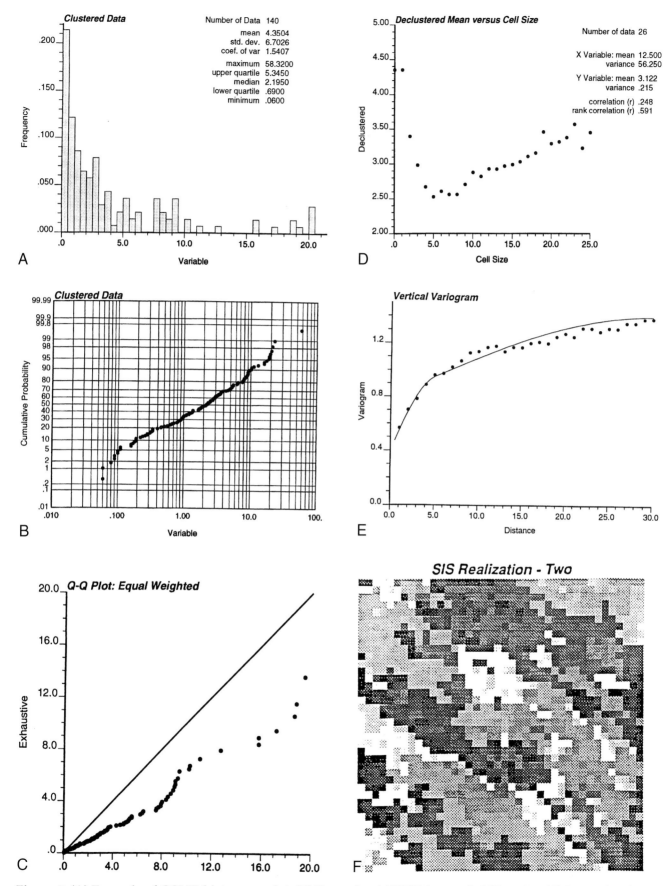

Figure 5. (A) Example of GSLIB histogram plot. (B) Example of GSLIB log probability plot. (C) Example of GSLIB Q-Q plot. (D) Example of GSLIB scatter plot. (E) Example of GSLIB experimental and fitted-model variogram. (F) Example of GSLIB gray-scale map plot.

GeoApplications can also provide the source code, written in FORTRAN (Microsoft 4.01), and a 600-page programmer's manual. The cost varies with the disk format requested: 5¼–in 1.2 MB is $86.50 ($78.50 plus $8 shipping), 3½–in 1.4 MB is $91.50 ($83.50 plus $8 shipping).

GeoApplications
P.O. Box 41082
Tucson, AZ 85717-1082
Telephone: (602) 323-9170 (Please call to confirm pricing.)
Fax: (602) 327-7752

Geo-EAS is also available through the International Ground Water Modeling Center for $50.00; specify order no. FOS 53 PC and whether high-density 5¼–in or 3½–in disks are required. Do not prepay; customers will be invoiced on receipt of order and orders shipped upon receipt of payment.

IGWMC USA
Institute for Ground-Water Research and Education
Colorado School of Mines
Golden, Colorado 80401-1887
Telephone: (303) 273-3103
Fax: (303) 273-3278

GEOPACK

GEOPACK is available from the Computer Oriented Geological Society (COGS). The program may be obtained on either 5¼–in 1.2 MB disks or 3½–in 1.4 MB diskettes (when ordering, please specify disk size and whether low- or high-density diskettes are required). The bound, 70-page user's manual is included. The list price for *Multi-Disk Set #8* is $50.00. COGS members are eligible for a 20% discount.

Computer Oriented Geological Society
P.O. Box 370246
Denver, CO 80237
Telephone: (303) 751-8553 (Please call to confirm current pricing.)

GEOPACK may also be obtained directly from the EPA if you send four formatted 1.2 MB 5¼–in diskettes.

"GEOPACK"
Robert S. Kerr Environmental Research Laboratory
Office of Research and Development
U.S. Environmental Protection Agency
Ada, OK 74820

GEOPACK is available from the International Ground Water Modeling Center for $50.00; specify order no. FOS 54 PC and whether 5¼–in or 3½–in disks are needed. Do not prepay; customers are invoiced on receipt of the order and orders are shipped upon receipt of payment.

IGWMC USA
Institute for Ground-Water Research and Education
Colorado School of Mines
Golden, Colorado 80401-1887

Telephone: (303) 273-3103
Fax: (303) 273-3278

Geostatistical Toolbox

Geostatistical Toolbox is available from the Computer Oriented Geological Society (COGS) as *Multi-Disk Set #11*. The program may be obtained on either 5¼–in or 3½–in disks in either high or low density. COGS charges $20.00 for the cost of duplicating and distributing the program. COGS members are eligible for a 20% discount.

Computer Oriented Geological Society
P.O. Box 370246
Denver, CO 80237
Telephone: (303) 751-8553 (Please call to confirm current pricing.)

Geostatistical Toolbox may also be obtained from any of the four FSS International offices. Please send $40 and formatted diskettes with your request: one 3½–in 1.4 MB, two 3½–in 720 KB, two 5¼–in 1.2 MB, or four 360 KB disks. The addresses of the four FSS offices are as follows.

800 Millbank	245 Moonshine Circle
False Creek South	Reno, NV 89523
Vancouver, BC	U.S.A.
Canada V5Z 3Z4	
10 Chemin de Drize	P.O. Box 657
1256 Troinex	Eppling 2121
Switzerland	NSW, Australia

If the Turbo Pascal source code is requested, FSS will require a letter explicitly stating that the FSS source code, or modifications to the source code, will not be used in or incorporated into commercial software. FSS International has copyright protection of Geostatistical Toolbox executables and program source code.

GSLIB

GSLIB: Geostatistical Software Library and User's Guide by Clayton V. Deutsch and Andre Journel (ISBN 0-19-507392-4) is available from Oxford University Press for $49.95 plus $2.50 postage and handling.

Oxford University Press
Business and Customer Service
2001 Evans Road
Cary, NC 27513
Order by telephone: 1-800-451-7756

The book also may also be ordered or obtained through bookstores.

ACKNOWLEDGMENTS

I am indebted to the many reviewers of this manuscript for their clarifications and suggestions. Although every attempt has been made to ensure accurate, up-to-date information, the nature of software is such that this information will become dated. The sources for obtaining the software should be contacted prior to ordering or requesting the programs because these sources may discontinue distribution. My personal thanks go to the authors and developers of these geostatistical programs for making these tools available.

REFERENCES

Deutsch, C. V., and A. G. Journel, 1992, GSLIB: geostatistical software library and user's guide: New York, Oxford University Press, 336 p.

Englund, E., and A. Sparks, 1991, Geo-EAS 1.2.1 (geostatistical environmental assessment software) user's guide: U.S. EPA Document EPA/600/B-91/008, 128 p.

Froidevaux, R., 1990, Geostatistical Toolbox user's man-ual: Troinex, Switzerland, FSS International, 83 p.

Grundy, W. D., and A. T. Miesch, 1987, Brief descriptions of STATPAC and related statistical programs for the IBM personal computer: U.S. Geological Survey Open-File Report 87-411-A, 34 p.

Hohn, M. E., 1988, Geostatistics and petroleum geology: New York, Van Nostrand Reinhold, 264 p.

Isaaks, E. H., and R. M. Srivastava, 1989, Applied geostatistics: New York, Oxford University Press, 561 p.

Journel, A. G., and C. J. Huijbregts, 1978, Mining geostatistics: London, Academic Press, 600 p.

Olea, Ricardo A., ed., 1991, Geostatistical glossary and multilingual dictionary: New York, Oxford University Press, 177 p.

Yates, S. R., and M. V. Yates, 1990, Geostatistics for waste management: a user's manual for the GEOPACK (Version 1.0) geostatistical software system: U.S. EPA Document EPA/600/8-90/004, 70 p.

Selected Readings on Geostatistics

Jeffrey M. Yarus

In compiling a bibliography on geostatistics for those interested in studying the topic, I am faced with selecting only a few of the many excellent articles and books that have been published. The small set chosen represents those I have found particularly helpful to me in my endeavor to understand geostatistics. Admittedly, the subject is notorious for its jargon, and most papers tend to be riddled with mathematical notation. That most geostatistical papers are written that way should be viewed by the reader as a symptom of severe psychological trauma. Authors of such papers usually live in high-stress environments and need to be shown more than the average amount of kindness and patience. Lurking behind the madness, however, are informational gems. The reader is encouraged to be tenacious when reading, and not fall prey to the feelings of tentativeness and xenophobia that often occur when encountering the strange and unusual.

Although all of the references listed here are worth perusing, the following discussion outlines an initial reading list that may be of interest to beginners. For the newcomers to geostatistics, I recommend reading the chapters on semivariograms and kriging in John Davis' book, *Statistics and Data Analysis in Geology*. Davis covers the rudiments of geostatistics, and his descriptions are lucid and perhaps more easily understood by most geologists. With Davis in mind, the sequence of papers published by the *Engineering and Mining Journal* in 1979 would be next on the list, in particular, articles by Booker on kriging, Clark on the semivariogram, and Journel on simulation (these articles are in the reference list). Keep in mind that geostatistics had its origin in mining, and that this series has a mining orientation. The article by D. G. Krige, "A Statistical Approach to Some Basic Valuation Problems on the Witwatersrand," published in 1951 is what started it all (Krige = kriging).

The textbooks by Clark and David are very much concerned with mining and perhaps not as interesting to the petroleum geoscientist. However, they are excellent resources for addressing specific issues as they arise. More practical texts are Hohn's *Geostatistics and Petroleum Geology*, and Isaaks' *An Introduction to Applied Geostatistics*. Both of these do a reasonable job of minimizing the mathematics and addressing problems associated with petroleum exploration and production.

I strongly recommend reading Cressie's article on "The Origins of Kriging." He does an excellent job unmasking the mystery behind geostatistics and its relationship to the more classic forms of statistics, such as regression analysis. Finally, Englund's paper, "A Variance of Geostatisticians," is a must. It emphasizes that there is as much art in the geostatistical process as there is science!

For further reading on specific topics addressed in this volume, please review the references cited at the end of each paper. A list of general references follows.

BOOKS

Agterberg, F. P., 1974, Geomathematics: New York, Elsevier Scientific Publishing, 596 p.

Clark, I., 1979, Practical geostatistics: London, Applied Science Publishers, 129 p.

David, M., 1977, Geostatistical ore reserve estimation: New York, Elsevier Scientific Publishing, 364 p.

Davis, J. C., 1986, Statistics and data analysis in geology, 2nd ed.: New York, John Wiley, 646 p.

Deutsch, C. V., and Journel, A. G., 1992, GSLIB: geostatistical software library and user's guide: New York, Oxford University Press, 340 p.

Isaaks, E. H., and R. M. Srivastava, 1989, An introduction to applied geostatistics: New York, Oxford University Press, 561 p.

Journel, A. G., and C. J. Huijbregts, 1978, Mining geostatistics: Orlando, Florida, Academic Press, 600 p.

Hohn, M. E., 1988, Geostatistics and petroleum geology: New York, Van Nostrand Rienhold, 264 p.

PAPERS

Abry, C. G., 1975, Geostatistical model for predicting oil: Tatum Basin, New Mexico: AAPG Bulletin, v. 59, p. 2111–2127.

Agterberg, F., 1976, New problems at the interface between geostatistics and geology, *in* M. Guarascio, M. David, and C. Huijbregts, eds., Advanced geostatistics in the mining industry: Dordrecht, Holland, D. Reidel , p. 403–421.

Booker, P. I., 1979, Kriging: Engineering and Mining Journal, v. 180, no. 8, p. 148–153.

Chiles, J. P. and P. Delfiner, 1977, Conditional simulations: a new Monte Carlo approach to probabilistic evaluation of hydrocarbon in place: Fountainebleau, France: Ecole de Mines de Paris, Notes et etudes, N526, Centre de Geostatistique, 30 p.

Clark, I., 1979, The semivariogram—part 1: Engineering and Mining Journal, v. 180, no. 7, p. 90–94.

Clark, I., 1979, The semivariogram—part 2: Engineering and Mining Journal, v. 180, no. 8, p. 90–97.

Cressie, N., 1990, The origins of kriging: Mathematical Geology, v. 22, no. 3, p. 239–252.

Cressie, N., and D. M. Hawkins, 1980, Robust estimation of the semivariogram, I: Mathematical Geology, v. 12, no. 2, p. 115–125.

DaCosta, A. J., 1985, A new approach to the characterization of reservoir heterogeneity based on the geomathematical model and kriging technique: SPE 14275, 60th Annual Technical Conference and Exhibition, Proceedings.

Dowds, J. P., 1972, Geostatistics aid gas, oil discovery and exploration: World Oil, v. 75, p. 57–58, 60.

Englund, E. J., 1990, A variance of geostatisticians: Mathematical Geology, v. 22, no. 4, p. 417–455.

Isaaks, E. H., and R. M. Srivastava, 1988, Spatial continuity measures for probabilistic and deterministic geostatistics: Mathematical Geology, v. 20, no. 4, p. 313–341.

Journel, A. G., 1974, Geostatistics for conditional simulation of ore bodies: Economic Geology, v. 69, p. 673–687.

Journel, A. G., 1979, Geostatistical simulation: methods for exploration and mine planning: Engineering and Mining Journal, v. 180, no. 12, p. 86–91.

Journel, A. G., 1983, Non-parametric estimation of spatial distributions: Mathematical Geology, v. 15, no. 3, p. 445–468.

Journel, A. G., 1988, Non-parametric geostatistics for risk and additional sampling assessment, in L. Keith, ed., Principles of environmental sampling:

American Chemical Society, p. 45–72.

Krige, D. G., 1951, A statistical approach to some basic valuation problems on the Witwatersrand: Journal of the Chemical, Metallurgical and Mining Society of South Africa, no. 52, 1951, p. 119–139.

Krige, D. G., 1966, Two-dimensional weighted average trend surfaces for ore valuation, in Proceedings of the symposium on mathematical statistics and computer applications in ore valuation: South African Institute of Mining and Metallurgy, Johannesburg, South Africa, p. 13–38.

Matheron, G. F., 1963, Principles of geostatistics: Economic Geology, v. 58, p. 1246–1266.

Matheron, G. F., 1970, La théorie des variables régionalisées et ses applications: Tech. Rep. Fascicule 5, Les Cahiers du Centre de Morphologie Mathématique de Fontainebleau, Ecole Supérieur des Mines de Paris.

Olea, R. A., 1977, Measuring spatial dependence with semivariograms: Lawrence Kansas, Kansas Geological Survey, Series on Spatial Analysis no. 3, 29 p.

Royle, A. G., 1979, Why geostatistics?: Engineering and Mining Journal, v. 180, no. 5, p. 92–101.

Srivastava, R. M., 1987, Minimum variance or maximum profitability: CIM Bulletin, v. 80, p. 63–68.

Srivastava, R. M., and H. Parker, 1988, Robust measures of spatial continuity, in M. E. A. Armstrong, ed., Third International Geostatistics Conference: Dordrecht, Holland, D. Reidel.

Verly, G., M. David, A. G. Journel, and A. Marchechal, eds., 1983, Geostatistics for natural resource characterization: NATO Advanced Study Institute, South Lake Tahoe, California: Dordrecht, Holland, D. Reidel.

Index